# The Alpheidae from China Seas

Zhong-li Sha • Yan-rong Wang
Dong-ling Cui

# The Alpheidae from China Seas

## Crustacea: Decapoda: Caridea

Springer

Zhong-li Sha
Institute of Oceanology
Chinese Academy of Sciences
Qingdao, Shandong, China

Yan-rong Wang
Institute of Oceanology
Chinese Academy of Sciences
Qingdao, Shandong, China

Dong-ling Cui
Institute of Oceanology
Chinese Academy of Sciences
Qingdao, Shandong, China

ISBN 978-981-15-0647-5 ISBN 978-981-15-0648-2 (eBook)
https://doi.org/10.1007/978-981-15-0648-2



This Springer imprint is published by the registered company Springer Nature Singapore Pte Ltd.
The registered company address is: 152 Beach Road, #21-01/04 Gateway East, Singapore 189721, Singapore

# Acknowledgments

We would like to express our indebtedness to the late Prof. Dr. Liu Rui-yu, who was a pioneer in studying crustacean of China and left us a great deal of valuable reference material of the alpheid shrimps. These large collections of reference material were obtained from several expeditions. Therefore, we are deeply indebted to all the collectors, including Chen Mu, Cheng Li-ren, Cui Yu-hang, Ding Lan-ping, Fan Zhen-gang, Fang Xiu-bin, Jiang Jin-xiang, Li Huan-lin, Li Jin-he, Li Xin-zheng, Liang Guan-sheng, Liao Yu-lin, Liu Rui-yu, Liu Ji-xing, Liu Xi-xing, Ma Xiu-tong, Volker Neumann, Ren Xian-qiu, Ruan Wen-guang, Shen Shou-peng, Sun Ke-zhi, Sun Fu-zeng, Tang Zhi-can, Wang Yong-liang, Wang Yong-qiang, Wu Bao-ling, Wu Xu-wen, Xu Feng-shan, Xu Jie-shan, Zhang Bao-lin, Zhang Jun-long, Zhang Wei-quan, and Zheng Feng, who offered us abundant specimens of alpheids to examine. Two of the collectors, Ren Xian-qiu and Wang Yong-liang, although retired from active systematic research, graciously gave close attention to the draft of this report and considerably enhanced its value. The report has also been materially improved by a detailed review by our colleagues Cheng Jiao, Sun Shao'e, and Yuan Zi-ming. We also thank all the workers in the Marine Biological Museum, CAS. Although the present report is not perfect, as a young team, we wish it would be helpful for the further research of the crustacean.

# Contents

# 1 Instruction

As one of the most diverse groups of marine decapod crustaceans, the Alpheidae contain more than 700 species within 48 established genera (e.g. De Grave & Fransen, 2011; Komai & Fujita, 2018). Alpheid shrimps represent one of the most abundant decapods in tropical and subtropical areas, which are commonly found in shallow-waters, especially on coral reefs (e.g. Banner & Banner, 1966; Felder & Chaney, 1979; Liu & Lan, 1980). Few alpheids are also found in hypohaline environments, such as mangrove and estuary. About 16 species within Alpheidae were also reported living in freshwater, brackish or marine cave and anchialine pools (Anker, 2008; Komai & Fujita, 2018). Moreover, small number of species of genus *Alpheopsis* and *Alpheus* live in deep-waters (Anker, 2017; Komai & Ohtomi, 2018; Ramos-Tafur, 2018). Most alpheids are permanent association with sessile marine invertebrates, particularly sponges, corals and crinoids, and gobiid fishes (e.g. Banner & Banner, 1975; Bruce, 1976; Duffy, 1992; Karplus & Thompson, 2011). Social monogamy is the common mating system employed by many crustaceans in order to defend and share a specific microhabitat or refuge (Correa & Thiel, 2003; Thiel & Baeza, 2001), which is also frequently observed among alpheid shrimps (Pescinelli et al., 2018). Interestingly, Duffy (1996) reported the first case of eusociality in a sponge-dwelling shrimp, *Synalpheus regalis* Duffy, 1996. That shrimp colonies consist of only one to few reproductive females and tens to hundreds of cohabiting genetically related, non-breeding colony males (Duffy & Macdonald, 1999). The cooperative colonies are always competition for sponge hosts (Duffy, 2003), and the major chela is a powerful weapon, which is used in communication and combat (Nolan & Salmon, 1970; Duffy et al., 2002). When the snapping shrimps rapidly close their claws, they direct a powerful water jet at a nearby opponent and generate a loud noise (Knowlton & Moulton, 1963; Versluis et al., 2000). The biological noises of oceans are also caused by the snapping shrimp, except for the fish, dolphins and whales (Versluis et al., 2000). Moreover, some of the alpheid shrimps, such as *Alpheus japonicus* Miers, 1879 and *Alpheus digitalis* De Haan, 1844, also have economic values (Liu & Lan, 1980). As one of the most species-rich groups, the Alpheidae present a particular challenge for taxonomists. The alpheid shrimps often contain cryptic taxa that are difficult to be identified using traditional, morphology-based criteria (Knowlton, 1986, 1993, 2000; Hebert et al., 2003; Witt et al., 2006; Barber & Boyce, 2006; Mathews, 2006). Within the 48 established genera, *Alpheus* is the most species-rich genus followed by *Synalpheus*, and the other genera of Alpheidae are consisted of not more than 60 species (e.g. De Grave & Fransen, 2011). Both *Alpheus* and *Synalpheus* are divided into several different taxonomic groups (e.g. Coutière, 1899, 1905; Banner & Banner, 1975; Ríos & Duffy,


Z.-l. Sha et al., *The Alpheidae from China Seas*,
https://doi.org/10.1007/978-981-15-0648-2_1

2007; Anker & De Grave, 2008). The history of taxonomic research on the Alpheidae is listed as following.

*Alpheus* Fabricius, 1798 contains 311 described species now (updated by Komai & Ohtomi, 2018). *Alpheus* was ever confused with *Crangon* Fabricius, 1798 for many decades, until the International Commission officially accepted the name *Alpheus* and the family name Alpheidae in 1955 (see detail in Banner & Banner, 1982). Coutière (1899) divided *Alpheus* into five species groups (*A. brevirostris*, *A. crinitus*, *A. edwardsii*, *A. macrochirus* and *A. megacheles*), then the species group *A. crinitus* was further divided into three subgroups: *A. crinitus*, *A. insignis* and *A. obesomanus* by Coutière (1905). However, with more and more species being discovered, the redefinition of the groups was requested. Banner & Banner (1966) elevated the above three subgroups to the status of group, and re-divided the *Alpheus* into the following 7 species groups: *A. brevirostris*, *A. crinitus*, *A. diadema*, *A. edwardsii*, *A. macrochirus*, *A. megacheles* and *A. obesomanus* with accepting the suggestion of De Man (1911). Presently, the redefined species groups (*A. brevirostris*, *A. crinitus*, *A. diadema*, *A. edwardsii*, *A. macrocheles*, *A. obesomanus* and *A. sulcatus*) have be well-accepted (e.g. Banner & Banner, 1982, 1985; Chace, 1988; Bruce, 1990), which are also adopted herein. Same as *Alpheus*, *Synalpheus* was also divided into six species groups: *S. brevicarpus*, *S. biunguiculatus* (later changed to *S. coutierei*), *S. comatularum*, *S. Laevimanus* (later changed to *S. gambarelloides*), *S. neomeris*, and *S. paulsoni* by Coutière (1908, 1909) and Banner & Banner (1975). However, only three of these traditionally recognized morphology-based species group (*S. brevicarpus*, *S. comatularum*, and *S. gambarelloides* groups) were supported as monophyletic using molecular analyses (Hultgren et al., 2014). Therefore, the *Synalpheus* is not divided into species groups in the present report.

The anatomical parts used to identify of the alpheid shrimps in this study are shown in Figs. 1.1 and 1.2 and discussed as following. **The anterior part of carapace:** the rostrum is present or absent, if present it's usually unarmed with serrated teeth both on dorsal and ventral margins (species in *Salmoneus* sometime bearing apical teeth on ventral margin); the rostral carina is well-developed in some species and lacking in the other species; the eyes are usually covered by the carapace or part of them are visible in dorsal and lateral view, but species in *Automate*, *Bermudacaris* and *Stenalpheops* with the eyes are exposed completely; Not all alpheid shrimps have the orbital tooth; the extra-corneal and supra-corneal tooth develop in some species of *Arete*, *Athanas* and *Rugathanas*; the pterygostomian tooth is present or absent. **The antennular peduncle:** it consists of three segments; the basal segment has a shield, the stylocerite, which is usually not overreaching the end margin of the second antennular segment. **The antennae:** the basicerite is visible both in dorsal and lateral view, and bearing strong distodorsal or distolateral tooth; the scaphocerite with strong lateral teeth distally is usually well-developed and beyond the antennular peduncle; the length of the cylindrical carpocerite is often an aid in diagnosis. **The mouth parts:** The mandible usually consists of molar process, incisor process and two-segmented palp, but the palp is lacking in the *Batella* and both the molar process and palp are lacking in the *Prionalpheus*; the remaining parts of the mouth are not showing distinctly differences between alpheid shrimps. **The first pereiopods (chelipeds):** the chelipeds are larger and stronger than the following four pairs of pereiopod, and most alpheid shrimps, especially species of *Alpheus* and *Athanas* with the chelipeds vary with age and sex; chelipeds are usually asymmetrical in size and shape except the species in genus *Stenalpheops* and *Thuylamea*, which have subequal and symmetrical chelipeds; the chela is often carrying the dactylus dorsally, but the dactylus of species in *Aretopsis* and *Bermudacaris* are situated ventrally; The chelipeds, especially the major cheliped, are probably carrying the most important single character for the identification of alpheid shrimps. **The second pereiopod:** they are chelate; the carpus often consists of 3–5 segments, and the ratio of these segments is taxonomic significance. **The following three**

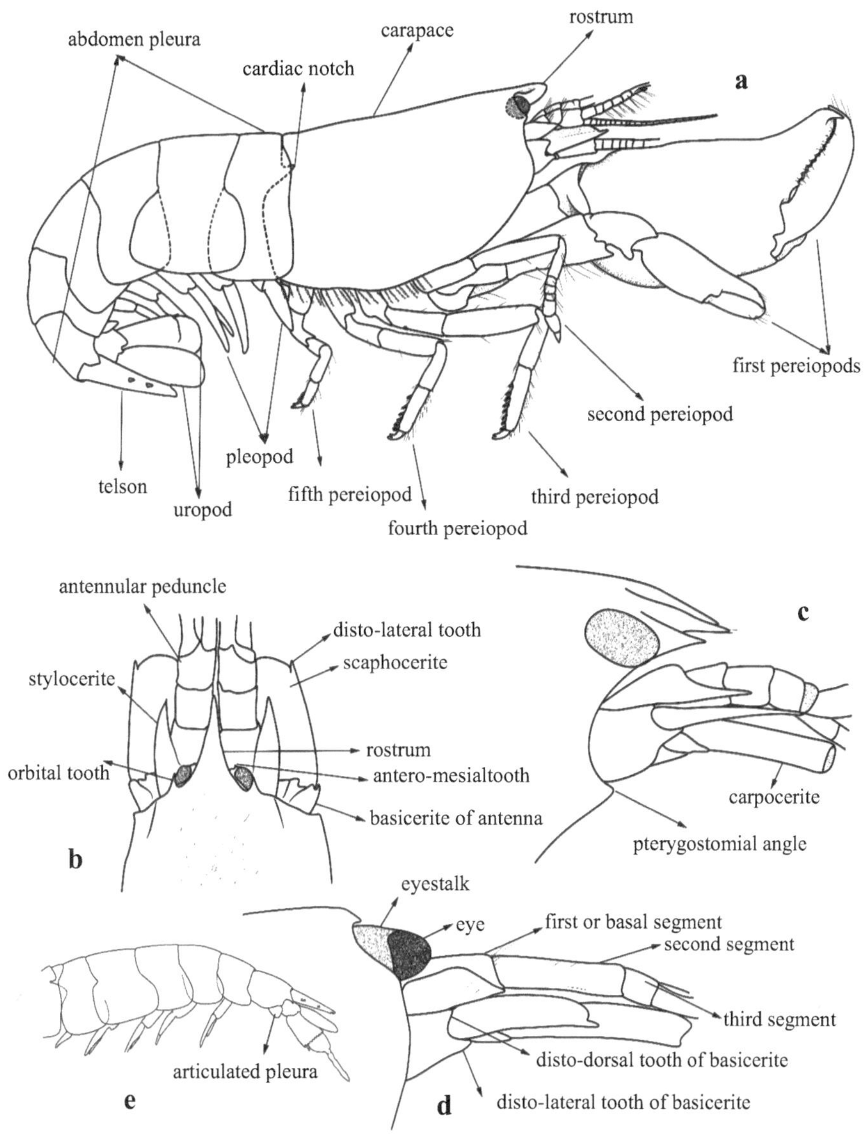


**Fig. 1.1** Schematic drawing of alpheid shrimp: (**a**) entire animal in lateral view; (**b**) anterior part of carapace in dorsal view; (**c**) and (**d**) anterior part of carapace in lateral view; (**e**) abdomen pleura in lateral view

**pereiopods:** third pereiopods are not chelate; the dactylus is simple or biunguiculate or even triunguiculate, and is an important identified character; the proportions of each articles and all spines or teeth situated on each articles are taxonomic significant. The fourth and fifth pereiopods are both similar to third pereiopod, and are not ordinarily used in definition of the species. The branchial formula is useful at the generic level in alpheid shrimps. **The abdominal segments:** the sixth abdominal segment bearing triangular articulate plate is one of most important characters in generic level's definition. **The pleopods:** the endopod of second pleopod is often bearing appendix interna and masculina in males while appendix interna only in females. However, species of *Synalpheus* and *Automate* are only bearing appendix interna both in males and females and some species of *Salmoneus* have appendix interna and mascu-

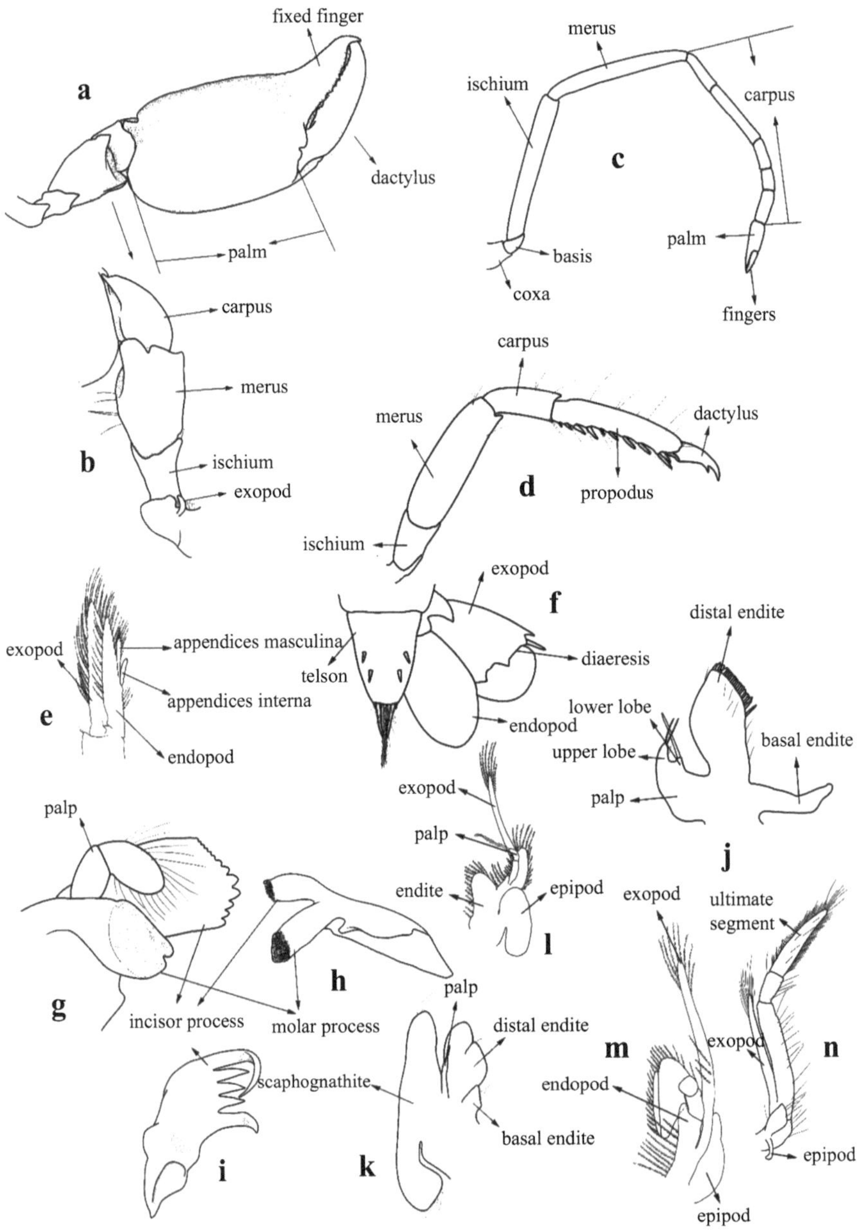


**Fig. 1.2** Schematic drawing of alpheid shrimp: (**a**) and (**b**) major cheliped; (**c**) second pereiopod; (**d**) third pereiopod; (**e**) second pleopod; (**f**) telson and right uropod; (**g–i**) mandible; (**j**) maxillule; (**k**) maxilla; (**l**) first maxilliped; (**m**) second maxilliped; (**n**) third maxilliped

lina in both males and females. **Uropods:** the exopod is articulated, and is extending as a tooth laterally and bearing a contiguous spine; the endopod is rounded; the uropods are usually less taxonomic significant. **The telson:** the telson is subtriangular and truncate distally; dorsal surface has two pairs of spines; posterior margin is armed with two pairs of posterolateral spines, and is straight, convex or notched between those posterolateral spines; the location of the dorsal spines and the shape of the posterior margin are important characters to define species.

Since the twentieth century, there are lots of taxonomic works on alpheid shrimps that has been done. Coutière (e.g. 1899, 1905, 1908, 1909) who established many genera and described many species, made great contribution to Alpheidae. De Man (1910, 1911) reported the alpheid shrimps of Indonesia through the Siboga Expedition, and described many species. Banner & Banner (e.g. 1966, 1975) has done a series of regional works from the West-Pacific and Indian Oceans, and also recorded the habitats of the shrimps. Miyake (1998) and Miya (e.g. 1997) reported the alpheids from Japan waters. Recently, Komai (e.g. 2009, 2011, 2015) and Anker (e.g. 2000, 2001, 2007, 2008, 2010, 2011a, 2011b, 2012, 2017) did great contribution to the Alpheidae for establishing many genera and describing many species. Furthermore, the taxonomic works of the Alpheidae from China seas are listed as following. Yu (1935) described 6 species of the genus *Alpheus* and two of them were new to science. Liu (1995) reported 4 species of the genus *Alpheus*. Liu & Lan (1980) described and illustrated 18 species of the genus *Alpheus* from Xisha Islands. Jeng & Chang (1985) and Jeng (1997) reported the alpheid shrimps from Taiwan. Bruce (1990) reported the alpheid shrimps from Hong Kong. Sha & Liu (2007, 2008) recorded the species of *Athanas* and *Arete*. Cui & Sha (2014) described one newly recorded species of *Alpheus* from Beibu Gulf. Wang & Sha (2015–2017) did some works of the *Automate*, *Salmoneus*, *Stenalpheops*, *Synalpheus* and *Thuylamea*. However, a systematic morphological research on the Alpheidae is needed. Based on more than 2000 specimens collected from China seas, 145 species belonging to 16 genera of the Alpheidae are described and illustrated herein.

## 1.1 Material and Methods

The present more than 2000 specimens were collected from China seas, including the Bohai Sea, the Yellow Sea, the East and South China Sea mainly by Agassiz trawl. Specimens were collected by the National Comprehensive Oceanographic Survey (1958–1960), the China-Vietnam Comprehensive Oceanographic Survey of the Tonkin Gulf (1959–1960, 1962), the China-Japan Joint Investigation of the Hainan Island (1997), the China-Germany Joint Investigation of the Hainan Island (1990, 1992), the CAS Xisha Islands Marine Biological Investigation (1956–1985), the CAS Nansha Islands Multi-disciplinary Investigation (1985, 1987–2000), and so on. The specimens were immediately fixed in 10% formalin and then deposited in the Marine Biological Museum, Chinese Academy of Sciences (MBMCAS), Qingdao, China, and preserved in the 75% ethanol now. Specimens were examined and dissected under a stereomicroscope (Nikon SMZ1500, Japan). Line drawings were completed in the Adobe Photoshop CS6 with graphics tablet. Carapace length (CL) was measured from the posterior margin of the orbit to the posterior margin of the carapace and is used herein as an indication of specimen size. All measurements are in millimeters (mm).

## 1.2 Zoogeographic Summary

145 species within 16 genera of the Alpheidae from the China seas are descripted in the present report, including 3 species within 2 genera in the Bohai Sea; 21 species within 4 genera in the Yellow Sea; 23 species within 6 genera in the East China Sea; 132 species within 13 genera in the South China Sea and 27 species within 8 genera in Taiwan (Fig. 1.3).

The Bohai and Yellow Sea: is a marginal sea in the northwestern Pacific Ocean, semi-enclosed by China and the peninsula of Korea. The Bohai and Yellow Sea are each with an average depth of 18 and 45.3 m, respectively. The macro faunal communities in this area are affected by several factors such as: Several large rivers flowing into this area, The Yellow Sea Cold Water Mass, the Yellow Sea Warm Current, etc.

The East China Sea: has the most extensive continental shelf in the northwestern Pacific

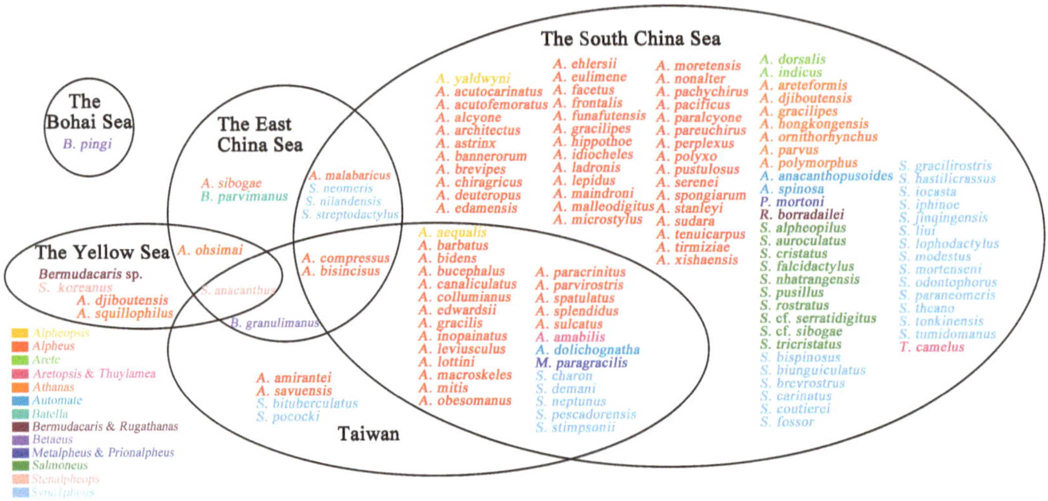


**Fig. 1.3** The alpheid shrimps in each part of China seas, except for the following 15 species: *Alpheus brevicristatus* and *Alpheus digitalis* found from the whole China seas; *Alpheus euphrosyne, Alpheus hoplocheles* and *Alpheus strenuus strenuus* found from the whole China seas except the Bohai Sea; *Alpheus balaenodigitus, Alpheus diadema, Alpheus leptocheles, Alpheus pacificus, Athanas dimorphus* and *Athanas japonicus* found from both the Yellow Sea and the South China Sea; *Alpheus japonicus, Alpheus malabaricus* and *Alpheus rapacida* found from the Yellow Sea, the East and South China Sea; *Alpheus lobidens* found from the Yellow Sea, the South China Sea and Taiwan

Ocean and covers an area of 7.7 × 10 km². Its hydrological characteristics are mainly influenced by the coastal water, the Yangtze River-diluted water and the Kuroshio Current.

The South China Sea: is the largest marginal sea of the northwest Pacific Ocean, with an average depth of 1212 m and a maximum depth of 5377 m.

Taiwan: the north part of it belongs to the East China Sea, and the south part of it belongs to the South China Sea according to the Scheme of the World Maine Biogeographic Regions shown in Liu (2008).

The Sampling stations of each alpheid genus are shown in Fig. 1.4.

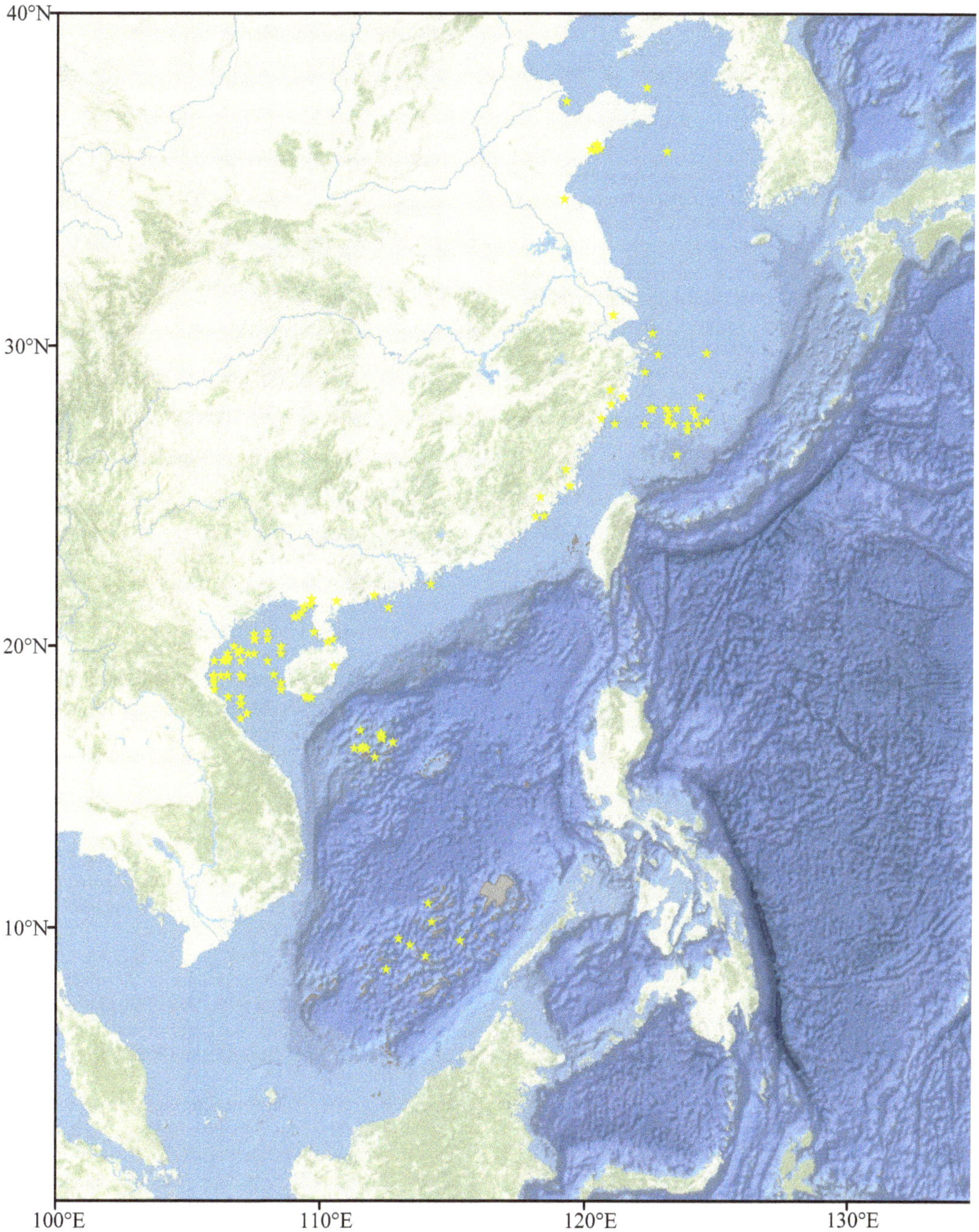


**Fig. 1.4** Location map of the sampling sites of the Alpheidae from China seas

## References


Anker A (2000) Taxonomical problems of the goby-associated species of *Alpheus* (Decapoda, Alpheidae). I.O.P. Diving News 11:2–7. (in Japanese, with English abstract)

Anker A (2001) Two new species of snapping shrimps from the Indo-Pacific, with remarks on colour patterns and sibling species in Alpheidae (Crustacea: Caridea). Raffles Bull Zool 49(1):57–72

Anker A (2007) New species and records of alpheid shrimps, genera Salmoneus Holthuis and Parabetaeus Coutière, from the tropical western Atlantic (Decapoda, Caridea). Zootaxa 1653:21–39

Anker A (2008) A worldwide review of stygobiotic and stygophilic shrimps of the family Alpheidae (Crustacea, Decapoda, Caridea). Subterranean Biol 6:1–16

Anker A (2010) The shrimp genus *Salmoneus* Holthuis, 1955 (Crustacea, Decapoda, Alpheidae) in the tropical western Atlantic, with description of five new species. Zootaxa 2372:177–205

Anker A (2011a) Three new species of the alpheid shrimp genus *Salmoneus* Holthuis, 1955 (Crustacea, Decapoda) from the tropical western Pacific. Zootaxa 2839:67–84

Anker A (2011b) Two new species of *Salmoneus* Holthuis, 1955 with a deep dorsal depression on the carapace (Crustacea, Decapoda, Alpheidae). Zootaxa 3041:39–50

Anker A (2012) Revision of the western Atlantic members of the *Alpheus armillatus* H. Milne Edwards, 1837 species complex (Decapoda, Alpheidae), with description of seven new species. Zootaxa 3386(1):109

Anker A (2017) Strongly carinate species of Alpheopsis Coutière, 1897 of the tropical Atlantic and eastern Pacific, with redescription of *A. trigona* (Rathbun, 1901) and description of three new species (Malacostraca: Decapoda: Alpheidae). Zootaxa 4277(2):199–227

Anker A, De Grave S (2008) *Zuzapheus* Ríos and Duffy, 2007: a junior synonym of Synalpheus Bate, 1888 (Decapoda: Alpheidae). J Crustac Biol 28:735–740

Anker A, Komai T (2004) Descriptions of two new species of alpheid shrimps from Japan and Australia, with notes on taxonomy of *Automate* De Man, *Coronalpheus* Wicksten and *Bermudacaris* Anker and Iliffe (Crustacea: Decapoda: Caridea). J Nat Hist 38(15):1895–1914

Anker A, Marin IN (2006) New records and species of Alpheidae (Crustacea: Decapoda) from Vietnam. Part I. Genus *Salmoneus* Holthuis, 1955. Raffles Bull Zool 54(2):295–319

Anker A, Pachelle PPG, De Grave S, Hultgren KM (2012) Taxonomic and biological notes on some Atlantic species of the snapping shrimp genus *Synalpheus* Spence Bate, 1888 (Decapoda, Alpheidae). Zootaxa 3598:1–96

Banner AH, Banner DM (1966) The alpheid shrimp of Thailand. Siam Soc Monogr Ser 3:1–168

Banner DM, Banner AH (1975) The alpheid shrimp of Australia. Part 2: the genus *Synalpheus*. Rec Aust Mus 29(12):267–389

Banner DM, Banner AH (1982) The alpheid shrimp of Australia part III: the remaining alpheids, principally the genus Alpheus, and the family Ogyrididae. Rec Aust Mus 34(1):1–357

Banner DM, Banner AH (1985) The alpheid shrimp of Indonesia, based upon J. G. De Man's "the Decapoda of the Siboga expedition, part II, family Alpheidae." (1911). Mar Res Indones 25:1–79

Barber P, Boyce SL (2006) Estimating diversity of indo-pacific coral reef stomatopods through DNA barcoding of stomatopod larvae. Proc R Soc B:1–9

Bruce AJ (1976) A report on a small collection of shrimps from the Kenya National Marine Parks at Malindi, with notes on selected species. Zoologische Verhandelingen, Leiden 145:3–72

Bruce AJ (1990) Additions to the marine shrimp fauna of Hong Kong. In: Morton B (ed) Proceedings of the Second International Marine Biological Workshop: The Marine Flora and Fauna of Hong Kong and Southern China, Hong Kong, 1986. Hong Kong, Hong Kong University Press, pp 611–648

Chace FA Jr (1988) The caridean shrimps (Crustacea: Decapoda) of the "Albatross" Philippine expedition, 1907–1910, part 5: family Alpheidae. Smithsonian Contributions to Zoology 466:1–99

Correa C, Thiel M (2003) Mating systems in caridean shrimp (Decapoda: Caridea) and their evolutionary consequences for sexual dimorphism and reproductive biology. Rev Chil Hist Nat 76:187–203

Coutière H (1899) Les "Alpheidae" morphologie externe et interne, formes larvaires, bionomie. Theses presenteesala Faculte des Sciences de Paris, Ser. A, No. 321 No. d'ordre 980. 559 pp

Coutière H (1905) Les Alpheidae. In: Stanley Gardiner J (ed) The fauna and geography of the Maldive and Laccadive archipelagoes 2(4):852–921, Plates 70–87

Coutière H (1908) Sur quelques nouvelles espèces d'Alpheidae. Bull Société philomathique de Paris 10(9):191–216

Coutière H (1909) The American species of snapping shrimps of the genus Synalpheus. Proc U S Natl Mus Smithsonian Inst Washington 36(1659):1–93

Cui DL, Sha ZL (2014) A new record of a species of *Alpheus* (Decapoda: Caridea: Alpheidae) from China. Chin J Oceanol Limnol 32(5):1123–1127

De Grave S, Fransen CHJM (2011) Carideorum catalogus: the recent species of the dendrobranchiate, stenopodidean, procarididean and caridean shrimps (Crustacea: Decapoda). Zoologische Mededelingen. Leiden 89(5):195–589

De Man JG (1910) Diagnoses of new species of macrurous decapod Crustacea from the «Siboga–Expedition», V. Tijdschrift der Nederlandsche Dierkundige Vereeniging 11: 287–319

De Man JG (1911) The Decapoda of the Siboga Expedition, Part II Family Alpheidae. Siboga Expedition Monogr 39a1:18–464, Plates XXIII

Duffy JE (1992) Host use patterns and demography in a guild of tropical sponge-dwelling alpheid shrimp. Biol J Linnean Soc Lond 58:307–324

Duffy JE (1996) *Synalpheus regalis*, new species, a sponge-dwelling shrimp from the Belize barrier reef, with comments on host specificity in *Synalpheus*. J Crustac Biol 16(3):564–573

Duffy JE (2003) In: Kikuchi T, Azuma N, Higashi S (eds) The ecology and evolution of eusociality in sponge-dwelling shrimp. Genes, behavior and evolution in social insect. Hokkaido University Press, Sapporo, pp 217–252

Duffy JE, Macdonald KS (1999) Colony structure of the social snapping shrimp *Synalpheus filidigitatus* in Belize. J Crustac Biol 19(2):283–292

Duffy JE, Morrison CL, Macdonald KS (2002) Colony defense and behavioral differentiation in the eusocial shrimp *Synalpheus regalis*. Behav Ecol Sociobiol 51:488–495

Felder DL, Chaney AH (1979) Decapod fauna of Seven and One-Half Fathom Reef, Texas: species composition, abundance and diversity. Contrib Mar Sci 22:1–29

Hebert PD, Cywinska A, Ball SL, Dewaard JR (2003) Biological identifications through DNA barcodes. Proc Biol Sci 270:313–321

Hultgren KM et al (2014) Phylogenetic relationships within the snapping shrimp genus Synalpheus (Decapoda: Alpheidae). Mol Phylogenet Evol 77:116–125

Jeng MS (1997) Studies on the land and aquatic decapod crustacean fauna of the Kenting National Park (II). Conservation Research Report, 96. Kenting National Park, Taiwan

Jeng MS, Chang KH (1985) Snapping shrimps (Crustacea: Decapoda: Alpheidae) of Taiwan. Bull Inst Zool Acad Sinica 24:241–256

Karplus L, Thompson A (2011) The partnership between gobiid fishes and burrowing alpheid shrimps. In: Patzner RA, Van Tassell JL, Kovacic M, Kapoor BG (eds) The biology of gobies. Science Publishers, Boca Raton, pp 559–608

Komai T (2009) A new species of the alpheid shrimp genus *Salmoneus* (Decapoda, Caridea) from the Ryukyu Islands, Japan, associated with a callianassid ghost shrimp (Decapoda, Thalassinidea). Crustaceana 82:869–880

Komai T (2011) Deep-sea shrimps and lobsters (Crustacea: Decapoda: Dendrobranchiata and Pleocyemata) from the Sagami Sea and Izu Islands, Central Japan. Mem Natl Mus Nat Sci 47:279–337

Komai T (2015) A new species of the snapping shrimp genus *Alpheus* (Crustacea: Decapoda: Caridea: Alpheidae) from Japan, associated with the innkeeper worm Ikedosoma elegans (Annelida: Echiura: Echiuridae). Zootaxa 4058(1):101–111

Komai T, Fujita Y (2018) A new genus and new species of alpheid shrimp from a marine cave in the Ryukyu Islands, Japan, with additional record of Salmoneus antricola Komai, Yamada & Yunokawa, 2015 (Crustacea: Decapoda: Caridea). Zootaxa 4369(4):575–586

Komai T, Ohtomi J (2018) A new deep-sea species of the snapping shrimp genus Alpheus Fabricius, 1798 (Decapoda: Caridea: Alpheidae) from Kagoshima Bay, Japan. Zootaxa 4434(1):099–110

Knowlton N (1986) Cryptic and sibling species among the decapod Crustacea. J Crustac Biol 6(3):356–363

Knowlton N (1993) Sibling species in the sea. Annu Rev Ecol Syst 24:189–216

Knowlton N (2000) Molecular genetic analyses of species boundaries in the sea. Hydrobiologia 420:73–90

Knowlton RE, Moulton JM (1963) Sound production in the snapping shrimps Alpheus (Crangon) and Synalpheus. Biol Bull 125:311–331

Liu RY (1995) Economic shrimps and prawns of Northern China. Marine Biological Institute of Academy Science, Beijing, i–iii+73pp. (in Chinese)

Liu RY, Lan JY (1980) On a collection of the genus Alpheus (Crustacea Decapoda) from the Xisha Islands, Guangdong Province, China. Studia Marina Sinica 17:77–115. (in Chinese)

Mathews LM (2006) Cryptic biodiversity and phylogeographical patterns in a snapping shrimp species complex. Mol Ecol 15:4049–4063

Miya Y (1997) *Stenalpheops anacanthus*, new genus, new species (Crustacea, Decapoda, Alpheidae) from the Seto Inland Sea and the Sea of Ariake, South Japan. Bulletin of the Faculty of Liberal Arts. Nagasaki University. Nat Sci 38(1):145–161

Miyake S (1998) Japanese Crustacean Decapods and Stomatopods in Color, Vol. I. Macrura, Anomura and Stomatopoda. Second Printing, Hoikusha, Osaka, vii+261 pp. (in Japanese)

Nolan BA, Salmon M (1970) The behavior and ecology of snapping shrimp (Crustacea: Alpheus heterochaelis and Alpheus normanni). Forma et function 2:289–335

Pescinelli RA, Almeida AO, Costa RC (2018) Population structure, relative growth and morphological sexual maturity of the snapping shrimp *Alpheus brasileiro* Anker, 2012 (Caridea: Alpheidae) from the south-eastern coast of Brazil. Marine Biology Research 14(6):610–620

Ramos-Tafur GE (2018) *Alpheus luiszapatai*, a new species of rare and colorful deep water alpheid shrimp (Crustacea: Decapoda: Alpheidae) from Arusí, Chocó Department, Pacific Coast of Colombia. Zootaxa 4403(3):540–556

Ríos R, Duffy JE (2007) A review of the sponge-dwelling snapping shrimp from Carrie Bow Cay, Belize, with description of *Zuzalpheus*, new genus, and six new species (Crustacea: Decapoda: Alpheidae). Zootaxa 1602:1–89

Sha ZL, Liu JY (2007) Study on Alpheidae (Crustacea, Decapoda) of China seas, genus *Athanas* Lecah. Acta Zootaxon Sinica 32(4):749–755

Sha ZL, Liu JY (2008) Studies on genus *Arete* Stimpson of Alpheidae (Crustacea, Decapoda) from China seas. Acta Zootaxon Sin 33(1):7–9

Thiel M, Baeza JA (2001) Factors affecting the social behavior ofcrustaceans living symbiotically with other marine invertebrates: a modeling approach. Symbiosis 30:163–190

Versluis M, Schmitz B, von der Heydt A, Lohse D (2000) How snapping shrimp snap: through cavitation bubbles. Science 285:2114–2117

Wang YR, Sha ZL (2015) A review of the genus *Synalpheus* (Crustacea: Decapoda: Caridea: Alpheidae) from China seas. Zool Syst 40(4):357–435

Wang YR, Sha ZL (2016a) A review of the genus *Alpheopsis* (Decapoda, Caridea, Alpheidae) from the seas of China. Crustaceana 89(9):1099–1113

Wang YR, Sha ZL (2016b) Redescription of the type species of the genus *Thuylamea* Nguyên, 2001 (Decapoda, Caridea, Alpheidae), based on a female specimen. Crustaceana 89(10):1185–1191

Wang YR, Sha ZL (2016c) Taxonomic Notes on the genus *Salmoneus* Holthuis, 1955 (Decapoda, Caridea, Alpheidae) from the South China Sea. Crustaceana 89(14):1585–1609

Wang YR, Sha ZL (2017a) Description of two new species of the genus *Automate* De Man, 1888 (Crustacea: Decapoda: Caridea) from the South China Sea. Zootaxa 4238(1):030–042

Wang YR, Sha ZL (2017b) The caudal appendix as an important character to identify various species in the genus *Stenalpheops* (Decapoda, Alpheidae). Crustaceana 90(13):1615–1640

Witt JD, Threloff DL, Hebert PD (2006) DNA barcoding reveals extraordinary cryptic diversity in an amphipod genus: implications for desert spring conservation. Mol Ecol 15:3073–3082

# 2 Taxonomy of Alpheidae from China Seas

Key to genus of Alpheidae from China seas

1. Carapace with ventral margins bearing one large notch on each side .......................................... *Stenalpheops*
-. Carapace with ventral margins not bearing one notch on each side .......................................... 2
2. Posterior surface of carapace bearing one backwards large tooth medially .......................................... *Thuylamea*
-. Posterior surface of carapace smooth, without large tooth .......................................... 3
3. Mandible without 2-segmented palp .......................................... 4
-. Mandible with 2-segmented palp .......................................... 5
4. Mandible with molar process absent .......................................... *Prionalpheus*
-. Mandible with molar process present .......................................... *Batella*
5. Sixth abdominal segment with articulate plate on posterolateral margin .......................................... 6
-. Sixth abdominal segment without articulate plate on posterolateral margin .......................................... 11
6. Chelipeds with dactylus situated ventrally .......................................... *Aretopsis*
-. Chelipeds with dactylus situated dorsally .......................................... 7
7. Carapace with rostrum and orbital teeth both absent on anterior margin .......................................... *Betaeus*
-. Carapace at least with rostrum present on anterior margin .......................................... 8
8. Eyes covered by carapace .......................................... *Alpheopsis*
-. Eyes visible in dorsal and lateral view .......................................... 9
9. Major chela in male with margins of palm rugged .......................................... *Rugathanas*
-. Major chela in male with margins of palm smooth or bearing tubercles .......................................... 10
10. Second pereiopod with carpus 4-segmented .......................................... *Arete*
-. Second pereiopod with carpus 5-segmented .......................................... *Athanas*
11. Second pleopod without appendix masculina in male .......................................... 12
-. Second pleopod with appendix masculina in male .......................................... 13
12. Carapace without orbital teeth and pterygostomial teeth .......................................... *Automate*
-. Carapace with orbital teeth and pterygostomial teeth .......................................... *Synalpheus*
13. Eyestalk visible in dorsal and lateral view .......................................... *Bermudacaris*
-. Eyestalk invisible in dorsal and lateral view .......................................... 14
14. Mandible with incisor process expand, bearing distally more than 10 irregular teeth .......................................... *Metalpheus*
-. Mandible with incisor process not expand, bearing distally not more than 10 irregular teeth .......................................... 15
15. Major chela with cutting edges of dactylus bearing plunger .......................................... *Alpheus*
-. Major chela with cutting edges of dactylus not bearing plunger .......................................... *Salmoneus*



Z.-l. Sha et al., *The Alpheidae from China Seas*,
https://doi.org/10.1007/978-981-15-0648-2_2

## 2.1 Genus *Alpheopsis* Coutière, 1897

**Diagnosis** Rostrum present. Orbital teeth present or absent. Cornea of eyes always visible in dorsal and lateral view. Antennular peduncle short and stout. Scaphocerite usually broad, lateral tooth usually longer than blade. First pereiopod asymmetry, Fingers compressed, cutting edges without teeth or with simple arrangement of exactly fitting teeth; palm subcylindrical or with lines and depressions; carpus cup-shaped; merus roughly triangular. Second pereiopod with carpus composed by 3–5 segments. Third to fifth pereiopods robust, dactylus simple or biunguiculate; merus without spines on ventral margin. Sixth abdominal segment without articulated plate. Posterior margin of telson convex (Banner & Banner, 1953).

Key to species of the genus of *Alpheopsis* from China seas

1. Major chela subcylindrical; Stylocerite not reaching to end margin of second antennular segment ......*A. aequalis*

-. Major chela with mesial face flattened, lateral face rounded; Stylocerite reaching to end margin of third antennular segment ...........................................................................................................*A. yaldwyni*

### 2.1.1 *Alpheopsis aequalis* Coutière, 1897 (Figs. 2.1 and 2.2)

*Alpheopsis aequalis* Coutière, 1897b: 382; Wang & Sha, 2016a: 1100, figs. 1, 2.

*Alpheopsis equalis* –. Banner, 1953: 15, fig. 4a–o; Banner AH & Banner DM, 1966b: 33, fig. 6; Banner DM & Banner AH, 1973: 342, fig. 16.

*Alpheopsis equalis* var. *truncatus* Coutière, 1903: 89, Figs 37–38.

*Alpheopsis consobrinus* De Man, 1910: 305.

**Material Examined** MBM 284426, 1 ovigerous♀, CL 3.9 mm, TL 9.7 mm, Chenhang (Dengqing) Island, Xisha Islands, Xu Jie-shan & Fan Zhen-gang Coll., 4 May 1958; MBM 158597, 1♂, CL 4.1 mm, TL 10.2 mm, 1♀, CL 2.7 mm, TL 7.6 mm, Yongshu Reef, Nansha Islands, Li Xin-zheng Coll., 18 May 1993.

**Description** Carapace smooth, glabrous. Rostrum acute, reaching about 1/2 length of visible part of first segment of antennular peduncle, about as long as wide at base, termination in lateral view acute. Orbital teeth absent. Pterygostomial angle produced into acute tooth. Cardiac notch well developed. Eyes well developed, cornea invisible dorsally, partly visible laterally.

Endopod of second pleopod with appendices interna and masculina in male, appendix interna only in female. Telson about 1.7 times as long as wide at base dorsal surface with two pair of spines (as show in Fig. 11, left side with 3 spines was unusual) situated at some distance from lateral margin, anterior and posterior pair situate about 2/5 and 7/10 length of telson, respectively; posterior margin rounded, bearing two pairs of slender posterolateral spines, mesial about twice as long as lateral; anal tubercles absent. Uropods slightly exceeding telson; diaeresis with one strong acute spine; distolateral spine long, slender, reaching posterior margin of exopod.

Antennular peduncle moderately slender; first segment longest; second segment slightly longer than wide; stylocerite reaching about 3/4 length of second segment of antennular peduncle, acute distally; lateral flagellum biramous, shorter ramus composed of at least three segments bearing tufts of aesthetascs; proximal fused portion composed of at least four segments. Basicerite bearing strong ventrolateral tooth; scaphocerite oval, anterior margin of blade rounded; distolateral tooth strong, acute, reaching beyond anterior margin of blade, reaching about end of third segment of antennular peduncle; carpocerite moderately slender, overreaching end of third segment of antennular peduncle.

Mandible with incisor process bearing distally 10 rounded irregular teeth, with fourth and fifth larger than others, molar process typical for genus. Maxillule with two endites, distal endite with numerous large spine-like setae on inner margin, basal endite with long setae on distal margins; palp bilobed, lower lobe with one robust plumatus seta, upper lobe rounded. First and

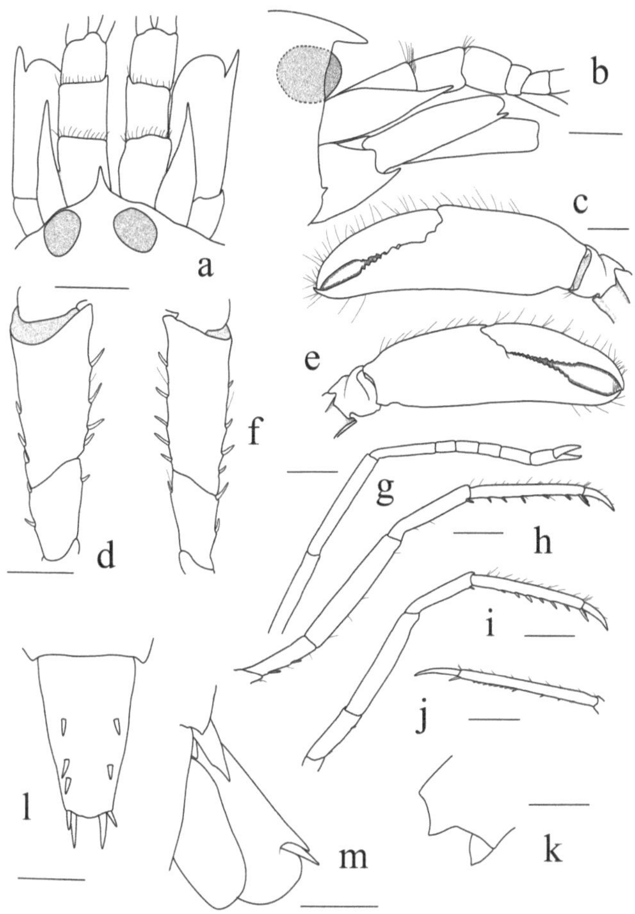


**Fig. 2.1** *Alpheopsis aequalis* Coutière, 1897. MBM 158597, male, CL 4.1 mm, **a**, anterior region, dorsal view; **b**, anterior region, lateral view; **c**, major chela; **d**, same, merus and ischium; **e**, minor chela, **f**, same, merus and ischium; **g**, right second pereiopod; **h**, right third pereiopod; **i**, left fourth pereiopod; **j**, dactylus and propodus of left fifth pereiopod; **k**, articulated pleura of sixth abdominal segment; **l**, telson; **m**, right uropod (scale bar = 0.5 mm. **c** and **e** same scale bar; **d** and **f** same scale bar)

second maxilliped normal. Third maxilliped moderately stout; ultimate segment very setose, distally tapering, bearing 2 subapical spines; arthrobranch normally developed.

First chelipeds nearly symmetrical in shape and size. Chela subcylindrical, without grooves on surface; cutting edge of dactylus and pollex bearing teeth on about 3/5 length, distal 2/5 length without teeth, stiff setae along full length of cutting edge of fingers, fingers tips corneous, crossing when closed; palm nearly as long as fingers. Carpus with outstanding flange into which proximal end of propodus fit, with slight constriction before this flange. Merus with ventromesial margin bearing 4 or 5 large spines, and dorsal margin bearing 3 or 4 spines. Ischium with ventral margin bearing 1 spine, and dorsal margin bearing 1 distal spine and 1 medial spine.

Second pereiopod slender, fingers slightly longer than palm. Carpus 5-segmented, length ratio subequal to 2: 0.6: 0.7: 0.7: 1.

Third pereiopod slender, with dactylus simple, gradually curved, about 0.3 times as long as propodus, with setae. Propodus slightly shorter than merus, armed with 5 spines along ventral margin, and distally one pair spines. Carpus about 3/4 length of merus, with distal 1 spine on ventral margin. Merus more than twice length of ischium, unarmed, about 8 times as long as wide at base. Ischium armed with 2 spines on ventrolateral margin

Fourth and fifth pereiopod similar to third pereiopod, bur more slender. Fifth pereiopod

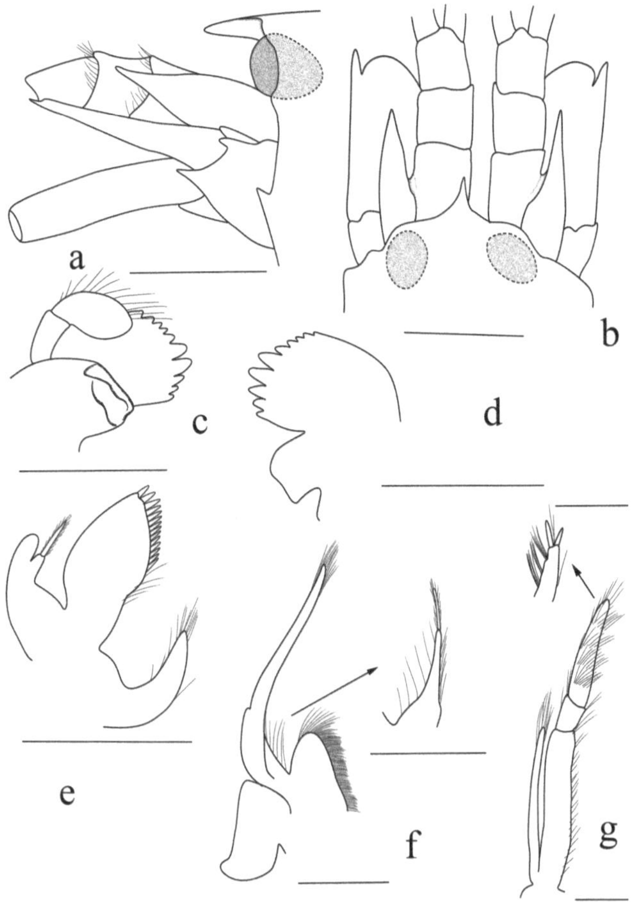


**Fig. 2.2** *Alpheopsis aequalis* Coutière, 1897. MBM 158597, female, CL 2.7 mm, **a**, anterior region, lateral view; **b**, same, dorsal view. MBM 158597, male, CL 4.1 mm, **c**, **d**, left mandible; **e**, left maxillule; **f**, left first maxilliped; **g**, left third maxilliped (scale bar = 0.5 mm)

with dactylus more slender than that of third and fourth pereiopods. Propodus with 3 small spine and at least 8 rows of short setae along ventral margin.

**Remarks** *Alpheopsis aequalis* is originally described by Coutière (1897b) in a very brief way and without illustrations. Coutière (1903) illustrated the anterior part of the carapace. Wide variation has found in its morphological characters (De Man, 1910; Armstrong, 1941; Banner, 1953; Banner & Banner, 1973). Therefore, Anker et al. (2005) stated that *A. aequalis* might be a species complex, but the present material is not enough to prove it. However, the type-specimens have gone missing from the Paris museum. The neotype should be selected and described from the type locality for the wide variation of this species. The present specimens are morphologically similar to each other, and are more similar to the descriptions of De Man (1910) and Banner & Banner (1966b) than to other re-descriptions. Additionally, the differences between the present material and the descriptions published previously (Banner, 1953; Banner & Banner, 1966b, 1973) are listed below: the longer rostrum shown by Banner & Banner (1973, fig. 16f) not found in the present specimens, and the rostrum not overreaching end margin of the first antennular segment in the present material; the longer stylocerite shown by Banner & Banner (1973, fig. 16a) not found in the present specimens, and the stylocerite of the present specimens not overreaching end margin of the second antennular segment; the pterygostomial angle well developed into an acute tooth in the present specimens. Additionally, Armstrong (1941) discusses the variation of the pterygostomial angle. The rounded pterygostomial angle as figured by Armstrong (1941) and Banner (1953) is not found in the present specimens. Moreover, the mouthparts of the species are now dissected and illustrated. There are some differences found between this species and other species of the genus (Chace, 1972; Anker et al., 2005). The incisor process bears 10 rounded irregular teeth distally, while it only bears 6 teeth in *A. labis* Chace, 1972 (cf. Chace, 1972, fig. 15g), and 7 teeth in *A. azorica* Anker, d'Udekem d'Acoz & Poddoubtchenko, 2005a and *A. africana* Holthuis, 1952 (cf. Anker et al., 2005). The palp of the mandible is usually 2-segmented in the genus, while it is described as 3-segmented (Chace, 1972, fig. 15g) in *A. labis* (with broadly rounded distal segment). Considering the fact that molecular evidence is not available, the present specimens are treated as *A. aequalis*.

**Distribution** Xisha Islands and Nansha Islands, the South China Sea; from Red Sea and Eastern Africa to Hawaii; Society Islands (Bora Bora, Tahiti); Easter Island; Savaii, eastern reef. intertidal to 80 m depth.

### 2.1.2 *Alpheopsis yaldwyni* Banner and Banner, 1973 (Figs. 2.3, 2.4, and 2.5)

*Alpheopsis yaldwyni* Banner DM & Banner AH, 1973: 344–346, fig. 17; 1985: 9, fig. 1; Wang & Sha, 2016a: 1105, figs. 3–5.

**Material Examined** MBM 284428, 1♂, CL 4.6 mm, TL 11.4 mm, Chenhang (Dengqing) Island, Xisha Islands, Fan Zhen-gang & Xu Jie-shan Coll., 16 April 1958; MBM 158631, 1♂, CL 6.1 mm, TL 15.1 mm, Jinqing Island, Xisha Islands, 30 April 1958; MBM 158003, 2♀ (1 ovigerous), CL 4.6, 7.2 mm, TL 12.2, 16.5 mm, Yongxing (Wude) Island, Xisha Islands, 12–13 June 1958; MBM 284427, 1♀, CL 5.1 mm, TL 11.3 mm, Chenhang (Dengqing) Island, Xisha Islands, Fan Zhen-gang & Xu Jie-shan Coll., 4 May 1958.

**Description** Carapace smooth, glabrous. Rostrum acute, reaching from slightly beyond end of first segment to end of second segment of antennular peduncle, about as long wide at base or distinctly wider at base than long. Orbital teeth absent. Pterygostomial angle produced into acute tooth. Cardiac notch well developed. Eyes well developed, cornea invisible dorsally, partly visible or totally invisible laterally.

Endopod of second pleopod with appendices interna and masculina in male, appendix interna only in female; appendix masculina with distally long setae slightly longer than appendix interna. Telson about 1.3 times as long as wide at base; dorsal surface with two pairs of spines situated at some distance from lateral margin, anterior and posterior pair of spine inserted about 1/2 and 3/4 length of telson, respectively; posterior margin rounded, bearing two pair of slender posterolateral spines, mesial about 3 times as long as lateral. Uropods exceeding telson; diaeresis with two subacute teeth, one on each side of distolateral spine; distolateral spine long, slender, not reaching posterior margin of exopod.

Antennular peduncle moderately stout; first segment longest; second segment slightly wider than long; third segment near as long as second; stylocerite reaching near end of third segment of antennular peduncle, acute distally; lateral flagellum biramous, shorter ramus composed of at least three segments bearing tufts of aesthetascs; proximal fused portion composed of at least four segments. Basicerite bearing strong distolateral tooth, and rounded distodorsal tooth, shorter than distolateral tooth; scaphocerite oval, anterior margin of blade rounded; distolateral tooth strong, acute, reaching far beyond anterior margin of blade, slightly longer than end of third segment of antennular peduncle; carpocerite moderately slender, far overreaching end of third segment of antennular peduncle; flagellum relatively slender.

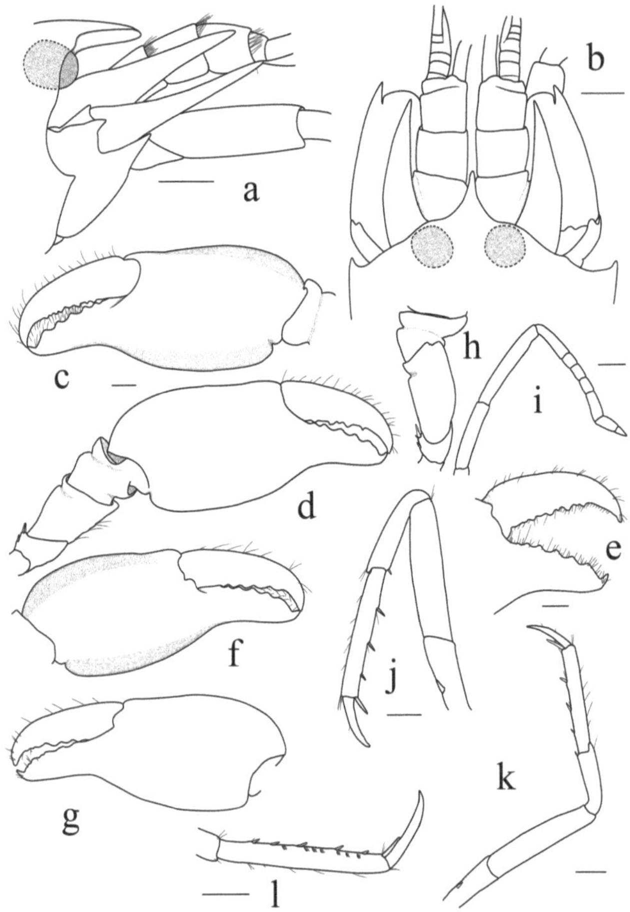


**Fig. 2.3** *Alpheopsis yaldwyni* Banner & Banner, 1973. MBM 158631, male, CL 6.1 mm, **a**, anterior region, lateral view; **b**, same, dorsal view; **c**, major chela, outer face; **d**, same, inner face; **e**, fingers of major chela; **f**, minor chela, outer face; **g**, same, inner face; **h**, carpus, merus and ischium; **i**, right second pereiopod; **j**, right third pereiopod; **k**, left fourth pereiopod; **l**, dactylus and propodus of left fifth pereiopod (scale bar = 0.5 mm. **c**, **d**, **f**, **g** and **h** same scale)

Mandible with incisor process bearing distally 13 rounded irregular teeth, with sixth and seventh larger than others, molar process typical for genus. Maxillule with two endites, distal endite with numerous large spine-like setae on inner margin, basal endite with long setae on distal margins; palp bilobed, lower lobe with one robust plumatus seta, upper lobe rounded. Maxilla with two endites, distal endite bilobed, both lobes with moderately short setae on inner margin; basal endite smaller, bearing long setae distally. First and second maxilliped normal. Third maxilliped moderately stout; ultimate segment very setose, distally tapering, tip bearing 2 spines; arthrobranch normally developed.

Frist cheliped only slightly asymmetrical in shape, but minor cheliped smaller than major cheliped in size. Major cheliped with fingers about 0.7 length of palm, cutting edges armed with irregular, not exactly fitting teeth, and densely short setae; tips corneous, crossing when closed. Palm without grooves, lateral face convex; mesial face completely flat; superior margin rounded. Carpus heavy, wider than long, distally bearing superior transverse constriction or groove and expanding into collar around base of palm. Merus stout, unarmed. Ischium stout, dorsal margin bearing 1 medial spine and one pair of spines. Minor cheliped with fingers about 0.7 length of palm, cutting edge armed with regular, exactly

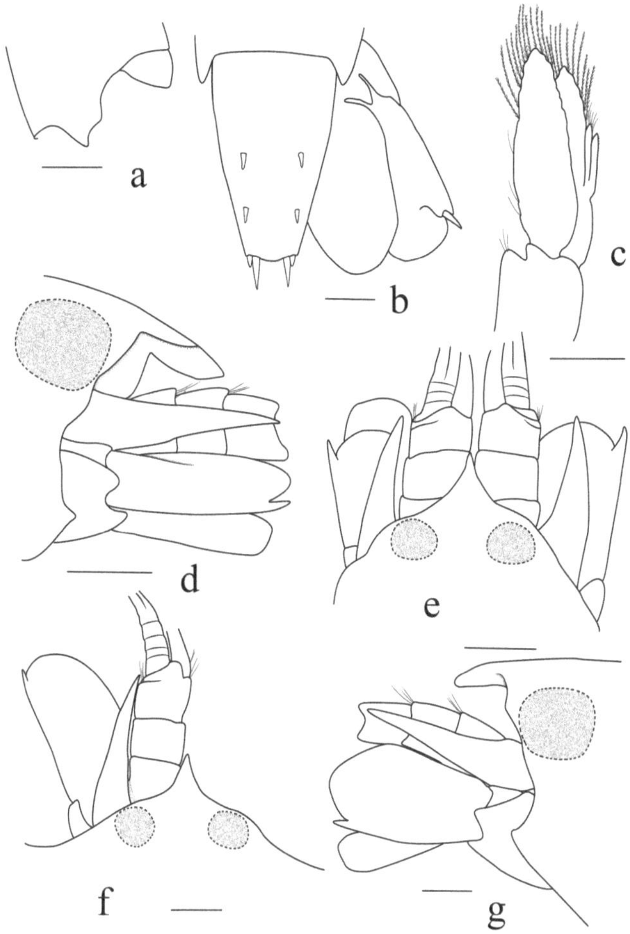


**Fig. 2.4** *Alpheopsis yaldwyni* Banner & Banner, 1973. MBM 158631, male, CL 6.1 mm, **a**, articulated pleura of sixth abdominal; **b**, telson and right uropod; **c**, left second pleopod. MBM 158003, female, CL 4.6 mm, **d**, anterior region, lateral view; **e**, same, dorsal view. MBM 158003, ovigerous female, CL 7.2 mm, **f**, anterior region, dorsal view; **g**, same, lateral view (scale bar = 0.5 mm)

fitting teeth, tips corneous, crossing when closed. Merus similar to that of major cheliped, but more slender. Ischium and carpus like that of major cheliped.

Second pereiopod with chela simple, palm slightly longer than fingers. Carpus 5-segmented, with length ratio subequal to 1.2: 0.4: 0.5: 0.5: 0.9. Ischium distinctly shorter than merus.

Third pereiopod with dactylus about 2/5 length of propodus, simple, slender, gradually curved, with short setae on superior margin. Propodus nearly as long as merus, armed with 3 smaller spines and one pair distal larger spines on ventral margin. Carpus slightly longer than 1/2 length of merus, distal inferior margin armed with usually 1 spine. Merus about 4.2 times as long as wide at base, unarmed. Ischium armed with 1 spine on ventral margin.

Fourth and pereiopod similar to third pereiopod, but fifth pereiopod more slender; dactylus about 1/2 length of propodus. Propodus armed with 5 smaller spines and distally one pair larger spines, and distal 2/3 length with at least 7 rows of short setae.

**Remarks** *A. yaldwyni* is a newly recorded species from China seas. This species is originally described by Banner & Banner (1973) from Australia. De Grave & Fransen (2011) and Anker et al. (2015) provide a colorful photograph of the

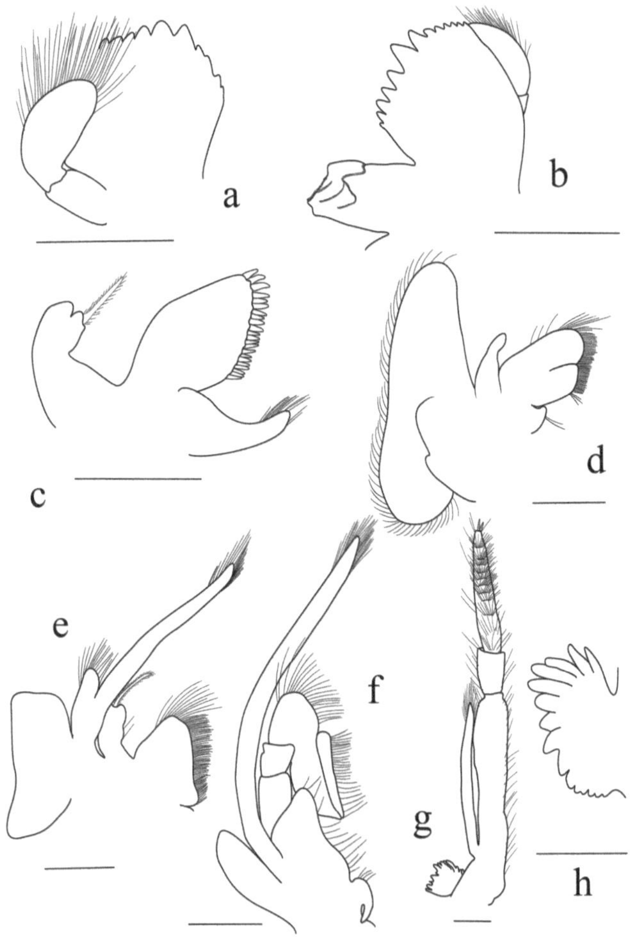


**Fig. 2.5** *Alpheopsis yaldwyni* Banner & Banner, 1973. MBM 158631, male, CL 6.1 mm, **a** & **b**, left mandible; **c**, left maxillule; **d**, left maxilla; **e**, left first maxilliped; **f**, left second maxilliped; **g**, third maxilliped; **h**, arthrobranch (scale bar = 0.5 mm)

species. The present specimens are generally agreed with the original description, except for the following characters: the rostrum varying from slightly overreaching the end of the first antennular segment to the end of the second antennular segment in the present specimens, while the rostrum only slightly beyond end margin of the first antennular segment in Banner & Banner (1973, fig. 17a, b), or only reached half-length of the second antennular segment in Banner & Banner (1985, fig. 1a); the rostrum nearly as long as wide at base in the present specimens, while the rostrum 1.5 times as long as wide at base in Banner & Banner (1973, fig. 17a); the basicerite of the antenna bearing a strong distodorsal tooth in the present specimens, while the basicerite not bearing distodorsal tooth in Banner & Banner (1973, fig. 17b; 1985, fig. 1a). *A. yaldwyni* is most similar to *A. aequalis* in *Alpheopsis*, but can readily be distinguished from *A. aequalis* by the following characters: the incisor process of the mandible bearing 13 rounded irregular teeth distally rather than 10 in *A. aequalis*; the rostrum slightly overreaching the end of the first segment of the antennular peduncle, while it not reaching that end in *A. aequalis*; the stylocerite reaching almost to end margin of the antennular peduncle, while the stylocerite not reaching end margin of the second antennular segment in *A. aequalis*; the mesial face of the major chela flattened, while the major chela sub-cylindrical in *A. aequalis*; the merus of the first chelipeds bearing no spines in all margins, while both the ventral and dorsal margin of the merus of the first chelipeds bearing moderately large spines in *A. aequalis*.

**Distribution** Xisha Islands, the South China Sea; Ambon (Rumah Tiga), Indonesia; Micronesia; Japan; Australia.

## 2.2 Genus *Alpheus* Fabricius, 1798

**Diagnosis** Rostrum usually present, triangular. Orbital hoods frequently with spiniform apex and usually demarked from rostral base by more or less pronounced depressions. Carapace with at least slight rostral carina; pterygostomian angle rounded. Eyes are completely enclosed by orbital hoods except on ventral side. Antennules usually with basal peduncular segment and stylocerite reduced. Antennal scale with blade frequently shorter than lateral spine. First chelipeds very asymmetrical. Large chela variable form, surfaces smooth or with grooves; dactylus usually with pistonlike process, which fits into cavity on fixed finger; carpus hemispherical; merus triangular in section. Small chela sometimes showing marked sexual dimorphism. Second pereiopods having carpus composed of five segments. Third to five pereiopods similar in shape; dactylus simple or biunguiculate. Sixth abdominal somite without articulated plate. Telson usually having the dorsal surface with two pair spines, with posterior margin convex (Banner & Banner, 1966).

Key to species groups of the genus *Alpheus*

1. Dorsal surface of palm of large chela with transverse groove near distally, which extending into triangular and rectangular depressions on the mesial and lateral face of the palm, respectively ........................... Edwardsii Group
-. Dorsal surface of palm of large chela with or without transverse groove near distally, if with, the groove not extending into triangular and rectangular depressions on the mesial and lateral face of the palm, respectively ....... 2
2. Palm of large chela twisted, and with 3 heavy longitudinal ridges and grooves............................................ .......................................................................................................................... Macrocheles Group
-. Palm of large chela without 3 heavy longitudinal ridges and grooves ........................................................... 3
3. Large chela having palm only with slightly longitudinal groove, and dactylus of the form of a single to double-headed hammer; Second pereiopods at times asymmetrical development .............................. Obesomanus Group
-. Large chela having palm with or without groove, and dactylus without single or double-headed hammer; Second pereiopods symmetrical development.......................................................................................... 4
4. Large chela always distinctively compressed................................................................ Brevirostris Group
-. Large chela not distinctively compressed ........................................................................................ 5
5. Palm of large chela smooth, without grooves ..................................................................... Crinitus Group
-. Palm of large chela with grooves ................................................................................................... 6
6. Longitudinal grooves absent on the palm of large chela ...................................................... Diadema Group
-. Transverse grooves absent on the palm of lager chela............................................................ Sulcatus Group

**Brevirostris Group (Fig. 2.6)**

Orbital teeth absent. Large chela always compressed, more or less quadrangular in section; dorsal face of palm with or without transverse groove proximal to dactylus. Dactylus of third pereiopod always not biunguiculate; merus usually unarmed (Banner & Banner, 1982).

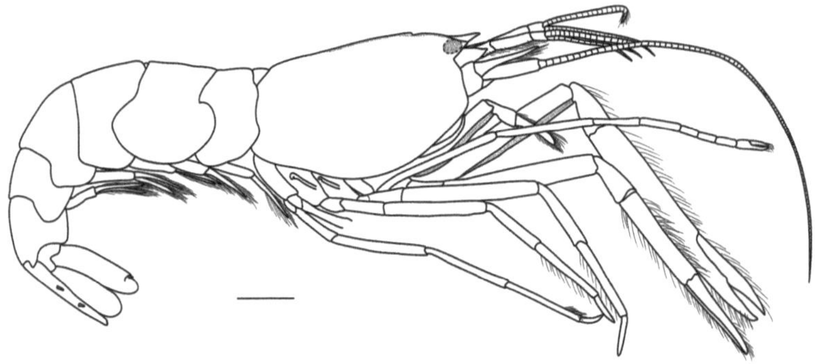

**Fig. 2.6** *Alpheus acutocarinatus* De Man, 1909. entire animal of Brevirostris Group (scale bar = 2 mm)

Key to species of Brevirostris Group from China seas

1. Rostral carina with one tooth medially ........................................ 2
-. Rostral carina with one tooth medially ........................................ 3
2. Major chela with palm bearing one longitudinal groove on lateral face ........................................ *A. sibogae*
-. Major chela with palm not bearing one longitudinal groove on lateral face ........................................ *A. acutocarinatus*
3. Body strongly compressed ........................................ *A. compressus*
-. Body not unusually compressed ........................................ 4
4. Palm of major chela bearing transverse groove sub-distally on dorsal margin ........................................ 5
-. Palm of major chela not bearing transverse groove on dorsal margin ........................................ 9
5. Rostrum short, only slightly overreaching anterior margins of orbital hoods ........................................ *A. barbatus*
-. Rostrum reaching to end margin of first antennular segment ........................................ 6
6. Second pereiopod with first carpal segment distinctly short than second segment ........................................ *A. savuensis*
-. Second pereiopod with first carpal segment sub-equal to or longer than second segment ........................................ 7
7. Fingers of minor chela sub-equal to palm ........................................ *A. moretensis*
-. Fingers of minor chela distinctly longer than palm ........................................ 8
8. Antennular peduncle stout, with second segment about 1.6 times as long as wide ........................................ *A. djiboutensis*
-. Antennular peduncle slender, with second segment about 3 times as long as wide ........................................ *A. brevicristatus*
9. Major and minor chelipeds strongly compressed, ventral margin knife-shaped ........................................ *A. leptocheles*
-. Major and minor chelipeds not unusually compressed, ventral margin rounded ........................................ 10
10. Merus of major and minor chelipeds armed with slender, acicular spines on ventral margins ........................................ 11
-. Merus of major and minor chelipeds armed with normal small spines or unarmed ........................................ 13
11. Major and minor chelipeds with merus bearing large acute tooth on dorsal margin sub-distally... ........................................ *A. macroskeles*
-. Major and minor chelipeds with merus not bearing acute tooth on dorsal margin sub-distally ........................................ 12
12. Third pereiopod with propodus only bearing setae on ventral margin; carpus of second pereiopod with first segment distinctly longer than second segment ........................................ *A. nonalter*
-. Third pereiopod with propodus bearing 5–8 spines; carpus of second pereiopod with first segment distinctly shorter than second segment ........................................ *A. pustulosus*

13. Minor chela with fingers sub-equal to palm........................................*A. tenuicarpus*
-. Minor chela with fingers distinctly longer than palm........................................14
14. Major chela with palm flattened on dorsal surface, bearing two longitudinal ridges........................................15
-. Major chela with palm sharped dorsally, not flat........................................*A. lepidus*
15. Major chela with palm bearing longitudinal ridge on lateral face........................................*A. digitalis*
-. Major chela with palm not bearing longitudinal ridge on lateral face........................................*A. rapacida*

### 2.2.1 *Alpheus acutocarinatus* De Man, 1909 (Fig. 2.7)

*Alpheus acutocarinatus* De Man, 1909a: 104; 1911: 401, fig. 94; Banner AH & Banner DM, 1966b: 120, fig. 43; Banner DM & Banner AH, 1982, 151, fig. 44; Chace, 1988: 14.

**Material Examined** CL 7.0–9.5 mm, TL 18.5–26.3 mm. MBM 159355, 5♂♂1♀, Sanya Bay, 23 May 1992; MBM 158844, 1♂, Nansha Islands, Tang Zhi-can Coll., 29 July 1988; MBM 158601, 6♀♀4♂♂, Sanya Bay, 23 May 1992; MBM 158484, 5♀♀3♂♂, Sanya Bay, 22 May 1992; MBM 158501, 7♀♀2♂♂, Sanya Bay, 23 May 1992; MBM 210152, 4♂♂, Beibu Gulf, Sun Fu-zeng Coll., 22 April 1962; MBM 210158, 1♂, Beibu Gulf, Sun Fu-zeng Coll., 12 December 1959; MBM 210147, 1♂, Beibu Gulf, Sun Fu-zeng Coll., 23 January 1962; MBM 210142, 5 larva, Beibu Gulf; MBM 210169, 1♂, Beibu Gulf, 21 April 1960; MBM 210140, 2♂♂, Beibu Gulf, 19 April 1962; MBM 210134, 3♂♂, Beibu Gulf, Wang Yong-liang Coll., 15 April 1962; MBM 210167, 3♂♂, Beibu Gulf, Sun Fu-zeng Coll. 22 April 1962; MBM 210177, 1♂, Beibu Gulf, Zhang Bao-lin Coll., 15 November 1960; MBM 210168, 1♂, Beibu Gulf, Sun Fu-zeng Coll., 11 February 1960; MBM 210160, 1♂, Beibu Gulf, Sun Fu-zeng Coll., 8 February 1960; MBM 210182, 1♂, Beibu Gulf, Sun

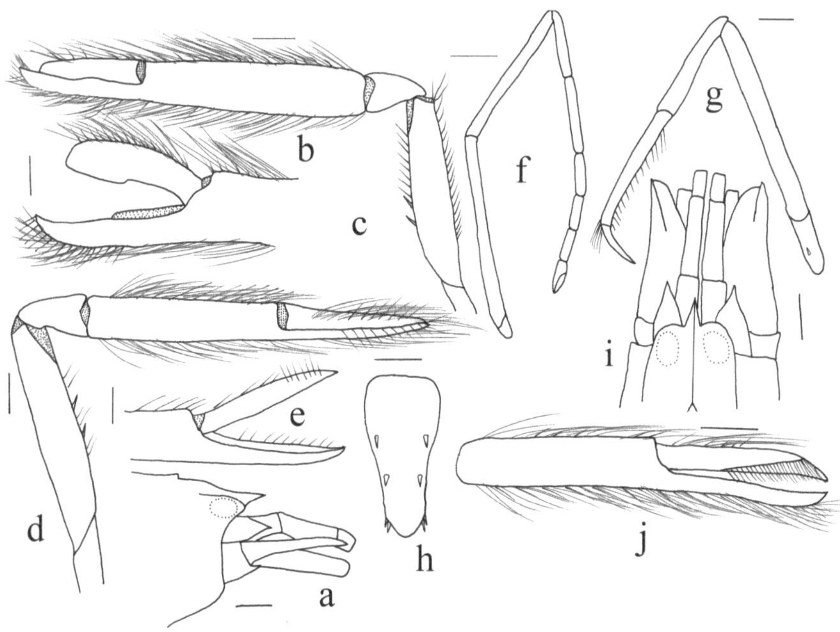


**Fig. 2.7** *Alpheus acutocarinatus* De Man, 1909. MBM 158484, female, **a**, anterior part of carapace, lateral view; **b**, major cheliped; **c**, same, fingers; **d**, minor cheliped; **e**, same, fingers; **f**, second pereiopod; **g**, third pereiopod; **h**, telson. MBM 210158, male, **i**, anterior part of carapace, dorsal view; **j**, minor chela (scale bar = 1 mm)

Fu-zeng Coll., 19 February 1960; MBM 210133, 4♀♀, Beibu Gulf, Sun Fu-zeng Coll., 16 February 1960; MBM 210170, 1♀1♂, Beibu Gulf, Sun Fu-zeng Coll., 21 April 1960; MBM 210171, 1♂, Beibu Gulf, Sun Fu-zeng Coll., 16 February 1960; MBM 210166, 1♂, Beibu Gulf, 10 February 1960; MBM 210157, 2♂♂, Beibu Gulf, Ma Xiu-tong Coll., 3 December 1960; MBM 210136, 1♂, Beibu Gulf, 15 January 1962; MBM 210179, 1♂, Beibu Gulf, Sun Fu-zeng Coll., 17 February 1960; MBM 210181, 1♀1♂, Beibu Gulf, Sun Fu-zeng Coll., 23 January 1962; MBM 210144, 3♀♀, Beibu Gulf, 22 December 1959; MBM 210172, 3♂♂1♀, Beibu Gulf, Sun Fu-zeng Coll., 22 December 1959; MBM 210138, 4 larva, Beibu Gulf, 14 December 1959; MBM 210151, 1♂, Beibu Gulf, Sun Fu-zeng Coll., 19 December 1959; MBM 210163, 2♂♂, Beibu Gulf, Zhang Wei-quan Coll., 12 April 1960; MBM 210137, 1♀, Beibu Gulf, Sun Fu-zeng Coll., 27 August 1962; MBM 210159, 2♀♀, Beibu Gulf, Zhang Bao-lin Coll., 23 November 1960; MBM 210150, 5♂♂1♀, Beibu Gulf, Sun Fu-zeng Coll., 24 April 1962; MBM 210153, 1♂, Beibu Gulf, 22 January 1962; MBM 210180, 1♂, Beibu Gulf, Sun Fu-zeng Coll., 11 April 1962; MBM 270031, 1 larva, the East China Sea, 28 August 1976; MBM 270044, 6♂♂5♀♀, the South China Sea, Liu Xi-xing Coll., 10 April 1960; MBM 270189, 1 larva, Beibu Gulf, Fan Zhen-gang Coll., 20 April 1960; MBM 270186, 1 larva, Beibu Gulf, Wang Yong-liang Coll., 14 April 1960; MBM 270184, 1 larva, Beibu Gulf, Sun Fu-zeng Coll., 12 December 1959; MBM 270173, 1♀, Beibu Gulf, Ma Xiu-tong Coll., 4 December 1959; MBM 270573, 1 larva, Beibu Gulf, 18 August 1962; MBM 270486, 1♂, Beibu Gulf, Fan Zhen-gang Coll., 18 April 1960; MBM 210139, 1 larva, Beibu Gulf, Liang Guan-sheng Coll., 11 December 1959; MBM 210105, 1♂, Beibu Gulf, Fan Zhen-gang Coll., 20 April 1960.

**Description** Rostrum triangular, sharp, reaching 2/3 length of first antennular segment; rostral carina sharp, with a tooth medially, extending to about 2/3 length of carapace; orbital hoods produced anteriorly, unarmed; orbitorostral grooves deep; pterygostomial corner rounded.

Telson about 3 and 4 times as long as anterior and posterior width, respectively; dorsal surface with two pairs of spines, inserted approximately at 1/3 and 2/3 length of telson, respectively; posterior margin strongly convex, beyond longer posterolateral spines.

Antennular peduncle with distally acute stylocerite, hardly reaching to distal margin of first antennular segment; second antennular segment about 1.5 times as long as first segment, and 2 times as long as third segment. Antenna with lateral margin of scaphocerite slightly concave, not reaching to distal margin of third antennular segment; antennal scale with distolateral tooth about as long as blade. Carpocerite reaching to distal margin of second antennular segment.

Major cheliped overreaching antennular peduncle by carpus and chela, with chela compressed, about 6 times as long as wide, dorsal and ventral margins fringed with long setae. Dactylus slightly longer than 1/2 length of palm, slightly shorter than or as long as fixed finger distally acute or rounded, fingers crossed when closed. Palm without groove, surfaces lightly granular. Carpus cup-shaped. Merus slender, ventromesial margin bearing 2–3 spines and acute tooth distally. Ischium unarmed. Minor cheliped slender, not showing distinctly sexual dimorphism, chela nearly as long as major chela, dorsal and ventral margins fringed with long setae. Fingers slightly shorter than palm; palm without groove, surfaces lightly granular. Carpus cup-shaped. Merus elongate, ventromesial margin bearing 2–4 spines and acute tooth distally. Ischium unarmed.

Second pereiopod slender, overreaching antennular peduncle by distal of merus, carpus and chela; carpus 5-segmented, segmented ratio subequal to 5–9: 10: 4–5: 4–5: 4.

Third pereiopod slender, dactylus spatulate, ventral face slightly excavate, shorter than 1/2 length of propodus. Ventral margin of propodus without spines, but long setae; carpus slightly shorter than propodus. Merus distinctly longer than propodus, unarmed. Ventral margin of ischium with 1 spine. Fourth and fifth pereiopod similar to third one, but shorter and slender, ventral margin of propodus of fifth pereiopod bearing transverse rows of setae; ventral margin of ischium without spine.

**Remarks** *A. acutocarinatus* is distinguished from other species in Brevirostris group by one acute tooth situated on the middle line of the carapace. This species is most similar to *A. sibogae* according to the acute tooth on the carapace, but it differs from *A. sibogae* by the lateral face of the major chela lacking the longitudinal depression.

The present specimens are generally similar to the descriptions of Banner & Banner (1982) and Chace (1988). However, the materials is slight variations as following: the dactylus of the major chela distinctly shorter than the fixed finger or subequal to it as shown in Banner & Banner (1982, fig. 44d); the merus of major and minor cheliped having 1–4 spines on the ventromesial margin.

**Distribution** The South China Sea; Madagascar; Vietnam; Thailand; Philippines; Indonesia; southern Queensland; Australia. Depth: 20–72 m.

### 2.2.2 *Alpheus barbatus* Coutière, 1897 (Fig. 2.8)

*Alpheus barbatus* Coutière, 1897c: 235; 1899: figs. 279, 280; De Man, 1911: 387, fig. 88; Banner DM & Banner AH, 1978: 220; 1982: 163, fig. 49; Chace, 1988: 16.

**Material Examined** This species was reported by Banner & Banner (1978) and Liu (2008) from the South China Sea, but not found in the present specimens.

**Description** Rostrum short, triangular, slightly overreaching anterior margin of orbital hoods; rostral carina extending to posterior margin of orbital hoods; orbital hoods unarmed; orbitorostral grooves shallow; pterygostomial corner rounded.

Telson elongate, 2.7 times as long as posterior marginal wide; dorsal surface with two pairs of small spines, all inserted posterior 1/3 length of telson; posterior margin nearly straight, posterolateral spines beyond posterior margin.

Antennular peduncle stout, with distally rounded stylocerite almost reaching distal margin of first antennular segment; three antennular segments nearly equal in length. Antenna with distolateral tooth of scaphocerite distinctly longer than blade, which overreaching antennular peduncle. Carpocerite overreaching distolateral spine of scaphocerite.

Major cheliped with chela compressed, fingers about 1/2 length of palm, tip rounded. Dorsal margin of palm bearing narrow and curving transverse groove nearly distal; ventral margin rounded without groove; mesial face with shallow longitudinal groove extending from near middle length of palm to middle length of fixed finger. Merus 1.9 times as long as wide, ventrolateral and ventromesial margin distally rounded, without spines. Ischium unarmed. Minor cheliped not showing sexual dimorphic, fingers 1.7 times longer than palm, cutting margin of proximal half-length of dactylus fringed with dense setae, and medial face of both fingers fringed with setae that obscure fingers almost to tips. Surfaces of palm without groove. Merus like that of major cheliped. Ischium unarmed.

Carpus of second pereiopod 5-segmented, segmented ratio subequal to 10: 5: 2: 2: 5.

Third pereiopod stout, dactylus simple, conical, 0.4 times as long as propodus. Propodus 0.7 times as long as merus, ventral margin bearing 5 and distal one pair spines. Carpus slightly shorter than propodus. Merus 3 times as long as wide, unarmed. Ischium armed with one spine on ventral margin (Banner & Banner 1982).

**Distribution** North of the South China Sea; Red Sea and eastern Africa to Philippines; Indonesia; Queensland; Australia. Depth: intertidal to 10 m.

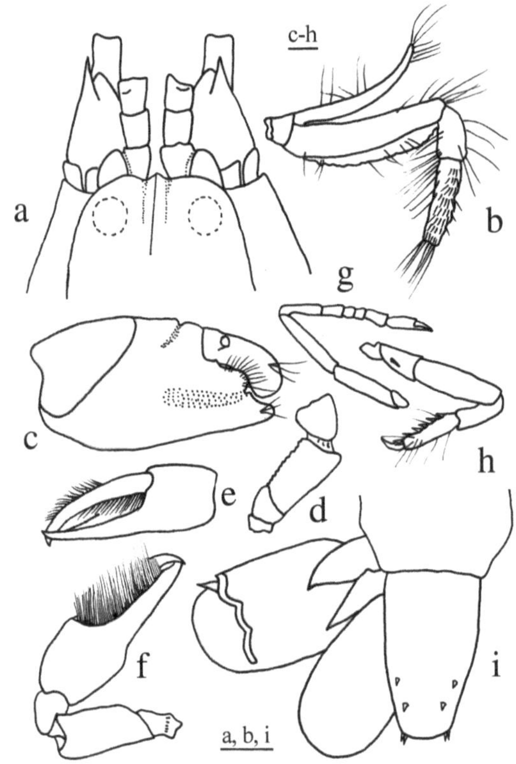


**Fig. 2.8** *Alpheus barbatus* Coutière, 1897 (copy from Banner & Banner, 1982). Female, **a**, anterior part of carapace, dorsal view; **b**, third maxilliped; **c**, major chela, lateral view; **d**, merus of major cheliped; **e**, minor chela, lateral view; **f**, minor cheliped, mesial view; **g**, second pereiopod; **h**, third pereiopod; **i**, telson (scale bar = 1 mm)

### 2.2.3 *Alpheus brevicristatus* De Haan, 1844 (Fig. 2.9)

*Alpheus brevicristatus* De Haan, 1844 (in De Haan, 1833–1850): Pl. 45, fig. 1 (1844); De Man, 1909b: 158; Jeng & Chang, 1985: 247, fig. 17; Bruce, 1990: 610; Yang & Kim, 1998: 187–193, fig. 1–4

*Alpheus kingsleyi* Miers, 1879: 54.

**Material Examined** CL 10.3–14.5 mm, TL 26.4–37.7 mm. MBM 159116, 1♂, Cangkou, Qingdao, 25 March 1955; MBM 159054, 2♂♂1♀, Cangkou, Qingdao, 31 January 1956; MBM 233272, 1♂ Fujian, 17 April 1984; MBM 159130, 13 larva, the Yellow Sea, 28 June 1957; MBM 158963, 11 larva, Xia Men, Fu Jian, 12 April 1982; MBM 158139, 2♂♂, Xiying, Guangdong, 3 December 1954; MBM 158035, Fujian, 4 May 1982; MBM 158046, 8 larva, the Yellow Sea, 27 October 1958; MBM 157990, 1♀, Dadonghai, Sanya, Ren Xian-qiu Coll., 24 November 1997; MBM 159230, 1♂, the Yellow Sea, 1958–1959; MBM 157979, Fu Jian, 29 April 1982; MBM 157890, 4♂♂1♀, Hainan, Liu Xi-xing, 24 May 1958; MBM 157758, 2♀♀2♂♂, 2 May 1957; MBM 157852, 3♀♀3♂♂, Cangkou, Qingdao, 3 June 1960; MBM 157817, 1♂, Cangkou, Qingdao, 1 May 1960; MBM 157813, 1♂, Qingdao, 14 November 1950; MBM 157791, 1♀1♂, the Yellow Sea, 24 January 1956; MBM 158518, 2 larva, the Yellow Sea; MBM 157829, 1♀, Cangkou, Qingdao, 20 August 1958; MBM 159986, 2 larva, Cangkou, Qingdao, 6 July 1957; MBM 159974, 26 larva, Xuejiadao, Qingdao, 12 September 1954; MBM 159232, 13♂♂♂4♀♀, 24 April 1984; MBM 158489, 1♂, Shanwei, Guangdong, April 1963; MBM 158731, 1♀, Xuejiadao, Qingdao, 30 July 1951; MBM 158482, Mangrove in Qukou, Liu

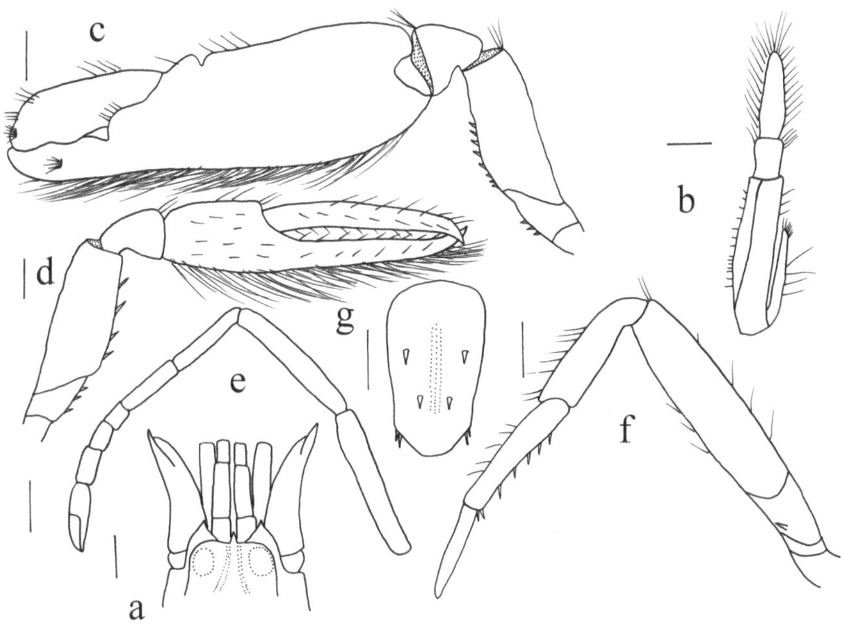


**Fig. 2.9** *Alpheus brevicristatus* De Haan, 1844. MBM 270092, female, **a**, anterior part of carapace; **b**, third maxilliped; **c**, major cheliped; **d**, minor cheliped; **e**, second pereiopod; **f**, third pereiopod; **g**, telson (scale bar = 1 mm)

Xi-xing Coll. 4 January 1992; MBM 158727, 1♀, Qingdao, 17 April 1951; MBM 158725, 1♂, Huiquan Bay, Qingdao, 18 June 1952; MBM 157905, 1♂, Jimen, Xiamen, 7 April 1982; MBM 157945, 4♀♀2♂♂, Fujian, Fan Zhen-gang & Xu Jie-shan Coll. 18 April 1957; MBM 157858, Fujian, Chen Mu Coll., 26 June 1975; MBM 158875, Xiamen, Fujian, 8 April 1982; MBM 233271, 1♀, Pingtan, Fujian, 1 May 1984; MBM 270091, 2♀♀, Xuejian Island, Qingdao, Ren Xian-qiu Coll., 23 October 1986; MBM 270092, 1♀, Xuejiadao, Qingdao, Ren xian-qiu Coll., 13 June 1987; MBM 270019, 2♂♂, Xuejiadao, Qingdao, 16 March 1987; MBM 270073, 1♂, Wenchang, Hainan, 22 July 1957; MBM 158108, 9 larva, Hainan, 1 January 1955.

**Description** Rostrum acute, triangular, reaching to about middle length of visible part of first antennular segment; rostral carina extending to posterior margin of orbital hoods; orbital hoods without tooth; orbitorostral grooves narrow and deep; pterygostomial corner rounded.

Telson about 2 times as long as anterior marginal wide; dorsal surface bearing two pairs of spines, approximately inserted 1/3 and 2/3 length of telson, respectively; posterior margin strongly convex, beyond posterolateral spines.

Antennular peduncle with distally acute stylocerite, hardly reaching distal margin of first antennular segment; visible part of first antennular segment nearly equal to third segment; second antennular segment longest, 2.5 times as long as third segment. Antenna with lateral margin of scaphocerite concave, distinctly overreaching antennular peduncle; antennal scale with distolateral tooth longer than blade. Carpocerite reaching distal margin of antennular peduncle.

Chelipeds showing sexual dimorphic: in male dactylus of major and minor chela distinctly longer fixed finger, in female dactylus of major and minor chela as long as fixed finger; in male chela balaeniceps. Major cheliped with chela compressed, about 4 times as long as wide, ventral margin fringed with long setae; fingers shorter than palm; dorsal margin of palm bearing narrow transverse groove proximal distally, faces without groove. Carpus cup-shaped. Ventromesial margin of merus bearing 5–7 spines, distally rounded. Ischium with 1–3 spines on ventral margin. Minor cheliped with chela about 5 times as long as wide; fingers about 2 times as long as palm, cutting margins bearing setae; ventral margin of chela fringed with long setae. Carpus cup-shaped. Merus like that of major cheliped. Ventral margin of ischium bearing 1–3 spines.

Second pereiopod with carpus 5-segmented, segmented ratio subequal to 9: 8: 3: 3: 4.

Third pereiopod stout, dactylus spatulate, longer than 1/2 length of propodus. Propodus nearly as long as carpus, ventral margin bearing 4–6 and distal one pair spines. Merus longer than propodus, unarmed. Ischium bearing 1 spine on ventral margin. Fourth and fifth pereiopods similar to third, but fifth pereiopod slender than third and fourth, ventral margin of propodus bearing only transverse rows of short setae; ventral margin of ischium without spine.

**Remarks** *A. brevicristatus* is widely distributed in China seas, but has been rarely recorded and illustrated. Jeng & Chang (1985) has shown the colorful photograph of this species. Yang & Kim (1998) has only illustrated the zoeal stages of this species from Korean. The present specimens differ from the description of Miers (1879) from Japan by the fingers of the major distinctly shorter than the palm, rather than nearly as long as the palm. Moreover, the present specimens show sexual dimorphism: in the male, the dactylus of major and minor chela is distinctly longer than the fixed finger and the minor chela is balaeniceps; in the female, the dactylus of major and minor chela is nearly as long as the fixed finger and the minor chela is not balaeniceps.

**Distribution** The Yellow Sea, the East Sea, and the South China Sea; Korea; Japes.

### 2.2.4 *Alpheus compressus* Banner & Banner, 1981 (Fig. 2.10)

*Alpheus compressus* Banner AH & Banner DM, 1981: 227, fig. 3; Chace, 1988: 19.

**Material Examined** CL 7.1 mm, TL 19.5 mm. MBM 270574, 1♀, Beibu Gulf, Sun Fu-zeng Coll. 9 April 1962; MBM 270575, 1♀, Beibu Gulf, 26 August 1962; MBM 270548, 1♀, Beibu Gulf, Fan Zhen-gang Coll., 7 July 1960; MBM 270553, 1 larva, Sun Fu-zeng Coll., 14 December 1959; MBM 270272, 1 larva, Beibu Gulf, Liu Xi-xing Coll., 12 April 1960; MBM 270554, 1 larva, Beibu Gulf, Fan Zhen-gang Coll. 19 April 1960; MBM 270550, 1♂, Beibu Gulf, Sun Fu-zeng Coll. 12 April 1960; MBM 270551, 1♂, Beibu Gulf, Zhang Bao-lin Coll., 9 November 1960; MBM 270549, 1♀, Beibu Gulf, Sun Fu-zeng Coll., 21 April 1960; MBM 270547, 1♀, Beibu Gulf, Liang Guan-sheng & Ruan Wen-guang, 14 December 1959.

**Description** Rostrum triangular, distally acute, reaching from 1/2 length to distal margin of visible part of first antennular segment; rostral carina inconspicuous; orbital hoods without tooth; orbitorostral grooves shallow; pterygostomial corner rounded.

Telson about 3 and 4.5 times as long as anterior and posterior width, respectively; dorsal surface bearing two pairs of spines, approximately inserted 2/5 and 7/10 length of telson, respectively; posterior margin convex, not overreaching posterolateral spines.

Antennular peduncle with distally acute stylocerite, slightly overreaching distal margin of first antennular segment; visible part of first antennular segment slightly longer than third segment; second segment longest, about 2 times as long as third segment. Antenna with lateral margin of scaphocerite nearly straight, only reaching 1.2 length of third antennular segment; antennal scale with distolateral tooth nearly subequal to

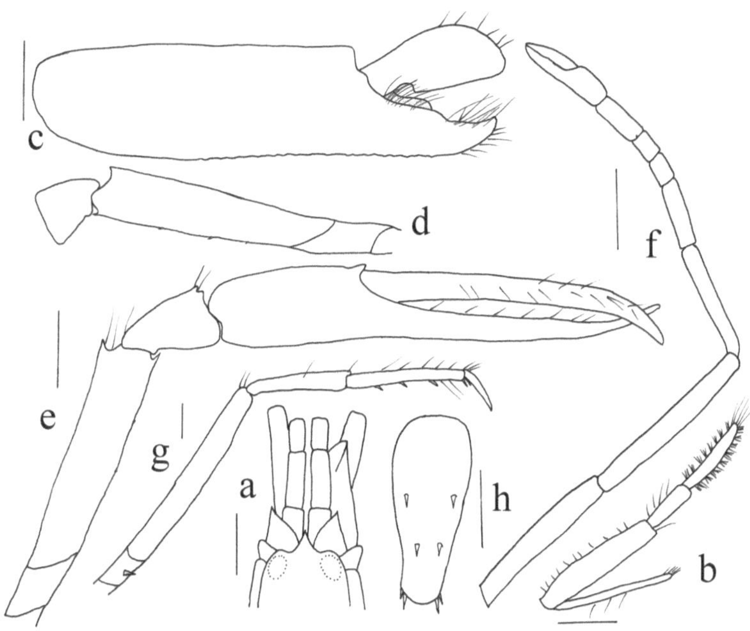


**Fig. 2.10** *Alpheus compressus* Banner & Banner, 1981. MBM 270547, female, **a**, anterior part of carapace, dorsal view; **b**, third maxilliped; **c**, major chela, lateral view; **d**, carpus to ischium of major cheliped; **e**, minor cheliped mesial view; **f**, second pereiopod; **g**, third pereiopod; **h**, telson (scale bar = 1 mm)

blade. Carpocerite overreaching distal margin of antennular peduncle.

Major cheliped moderately compressed, fingers about 1/3 length of palm; dactylus almost straight in lateral view, cutting margin of dactylus only bearing poorly developed plunger; faces of palm without any groove, ventral margin slightly granular. Carpus cup-shaped. Merus slender, dorsal margin with acute triangular tooth distally; ventromesial margin bearing 2 spines and distally one acute tooth. Ischium unarmed. Minor cheliped compressed, moderately slender, with chela about 6 times as long as wide; fingers distally acute, about 2 times as long as palm, crossed when closed with gap medially; distal margin of palm bearing one acute tooth on mesial face, flanking dactylus. Carpus cup-shaped, distal margin with 2 teeth. Merus like that of major cheliped. Ischium unarmed.

Second pereiopod with carpus 5-segmented, segmented ratio subequal to 10: 5: 2: 2: 4.

Third pereiopod moderately slender, dactylus simple, slightly shorter than 1/2 length of propodus. Propodus slightly longer than carpus, ventral margin bearing 2–3 and distal one pair spines. Merus about 2 times as long as carpus, unarmed. Ischium armed with 1 spine on ventral margin.

**Remarks** *A. compressus* is readily distinguished from other species in *Alpheus* by the extremely compressed body. Compared with the original description, the present specimens not bear the slender rostrum. The rostrum of present specimens is only slightly overreaching half-length of the first antennular segment, rather than hardly reaching to the end margin of the first antennular segment as shown in Banner & Banner (1981, fig. 3a). The scaphocerite is not reaching to the half-length of third antennular segment rather than reaching the end margin of that segment showed in Banner & Banner (1981, fig. 3a). The length of the rostrum and the scaphocerite is a variable character in the genus. So the length of the rostrum and the scaphocerite is maybe variable in *A. compressus*. Additionally, Banner & Banner (1981, fig. 3r) has described the paratype bearing two small orbital teeth. However, the orbital teeth are not found in the present specimens. Banner &

Banner (1981) regarded the presence or absence of the orbital teeth as a variable character of *A. compressus*.

**Distribution** The East Sea, the South China Sea and Taiwan; Thailand; Philippines; Indonesia. Depth: 14–280+ m.

### 2.2.5 *Alpheus digitalis* De Haan, 1844 (Fig. 2.11)

*Alpheus digitalis* De Haan, 1844 (in De Haan, 1833–1850): 178, pl. 45, fig. 4; Coutière, 1898c: 249, fig. 2; 1899: 230; Holthuis & Sakai, 1970: 292; Yamaguchi & Baba, 1993: 224, figs. 45a, 45b (in part); Cha et al., 2001: 87, unnumbered figs (in part).

*Alpheus rapax*–. De Haan, 1844 (in De Haan, 1833–1850): 177, Pl. 45, fig. 2 (in part); De Man, 1888a: 264; Ortmann, 1890: 481; Kim, 1977: 241, Pl. 23, 42a, b, figs. 96, 97 (Not *Alpheus rapax* Fabricius, 1798).

*Alpheus distinguendus* De Man, 1909b: 155, Pl. 7, figs. 9–14 (in part); Motoh, 1972: 39, Pl. 7, figs. 2, 3 (in part); Sakamoto & Hayashi, 1977: 1263 (list); Banner DM & Banner AH, 1982: 157, fig. 47; Miyake, 1998: 43, Pl. 15, fig. 6.

*Alpheus bevicristatus*–. Igarashi, 1969: 4, Pl. 4, fig. 10; Kim, 1977: Pl. 50, fig. 41 (Not *Alpheus brevicaristatus* De Haan, 1844).

Not *Alpheus digitalis*–. Hayashi, 1998a: 213, figs. 348d, k, 350d, 351d (*Alpheus longiforceps* Hayashi & Nagata, 2002).

**Material Examined** CL 7.5–10.7 mm, TL 18.3–32.1 mm. MBM 159368, 2♀♀4♂♂, Xinying, Hainan, 2 April 1992; MBM 159365, 1♀, Xinying, Hainan, 7 April 1992; MBM 159485, 1♂2♀♀, the East China Sea, Tang Zhican Coll., 21 July 1976; MBM 15935, 6♂♂7♀♀, Sanya Bay, Hainan, 23 March 1992; MBM 158584, 1♂, the East China Sea, 18 September 1976; MBM 159164, 1♀, Sanya Bay, Hainan, 26 November 1997; MBM 158225, 2♂♂, Shanwei, Guangdong, 5 February 1956; MBM 159371,

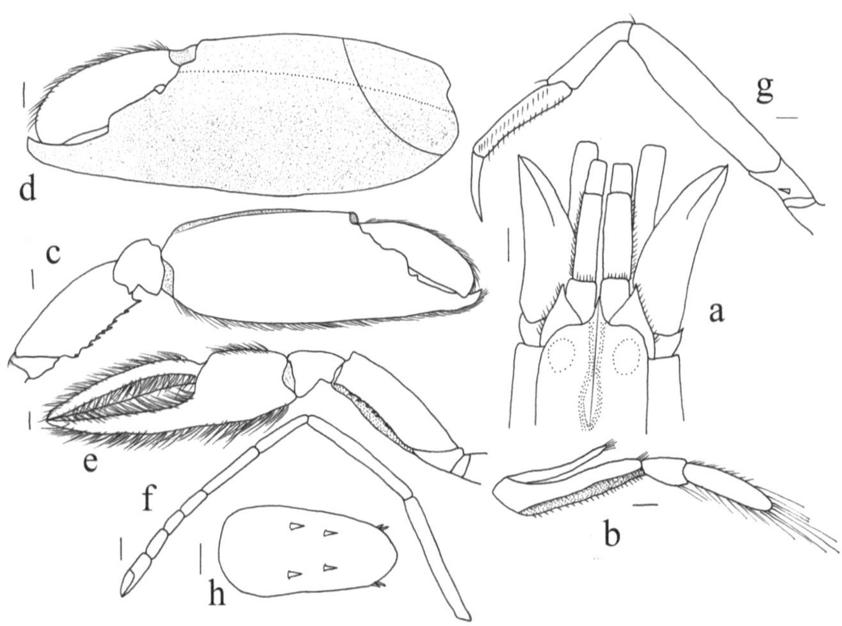


**Fig. 2.11** *Alpheus digitalis* De Haan, 1844. Male, **a**, anterior part of carapace, dorsal view; **b**, third maxilliped; **c**, major cheliped, mesial view; **d**, same, chela, lateral view; **e**, minor cheliped, mesial view; **f**, second pereiopod; **g**, third pereiopod; **h**, telson (scale bar = 1 mm)

4♂♂8♀♀, Hainan, 12 December 1955; MBM 157947, 1♀3♂♂, Fujian, 1 June 1974; MBM 159252, 1♀, the East China Sea, Xu Feng-shan Coll., 21 February 1962; MBM 159529, 3♂♂2♀♀, the Yellow Sea, 9 March 1962; MBM 158954, 1♂1♀, Beibu Gulf, 22 April 1960; MBM 157931, 2♂♂, Longxu Island, Rongcheng, 6 May 1951; MBM 157928, 1♀, Nvgukou, Qingdao, 18 September 1936; MBM 159156, 1♂, Shanghai, Sun Fu-zeng, 16 September 1985; MBM 157877, 1♂, Qingdao, 26 May 1951; MBM 159188, 1♂; MBM 158118, 1♂, Cangkou, Qingdao, 20 November 1956; MBM 159237, 15 larva, Nanji Island, 24 October 1984; MBM 107682, 1♀, Sanya Bay, Hainan, 23 February 1992; MBM 159379, 20 larva, Guangdong, 31 March 1956; MBM 159246, 1♀, Hong Kong, Chen Chuan-chao Coll., 1 March 1999; MBM 138275, 1; MBM 159979, 24♀♂, Shilaoren, Qingdao, 10 July 1956; MBM 270795, 99♀♂, the Yellow Sea, 19 June 2007; MBM 158617, 2♂♂, 7 April 1992; MBM 159354, 1♂, Sanya Bay, Hainan, 18 November 1990; MBM 270797, 28♀♂, the Yellow Sea,21 June 2007; MBM 270706, 2♀♀1♂, the Yellow Sea, 27 June 2007; MBM 270707, 1♂1♀, the Yellow Sea, 22 June 2007; MBM 270708, 1♂, the Yellow Sea, 28 June 2007; MBM 270096, 2♂♂, Zhejiang, 7 March 1954; MBM 270098, 1♂3♀♀, the Bohai Sea, 24 April 1959; MBM 270786, 1♂1♀, the Yellow Sea, 22 June 2007; MBM 270104, 2♀♀, the Bohai Sea, 25 October 1959; MBM 270100, 4♂♂2♀♀, the South China Sea, Chen Mu Coll., 24 April 1960; MBM 270009, 1♀, Zhapo, Guangdong, 15 November 1954; MBM 270029, 1♂1♀, the Bohai Sea, 22 July 1979; MBM 270060, 4♂♂4♀♀, Zhejiang; MBM 270699, 2♂♂, the Yellow Sea, 21 June 2007; MBM 270794, 3♂♂3♀♀, Xinying, Hainan, 1 April 2007; MBM 159370, 1♂, Hainan, 12 December 1955; MBM 158490, 8♀, Sanya Bay, 23 March 1992.

**Description** Rostrum triangular, acute, hardly reaching distal margin of first antennular segment; rostral carina high and narrow, extending to middle length of carapace; orbital hoods without tooth; orbitorostral grooves deep; pterygostomial corner rounded.

Telson about 2.5 and 3 times as long as anterior and posterior width, respectively; dorsal surface bearing two pairs of spines, approximately inserted 2/5 and 3/5 length of telson, respectively; posterior margin strongly convex, beyond posterolateral spines.

Antennular peduncle with distally acute stylocerite, hardly reaching distal margin of first antennular segment; visible part of first antennular segment nearly as long as third segment; second segment longest, about 3 times as long as third segment; lateral margin of stylocerite and three segments fringed with row of short dense setae. Antenna with lateral margin of scaphocerite slightly concave, far overreaching antennular peduncle; antennal scale with acute distolateral tooth slightly longer than blade. Carpocerite as long as antennular peduncle.

Major cheliped compressed, with chela about 4.5 times as long as wide; fingers about 1/2 length of palm, dactylus fringed with short setae on dorsal margin, distally rounded, slightly shorter than fixed finger or nearly as long as it; plunger reduced. Palm with dorsal margin bearing two slightly oblique ridges, without transverse groove, between ridges flattened depression present, without transverse groove; lateral face with long longitudinal ridge, not extending to distal margin; dorsal and ventral margin fringed with row of long setae. Carpus cup-shaped. Merus 2.7–4 times as long as wide; ventromesial margin dentate, with 4–10 spines and distally bearing acute tooth; dorsal margin terminating in acute tooth. Ischium unarmed. Minor chela sexual dimorphism: chela of male slender with fingers about 2.2–2.8 times as long as palm, cutting edges bearing dense brush of long setae; chela of female with fingers about 1.6–1.7 times as long as palm, cutting edges only bearing row of few setae. Dactylus with dorsal margin bearing ridge and dense setae; cutting edges of fingers bearing dense long setae. Carpus cup-shaped. Merus like that of major cheliped, ventromesial margin with 5–8 spines. Ischium unarmed.

Second pereiopod with carpus 5-segmented, segmented ratio subequal to 5: 5: 2: 2: 2.

Third pereiopod moderately slender, dactylus spatulate, about 1/2 length of propodus. Propodus longer than carpus, ventral and lateral margin each bearing row of setae and ventral margin with 4–6 spinules covered by long setae. Merus about 2 times as long as carpus, unarmed. Ischium bearing 1 spine on ventral margin. Fourth and fifth pereiopod similar to third pereiopod, but more slender; but fifth pereiopod with ischium unarmed.

**Remarks** As the discussion of *A. digitalis* in Banner & Banner (1982, in the nominal species name *A. distinguendus*), this species is rather confused. This species is similar to *A. rapacida* in the Brevirostris species group. But it can be distinguished from *A. rapacida* by the following characters: the lateral face of the major chela bearing a longitudinal ridge in *A. digitalis*, and the ridge is lacking in *A. rapacida*; the palm of the minor chela in male is nearly as long as wide, but the palm is about twice as long as wide in *A. rapacida*.

**Distribution** From the Bohai Sea to the South China Sea; Mergui Archipelago; Singapore; Japan; Thailand; Vietnam.

### 2.2.6 *Alpheus djiboutensis* De Man, 1909 (Fig. 2.12)

*Alpheus djiboutensis* De Man, 1909b: 160, Pl. 7, fig, 17–24; Holthuis, 1958: 25, fig. 9; Banner DM & Banner AH, 1982: 180, fig. 55.

**Material Examined** MBM 159017, 1♂, CL 18.7 mm, TL 46.1 mm, the Yellow Sea, 8 May 1958.

**Description** Rostrum triangular, subacute, slightly overreaching distal margin of first antennular segment; rostral carina strong, sharp, extending to posterior margin of orbital hoods; orbital hoods without tooth; orbitorostral grooves moderately deep; pterygostomial corner rounded.

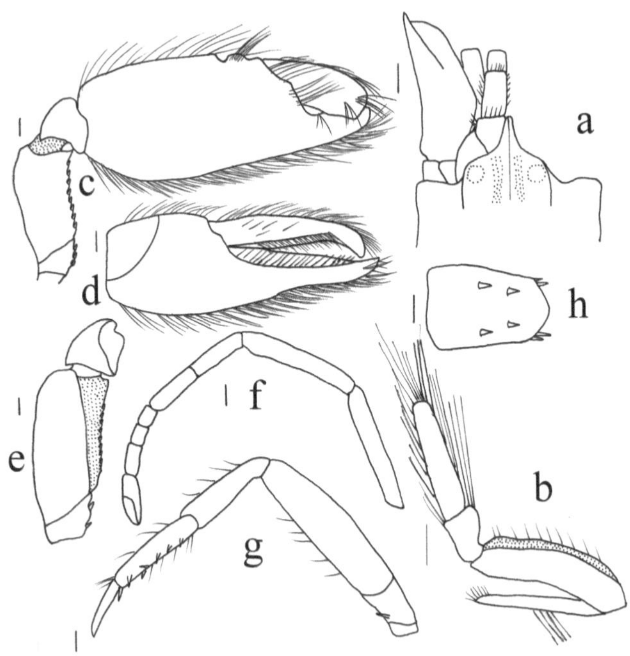


**Fig. 2.12** *Alpheus djiboutensis* De Man, 1909. MBM 159017, male, **a**, anterior part of carapace, dorsal view; **b**, third maxilliped; **c**, major cheliped, mesial view; **d**, minor chela, lateral view; **e**, carpus to ischium of minor cheliped, lateral view; **f**, second pereiopod; **g**, third pereiopod; **h**, telson (scale bar = 1 mm)

Telson about 1.8 and 2 times as long as anterior and posterior width, respectively; dorsal surface bearing two pairs of spines, approximately inserted 1/2 and 3/4 length of telson, respectively; posterior margin moderately convex, beyond posterolateral spines.

Antennular peduncle with distally subacute stylocerite, not reaching distal margin of first antennular segment; visible part of first antennular segment nearly as long as third segment; second segment longest, twice as long as third segment. Antenna with lateral margin of scaphocerite concave, far overreaching antennular peduncle; antennal scale with distolateral tooth longer than blade. Carpocerite reaching distal margin of antennular peduncle.

Major cheliped with chela about 2.5 times as long as wide; dorsal margin of palm slightly flattened, bearing two slightly oblique ridges terminated by moderately deep, narrow transverse groove nearly distal margin; dorsal and ventral margin of chela bearing row of setae; medial face slightly granulate. Carpus cup-shaped. Merus about 2.5 times as long as wide, ventromesial margin dentate, distal half-length bearing long setae and proximal half-length bearing several small spines. Ischium with ventral margin dentate, with or without spine. Minor chela of male "sub-balaeniceps", cutting edges of fingers bearing dense setae; dorsal margin of dactylus and ventral margin of fixed finger fringed with dense long setae. Palm nearly as long as wide; dorsal margin like that of major chela, but not bearing transverse groove; dorsal and ventral margins fringed with dense setae. Carpus cup-shaped. Merus like that of major cheliped, ventromesial margin with 5–8 spines. Ischium unarmed.

Second pereiopod with carpus 5-segmented, segmented ratio subequal to 5: 5: 2: 2: 2.

Third pereiopod moderately slender, dactylus spatulate, about 1/2 length of propodus. Propodus longer than carpus, ventral and lateral margin each bearing row of setae and ventral margin with 4–6 spines and distally one pair covered by long setae. Merus about 2 times as long as carpus, unarmed. Ischium bearing 1 spine on ventral margin. Fourth and fifth pereiopod similar to third pereiopod, but more slender; but fifth pereiopod with ischium unarmed.

**Remarks** The variations of *A. djiboutensis* were well discussed by Banner & Banner (1982). The present specimens are generally agreed with the re-description of Banner & Banner (1982), excepting for the following characters: blade of the scaphocerite far overreaching distal margin of the antennular peduncle, while it only reaching to that end in Banner & Banner (1982, fig. 55a, m); the spines on the ventral margin of propodus of third pereiopod arranged in single row in the present specimens rather than at least arranged in two rows in Banner & Banner (1982, fig. 55i, q). Moreover, no female specimen is found from China seas. Hence, the chela of female whether "sub-balaeniceps" or balaeniceps can not to be known.

**Distribution** The Yellow Sea; from the Red Sea across the Indo-Pacific to the Society Islands.

### 2.2.7 *Alpheus lepidus* De Man, 1908 (Fig. 2.13)

*Alpheus lepidus* De Man, 1908: 106; 1911: 397, Pl. XXI, fig. 92; 1922: 40, Pl. 4, fig, 17.

**Material Examined** CL 5.2 mm, TL 17.5 mm. MBM 270206, 1♀, Beibu Gulf, Shen Shou-peng Coll., 7 July 1960; MBM 270433, 1♂, Beibu Gulf, Sun Fu-zeng Coll., 11 April 1962; MBM 270290, 1 larva, Beibu Gulf, Fan Zhen-gang Coll., 14 May 1960; MBM 270297, 1♂, Beibu Gulf, Zheng Feng Coll., 19 October 1960; MBM 270234, 1♀1♂, Beibu Gulf, 11 October 1962; MBM 270432, 1♀, Beibu Gulf, Sun Fu-zeng Coll., 17 August 1962; MBM 270293, 1♂, Beibu Gulf, Fan Zhen-gang Coll., 18 April 1960; MBM 270289, 1♀, Beibu Gulf, Fan Zhen-gang Coll., 7 February 1960; MBM 270159, 1♂, Beibu Gulf, Sun Fu-zeng Coll., 11 February 1960; MBM 270296, 1♂, Beibu Gulf, Sun

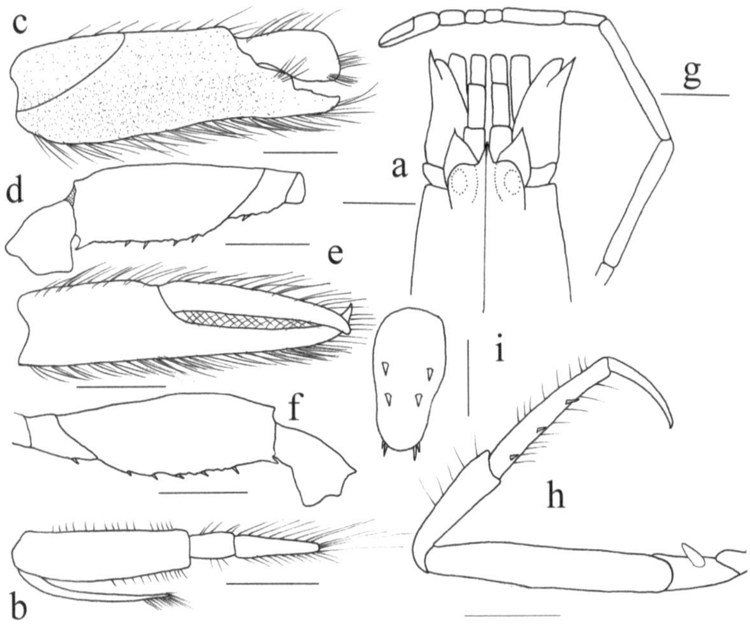


**Fig. 2.13** *Alpheus lepidus* De Man, 1908. MBM 270233, male, **a**, anterior part of carapace, dorsal view; **b**, third maxilliped; **c**, major chela, lateral view; **d**, carpus to ischium of major cheliped, mesial view; **e**, minor chela; **f**, carpus to ischium of minor cheliped; **g**, second pereiopod; **h**, third pereiopod; **i**, telson (scale bar = 1 mm)

Fu-zeng Coll., 12 July 1960; MBM 270235, 1♀, Beibu Gulf, 12 October 1962; MBM 270295, 1♂, Beibu Gulf, Sun Fu-zeng, 12 July 1960; MBM 270430, 1♀, Beibu Gulf, 15 April 1962; MBM 270153, 1♀, Beibu Gulf, Ren Xian-qiu Coll., 20 August 1962; MBM 270328, 1♀, Beibu Gulf, Shen Shou-peng Coll., 14 October 1960; MBM 270152, 1♂, Beibu Gulf, Fan Zhen-gang Coll., 6 July 1960; MBM 270434, 1♂, Beibu Gulf, Sun Fu-zeng Coll., 19 April 1962; MBM 270233, 1♀1♂, Beibu Gulf, Sun Fu-zeng Coll., 12 January 1962.

**Description** Rostrum triangular, acute, overreaching 1/2 length of first antennular segment; rostral carina strong, sharp, about extending to 1/2 length of carapace; orbital hoods without tooth; orbitorostral grooves moderately deep; pterygostomial corner rounded.

Telson about 2.5 and 3.5 times as long as anterior and posterior width, respectively; dorsal surface bearing two pairs of spines, approximately inserted 2/7 and 4/7 length of telson, respectively; posterior margin slightly convex, not beyond posterolateral spines.

Antennular peduncle with distally acute stylocerite, reaching distal margin of first antennular segment; visible part of first antennular segment as long as third segment; second segment about2 times as long as wide, and 1.3 times as long as third segment. Antenna with lateral margin of scaphocerite slightly concave, overreaching distal margin of antennular peduncle; antennal scale with distolateral tooth longer than blade. Carpocerite nearly reaching distal margin of antennular peduncle.

Major cheliped with chela compressed, about 3 times as long as wide. Fingers about 1/2 length of palm; dactylus slightly shorter than fixed finger. Palm with dorsal margin moderately sharp, without transverse groove nearly distal margin; mesial face lightly granular; ventral margin fringed with long setae. Carpus cup-shaped. Merus with proximal half-length of ventromesial margin armed with 4–5 spines, distally produced as small triangular acute tooth. Ischium with ventral margin dentate, bearing 1 small spine. Minor cheliped with chela more compressed, 5 times as long as wide; dorsal and ventral margins fringed with long setae. Fingers about 1.5 times as long as palm; Cutting edges bearing long setae; tips cross when closed with leaving a gape. Palm with faces lightly granular. Carpus cup-shaped. Merus and ischium like that of major cheliped.

Second pereiopod with carpus 5-segmented, segmented ratio subequal to 4: 8: 2: 2: 3.

Third pereiopod moderately slender, dactylus spatulate, about 2/3 length of propodus. Propodus longer than carpus, ventral margin bearing 3 spines. Merus longest, unarmed. Ischium armed with 1 strong spine on ventral margin. Fourth and fifth pereiopod similar to third pereiopod, but more slender; but fifth pereiopod with ischium unarmed.

**Remarks** *A. lepidus* is a newly recorded species from the China seas, and it is ever described by De Man (1909a, 1911, 1922) from Indonesia. The original description is simple and without illustration. The re-description by De Man (1911) is rather detailed with the line drawings. But, all the specimens collected by De Man (1911) are without the minor cheliped. Hence, a supplementary description is made with character of minor cheliped by De Man (1922) based on an ovigerous female specimen. The present specimens are well agreed with the previous descriptions, not showing variations.

**Distribution** The South China Sea; Indonesia.

### 2.2.8 *Alpheus leptocheles* Banner & Banner, 1975 (Fig. 2.14)

*Alpheus leptocheles* Banner AH & Banner DM, 1975: 261, fig. 1.

**Material Examined** CL 8.1–11.5 mm, TL 23.9–37.3 mm. MBM 158810, 1♀, Guanghai, Guangdong, 28 October 1954; MBM 158393, 1♂, the Yellow Sea, Liu Rui-yu Coll., 26 March 1956; MBM 270012, 2♂♂, Zhapo, Hainan, Liu Rui-yu Coll., 15 November 1954; MBM 270008, 10♂♂5♀♀, Zhapo, Hainan, 15 November 1954; MBM 157731, 25♀♂, Shangchuan Island, Guangdong, Liu Rui-yu & Wang Yong-liang Coll., 1 November 1954.

**Description** Rostrum acute, reaching 1/2 length of first antennular segment; rostral carina slightly rounded, extending to posterior margin of orbital hoods; orbital hoods without tooth; orbitorostral grooves moderately deep; pterygostomial corner rounded.

Telson about 3 and 4 times as long as anterior and posterior width, respectively; dorsal surface bearing two pairs of spines, approximately inserted 1/3 and 2/3 length of telson, respectively; posterior margin strongly convex, not beyond posterolateral slender spines.

Antennular peduncle with distally acute stylocerite, hardly reaching distal margin of first antennular segment; second antennular segment longest, about 3 times as long as wide, 1.5 times as long as first segment and 2.5 times as long as third segment. Antenna with lateral margin of scaphocerite nearly straight, slightly overreaching distal margin of antennular peduncle; antennal scale with distolateral tooth as long as blade. Carpocerite nearly as long as antennular peduncle.

Major chela showing sexual dimorphism: Chela of male strongly compressed; fingers very slim; dactylus dorsally carina, slightly shorter than fixed finger; palm widest at based of fingers, and tapering proximally; both dorsal and ventral margins thinned, knife-shaped; mesial and lateral faces granular; chela of female strongly compressed, approximately rectangular; dorsal and ventral margins of palm also knife-shaped. Carpus cup-shaped. Merus and ischium unarmed. Minor chela showing sexual dimorphism: chela of male highly compressed, 5–7 times as long as wide; dorsal and ventral margins fringed with dense setae; fingers 1.5–2.2 times as long as palm, tips crossing when closed with leaving a gape, cutting edges bearing rows of dense setae; mesial face of fixed finger bearing long setae; chela of female about 8 times as long as wide;

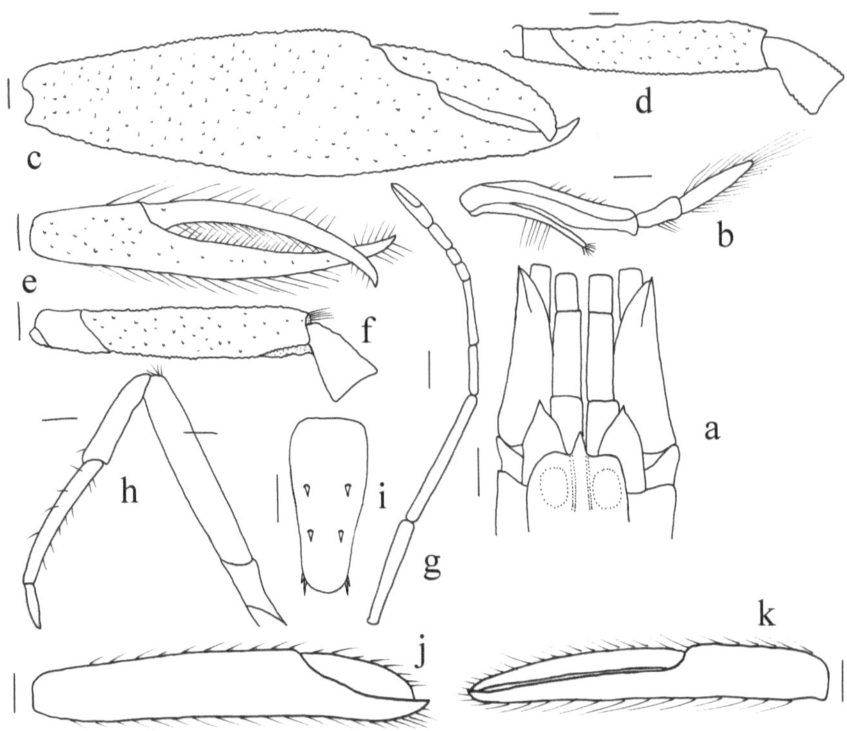


**Fig. 2.14** *Alpheus leptocheles* Banner & Banner, 1975. MBM 270012, male, **a**, anterior part of carapace, dorsal view; **b**, third maxilliped; **c**, major chela; **d**, carpus to ischium of major cheliped; **e**, minor chela; **f**, carpus to ischium of minor cheliped; **g**, second pereiopod; **h**, third pereiopod; **i**, telson. Female, **j**, major chela; **k**, minor chela (scale bar = 1 mm)

fingers less than 1.5 times as long as palm; cutting edges bearing short setae, not leaving a gape when closed; dorsal and ventral margins bearing fine setae. Carpus cup-shaped. Merus and ischium similar to that of major cheliped, unarmed.

Second pereiopod with carpus 5-segmented, segmented ratio subequal to 7: 9: 3: 3: 3.

Third pereiopod moderately slender, dactylus spatulate, about 1/2 length of propodus. Propodus only bearing long setae on ventral margin. Carpus shorter than propodus, dorsal margin slightly produced distally. Merus and ischium unarmed.

**Remarks** *A. leptocheles* is a newly recorded species from China seas, and it is originally described from Australia. This species is readily distinguished from other species in *Alpheus* by the extremely compressed major cheliped. The present specimens are generally agree with the original description, except for the following characters: the dactylus of male chela with tip more acute than the original description as shown in Banner & Banner (1975, fig. 1c), and it slightly shorter than the fixed finger rather than as long as the fixed finger; the telson with posterior margin strongly convex in the present specimens rather than nearly straight as shown in Banner & Banner (1975, fig. 1k).

**Distribution** The Yellow Sea and the South China Sea; New Guinea, Australia.

### 2.2.9 *Alpheus macroskeles* Alcock & Anderson, 1899 (Fig. 2.15)

*Alpheus macroskeles* Alcock & Anderson, 1899: Pl. 9, fig. 5; Banner DM & Banner AH, 1978: 224, fig. 1; Chace, 1988: 37, fig. 8.

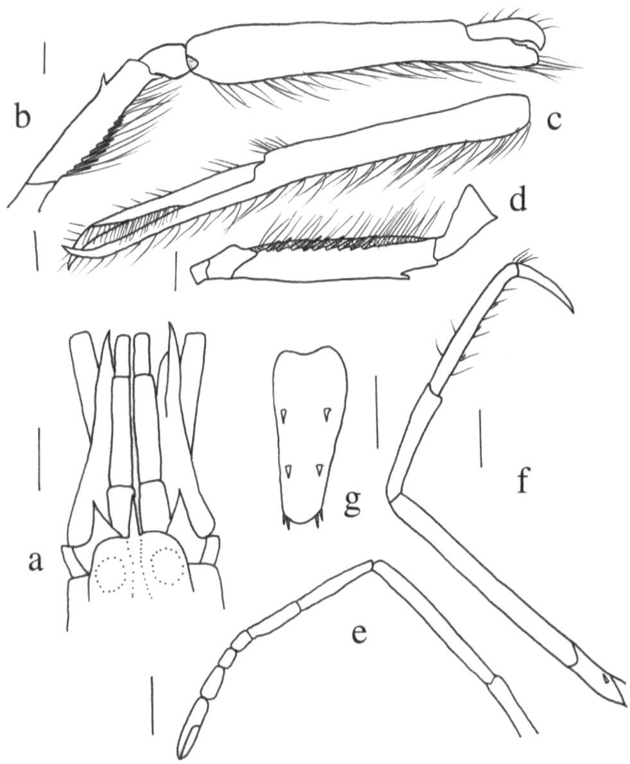


**Fig. 2.15** *Alpheus macroskeles* Alcock & Anderson, 1899. MBM 270015, male, **a**, anterior part of carapace, dorsal view; **b**, major cheliped, mesial view; **c**, minor chela, mesial view; **d**, carpus to ischium of minor cheliped, mesial view; **e**, second pereiopod; **f**, third pereiopod; **g**, telson (scale bar = 1 mm)

**Material Examined** MBM 270015, 1♂, CL 6.9 mm, TL 17.5 mm, the South China Sea, Wang Yong-liang Coll., 26 January 1959.

**Description** Rostrum acute, distinctly longer than wide at base, not reaching distal margin of first antennular segment; rostral carina blunt, feebly development, extending to posterior margin of orbital hoods; orbital hoods without tooth; orbitorostral grooves shallow; pterygostomial corner rounded.

Telson about 2.5 and 3.2 times as long as anterior and posterior width, respectively; dorsal surface bearing two pairs of spines, approximately inserted 1/3 and 2/3 length of telson, respectively; posterior margin slightly convex, not beyond posterolateral slender spines.

Antennular peduncle with distally acute stylocerite, not reaching distal margin of first antennular segment; second antennular segment longest, 3 times as long as first segment; third segment slightly shorter than visible part of first segment. Antenna with lateral margin of scaphocerite concave, slightly beyond distal margin of antennular peduncle; antennal scale with distolateral tooth far longer than blade. Carpocerite longer than antennular peduncle.

Major cheliped with chela subcylindrical; fingers about 1/4 length of palm; dactylus slightly longer or as long as fixed finger, with little developed plunger; surfaces of palm smooth, without groove. Carpus cup-shaped. Merus with dorsal margin bearing 1 large acute tooth sub-distally; ventromesial margin bearing 5 needlelike spines and distally acute tooth. Ischium with ventral margin dentate, armed with small spines. Minor cheliped with chela of male balaeniceps; fingers slightly shorter than palm. Merus and ischium similar to that of major cheliped.

Second pereiopod with carpus 5-segmented, segmented ratio subequal to 9: 6: 2: 2: 4.

Third pereiopod slender, dactylus sub-spatulate, 1/2 length of propodus. Propodus bearing 3 small spines and long setae on ventral margin. Merus longer than propodus, unarmed. Ischium armed with 1 spine on ventral margin.

**Remarks** *A. macroskeles* can be readily distinguished from other species in the Brevirostris group by merus of the first chelipeds armed with one acute sub-distal tooth. This species is originally described from the Bay of Bengal and the Andaman Sea. The original description is simple, and Banner & Banner (1978) has made a supplemental description and illustration of this species based on specimens from the South China Sea. The present one male specimen is also collected from the South China Sea. No female specimen is checked herein. Hence, the characters of female cannot be described. Based on the re-description of Banner & Banner (1978), the minor chela shows sexual dimorphism: chela of male balaeniceps, fingers shorter than palm; chela of female not balaeniceps, fingers nearly as long as palm.

**Distribution** The South China Sea, Taiwan; Indonesia; Philippines.

### 2.2.10 *Alpheus moretensis* Banner & Banner, 1982 (Fig. 2.16)

*Alpheus moretensis* Banner DM & Banner AH, 1982: 177, fig. 54.

**Material Examined** CL 7.9 mm, TL 20.1 mm. MBM 270501, 1♂, Beibu Gulf, Shen Shou-peng Coll., 11 July 1960; MBM 270391, 2♀♀2♂♂, Beibu Gulf, Phan Coll., 10 October 1962; MBM 270340, 2♀♀, Beibu Gulf, Sun Fu-zeng Coll., 13 January 1962; MBM 270191, 1♀, Beibu Gulf, Shen Shou-peng Coll., 17 October 1960-10-17; MBM 270190, 1♀, Beibu Gulf, Sun Fu-zeng Coll., 11 April 1960; MBM 270193, 2♀♀, Beibu Gulf, Fan Zhen-gang Coll., 7 February 1960.

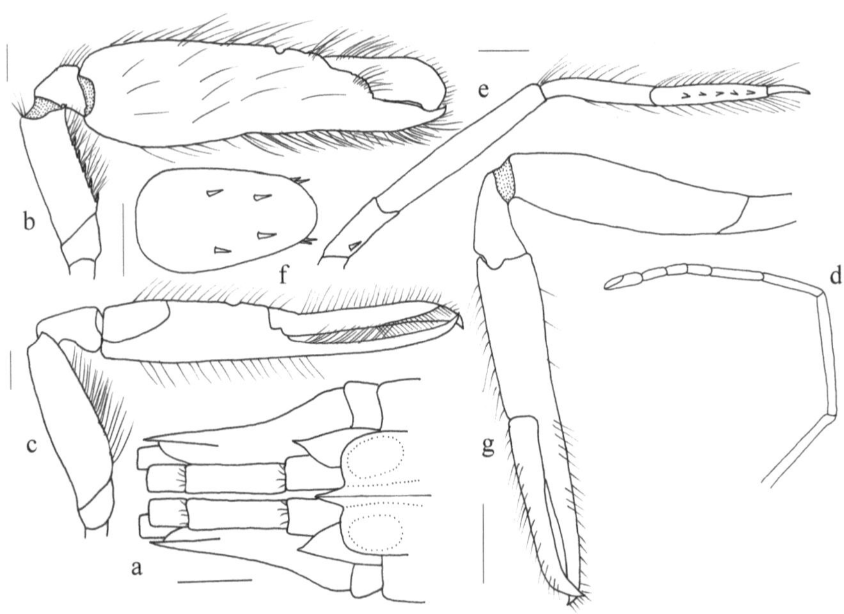


**Fig. 2.16** *Alpheus moretensis* Banner & Banner, 1982. MBM 270501, male, **a**, anterior part of carapace, dorsal view; **b**, major cheliped, mesial view; **c**, minor cheliped, lateral view; **d**, second pereiopod; **e**, third pereiopod; **f**, telson. MBM 270190, female, **g**, minor cheliped (scale bar = 1 mm)

**Description** Rostrum triangular, acute, overreaching 1/2 length of first antennular segment; rostral carina narrow and sharp, extending to middle length of carapace; orbital hoods inflated, without tooth; orbitorostral grooves narrow and deep; pterygostomial corner rounded.

Telson about 1.5 and 3 times as long as anterior and posterior width, respectively; dorsal surface bearing two pairs of spines, approximately inserted 2/5 and 7/10 length of telson, respectively; posterior margin strongly convex, beyond posterolateral spines.

Antennular peduncle with distally acute stylocerite, hardly reaching distal margin of first antennular segment; second antennular segment longest, about 2 times as long as visible part of first segment; third segment shorter than first segment. Antenna with lateral margin of scaphocerite slightly concave, reaching distal margin of antennular peduncle; antennal scale with distolateral tooth distinctly longer than narrow blade. Carpocerite nearly as long as antennular peduncle.

Major cheliped with chela about 3 times as long as wide. Fingers longer than 1/2 length of palm. Palm with dorsal margin bearing transverse groove near distally; dorsal and ventral margins bearing fine setae. Merus about 2.5 times as long as wide; ventromesial margin armed with 5–6 spines, distally produced to acute tooth. Ischium with ventral margin dentate, unarmed. Minor cheliped with chela showing sexual dimorphism: chela of male balaeniceps, cutting edge of fixed finger bearing dense plumatus setae, dorsal margin of palm bearing transverse groove sub-distally; chela of female not balaeniceps, palm smooth with dorsal margin not bearing transverse groove. Merus and ischium like those of major cheliped.

Second pereiopod with carpus 5-segmented, segmented ratio subequal to 5: 5: 2: 2: 2.

Third pereiopod slender, dactylus spatulate, about 1/3 length of propodus. Propodus shorter than merus, lateral margin armed with 5–6 spines and ventral margin with distally one pair spines. Merus unarmed. Ischium armed with 1 spine on ventral margin.

**Remarks** *A. moretensis* is a newly recorded species from the China seas, and it is originally described from Australia by Banner & Banner (1982). The present specimens are generally agreed with the original description, except for the following characters: the palm of major chela distinctly longer than fingers in the present specimens, rather than as long as the fingers shown in Banner & Banner (1982, fig. 54d); the palm of male minor chela nearly as long as the fingers in the present specimens, rather than distinctly shorter than the fingers shown in Banner & Banner (1982, fig. 54f); the ventral margin of ischium of first cheliped dentate in the present specimens, rather than smooth shown in Banner & Banner (1982, fig. 54e, f). As those characters are variable in the *Alpheus*, the present specimens are identified as *A. moretensis*. Additionally, the original description is only based on one male specimen. The female specimens are collected from the South China Sea. Hence, the characters of the female are supplementally described herein, and the minor chela shows sexual dimorphism as described above.

**Distribution** The South China Sea; Australia.

### 2.2.11 *Alpheus nonalter* Kensley, 1969 (Fig. 2.17)

*Alpheus nonalter* Kensley, 1969: 172, fig. 15; Banner AH & Banner DM, 1981: 232; Chace, 1988: 42, fig. 10.

**Material Examined** MBM 158848, 1♂ (with minor cheliped lacking), CL 7.1 mm, TL 20.3 mm, Nansha Islands, 23 September 1994.

**Description** Rostrum acute, reaching 2/3 length of visible part of first antennular segment; feebly developed rostral carina extending to posterior margin of orbital hoods; orbital hoods without tooth; orbitorostral grooves moderately shallow; pterygostomial corner rounded.

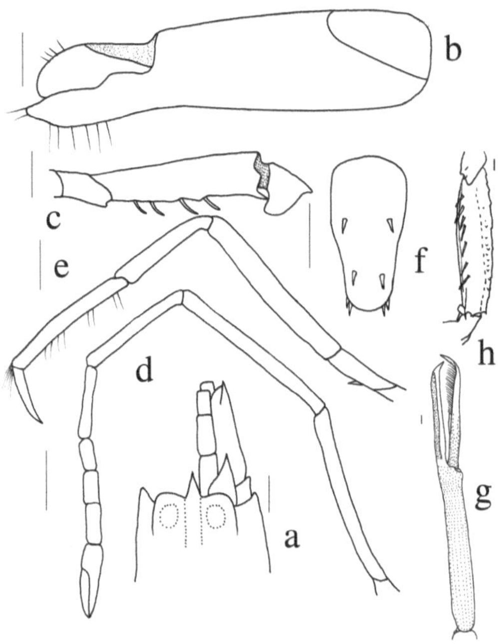


**Fig. 2.17** *Alpheus nonalter* Kensley, 1969. MBM 158848, **a**, anterior part of carapace, dorsal view; **b**, major chela; **c**, carpus to ischium of major cheliped; **d**, second pereiopod; **e**, third pereiopod; **f**, telson. Adapted from Chace (1988) **g**, minor chela; **h**, carpus to ischium of minor cheliped (scale bar = 1 mm)

Telson about 2.1 and 3.5 times as long as anterior and posterior width, respectively; dorsal surface bearing two pairs of spines, approximately inserted 3/7 and 4/5 length of telson, respectively; posterior margin slightly convex, not beyond posterolateral spines.

Antennular peduncle with distally acute stylocerite, reaching distal margin of first antennular segment; second antennular segment longest, about 1.5 times as long as first segment; third segment shorter than first segment. Antenna with lateral margin of scaphocerite nearly straight, overreaching distal margin of antennular peduncle; antennal scale with distolateral tooth distinctly longer than blade. Carpocerite slightly longer than antennular peduncle.

Major cheliped elongate, with chela about 5 times as long as wide. Fingers 1/3 length of palm; dactylus distinctly shorter than fixed finger; plunger feebly developed. Palm subrectangular, dorsal margin bearing two oblique slight ridges, between two ridges forming moderately flattened depression. Merus armed with 4 slender needle-like spines on ventromesial margin, distally produced as acute tooth. Ischium unarmed. Minor cheliped lost in the present material.

Second pereiopod with carpus 5-segmented, segmented ratio subequal to 14: 8: 4: 4: 5.

Third pereiopod slender, dactylus sub-spatulate, 1/3 length of propodus. Propodus shorter than merus, ventral margin only bearing long setae. Merus unarmed. Ischium armed with 1 spine on ventral margin.

**Remarks** *A. nonalter* is most similar to *A. macroskeles* in the Brevirostris group, but can be readily distinguished from *A. macroskeles* by the absence of sub-distal acute tooth on the dorsal margin of merus of the first cheliped. The present specimen is well agreed with the re-description made by Chace (1988). Unfortunately, the minor cheliped is losted in the present collections. Based on the description of Chace (1988), the minor chela shows sexual dimorphism: only chela of male balaeniceps.

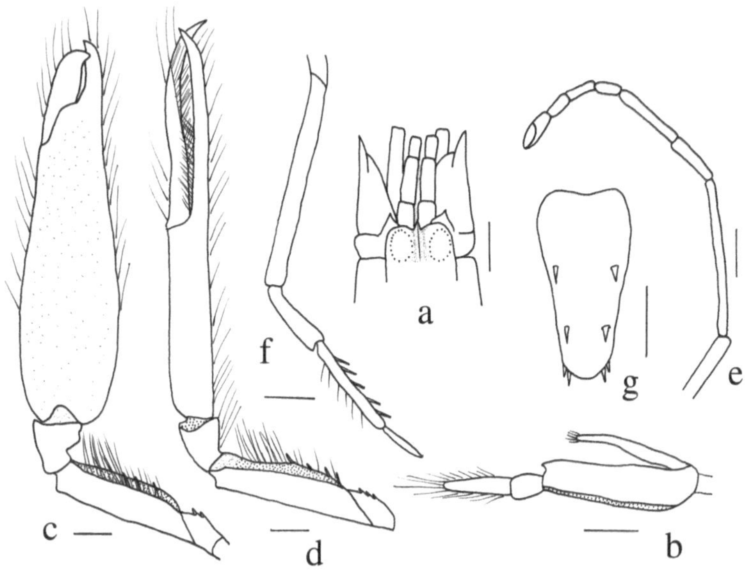


**Fig. 2.18** *Alpheus pustulosus* Banner & Banner, 1968. MBM 107682, male, **a**, anterior part of carapace; **b**, third maxilliped; **c**, major cheliped, mesial view; **d**, minor cheliped, lateral view; **e**, second pereiopod; **f**, third pereiopod; **g**, telson (scale bar = 1 mm)

**Distribution** Hong Kong, the South China Sea; Eastern Africa; Philippines.

## 2.2.12 *Alpheus pustulosus* Banner & Banner, 1968 (Fig. 2.18)

*Alpheus pustulosus* Banner AH & Banner DM, 1968a: 143, fig.2; 1981: 233; Chace, 1988: 49.

**Material Examined** CL 4.5 mm, TL 17.1 mm. MBM 107682, 6♂♂, Sanya Bay, 23 February 1992; MBM 210106, 1♂, Beibu Gulf, Fan Zhengang Coll., 25 April 1960-4-25.

**Description** Rostrum acute, reaching middle of visible part of first antennular segment; feebly developed rostral carina extending to posterior margin of orbital hoods; orbital hoods without tooth; orbitorostral grooves shallow; pterygostomial corner rounded.

Telson about 2.5 and 3.2 times as long as anterior and posterior width, respectively; dorsal surface bearing two pairs of spines, approximatcly inserted 1/3 and 2/3 length of telson, respectively; posterior margin slightly convex, not beyond posterolateral spines.

Antennular peduncle with distally acute stylocerite, hardly reaching distal margin of first antennular segment; visible part of first antennular segment shorter than second segment; second segment about 2.2–3 times as long as wide; third segment 1/2 length of second segment. Antenna with lateral margin of scaphocerite slightly concave, overreaching distal margin of antennular peduncle; antennal scale with distolateral tooth distinctly longer than narrow blade. Carpocerite slightly longer than antenna.

Major cheliped elongate with chela sub-rectangular, about 4 times as long as wide at base. Fingers about 1/4 length of palm. Palm without groove, surfaces lightly punctate; ventral face flattened, demarked from mesial and lateral face by almost sharp angles; dorsal face rounded. Carpus cup-shaped. Merus about 3.5 times as long as wide, ventral margin punctate; ventromesial margin bearing 4–7 slender spines, distally produced as acute tooth. Ischium with ventral margin dentate, armed with several slender spines. Minor cheliped elongate, chela of male balaeniceps (female unknown); fingers nearly as long as palm, cutting edge of fixed finger bearing short dense setae. Merus similar to that of major cheliped. Ischium bearing slender spines on ventral margin.

Second pereiopod with carpus 5-segmented, segmented ratio subequal to 10: 12: 3: 3–4: 4.

Third pereiopod slender, dactylus sub-spatulate, about 1/3 length of propodus. Propodus shorter than merus, ventral margin bearing 5–8 spines. Merus unarmed. Ischium armed with 1 spine on ventral margin.

**Remarks** *A. pustulosus* is originally described from Hong Kong, and then Chace (1988) reported it from Philippines. The present specimens are collected from the South China Sea and generally agreed with the original description, except for the slender minor chela with fingers nearly as long as palm rather than distinctly shorter than the palm as shown in Banner & Banner (1968a, fig. 2e). The type specimens of this species are all male. Unfortunately, the present collections are also male. Thus, the characters of the female are still unknown.

**Distribution** The South China Sea; Philippines.

### 2.2.13 *Alpheus rapacida* De Man, 1908 (Fig. 2.19)

*Alpheus rapacida De Man,* 1908*: 105; 1911: 394, fig. 91; Barnard,* 1950: 750, figs. 142a–f; Tiwari, 1963: 302, figs. 21, 22; Lewinsohn & Holthuis, 1964: 47, fig. 1; Banner AH & Banner DM, 1966b: 118, fig. 42; Banner DM & Banner AH, 1982: 160, fig. 48.

**Material Examined** CL 14.5 mm, TL 40.1 mm. MBM 158241, 2♂♂, the Yellow Sea, 15 March 1954; MBM 158040, 3♀♀2♂♂,Guanghai, Guangdong, Liu Rui-yu Coll., 28 October 1954; MBM 158098, 1♀, Ningbo, Zhejiang, 30 April 1953; MBM 157935, 2♀♀, Sanya, Hainan, 24 March 1955; MBM 158947, 4♂♂, Guangdong,

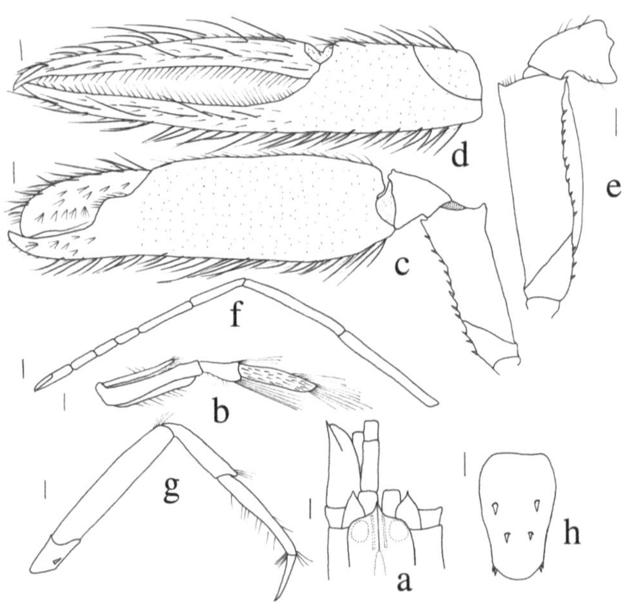


**Fig. 2.19** *Alpheus rapacida* De Man, 1908. MBM 158947, male, **a**, anterior part of carapace; **b**, third maxilliped; **c**, major cheliped, mesial view; **d**, minor chela, lateral view; **e**, carpus to ischium of minor cheliped; **f**, second pereiopod; **g**, third pereiopod; **h**, telson (scale bar = 1 mm)

Liu Rui-yu & Wang Yong-liang Coll., 26 March 1954; MBM 159453, 5♂♂; MBM 157731, 6♀♂, Shangchuan Island, Guangdong, Liu Rui-yu Coll., 1 November 1954-11-1; MBM 158257, 1♀1♂, the Yellow Sea, 12 March 1956; MBM 158263, 1♀6♂♂, Guanghai, Guangdong, 26 March 1956; MBM 158902, 2♀♀, Hainan, 23 May 1958-5-23; MBM 158959, 1♀, Hainan, Wang Yong-liang Coll., 26 June 1957; MBM 107680, 2♂♂, Qukou, Hainan, 3 November 1990; MBM 107681, 1♂, Qinglan, Hainan, 31 December 1991; MBM 159354, 2♂♂, Sanya Bay, 18 November 1990; MBM 270818, 2♀♀1♂, the South China Sea, Tang Zhi-can Coll., 10 April 1960; MBM 270148, 1♂, Beibu Gulf, Fan Zhen-gang Coll., 9 February 1960-2-9; MBM 270192, 1♀, Beibu Gulf, Ma Xiu-tong Coll., 7 December 1959; MBM 270444, 1larva, Beibu Gulf, Ren Xian-qiu Coll., 11 November 1960; MBM 270518, 1♀, Beibu Gulf, Sun Fu-zeng Coll., 23 April 1962; MBM 270507, 1♀, Beibu Gulf, 27 February 1960; MBM 270504, 1♂, Beibu Gulf, Ren Xian-qiu Coll., 14 November 1960; MBM 270523, 1♀1♂, Beibu Gulf, 15 December 1959; MBM 270359, 1♀, Beibu Gulf, 20 October 1960; MBM 270360, 1♀, Beibu Gulf, Ren Xian-qiu Coll., 13 November 1960; MBM 270494, 1♂, Beibu Gulf, 11 July 1960; MBM 270520, 1♀, Beibu Gulf, Sun Fu-zeng Coll., 27 August 1962-8-27; MBM 270522, 2♂♂, Beibu Gulf, Sun Fu-zeng Coll., 16 August 1962-8-16; MBM 270361, 1♀, Beibu Gulf, Shen Shou-peng Coll., 20 October 1960; MBM 210110, 1♀, Beibu Gulf, 15 April 1960; MBM 270129, 25♀♂, Zhapo, Guangdong, Liu Rui-yu & Wang Yong-liang Coll., 15 November 1954; MBM 159347, 2♀♀2♂♂, Sanya Bay, Wang Yong-liang Coll., 23 March 1992.

**Description** Rostrum acute, hardly reaching distal margin of first antennular segment; rostral carina extending to nearly middle length of carapace, with anteriorly knife-shaped and posteriorly rounded; orbital hoods without tooth; orbitorostral grooves deep; pterygostomial corner rounded.

Telson about 2 and 3 times as long as anterior and posterior width, respectively; dorsal surface bearing two pairs of spines, approximately inserted 2/5 and 3/5 length of telson, respectively; posterior margin strongly convex, beyond posterolateral spines.

Antennular peduncle with distally acute stylocerite, reaching distal margin of first antennular segment; visible part of first segment as long as third segment; second segment longest, about 3 times as long as third segment. Antenna with lateral margin of scaphocerite concave, far overreaching distal margin of antennular peduncle; antennal scale with distolateral tooth distinctly longer than narrow blade. Carpocerite slightly shorter than antennular peduncle.

Major cheliped compressed, with chela about 4 times as long as wide. Fingers about 2/5 length of palm; plunger of dactylus feebly developed. Palm with dorsal margin flattened, bearing two oblique slightly developed ridges, without transverse groove near distally; surfaces granular; dorsal and ventral margins fringed with fine setae. Carpus short, cup-shaped. Merus about 3.2 times as long as wide, surfaces slightly granular; ventromesial margin dentate, bearing 8–10 little spines, distally produced as small acute tooth; dorsal margin terminating in subacute to acute tooth. Ischium with ventral margin slightly dentate, bearing 3–4 little spines. Minor cheliped not showing sexual dimorphism. Fingers about 2.2 times as long as palm; faces and cutting edges bearing fine setae. Palm without any ridges and grooves; dorsal and ventral margins fringed with fine setae. Carpus cup-shaped. Merus and ischium similar to those of major cheliped.

Second pereiopod with carpus 5-segmented, segmented ratio subequal to 10: 10: 3: 3: 3.

Third pereiopod moderately stout, dactylus spatulate, about half length of propodus. Propodus without spines on ventral margin, only bearing long setae. Carpus about 2/5 length of merus. Merus unarmed. Ischium armed with 1 spine on ventral margin.

**Remarks** *A. rapacida* is a variable species as discussed by Banner & Banner (1982). The present specimens are more similar to the re-description made by Banner & Banner (1982) in the following characters: the second antennular peduncle much longer, with ratio of three antennular segments subequal to 1: 3: 1; the dactylus of major chela shorter than fixed finger. However, the rostrum is only reaching to middle length of the first antennular segment, not reaching to end margin of that segment as shown in Banner & Banner (1982, fig. 48b). The length of rostrum is same to the description made by De Man (1911). Those variations above have been taken as intraspecific variation by Banner & Banner (1982).

**Distribution** The Yellow Sea, the East China Sea and the South China Sea; South Africa; Red Sea; Israel; Singapore; Thailand; Vietnam; Indonesia; Australia; Hawaii.

### 2.2.14 *Alpheus savuensis* De Man, 1908 (Fig. 2.20)

*Alpheus savuensis* De Man, 1908: 110; 1911: 392; 1915: Pl. 20 fig. 90; Banner AH & Banner DM, 1966a: 180; Miya, 1984: 97; Banner DM & Banner AH, 1985: 32; Jeng & Chang, 1985: 247, figs. 18, 19; Jeng, 1997: 13, 28, fig. 23; Hayashi, 1998a: 216, figs. 348g, n, 350g, 351g.

**Material Examined** This species was reported by Liu (2008) from the South China Sea, but not found in the present specimens.

**Description** Rostrum acute, reaching 1/2 length of first antennular segment; rostral carina extending to nearly middle length of carapace, with anteriorly knife-shaped and posteriorly rounded; orbital hoods without tooth; orbitorostral grooves deep; pterygostomial corner rounded.

Telson about 2 and 3 times as long as anterior and posterior width, respectively; dorsal surface bearing two pairs of spines, approximately inserted 2/5 and 3/5 length of telson, respectively; posterior margin strongly convex, beyond posterolateral spines.

Antennular peduncle with distally acute stylocerite, reaching distal margin of first antennular segment; visible part of first segment as long as

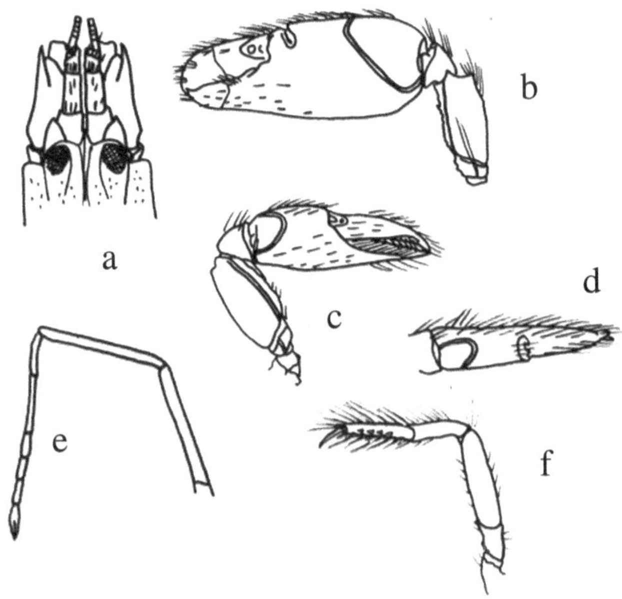


**Fig. 2.20** *Alpheus savuensis* De Man, 1908. (Adapted from De Man, 1908). **a**, anterior part of carapace, dorsal view; **b**, major cheliped, lateral view; **c**, minor cheliped, lateral view; **d**, minor chela, dorsal view; **e**, second pereiopod; **f**, third pereiopod

third segment; second segment longest, about 3 times as long as third segment. Antenna with lateral margin of scaphocerite concave, far overreaching distal margin of antennular peduncle; antennal scale with distolateral tooth nearly as long as narrow blade. Carpocerite slightly shorter than antennular peduncle.

Major cheliped compressed, with chela about 4 times as long as wide. Fingers about 2/5 length of palm; plunger of dactylus feebly developed. Palm with dorsal margin flattened, without transverse groove near distally; surfaces granular; dorsal and ventral margins bearing row of fine setae. Carpus short, cup-shaped. Merus about 3.2 times as long as wide, surfaces slightly granular; ventromesial margin dentate, bearing 8–10 little spines, distally produced as small acute tooth; dorsal margin terminating in subacute to acute tooth. Ischium with ventral margin slightly dentate, bearing 3–4 little spines. Minor cheliped not showing sexual dimorphism. Fingers about 22 times as long as palm. Palm without any ridges and grooves. Carpus cup-shaped. Merus and ischium similar to those of major cheliped.

Second pereiopod with carpus 5-segmented, segmented ratio subequal to 10: 10: 3: 3: 3.

Third pereiopod moderately stout, dactylus spatulate, about half length of propodus. Propodus without spines on ventral margin, only bearing long setae. Carpus about 2/5 length of merus. Merus unarmed. Ischium armed with 1 spine on ventral margin.

**Distribution** Taiwan; Fiji Islands; Indonesia.

### 2.2.15 *Alpheus sibogae* De Man, 1908 (Fig. 2.21)

*Alpheus sibogae* De Man, 1908: 107; 1911: 398, fig. 93.

**Material Examined** CL 7.9 mm, TL 20.5 mm. MBM 159240, 1♂, the East China Sea, Tang Zhi-can & Xu Jie-shan Coll., 5 June 1978; MBM 159224, 1♂, the East China Sea, Tang Zhi-can Coll., 18 September 1976; MBM 159243, 15♀♂, the East China Sea, 5 July 1976; MBM 159250, 2♂♂, the East China Sea, 15 August 1964.

**Description** Rostrum triangular, acute, longer than wide at base, reaching 2/3 length of first antennular segment; rostral carina sharp, extending to near middle length of carapace, with a tooth medially; orbital hoods inflated, without tooth; orbitorostral grooves deep; pterygostomial corner rounded.

Telson about 2.5 and 3 times as long as anterior and posterior width, respectively; dorsal surface bearing two pairs of spines, approximately inserted 1/3 and 2/3 length of telson, respectively; posterior margin slightly convex, not beyond posterolateral spines.

Antennular peduncle with distally acute stylocerite, not reaching distal margin of first antennular segment; visible part of first segment near two times as long as third segment; second segment longest, about 2.5 times as long as third segment. Antenna with lateral margin of scaphocerite concave, nearly as long as antennular peduncle; antennal scale with distolateral tooth distinctly longer than narrow blade. Carpocerite slightly longer than antennular peduncle.

Major cheliped compressed and elongate, with chela about 5.5 times as long as wide. Fingers about 4 times as long as palm; dactylus shorter than fixed finger. Palm without transverse groove on dorsal margin; lateral face with longitudinal depression. Carpus cup-shaped. Merus with dorsal margin dentate; ventromesial margin bearing 3–6 spines, distally produced as acute tooth. Ischium bearing 3–7 spines on ventral margin. Minor cheliped elongate, with chela about 7 times as long as wide; chela of male balaeniceps. Fingers slightly shorter than palm, cutting edge of fixed finger bearing dense plumose setae; tip crossing when closed not leaving gape. Palm without groove or ridges. Merus and Ischium similar to those of major cheliped.

Second pereiopod with carpus 5-segmented, segmented ratio subequal to 7: 8: 3: 3: 3.

Third pereiopod slender, dactylus spatulate, about 1/3 length of propodus. Propodus nearly as long as carpus, ventral margin bearing 5 and distally

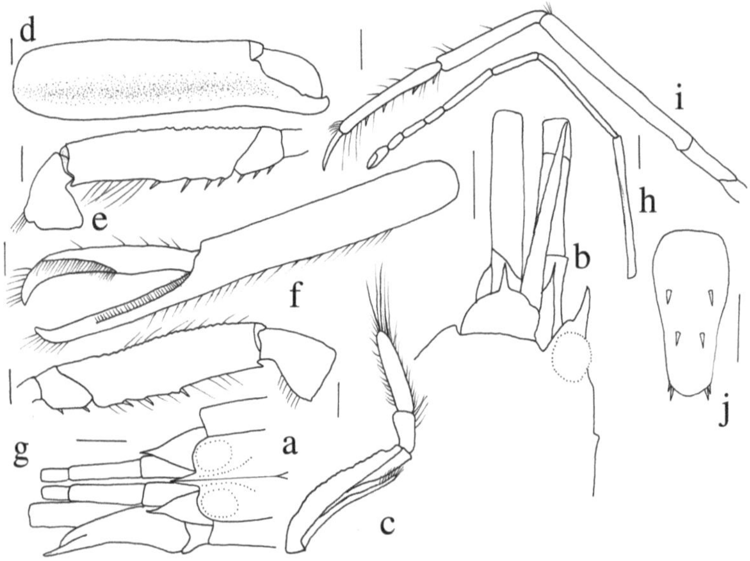


**Fig. 2.21** *Alpheus sibogae* De Man, 1908. MBM 159250, male, **a**, anterior part of carapace, dorsal view; **b**, same, lateral view; **c**, third maxilliped; **d**, major chela, lateral view; **e**, carpus to ischium of major cheliped, mesial view; **f**, minor chela, lateral view; **g**, carpus to ischium of minor cheliped, lateral view; **h**, second pereiopod; **i**, third pereiopod; **j**, telson (scale bar = 1 mm)

one pair spines. Merus 2 times as long as carpus, unarmed. Ischium without spine on ventral margin.

**Remarks** *A. sibogae* is readily distinguished from other species in the Brevirostris group by the major chela having narrow longitudinal deep depression on the lateral face of the palm. The present specimens are generally agreed with the description made by De Man (1911), except the following characters: the fingers of male minor chela distinctly shorter than palm, rather than nearly as long as palm as shown in De Man (1911, fig. 93d); ventromesial margin of ischium of major and minor cheliped armed with 3–7 small spines rather than unarmed as shown in De Man (1911, fig. 93c, d, f).

**Distribution** The East China Sea; Indonesia.

### 2.2.16 *Alpheus tenuicarpus* De Man, 1908 (Fig. 2.22)

*Alpheus tenuicarpus* De Man, 1908: 104; 1911: 381, fig. 84; Banner DM & Banner AH, 1981: 40; 1985: 32; Cui & Sha, 2014: 1123–1127, figs. 1, 2.

**Material Examined** CL 3.9 mm, TL 10.9 mm. MBM 270410, 1♀, Beibu Gulf, Shen Shou-peng Coll., 19 October 1960; MBM 210113, 1♂Beibu Gulf, Fan Zhen-gang 1960; MBM 270271, 1♀, Beibu Gulf, Fan Zhen-gang Coll., 22 April 1960; MBM 270147, 1♀, Beibu Gulf, Fan Zhen-gang & Ruan Wen-guang Coll., 14 February 1960; MBM 270145, 1♀, Beibu Gulf, Fan Zhen-gang Coll., 8 February 1960; MBM 270298, 1♀, Beibu Gulf, Ma Xiu-tong Coll., 5 December 1959; MBM 270572, 1♀, Beibu Gulf, Lam Coll., 10 October 1962; MBM 270485, 1♀, Beibu Gulf, Tang Zhi-can Coll., 4 December 1959; MBM 270409, 1♂, Beibu Gulf, Zhang Bao-lin Coll., 8 November 1960; MBM 270503, 1♀, Beibu Gulf, Fan Zhen-gang & Ruan Wen-guang Coll., 14 February 1960; MBM 270386, 1♂, Beibu Gulf, LamColl., 22 August 1962; MBM 270481, 2♀♀, Beibu Gulf, Fan Zhen-gang Coll., 10 February 1960; MBM 270267, 1♀, Beibu Gulf, Ma Xiu-tong Coll., 5 December 1959; MBM 270144, 1♀, Beibu Gulf, Fan Zhen-gang Coll., 6 July 1960; MBM 270469, 1♂, Beibu Gulf, Sun Fu-zeng Coll., 25 May 1960-5-25; MBM 270509, 1♀, Beibu Gulf, Sun Fu-Zeng Coll., 15 July 1960.

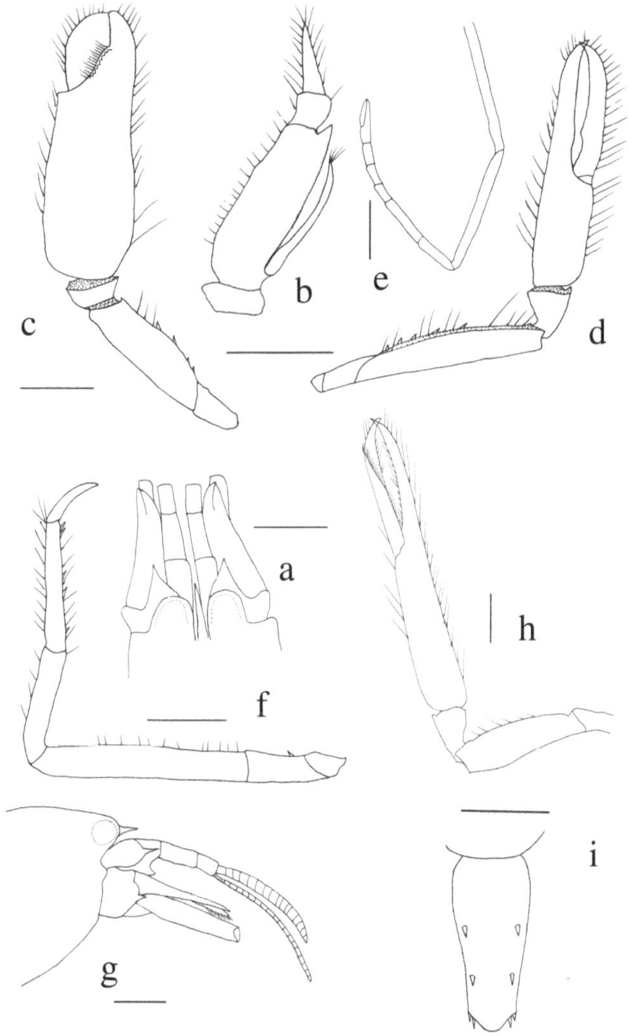


**Fig. 2.22** *Alpheus tenuicarpus* De Man, 1908. MBM 270410, female, **a**, anterior part of carapace, dorsal view; **b**, third maxilliped; **c**, major cheliped; **d**, minor cheliped; **e**, second pereiopod; **f**, third pereiopod. MBM 270113, male, **g**, anterior part of carapace, lateral view; **h**, minor cheliped; **i**, telson (scale bar = 1 mm)

**Description** Rostrum triangular, acute, reaching to middle of visible part of the first antennular article; rostral carina prominent, extending to posterior margin of orbital hoods; orbitorostral grooves deep; orbital hoods without tooth.

Telson about 2.5 and 3 times as long as anterior and posterior width, respectively; dorsal surface bearing two pairs of spines, approximately inserted 3/7 and 5/7 length of telson, respectively; posterior margin strongly convex, beyond posterolateral spines.

Antennular peduncle with distally acute stylocerite, nearly reaching distal margin of first antennular segment; first segment with ventromesial tooth, nearly reaching 1/2 length of second segment, visible part slightly shorter than second segment; second segment longest. 2 times as long as third segment. Antenna with lateral margin of scaphocerite slightly concave, reaching distal margin of antennular peduncle; antennal scale with distolateral tooth nearly as long as narrow blade. Carpocerite slightly longer than antennular peduncle.

Major cheliped with chela compressed, about 4 times as long as wide. Fingers slightly shorter than 1/2 length of palm; fixed finger bearing serrate teeth and setae on proximal half length. Palm nearly smooth, without any groove or ridge. Carpus short, cup-shaped. Merus with ventromesial margin bearing 5 spines and terminating in acute tooth. Ischium unarmed. Minor cheliped with chela showing sexual dimorphism: dactylus of male balaeniceps; fingers slightly shorter than palm; fingers of female longer than palm, tips crossing when closed. Carpus cup-shaped. Merus and ischium similar to those if major cheliped.

Second pereiopod with carpus 5-segmented, segmented ratio subequal to 2: 3: 1: 1: 1.

Third pereiopod moderately slender, dactylus spatulate, about 1/2 length of propodus. Propodus longer than carpus, ventral margin bearing 5 and distally one pair spines. Merus unarmed. Ischium with 1 spine on ventral margin.

**Remarks** *A. tenuicarpus* is described without characters of female. Cui & Sha (2014) has made a re-description of *A. tenuicarpus* with a supplementary description of female, and made a discussion of this species.

**Distribution** The South China Sea; Gulf of Aqaba; Indonesia.

### Crinitus Group (Fig. 2.23)

Rostrum usually reduced or absent; orbital hoods without tooth; Major cheliped with chela rounded in section, surfaces without groove or ridge. Minor chela of male often balaeniceps. Third pereiopod with dactylus simple or biunguiculate; merus usually unarmed (Banner & Banner, 1982).

Key to species of Crinitus Group from China seas

1. Antennal scale with blade reduced, not reaching to distal margin of second antennular segment ......................2
-. Antennal scale with blade normally developed, at least reaching to middle length of third antennular segment....4
2. Minor chela with fingers distinctly longer than palm ......................................................................3
-. Minor chela with fingers slightly shorter than palm ............................................................ *A. spongiarum*
3. Fingers of minor chela spatulate; third pereiopod with merus rounded distally on ventral margin.....
.......................................................................................................................*A. xishaensis*
-. Fingers of minor chela tapering; third pereiopod with merus bearing acute tooth distally on ventral margin
.......................................................................................................................*A. eulimene*
4. Third pereiopod armed with spines on ventral margin.................................................................. *A. alcyone*
-. Third pereiopod unarmed with spines on ventral margin..............................................................................5
5. Sixth abdominal segment with posterior margin bearing teeth medially ........................................................
.....................................................................................................................*A. paralcyone*
-. Sixth abdominal segment with posterior margin smooth medially, not bearing teeth ......................................
.................................................................................................................................................6
6. Rostrum absent.............................................................................................................. *A. frontalis*
-. Rostrum small, acute.......................................................................................................................7
7. Minor chela not showing sexual dimorphism, with fingers tapering...............................................................8
-. Minor chela showing sexual dimorphism: fingers of male spatulate, dactylus bearing two rows of plumose setae on lateral margins; fingers of female tapering ........................................................................................9
8. Uropod with only endopod bearing spinules on posterolateral margin ........................................... *A. stanleyi*
-. Uropod with both endopod and exopod bearing spinules on posterolateral margin .........................................
.......................................................................................................................*A. brevipes*
9. Carpus of second pereiopod with first segment distinctly shorter than second segment.................*A. bucephalus*
-. Carpus of second pereiopod with first segment distinctly longer than second segment ....................................
..................................................................................................................... *A. pachychirus*

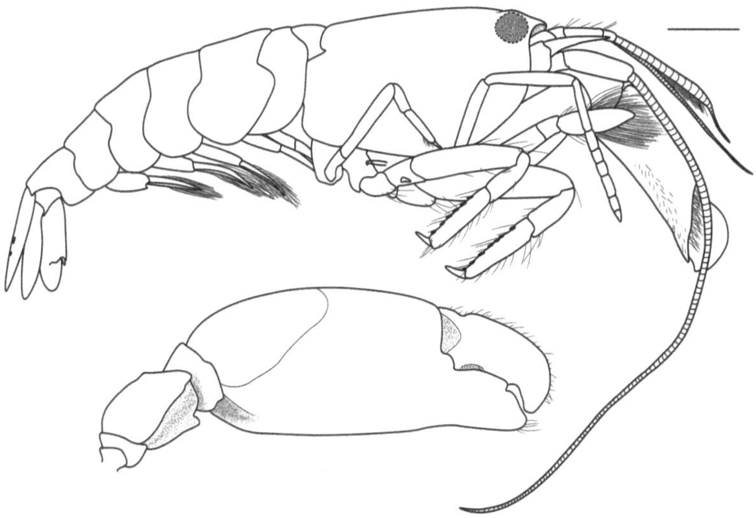

**Fig. 2.23** *Alpheus frontalis* H. Milne Edwards, 1837. entire animal of Crinitus Group (scale bar = 2mm)

## 2.2.17 *Alpheus alcyone* De Man, 1902 (Fig. 2.24)

*Alpheus alcyone* De Man, 1902: 870, Pl. 27, fig. 61; Nobili, 1906b: 32; De Man, 1911: 351; Banner AH & Banner DM, 1966b: 107, fig. 36; Miya, 1974: 144, Pl. 26; Liu & Lan, 1980: 100, fig. 12; Banner DM & Banner AH, 1982: 110, fig. 29; Chace, 1982: 15.

*Alpheus aculeipes* Coutière, 1905: 892, Pl. 79, fig. 31.

*Alpheus crinitus*–. Bate, 1888: 548, Pl. 98, fig. 2; Zehntner, 1849: 206 (Not *A. crinitus* Dana, 1852).

**Material Examined** CL 3.9–5.1 mm, TL 9.5–12.9 mm. MBM 159039, 1♂1♀, East Island, Xisha Islands, 28 May 1980; MBM 159038, 1♂, Yongxing Island, Xisha Islands, 11 June 1980; MBM 159037, 1♂, Shanhu Island, Xisha Islands, 19 May 1980; MBM 159040, 24♂♂6♀♀, Shanhu Island, Xisha Islands, 19 May 1980; MBM 158033, 2♀♀, Xisha Islands, Fan Zhen-gang & Xu Jie-shan Coll., 28 April 1958-4-28; MBM 159001, 8♀♀15♂♂, Guangjin Island, Xisha Islands, Chen Mu Coll., 23 May 1981; MBM 158100, 3♂♂1♀, Shi Island, Xisha Islands, 13 May 1957; MBM 158116, 1♂, Yongxing Island, Xisha Islands, Fan Zhen-gang & Xu Jie-shan Coll., 15 May 1958; MBM 159215, 1♂, Nansha Islands, 16 April 1999; MBM 157985, 1♂, Tree Island, Xisha Islands, Fan Zhen-gang Coll., 4 May 1958; MBM 157909, 2♂♂2♀♀, Xisha Islands, 29 May 1990; MBM 159041, 5♀♀6♂♂, Yongxing Island, Xisha Islands, 11 June 1980; MBM 158993, 3♂♂, Shi Island, Xisha Islands, Chen Mu Coll., 2 June 1981; MBM 158109, 2♂♂, Dengqing Island, Xisha Islands, 11 April 1958; MBM 158221, 1♂, Xincun, Hainan, 9 July 1957; MBM 158122, 1♂, Jinqing Island, Xisha Islands, Fan Zhen-gang & Xu Jie-shan Coll., 1 May 1958; MBM 159044, 1♂2♀♀, Quanfu Island, Xisha Islands, 27 March 1975; MBM 158657, 2♀♀1♂, Yezhu Island, Yalong Bay, 19 March 1992-3-19; MBM 157735, 1♂1♀, Xinying, Hainan, Wang Yong-liang Coll., 23 May 1955-5-23; MBM 158274, 1♂, Zhongjian Island, Xisha Islands, 12 May 1975; MBM 157718, 2♂♂, Shi Island, Xisha Islands, Wang Yong-liang Coll., 13 May 1957-5-13; MBM 157973, 13♀♂, Jinqing Island, Xisha Islands, 25 April 1958; MBM 158319, 1♀, Yongxing Island, Xisha Islands, Chen Mu Coll., 31 May 1981; MBM 159306-9, 1♂, Yezhu Island, Yalong Bay, Wang Yong-liang Coll., 22 March 1992; MBM 157968, 1♂1♀, Yongxing Island, Xisha Islands, 13 April 1957; MBM 158728, 2♀♀1♂, Jinqing Island, Xisha Islands, 25 April 1958; MBM 158578, 2♀♀,

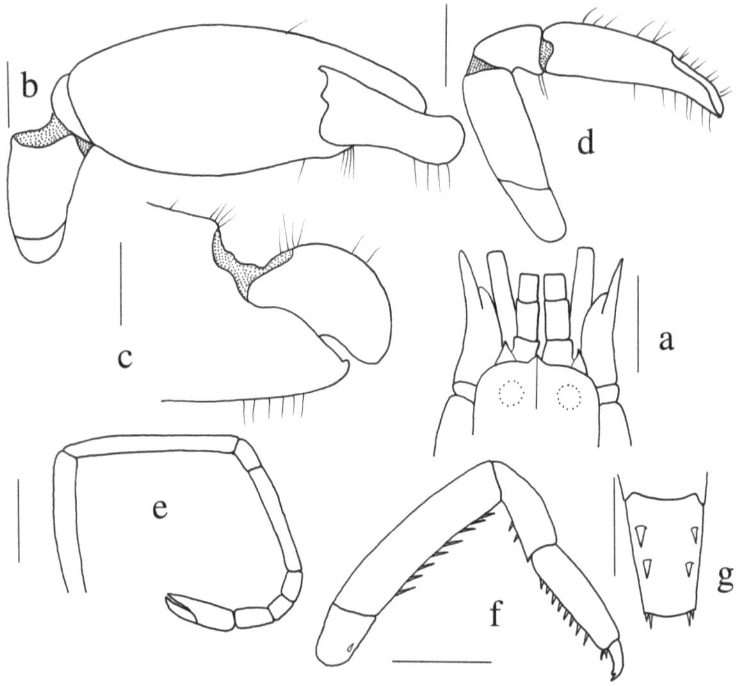


**Fig. 2.24** *Alpheus alcyone* De Man, 1902. MBM 158112, male, **a**, anterior part of carapace, dorsal view; **b**, major cheliped; **c**, same, fingers; **d**, minor cheliped; **e**, second pereiopod; **f**, third pereiopod; **g**, telson (scale bar = 1 mm)

Zhubi Reef, Nansha Islands, 30 March 1993; MBM 158891, 1♂1♀, Yongxing Island, Xisha Islands, Fan Zhen-gang & Xu Jie-shan Coll., 20 March 1958; MBM 158112, 2♀♀3♂♂, Nansha Islands, 28 April 1958; MBM 158717, 3♀♀7♂♂, Nansha Islands, Fan Zhen-gang & Xu Jie-shan Coll., 28 April 1958; MBM 158274, 1♂1♀, Yongxing Island, Xisha Islands, 7 June 1975; 04II71AD-28, 1♀, Zhubi Reef, Nansha Islands, 11 May 2004; 04II76AD-27, 2♀♀, Zhubi Reef, Nansha Islands, 12 May 2004; 04II66AD-16, 1♀, Zhubi Reef, Nansha Islands, 10 May 2004; MBM 233238, 4♀♂, Lingyang Reef, Xisha Islands, Liu Xi-xing Coll., 28 March 1975.

**Description** Rostrum small, triangular, only slightly overreaching orbital margin; rostral carina sharp, extending to posterior margin of orbital hoods; orbitorostral grooves moderately shallow; orbital hoods without tooth.

Telson about 2.5 and 3 times as long as anterior and posterior width, respectively; dorsal surface bearing two pairs of large spines, approximately inserted 1/4 and 1/2 length of telson, respectively; posterior margin only slightly convex, not beyond posterolateral spines. Uropod with lateral distal margin of endopod bearing a few spines.

Antennular peduncle with distally acute stylocerite, not reaching distal margin of first antennular segment; visible part of first antennular segment as long as third segment; second segment twice as long as wide, about 1.5 times as long as third segment. Antenna with lateral margin of scaphocerite concave, nearly as long as carpocerite; antennal scale with distolateral tooth far overreaching narrow blade, blade reaching about middle length of third antennular segment. Carpocerite far overreaching antennular peduncle.

Major cheliped with chela cylindrical, about 2.5 times as long as wide. Fingers about 1/2 length of palm; dactylus distinctly longer than fixed finger. Palm smooth, without any groove and ridge. Carpus short, cup-shaped. Merus stout, dorsal and ventromesial margin terminating in acute tooth. Ischium unarmed. Minor cheliped with chela not showing sexual dimorphism, chela about 4 times as long as wide. Dorsal margin and ventromesial margin not produced as acute tooth distally. Ischium unarmed.

Second pereiopod with carpus 5-segmented, segmented ratio subequal to 5: 16: 3: 3: 5.

Third pereiopod moderately stout, dactylus biunguiculate, slender, about 1/5 length of propodus. Propodus longer than carpus, ventral margin bearing 6–8 and distally one pair spines. Carpus about 1/2 length of merus, ventral margin bearing 1–4 spines. Merus about 3 times as long as wide, ventromesial margin bearing 3–12 spines; ventrolateral margin terminating in acute tooth. Ischium with one spine on ventral margin.

**Remarks** *A. alcyone* is readily distinguished from other species in the Crinitus group by the third pereiopod with ventromesial margin armed with 3–12 spines. The present specimens are generally agreed with the re-description made by Banner & Banner (1982), but have slight variation in the following characters: the antennal scale with blade reaching to middle length of third antennular segment rather than hardly reaching to end margin of second antennular segment as shown in Banner & Banner (1982, fig. 29b); the scaphocerite far overreaching distal margin of antennular peduncle rather than only reaching to distal margin of that end as shown in Banner & Banner (1982, fig. 29b). Moreover, the number of the spines on carpus and merus of third pereiopod is variable as described above.

**Distribution** The South China Sea, Taiwan; Red Sea; East Africa; Persian Gulf; Maldives; Ceylon; Malaysia; Thailand; Indonesia; Australia; Philippines; Japan; Marshall; Caroline; Fiji; Tonga; Samoa Islands.

### 2.2.18 *Alpheus brevipes* Stimpson, 1860 (Fig. 2.25)

*Alpheus brevipes* Stimpson, 1860: 30; Banner AH & Banner DM, 1966a: 177; 1968b: 285; Liu & Lan, 1980: 99, fig. 11.

*Crangon brevipes* –. Banner AH, 1953: 103, figs. 35a–j, 36a–e, 37.

**Material Examined** MBM 159042, 1♀, CL 5.9 mm, TL 18.5 mm, East Island, Xisha Islands, 28 May 1980.

**Description** Rostrum small, triangular, only slightly overreaching orbital margin; rostral carina extending to posterior margin of orbital hoods; orbitorostral grooves narrow and shallow; orbital hoods without tooth.

Telson about 1.5 and 2.5 times as long as anterior and posterior width, respectively; dorsal surface bearing two pairs of large spines, approximately inserted 1/3 and 2/3 length of telson, respectively; posterior margin only slightly convex, not beyond posterolateral spines, bearing few spinules between posterolateral spines. Uropod with lateral distal margin of endopod and posterior margin of exopod each bearing a few spines.

Antennular peduncle with distally acute stylocerite, not reaching distal margin of first antennular segment; visible part of first antennular segment slightly shorter than third segment; second segment longest; about twice as long as third segment. Antenna with lateral margin of scaphocerite slightly concave, slightly overreaching distal margin of antennular peduncle; antennal scale with distolateral tooth far overreaching narrow blade. Carpocerite longer than antennular peduncle.

Major cheliped with chela cylindrical, about 2.6 times as long as wide. Fingers about 1/3 length of palm. Palm smooth, without any groove and ridge. Carpus short, cup-shaped. Merus and ischium unarmed. Minor cheliped slender, with chela not showing sexual dimorphism. Fingers tapering. Carpus cup-shaped. Merus and ischium similar to those of major cheliped.

Second pereiopod with carpus 5-segmented, segmented ratio subequal to 18: 19: 7: 8: 13.

Third pereiopod moderately stout, dactylus simple. Propodus slightly longer than carpus,

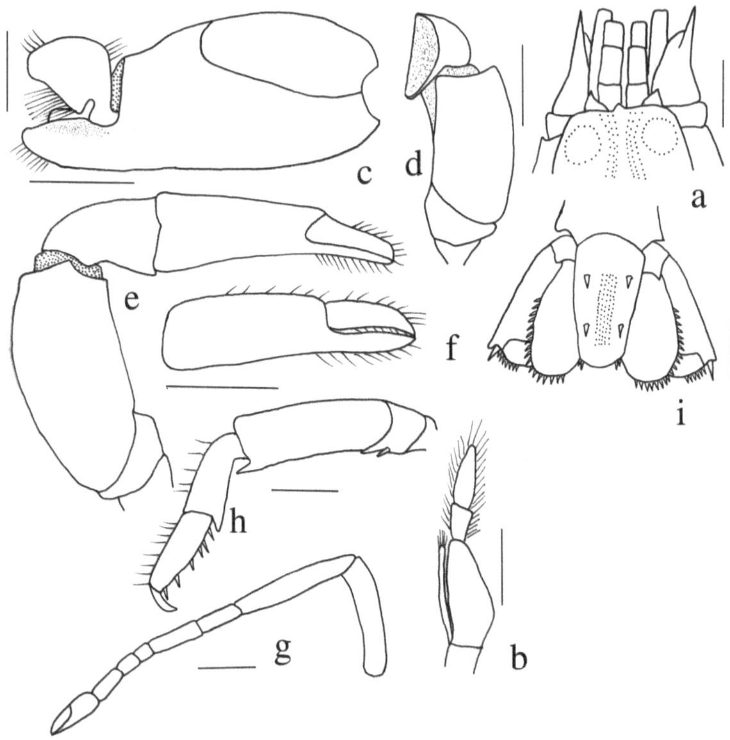


**Fig. 2.25** *Alpheus brevipes* Stimpson, 1860. MBM 159042, female, **a**, anterior part of carapace, dorsal view; **b**, third maxilliped; **c**, major chela, lateral view; **d**, carpus to ischium of major cheliped, lateral view; **e**, minor cheliped, lateral view; **f**, same, minor chela, lateral view; **g**, second pereiopod; **h**, third pereiopod; **i**, uropods and telson (scale bar = 1 mm)

ventral margin bearing 7 spines; lateral margin bearing 4 accessary spines. Merus about 3.3 times as long as wide, ventral margin terminating in acute tooth. Ischium armed with one spine on ventral margin.

**Remarks** *A. brevipes* is readily distinguished from other species in Crinitus group by the posterior and/or lateral margin of endopod and exopod of both having spinules.

**Distribution** The South China Sea; Hawaiian Islands.

### 2.2.19 *Alpheus bucephalus* Coutière, 1905 (Fig. 2.26)

*Alpheus bucephalus* Coutière, 1905: 890, Pl. 78, fig. 29; Banner AH, 1957: 201; Banner AH & Banner DM, 1966b: 110, fig. 38; Banner DM & Banner AH, 1982: 120, figs. 23d–f, 32; Liu & Lan, 1980: 97, fig. 10; Chace, 1988: 17.

*Alpheus consobrinus* De Man, 1908: 101; 1911: 360.

*Crangon bucephalus* Edmondson, 1925: 14.

*Alpheus crinitus* –. Coutière, 1900: 413 (Not *A. crinitus* Dana, 1852).

**Material Examined** CL 3.2–5.1 mm, TL 8.1–13.5 mm. MBM 159047, 1♂, Zhaoshu Island, Xisha Islands, 11 May 1980; MBM 159048, 2♂♂1♀, Zhaoshu Island, Xisha Islands, 28 May 1980; MBM 159051, 5♂♂, Shi Island, Xisha Islands, 10 May 1980; MBM 159043, 2♂♂, Yongxing Island, Xisha Islands, 9 May 1980; MBM 159053, 7♀♂, Yongxing Island, Xisha Islands, 11–13 June 1980; MBM 158078, 1♀, Sanya, Hainan, 20 November 1997; MBM 159049, 3♀♀2♂♂, Jinying Island, Xisha Islands,

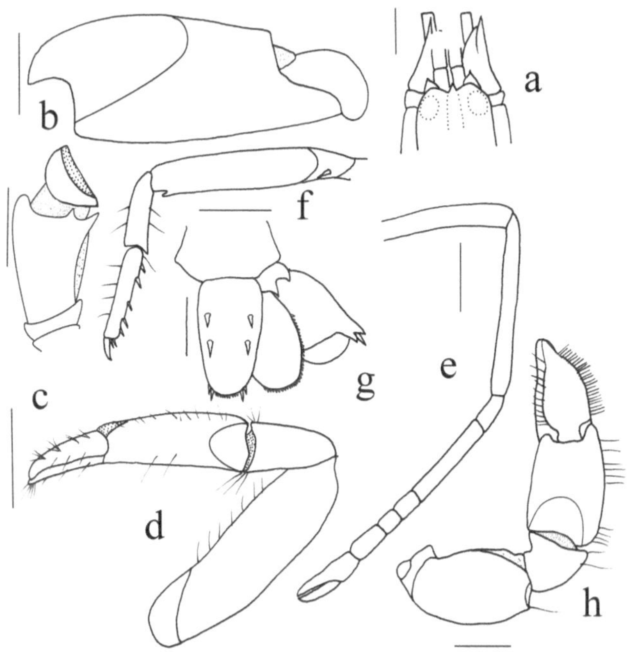


**Fig. 2.26** *Alpheus bucephalus* Coutière, 1905. MBM 158272, female, **a**, anterior part of carapace, dorsal view; **b**, major chela, lateral view; **c**, carpus to ischium of major cheliped, lateral view; **d**, minor cheliped, lateral view; **e**, second pereiopod; **f**, third pereiopod; **g**, right uropod and telson. Male, **h**, minor cheliped, lateral view (scale bar = 1 mm)

15 May 1980; MBM 158814, 1♂, Jinqing Islalnd, Xisha Islands, 1 May 1958; MBM 158851, 1♂, Xincun, Hainan, 25 December 1955; MBM 159052, 1♀, Shanhu Island, 19–21 May 1980; MBM 157861, 1♂, Jinqing Island, Xisha Islands, 19 May 1981-5-19; MBM 158006, 4♀♀2♂♂, Yongxing Island, Xisha Islands, 15 May 1958; MBM 158208, 2♂♂, Dengqing Island, Xisha Islands, 12 April 1958; MBM 157857, 1♀2♂♂, Xincun, Hainan, 18 April 1958; MBM 157886, 1♂1♀, Shanhu Island, Xisha Islands, 23 March 1958; MBM 158759, 1♀, Dengqing Island, Xisha Islands, 11 April 1958; MBM 158993, 2♂♂1♀, Zhubi Reef, 10 April 1999; MBM 158589, 1♀, 23 April 1958; MBM 158654, 1♀, Yufang Reef, Nansha Islands, 7 May 1990; MBM 158308, 1♀, Shi Island, Xisha Islands, 8 May 1975; MBM 159278, 1♂1♀, Xincun, Hainan, Wang Yong-liang Coll., 27 March 1992; MBM 158272, 3♂♂1♀, Jinyin Island, Xisha Islands, 24 May 1975; MBM 158570, 1♂1♀, Nansha Islands, 21 May 1993; MBM 158728, 2♂♂6♀♀, Jinqing Island, Xisha Islands, 25 April 1958; MBM 158603, 3♀♀1♂, Nansha Islands, 18 May 1993; MBM 159263, 1♀, Nansha Islands, 17 April 1999; MBM 233264, 2♀♀1♂, Lingyang Reef, 28 March 1975; MBM 233236, 1♀, Jinying Island, Xisha Islands, Wang Yong-liang Coll., 26 May 1975-5-26; MBM 158651, 1♀, Yongshu Reef, 16 May 1993; 04II67AD-4, 2♀♀1♂, Zhubi Reef, Nansha Islands, 10 May 2004; 04II84AD-31, 2♀♀, Zhubi Reef, Nansha Islands, 14 May 2004; 04II82AD-12, 1♀, Zhubi Reef, Nansha Islands 14 May 2004.

**Description** Rostrum triangular, reaching nearly 1/2 length of first antennular segment; rostral carina high and narrow, extending to posterior margin of orbital hoods; orbitorostral grooves deep; orbital hoods without tooth.

Telson about 1.1 and 2.2 times as long as anterior and posterior width, respectively; dorsal surface bearing two pairs of moderately large spines, approximately inserted 1/4 and 1/2 length of telson, respectively; posterior margin only slightly convex, not beyond posterolateral spines, bearing few spinules between posterolateral spines. Uropod with lateral distal margin of endopod and posterior margin of exopod each bearing a few spines.

Antennular peduncle with distally acute stylocerite, not reaching distal margin of first antennular segment; visible part of first antennular segment nearly as long as third segment; second segment longest; about 1.5 times as long as third segment. Antenna with lateral margin of scaphocerite slightly concave, slightly overreaching distal margin of antennular peduncle; antennal scale with distolateral tooth far overreaching narrow blade. Carpocerite longer than antennular peduncle.

Major cheliped with chela sub-cylindrical, about 2.5 times as long as wide. Fingers about 1/2 length of palm, dactylus distinctly longer than fixed finger. Palm smooth, without any groove and ridge. Carpus short, cup-shaped. Merus stout, ventromesial margin usually bearing sub-terminal rounded or acute tooth. Ischium unarmed. Minor cheliped showing sexual dimorphism: chela of male stout, dactylus balaeniceps, fingers as long as palm, tips rounded; chela of female more slender, dactylus never balaeniceps, fingers about 0.7 times as long as palm, tip tapering. Carpus cup-shaped. Merus and ischium similar to those of major cheliped.

Second pereiopod with carpus 5-segmented, segmented ratio subequal to 4: 11: 2: 2: 4.

Third pereiopod moderately stout, dactylus simple. Propodus longer than carpus, ventral margin bearing 9 spines (not in single row) and distally one pair. Carpus about 1/2 length of merus. Merus with ventral margin terminating in large acute tooth. Ischium bearing one spine on ventral margin.

**Remarks** *A. bucephalus* widely distributed in the Indian Ocean and the Pacific. The present specimens are all collected from the South China Sea, and are generally agreed with the re-description made by Banner & Banner (1982). The variations of this species has been discussed by Banner (1957) are also found in the present collections.

**Distribution** The South China Sea, Taiwan; Red Sea and eastern Africa to Japan; Philippines; Indonesia; Australia; Pacific Islands to Line and Society Islands.

### 2.2.20 *Alpheus eulimene* De Man, 1909 (Fig. 2.27)

*Alpheus eulimene* De Man, 1909a: 101; 1911: 364, Pl. 16, fig. 76; Banner AH, 1956: 356; Miya, 1974: 146, Pl. 27; Bruce, 1976: 47. Banner DM & Banner AH, 1982: 105, fig. 27; Chace, 1988: 26.

**Material Examined** CL 3.1 mm, TL 7.9 mm. MBM 158412, 1♀, Banyue Reef, Nansha Islands, 28 September 1994; 04II54AD-22, 1 larva, Zhubi Reef, Nansha Islands, 6 May 2004; 04II56AD-43, 1♀1♂, Zhubi Reef, Nansha Islands, 6 May 2004; 04II77AD-20, 1♀, Zhubi Reef, Nansha Islands, 12 May 2004; 04II58AD-14, 1 larva, Zhubi Reef, Nansha Islands, 8 May 2004; 04II76AD-27, 1 larva, Zhubi Reef, Nansha Islands, 12 May 2004.

**Description** Rostrum triangular, small, about reaching 1/5 length of first antennular segment; anterior margin of carapace projecting; rostral carina sharp, about extending to middle length of carapace; orbitorostral grooves moderately deep; orbital hoods without tooth.

Telson about 1.7 and 3.4 times as long as anterior and posterior width, respectively; dorsal surface bearing two pairs of moderately large spines, approximately inserted 1/3 and 2/3 length of telson, respectively; posterior margin only nearly straight, not beyond posterolateral spines.

Antennular peduncle with distally acute stylocerite, not reaching distal margin of first antennular segment; visible part of first antennular segment nearly as long as third segment; second segment about 1.5 times as long as third segment. Antenna with lateral margin scaphocerite concave, slightly overreaching antennular peduncle; blade of antennal scale reduced, only reaching

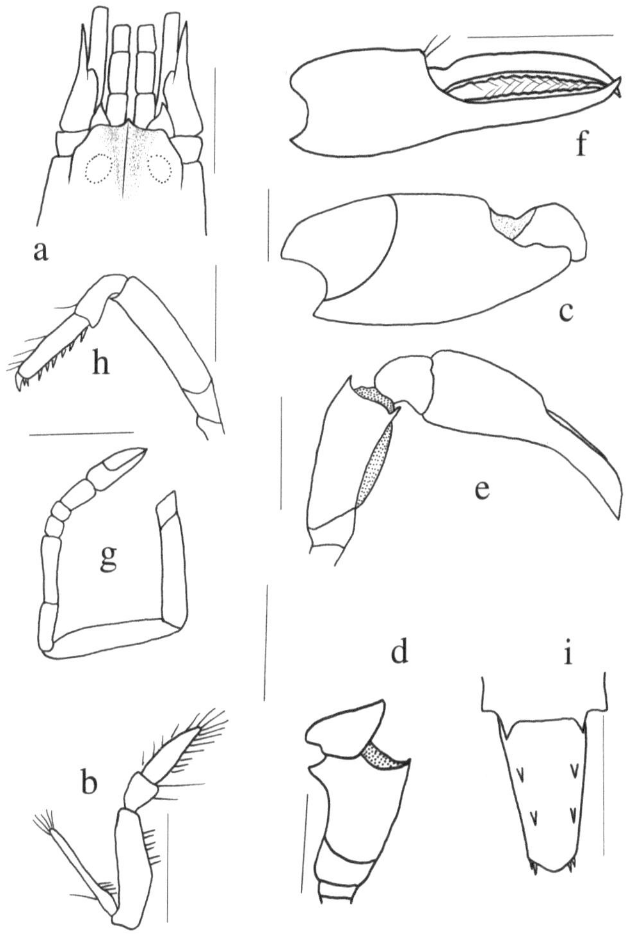


**Fig. 2.27** *Alpheus eulimene* De Man, 1909. 04II58AD-14, male, **a**, anterior part of carapace, dorsal view; **b**, third maxilliped; **c**, major chela, lateral view; **d**, carpus to ischium of major cheliped, mesial view; **e**, minor cheliped, mesial view; **f**, minor chela, mesial view; **g**, second pereiopod; **h**, third pereiopod; **i**, telson (scale bar = 1 mm)

middle length of second antennular peduncle. Carpocerite distinctly longer than antennular peduncle.

Major cheliped with chela about 2.5 times as long as wide. Fingers about 1/3 length of palm, dactylus distinctly longer than fixed finger. Palm gradually widened distally; surfaces without any grooves and ridges. Carpus cup-shaped. Merus stout, dorsal margin rounded distally; ventromesial margin bearing sub-terminal acute tooth. Ischium unarmed. Minor cheliped with chela not showing sexual dimorphism. Fingers distinctly longer than palm, slender, curved laterally; cutting edges bearing slender short setae. Palm wider than fingers; surfaces without any grooves and ridges. Carpus cup-shaped. Merus slender than that of major cheliped, dorsal margin terminating in acute tooth; ventromesial margin bearing an acute terminal tooth.

Second pereiopod with carpus 5-segmented, segmented ratio subequal to 10: 14: 4: 4: 7.

Third pereiopod moderately stout, dactylus simple or bearing a small acute tooth on ventral margin. Propodus longer than carpus, ventral margin bearing 6 and distally one pair spines. Carpus about 1/2 length of merus, ventral margin produced distally, with or without spines. Merus with ventral margin produced as acute tooth distally. Ischium unarmed.

**Remarks** Banner & Banner (1982) have discussed the difference between *A. eulimene* and its allied species in Crinitus group. *A. eulimene* is readily recognized by the reduced blade of the scaphocerite. The present specimens are generally agreed with the re-description made by Banner & Banner (1982).

**Distribution** The South China Sea; Maldive Islands; Kenya; Malaysia; Australia; Indonesia; Philippine; Japan; Mariana Islands.

### 2.2.21 *Alpheus frontalis* H. Milne Edwards, 1837 (Fig. 2.28)

*Alpheus frontalis* H. Milne Edwards, 1837: 356; Ortmann, 1890: 488; Tiwari, 1963: 294, fig. 15; Fishelson, 1966: 98, figs. 1–3; Banner AH & Banner DM, 1968b: 286; Miya, 1974: 135, Pl. 23 figs. d–g; Banner DM & Banner AH, 1982: 99, figs. 23l, m, 25; Chace, 1988: 28.

*Alpheus latifrons* A. Milne Edwards, 1873: 87; De Man, 1888b: 521, Pl. 22, fig. 4; 1890: 119, Pl. 6, fig. 15.

*Betaeus utricola* Richters, 1880: 164, Pl. 17, figs. 34, 35.

**Material Examined** CL 6.5 mm, TL 17.8 mm. MBM 159417, 2♂♂, Yongxing Island, Xisha Islands, 11 June 1980; MBM 159415, 1♀1♂, Shanhu Island, Xisha Islands, 22 May 1980; MBM 157941, 2♀♀1♂, Guangjin Island, Xisha Islands, 21 May 1981; MBM 158355, 1♂, Nanping Reef, Nansha Islands, 2 August 1988; MBM 159412, 1♀, Yezhu Island, Sanya Bay, Neumenn Coll., 27 March 1992-3-27; MBM 157924, 1♂, Yongxing Island, Xisha Islands, 18 May 1981-5-18; MBM 158272, 1♂, Jinyin Island, Xisha Islands, Ren Xian-qiu Coll., 24 May 1975; MBM 159418, 136♀♂, Shanhu Island, Xisha Islands, 19 May 1980; 04II58AD-14, 2♀♀1♂, Zhubi Reef, Nansha Islands, 8 May 2004.

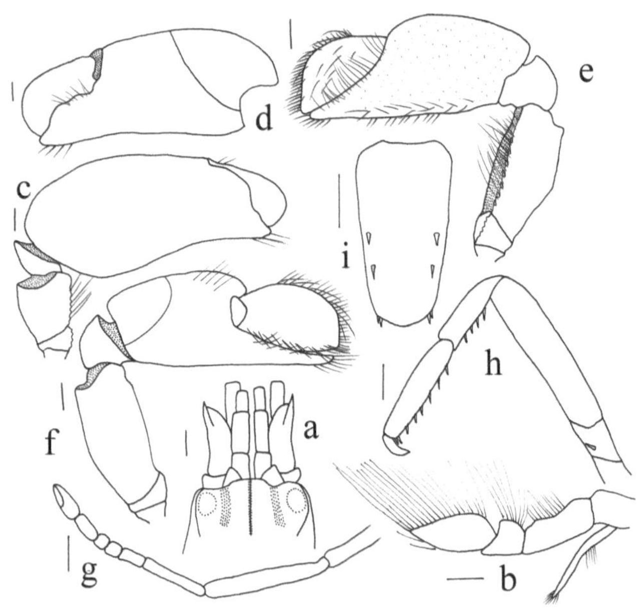


**Fig. 2.28** *Alpheus frontalis* H. Milne Edwards, 1837. MBM 157924, male, **a**, anterior part of carapace, dorsal view; **b**, third maxilliped; **c**, major cheliped, mesial view; **d**, major chela, lateral view; **e**, minor cheliped, mesial view; **f**, same, lateral view; **g**, second pereiopod; h, third pereiopod; **i**, telson (scale bar = 1 mm)

**Description** Rostrum absent; anterior margin of carapace extending as vaulted shelf-like projection between orbits; rostral carina rounded, extending to posterior margin of orbital hoods; orbital hoods without tooth.

Telson about 2.2 and 2.5 times as long as anterior and posterior width, respectively; dorsal surface bearing two pairs of spines, approximately inserted 1/2 and 2/3 length of telson, respectively; posterior margin only slightly convex, not beyond posterolateral small spines.

Antennular peduncle with distally rounded stylocerite, not reaching distal margin of first antennular segment; visible part of first antennular segment slightly longer than third segment; second segment about twice as long as third segment. Antenna with lateral margin of scaphocerite slightly concave, overreaching distal margin of antennular peduncle; antennal scale with distolateral tooth longer than blade. Carpocerite longer than antennular peduncle.

Major cheliped with chela sub-cylindrical, about 2.5 times as long as wide. Fingers about 1/3 length of palm; dactylus with tip curved, slightly longer than fixed finger when closed. Palm without any grooves and ridges. Carpus cup-shaped. Merus stout, ventromesial with irregular teeth and long setae. Ischium unarmed. Minor cheliped with chela showing sexual dimorphism: chela of male about 3 times as long as wide, fingers slightly shorter than palm, dactylus balaeniceps; chela of female slender, fingers 1/2 length of palm. Carpus cup-shaped. Merus similar to that of major cheliped, but more slender. Ischium unarmed.

Second pereiopod with carpus 5-segmented, segmented ratio subequal to 12: 5: 3: 3: 5.

Third pereiopod stout, dactylus simple, curved. Propodus longer than carpus, ventral margin bearing 9 spines and distally one pair. Carpus about 1/2 length of merus, ventral margin bearing 2–4 spines, distally terminating in acute tooth. Merus about 3.3 times as long as wide, unarmed. Ischium armed with one spine on ventral margin.

**Remarks** *A. frontalis* is readily recognized in Crinitus group by the absence of rostrum. The variation of the anterior margin of the carapace has been shown in De Man (1911, fig. 79, 79a, b). The present specimens also show the variation of the shape of the anterior margin of the carapace, but are generally agreed with the previous descriptions.

**Distribution** The South China Sea; Red Sea and eastern Africa to Japan; Philippines; Indonesia; Australia; across Pacific to the Society Islands, but not Hawaii.

### 2.2.22 *Alpheus pachychirus* Stimpson, 1860 (Fig. 2.29)

*Alpheus pachychirus* Stimpson, 1860: 30; Ortmann, 1890: 489, Pl. 36, fig. 17a, k; Borradaile, 1898: 1013; De Man, 1902: 875; 1911: 336; Coutière, 1905: 899; 1921: 425; Tattersall, 1921: 380; Banner AH, 1956: 3556; 1957: 203; 1958; 167; Banner AH & Banner DM, 1966a: 176; 1968b: 283; Miya, 1974: 133, Pl. 23, fig. a–c; Liu & Lan, 1980: 95, fig. 9; Banner DM & Banner AH, 1982: 102, fig. 13j, k, 26; Chace, 1988: 45.

**Material Examined** CL 5.1–6.2 mm, TL 13.3–16.5 mm. MBM 157941, 9♀♀3♂♂, Guangjin Island, Xisha Islands, 21 May 1981; MBM 107625, 1♂, Zhaoshu Island, Xisha Islands, 11 May 1980-5-11; MBM 157861, 4♀♀10♂♂, Jinqing Island, Xisha Islands, Chen Mu Coll., 19 May 1981; MBM 157909, 3♂♂, Xinyi Reef, Nansha Islands, 29 May 1990-5-29; MBM 157900, 3♂♂1♀, Yongshu Reef, Nansha Islands, Ren Xian-qiu Coll., 11 April 1999; MBM 158852, 1♂, Daai Reef, Nansha Islands, 24 May 1990; MBM 159222, 1♀, Zhubi Reef, Nansha Islands, 10 April 1999-4-10; MBM 159414, 2♂♂1♀, Jinyin Island, Xisha Islands, 15 May 1980; MBM 158983, 3♂♂, Yongxing Island, Xisha Islands, 21 March 1980; MBM 158654, 3♀♀2♂♂, Yufang Reef, Nansha Islands, 4 May

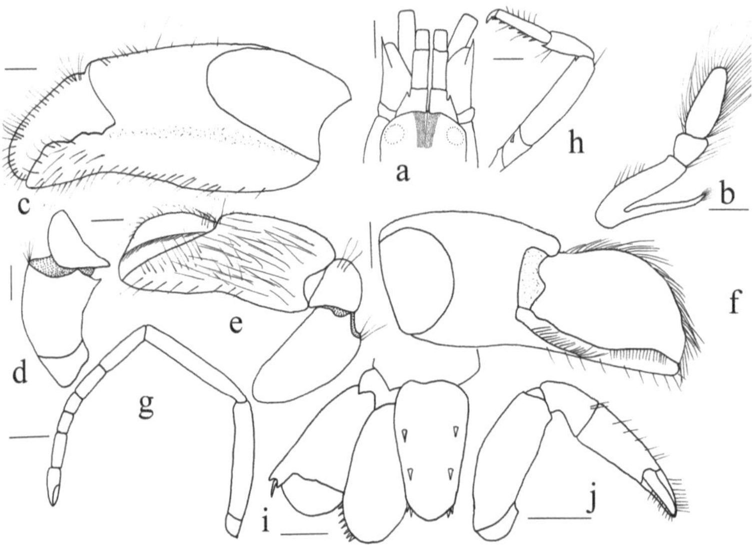


**Fig. 2.29** *Alpheus pachychirus* Stimpson, 1860. MBM 158983, male, **a**, anterior part of carapace, dorsal view; **b**, third maxilliped; **c**, major chela, lateral view; **d**, carpus to ischium of major cheliped; **e**, minor cheliped, mesial view; **f**, minor chela, lateral view; **g**, second pereiopod; **h**, third pereiopod; **i**, left uropod and telson . Female, **j**, minor cheliped (scale bar = 1 mm)

1990; MBM 158300, 1♀1♂, Xisha Islands, 13 May 1975; MBM 158637, 1♀, Chigua Reef, Nansha Islands, 19 May 1990; MBM 158768, 1♀, Nansha Islands, 25 May 1990; MBM 158627, 1♀1♂, Xinyi Reef, Nansha Islands, 29 May 1990; MBM 158625, 1♀, Haikou Reef, Nansha Islands, 3 May 1987-5-3, 04II77AD-20, 2♀♀2♂♂, Zhubi Reef, Nansha Islands, 12 May 2004; 04II60AD-6, 4♀♀, Zhubi Reef, Nansha Islands, 8 May 2004; 04II75AD-30, 22♀♂, Zhubi Reef, Nansha Islands, 12 May 2004; 04II69AD-32, 1♀, Zhubi Reef, Nansha Islands, 11 May 2004; 04II53AD-22, 1♀, Zhubi Reef, Nansha Islands, 4 May 2004; 04II72AD-19, 4 larva, Zhubi Reef, Nansha Islands, 11 May 2004; 04II58AD-14, 2♀♀, Zhubi Reef, Nansha Islands, 8 May 2004-5-8; 04II80AD-25, 19♀♂, Zhubi Reef, Nansha Islands, 13 May 2004; 04II66AD-16, 2 larva, Zhubi Reef, Nansha Islands, 10 May 2004; 04II76AD-27, 1♀, Zhubi Reef, Nansha Islands, 12 May 2004; 04II63AD-3, 2♀♀2♂♂, Zhubi Reef, Nansha Islands, 10 May 2004; 04II59AD-31, 1♂, Zhubi Reef, Nansha Islands, 8 May 2004; 04II83AD-27, 3♀♀1♂, Zhubi Reef, Nansha Islands, 14 May 2004; 04II52AD-21, 1♀1♂, Zhubi Reef, Nansha Islands, 4 May 2004-5-4; 04II79AD-28, 1♀, Zhubi Reef, Nansha Islands, 13 May 2004; 04II55AD-28, 1♂, Zhubi Reef, Nansha Islands, 6 May 2004; 04II84AD-31, 1♂, Zhubi Reef, Nansha Islands, 14 May 2004; MBM 233254, 1♀1♂, Xisha Islands, 28 March 1975.

**Description** Rostrum small, triangular, only slightly overreaching anterior margin of carapace; rostral carina not extending to posterior margin of orbital hoods; orbitorostral grooves broad and shallow; orbital hoods inflated, without tooth.

Telson about 2.2 and 2.8 times as long as anterior and posterior width, respectively; dorsal surface bearing two pairs of moderately large spines, approximately inserted 2/7 and 4/7 length of telson, respectively; posterior margin nearly straight, not beyond posterolateral small spines.

Antennular peduncle with distally acute stylocerite, not reaching distal margin of first antennular segment; visible part of first antennular segment slightly longer than third segment; second segment longest, about twice as long as third segment. Antenna with lateral margin of scaphocerite slightly concave, not reaching distal margin of antennular peduncle; antennal scale with distolateral tooth longer than blade. Carpocerite far overreaching antennular peduncle.

Major cheliped with chela cylindrical, about 2.5 times as long as wide. Fingers about 1/2 length of palm, dactylus as long as fixed finger. Palm smooth, without any grooves and ridges. Carpus cup-shaped. Merus stout, ventromesial margin terminating in acute tooth. Minor cheliped with chela showing sexual dimorphism: chela of male stout, mesial face bearing dense setae, fingers nearly as long as palm, dactylus spatulate, about 1.5 times as long as wide, cutting edge concave, distally acute, margin bearing setae; chela of female slender, fingers about 0.6 times as long as palm. Carpus cup-shaped. Merus slender, ventromesial margin without distally acute tooth. Ischium unarmed.

Second pereiopod with carpus 5-segmented, segmented ratio subequal to 7–8: 5: 2: 2: 4.

Third pereiopod stout, dactylus simple, curved. Propodus longer than carpus, ventral margin bearing 5–8 spines and distally one pair. Carpus about 1/2 length of merus, ventral margin terminating in acute tooth. Merus about 3.5 times as long as wide, ventral margin distally produced as acute tooth. Ischium with one spine on ventral margin.

**Remarks** *A. pachychirus* is a widely distributed species. The present specimens are generally agreed with the previous descriptions.

**Distribution** The South China Sea; Red Sea and eastern Africa to Indonesia; Philippines; Ryukyus; Australia, and Pacific Ocean islands as far eastward as Society Islands.

## 2.2.23 *Alpheus paralcyone* Coutière, 1905 (Fig. 2.30)

*Alpheus paralcyone* Coutière, 1905: 395, Pl. 30, fig. 34; Banner AH & Banner DM, 1966b: 108, fig. 37; Miya, 1974: 139, Pl. 24; Banner DM & Banner AH, 1982: 113, fig. 30; Chace, 1988: 46.

*Crangon paralcyone* Banner, 1953: 99, fig. 34.

*Crangon laysani* Edmondson, 1925:17, fig. 3.

**Material Examined** CL 4.9–7.1 mm, TL 10.5–19.3 mm. MBM 158851, 1♂, Xincun, Hainan, 25 December 1955; MBM 158646-7, 1♂, Yalong Bay, Hainan, Neumann Coll., 27 March 1992; MBM 158469, 2♀♀2♂♂, Xincun, Hainan, Wang Yong-liang Coll., 27 March 1992; MBM 159278, 2♀♀2♂♂, Xincun, Hainan, 27 March 1992-3-27; MBM 107651, 1♂, Yezhu Island, Yalong Bay, 8 January 1992; MBM 270175, 1♀, Beibu Gulf, Sun Fu-zeng Coll., 20 April 1960; MBM 270182, 1♀, Beibu Gulf, Sun Fu-zeng Coll., 10 December 1959; MBM 270533, 1♀1♂, Beibu Gulf, 15 April 1962; MBM 270495, 1♂, Beibu Gulf, 11 July 1960. MBM 270341, 1♀, Beibu Gulf, Sun Fu-zeng Coll., 10 April 1962-4-10; MBM 270343, 1♀, Beibu Gulf, Jiang Jin-xiang Coll., 15 April 1962; MBM 270342, 1♀, Beibu Gulf, Sun Fu-zeng Coll., 24 April 1962; MBM 270179, 1♀, Beibu Gulf, Sun Fu-zeng Coll., 17 February 1960; MBM 270174, 1♂, Beibu Gulf, Shen Shou-peng Coll., 20 October 1960; MBM 270500, 1♀, Beibu Gulf, 24 May 1960; MBM 270181, 1♀1♂, Beibu Gulf, 10 July 1960; MBM 270177, 1♂, Beibu Gulf, Shen Shou-peng Coll., 18 October 1960; MBM 270344, 1♂, Beibu Gulf, 18 August 1962; MBM 270178, 1♀, Beibu Gulf, Zhang Bao-lin Coll., 15 November 1960; MBM 270534, 1♀, Beibu Gulf, Sun Fu-zeng Coll., 24 April 1962; MBM 270531, 1♂, Beibu Gulf, 22 April 1962; MBM 270183, 1♂, Beibu Gulf, Fan Zhen-gang Coll., 4 December 1959; MBM 270176, 1♂, Beibu Gulf, Fan Zhen-gang Coll., 14 May 1960; MBM 270538, 1♂, Beibu Gulf, 12 October 1962.

**Description** Rostrum small, triangular, only slightly overreaching anterior margin of carapace; rostral carina extending to posterior margin of orbital hoods; orbitorostral grooves flat; orbital hoods slightly inflated, without tooth.

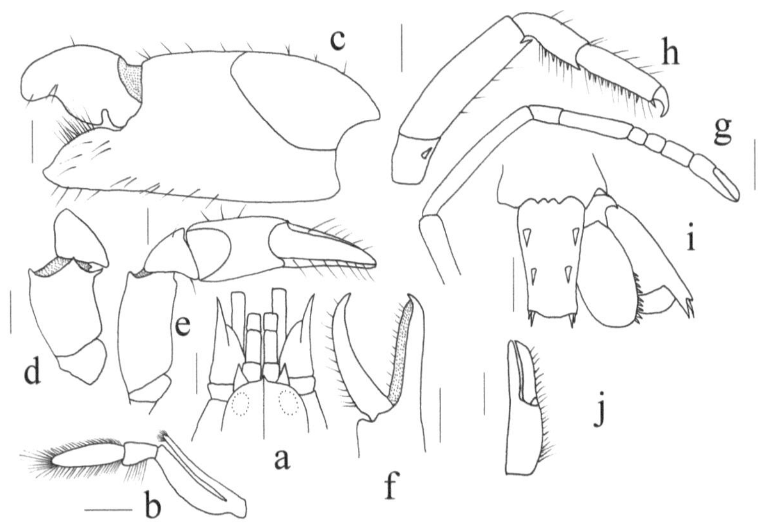


**Fig. 2.30** *Alpheus paralcyone* Coutière, 1905. MBM 159278, male, **a**, anterior part of carapace, dorsal view; **b**, third maxilliped; **c**, major chela, lateral view; **d**, carpus to ischium of major cheliped, mesial view; **e**, minor cheliped, lateral view; **f**, same, fingers; **g**, second pereiopod; **h**, third pereiopod; **i**, right uropod and telson;. Female, **j**, minor chela (scale bar = 1 mm)

Posterior margin of sixth abdominal segment bearing three teeth medially. Telson about 1.6 and 2 times as long as anterior and posterior width, respectively; dorsal surface bearing two pairs of moderately large spines, approximately inserted 1/4 and 1/2 length of telson, respectively; posterior margin only slightly convex, not beyond posterolateral small spines. Uropod with lateral margin of endopod bearing few spines.

Antennular peduncle with distally acute stylocerite, not reaching distal margin of first antennular segment; visible part of first antennular segment nearly as long as third segment; second segment twice as long as third segment. Antenna with lateral margin of scaphocerite concave, overreaching distal margin of antennular peduncle; antennal scale with blade reduced, only reaching middle length of third antennular segment. Carpocerite far overreaching antennular peduncle.

Major cheliped with chela cylindrical, about 3 times as long as wide. Fingers about 1/2 length of palm, slightly curved. Palm smooth, without any groove and ridge. Carpus cup-shaped. Merus stout, dorsal margin terminating in sub-acute tooth; ventromesial with distally large acute tooth. Ischium unarmed. Minor cheliped with chela showing slightly sexual dimorphism: chela of male stout, with fingers longer than palm, fingers broadened; chela of female slender, with fingers nearly as long as palm. Carpus cup-shaped. Merus similar to that of major cheliped, but more slender. Ischium unarmed.

Second pereiopod with carpus 5-segmented, segmented ratio subequal to 3: 6: 2: 2: 3.

Third pereiopod moderately stout, dactylus biunguiculate, ventral margin bearing small, acute unguis. Propodus longer than carpus, ventral margin bearing 14 spines and distal one pair. Carpus about 1/2 length of merus, ventral margin bearing 1–4 spines and distally acute tooth. Merus with ventral margin bearing distally acute tooth. Ischium armed with one spine on ventral margin.

**Remarks** *A. paralcyone* is readily distinguished from other species in Crinitus group by the posterior margin of sixth abdominal segment having three medial teeth. The present specimens are generally agreed with the previous descriptions, except for the rostrum distinctly beyond the anterior margin of carapace rather than only reaching to anterior margin of carapace as shown in Banner & Banner (1892, fig. 30b).

**Distribution** The South China Sea; Madagascar to Indonesia; Philippines; Japan; Australia, and Pacific islands to Hawaii.

## 2.2.24 *Alpheus spongiarum* Coutière, 1897 (Fig. 2.31)

*Alpheus spongiarum* Coutière, 1897cb: 236; Miya, 1974: 148, Pl. 28; Banner DM & Banner AH, 1982: 116, fig. 31; Chace, 1988: 54.
*Alpheus paraculeipes* Coutière, 1905: 849, Pl. 80, fig. 32; Pearson, 1905: 84; 1911: 356; Green, 1972: 67.

**Material Examined** CL 5.6 mm, TL 19.1 mm. MBM 158097, 3♀♀4♂♂, Xinying, Hainan, 13 January 1956; MBM 158883, 1♀, Nansha Islands, Fan Zhen-gang & Xu Jie-shan Coll., 28 April 1958; MBM 158223, 1♂, Weizhou Islandm Liu Rui-yu & Wang Yong-liang Coll., 25 December 1954; MBM 107678, 2♂♂1♀, Linchang Reef, Beibu Gulf, 31 December 1991; 04II74AD-20, 1♂, Zhubi Reef, Nansha Islands, 12 May 2004-5-12; 04II68AD-13, 1♂, Zhubi Reef, Nansha Islands, Sta. 9, 11 May 2004; MBM 158534, 1♂, Sanya Bay, 26 March 1992.

**Description** Rostrum small, triangular; orbitorostral margin almost straight; rostral carina sharp, about extending to middle length of carapace; orbitorostral grooves shallow; orbital hoods without tooth.

Telson about 2.2 and 3 times as long as anterior and posterior width, respectively; dorsal surface bearing two pairs of moderately large spines, approximately inserted 1/3 and 1/2 length of telson, respectively; posterior margin only slightly convex, not beyond posterolateral small spines. Uropod with posterolateral margin of endopod bearing few spines.

Antennular peduncle with distally acute stylocerite, not reaching distal margin of first antennular segment; visible part of antennular segment nearly as long as third segment; second segment about twice as long as third segment. Antenna with lateral margin of scaphocerite nearly straight, only reaching 1/2 length of third antennular segment; antennal scale with blade reduced,

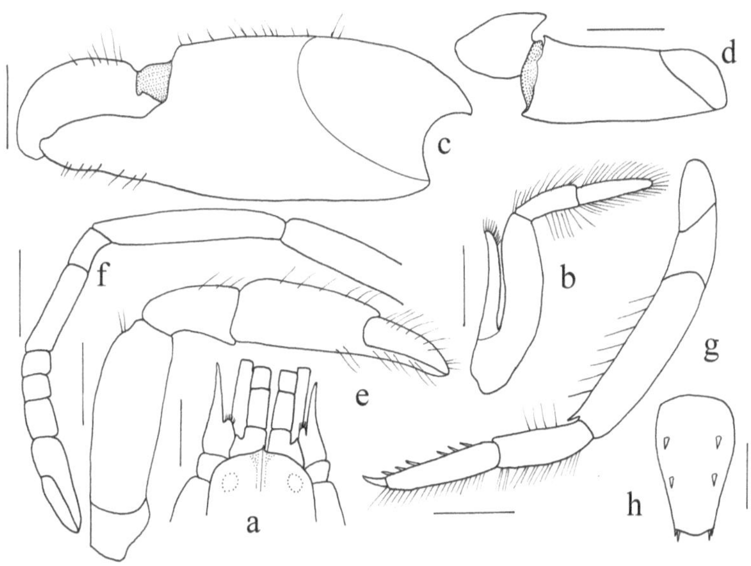


**Fig. 2.31** *Alpheus spongiarum* Coutière, 1897. MBM 158945, female, **a**, anterior part of carapace, dorsal view; **b**, third maxilliped; **c**, major chela, lateral view; **d**, carpus to ischium of major cheliped, mesial view; **e**, minor cheliped, lateral view; **f**, second pereiopod; **g**, third pereiopod; **h**, telson (scale bar = 1 mm)

only reaching 1/5 length of second antennular segment. Carpocerite beyond antennular peduncle.

Major cheliped with chela cylindrical, about 2.2 times as long as wide. Fingers about 1/2 length of palm, dactylus distinctly longer than fixed finger. Palm smooth, without any groove and ridge. Carpus cup-shaped. Merus stout, dorsal margin and ventromesial margin each terminating in acute tooth. Ischium unarmed. Minor cheliped with chela not showing sexual dimorphism, cylindrical. Fingers about 2/3 length of palm. Carpus cup-shaped. Merus slender, dorsal margin rounded distally; ventromesial margin terminating in acute tooth. Ischium unarmed.

Second pereiopod with carpus 5-segmented, segmented ratio subequal to 5: 11: 2: 2: 4.

Third pereiopod moderately stout, dactylus swelling. Propodus slightly longer than carpus, ventral margin bearing 5–7 and distally one pair spines. Carpus with ventral and dorsal margins both terminating in subacute tooth; ventral margin sometimes bearing 0–2 spines and long setae. Merus triangular in section, ventrolateral margin terminating in acute tooth, unarmed. Ischium unarmed.

**Remarks** *A. spongiarum* is similar to *A. eulimene* by having the reduced blade of scaphocerite. The present specimens are generally agreed with previous descriptions, except for the following characters: the lateral tooth of scaphocerite reaching to distal margin of the antennular peduncle rather than reaching to middle length of the third antennular segment as shown in Banner & Banner (1982, fig. 31a); the dactylus of third pereiopod only swelling, not biunguiculate as shown in Banner & Banner (1982, fig. 31h).

**Distribution** The South China Sea; Madagascar; Seychelles; Reunion; Gulf of Aden; Maldive and Laccadive Islands; Sri Lanka; Singapore; Japan; Philippines; Indonesia; Australia.

### 2.2.25 *Alpheus stanleyi* Coutière, 1908 (Fig. 2.32)

*Alpheus stanleyi* Coutière, 1908: 207; 1921: 423, Pl. 63, fig. 18; Banner DM & Banner AH, 1978: 230, fig. 2a–e; Chace, 1988: 55.
*Alpheus stanleyi* var. *dearmatus* De Man, 1910: 287; 1911: 367, Pl. 17, fig. 78.
*Alpheus cloudi* Banner, 1956: 352, fig. 16.

**Material Examined** This species was reported by Banner & Banner (1978) and Liu (2008) from the South China Sea, but not found in the present specimens.

**Description** Rostrum triangular, not reaching distal margin of first antennular segment; rostral carina extending to posterior margin of orbital hoods; orbitorostral grooves shallow; orbital hoods without tooth.

Telson slender, about 1.7 and 3.5 times as long as anterior and posterior width, respectively; dorsal surface bearing two pairs of small spines, approximately inserted 1/2 and 2/3 length of telson, respectively; posterior margin only slightly convex, not beyond posterolateral small spines. Uropod with posterolateral margin of endopod bearing few spines.

Antennular peduncle with distally acute stylocerite, not reaching distal margin of first antennular segment; visible part of first antennular segment slightly longer than third segment; second segment twice as long as third segment. Antenna with lateral margin of scaphocerite slightly concave, not reaching distal margin of antennular peduncle; antennal scale with distolateral tooth longer than blade. Carpocerite distinctly longer than antennular peduncle.

Major cheliped with chela sub-cylindrical, showing slightly sexual dimorphism: chela of male about twice as long as wide; chela of female 3 times as long as wide. Dactylus longer than fixed finger. Palm without any groove and ridge. Carpus cup-shaped. Merus stout, dorsal margin rounded distally; ventromesial margin terminat-

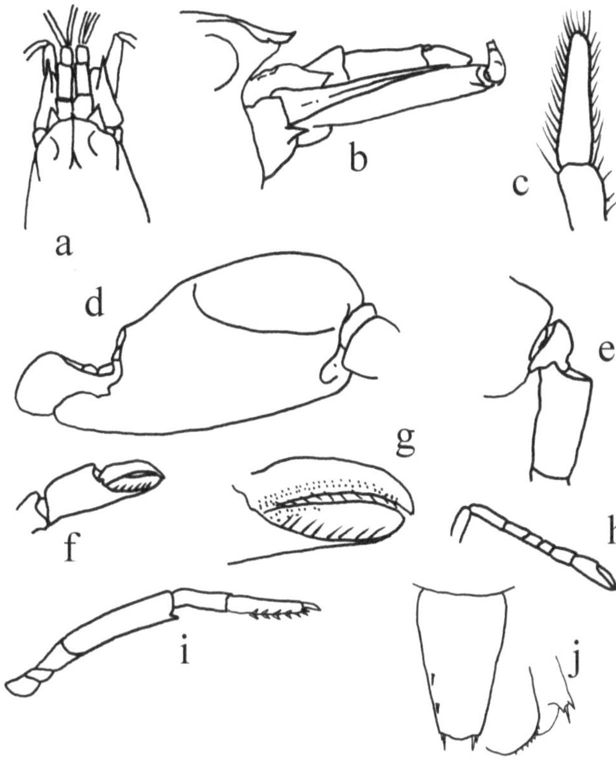


**Fig. 2.32** *Alpheus stanleyi* Coutière, 1908. (Adapted from Coutière, 1921). **a**, anterior part of carapace, dorsal view; **b**, same, lateral view; **c**, ultimate segment of third maxilliped; **d**, major chela; **e**, carpus and merus od major cheliped; **f**, minor chela; **g**, same, fingers; **h**, chela and carpus of second pereiopod; **i**, third pereiopod; **j**, posterior margin of right uropods and telson

ing in acute tooth or not. Ischium unarmed. Minor cheliped with chela not showing sexual dimorphism. Fingers nearly as long as palm.

Second pereiopod with carpus 5-segmented, segmented ratio subequal to 4: 2.8: 1.8: 1.8: 2.6.

Third pereiopod with dactylus variably biunguiculate. Propodus bearing 8 spines on ventral margin. Carpus about 1/2 length of merus. Merus with ventral margin terminating in acute tooth. Ischium with one spine on ventral margin.

**Distribution** The South China Sea; Western and central Indian Ocean; Philippines; Indonesia; Caroline; Fiji; Samoa Islands.

### 2.2.26 *Alpheus xishaensis* Liu & Lan, 1980 (Fig. 2.33)

*Alpheus xishaensis* Liu & Lan, 1980: 102, fig. 13.

**Material Examined** 4II68AD-13, 1♂, Zhubi Reef, Nansha Islands, Sta. 8, 11 May 2004; 04II54AD-22, 1 larva, Zhubi Reef, Nansha Island, 6 May 2004; MBM 233252, 1 larva, Zhaoshu Island, Xisha Islands, 11 May 1980-5-11.

**Description** Rostrum triangular, reaching 1/2 length of first antennular segment; rostral carina nearly extending to middle length of carapace; orbitorostral grooves moderately deep; orbital hoods inflated, without tooth.

Telson about 1.7 and 2.7 times as long as anterior and posterior width, respectively; dorsal surface bearing two pairs of moderately large spines, approximately inserted 1/3 and 2/3 length of telson, respectively; posterior margin only slightly convex, not beyond posterolateral small spines.

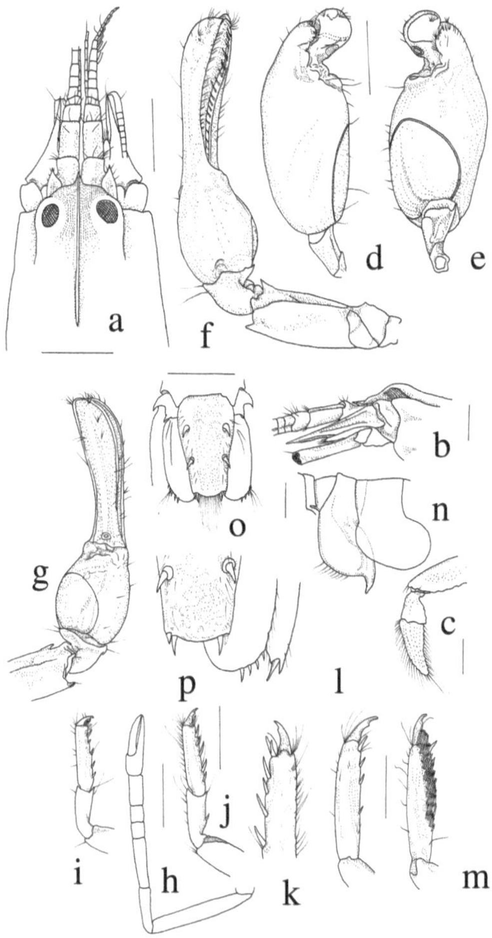


**Fig. 2.33** *Alpheus xishaensis* Liu & Lan, 1980. (Adapted from Liu, 1980). Holotype, **a**, anterior part of carapace, dorsal view; **b**, same, lateral view; **c**, third maxilliped; **d** & **e**, major cheliped; **f**, minor cheliped; **g**, minor chela; **h**, second pereiopod; **i**, third pereiopod; **j**, fourth pereiopod; **k**, dactylus and anterior part of propodus of fourth pereiopod; **l** & **m**, dactylus and propodus of fifth pereiopod, lateral and mesial view; **n**, first and second abdominal segments; **o**, uropods and telson; **p**, posterior margin of uropods and telson (scale bar: **a**–**h**, **k**–**n** = 1 mm; **i**, **j**, **o**, **p** = 0.5 mm)

Uropod with posterolateral margin of endopod bearing few spines.

Antennular peduncle with distally acute stylocerite, not reaching distal margin of first antennular segment; visible part of first antennular segment slightly longer than third segment; second segment about twice as long as third segment. Antenna with lateral margin of scaphocerite concave, slightly overreaching distal margin of antennular peduncle; antennal scale with blade reduced, only reaching 1/2 length of second antennular segment. Carpocerite nearly as long as antennular peduncle.

Major cheliped with chela sub-cylindrical, about twice as long as wide, slightly twist. Fingers with dactylus distinctly longer than fixed finger. Palm with dorsal margin bearing one transverse depression near distally, ventral margin and surfaces without groove and ridge. Carpus cup-shaped. Merus stout, about 1/4 length of chela, dorsal margin and ventromesial margin both terminating in acute tooth, unarmed. Ischium unarmed. Minor cheliped slender. Fingers about 1.5 times as long as palm, spatulate; cutting edges slightly concave; fixed finger with cutting edge bearing more than ten spines, gradually get bigger distally. Palm inflated, cylindrical, without any groove and ridge. Merus similar to that of major cheliped, but more slender. Ischium unarmed.

Second pereiopod with carpus 5-segmented, segmented ratio subequal to 4.5: 5: 2: 2: 3.

Third pereiopod moderately stout, dactylus biunguiculate, ventral margin bearing small, subacute unguis. Propodus distinctly longer than carpus, ventral margin bearing 5 and distally one pair spines. Carpus with dorsal and ventral margin both produced as acute tooth distally; ventral margin bearing one spine. Merus about 3.7 times as long as wide, unarmed. Ischium unarmed. Fourth and fifth pereiopod similar to third pereiopod, more slender, ventral margin of carpus without spine; propodus of fifth pereiopod bearing transverse rows of setae and 5 spines on ventral margin.

**Remarks** The differences between *A. xishaensis* and its similar species are discussed in Liu & Lan (1980).

**Distribution** The South China Sea.

**Diadema Group (Fig. 2.34)**
Orbital hoods always without tooth, at times projecting as vertical keels; orbitorostral grooves distinct. Carapace sometimes bearing teeth anteriorly. Major cheliped with chela cylindrical, dorsal margin bearing transverse groove nearly distal; without distinct longitudinal grooves. Minor cheliped with chela sometimes showing sexual dimorphism. Third pereiopod usually with dactylus simple; merus with or without spines on ventral margin (Banner & Banner, 1982).

Key to species of Diadema Group from China seas

1. Posterior portions of orbital hoods bearing flattened acute tooth ........ *A. bidens*
-. Dorsal face of carapace without teeth ........ 2
2. Major chela with palm bearing sub-terminal transverse groove on dorsal margin ........ 3
-. Major chela with palm not bearing sub-terminal transverse groove on dorsal margin ........ 5
3. Transverse groove on dorsal margin of major chela slightly and obliquely extending to mesial face as narrow depression ........ *A. ehlersii*
-. Transverse groove on dorsal margin of major chela not extending to mesial face ........ 4
4. Merus of major and minor chelipeds armed with spines on ventromesial margin ........ *A. diadema*
-. Merus of major and minor chelipeds unarmed with spines on ventromesial margin ........ *A. gracilipes*
5. Minor chela not showing sexual dimorphism; merus of major and minor chelipeds not bearing spines on ventromesial margin ........ *A. mitis*
-. Minor chela showing sexual dimorphism: chela of males at times balaeniceps; merus of major and minor chelipeds bearing small spines on ventromesial margin ........ *A. paracrinitus*

## 2.2.27 *Alpheus bidens* (Oliver, 1811) (Fig. 2.35)

*Palaemon bidens* Olivier, 1811: 663.

*Alpheus bidens*–. Milne-Edwards, 1837: 353, Pl. 24, figs. 11, 12; Coutière, 1899: 57, 274; De Man, 1911: 371, fig. 80; Banner, 1957: 203; Banner DM & Banner AH, 1982: 136, fig. 39; Chace, 1988: 17.

*Alpheus tridentatus* Zehntner, 1894: 204, Pl. 8, fig. 24.

*Alpheus praedator* De Man, 1908: 103; 1911: 373, fig. 81.

*Alpheus dissodontonotus* Stebbing, 1915: 83, Pl. 86.

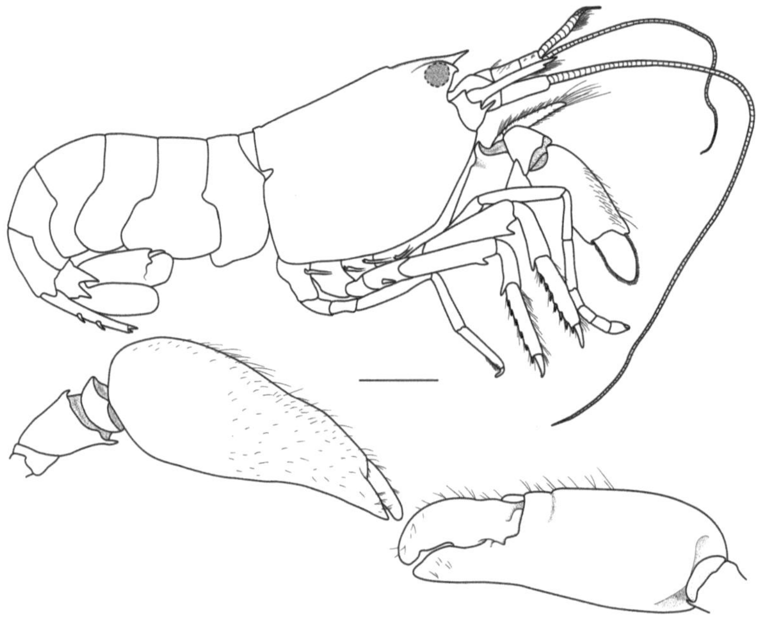

**Fig. 2.34** *Alpheus diadema* Dana, 1852. entire animal of Diadema Group (scale bar = 2 mm)

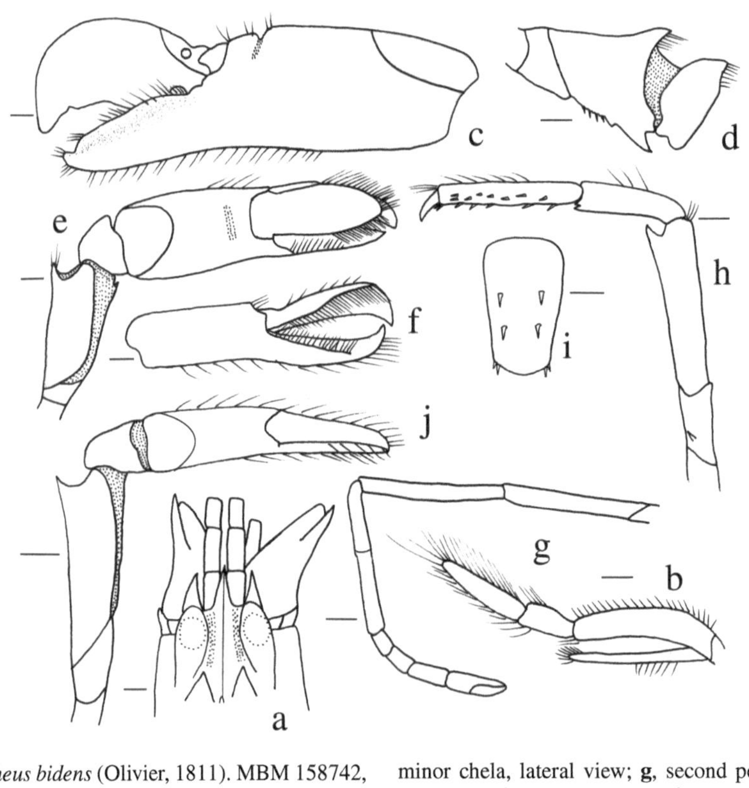


**Fig. 2.35** *Alpheus bidens* (Olivier, 1811). MBM 158742, male, **a**, anterior part of carapace, dorsal view; **b**, third maxilliped; **c**, major chela, lateral view; **d**, carpus to ischium, mesial view; **e**, minor cheliped, mesial view; **f**, minor chela, lateral view; **g**, second pereiopod; **h**, third pereiopod; **i**, telson. Female, **j**, minor cheliped, lateral view (scale bar = 1 mm)

**Material Examined** CL 12.5 mm, TL 30.1 mm. MBM 158742, 1♀1♂, Shalao, Hainan, 4 April 1992; MBM 107569, 1♀. Shalao, Hainan, 4 April 1992.

**Description** Rostrum triangular acute, reaching or overreaching distal margin of first antennular segment; rostral carina extending to posterior margin of orbital hoods, lies a small flattened area followed by a second dorsal carina; orbito-rostral grooves deep; orbital hoods inflated, without tooth; carapace with posterior surface of orbital hoods bearing acute tooth.

Telson about 1.6 and 2.4 times as long as anterior and posterior width, respectively; dorsal surface bearing two pairs of moderately large spines, approximately inserted 2/5 and 2/3 length of telson, respectively; posterior margin only slightly convex, not beyond posterolateral small spines.

Antennular peduncle with distally acute stylocerite, slightly overreaching distal margin of first antennular segment; visible part of first antennular segment as long as second segment; third segment shorter than second segment. Antenna with lateral margin of scaphocerite concave, far overreaching distal margin of antennular peduncle; antennal scale with distolateral tooth longer than blade. Carpocerite only reaching middle length of third antennular segment.

Major cheliped with chela sub-cylindrical, about 3.2 times as long as wide. Fingers about 2/3 length of palm, plunger feebly developed; lateral surface of fixed finger bearing slightly rounded longitudinal depression medially. Palm with dorsal surface bearing transverse groove sub-terminally. Merus stout, dorsal margin terminating in large subacute tooth; ventromesial margin slightly dentate, bearing small spines and with distally acute tooth; ventrolateral margin slightly dentate. Ischium unarmed. Minor cheliped with chela showing sexual dimorphism: fingers spatulate, dactylus of male balaeniceps, lateral face of fixed finger bearing row of stiff setae, fingers nearly as long as palm, dorsal surface of palm bearing transverse groove sub-terminally; in female fingers nearly as long as palm, dactylus not balaeniceps, fingers tapering, dorsal surface of palm smooth without transverse groove sub-terminally, mesial face of palm bearing long setae. Merus similar to that of major chela, but more slender. Ischium unarmed.

Second pereiopod with carpus 5-segmented, segmented ratio subequal to 10: 10: 3: 3: 5.

Third pereiopod moderately slender, dactylus simple. Propodus slightly shorter than merus, ventral surface bearing row of 7 spines with distally one pair and many randomly accessory spines. Carpus about 1/2 length of merus, dorsal margin terminating in sub-acute tooth; ventral margin terminating in two acute teeth. Merus with ventral margin terminating in one large acute tooth. Ischium armed with one spine on ventral margin.

**Remarks** *A. bidens* is readily distinguished from other species in Diadema group by the teeth on the posterior margin of orbital hoods. Chace (1988) has discussed the variability of the teeth on the first antennular segment, and the present specimens are all without the lateral tooth on the first antennular segment.

**Distribution** The South China Sea, Taiwan; Madagascar; Ryukyus; Philippines; Indonesia; Australia; Tasmania; Caroline; Marshall Islands.

### 2.2.28 *Alpheus diadema* Dana, 1852 (Fig. 2.36)

*Alpheus diadema* Dana, 1852b: 555, Pl. 35, fig. 7a–e; 1854: 23; Banner AH, 1956: 357; 1957: 203; 1959: 141, fig. 7; Holthuis, 1958: 24; Banner AH & Banner DM, 1964: 91; 1966a: 179; 1967: 276; 1968b: 287; Miya, 1974: 152, Pl. 29; Liu & Lan, 1980: 91, fig. 7; Banner DM & Banner AH, 1982: 140, fig. 40; Chace, 1988: 23.

*Alpheus insignis* Heller, 1861: 269, Pl. 2, figs. 17, 18; De Man, 1902: 846, Pl. 26, fig. 60a, b; 1911: 377; Coutière, 1905: 899; 1921: 426; Nobili, 1907: 5; Tattersall, 1921: 380.

*Crangon diadema* Banner, 1953: 118, fig. 43.

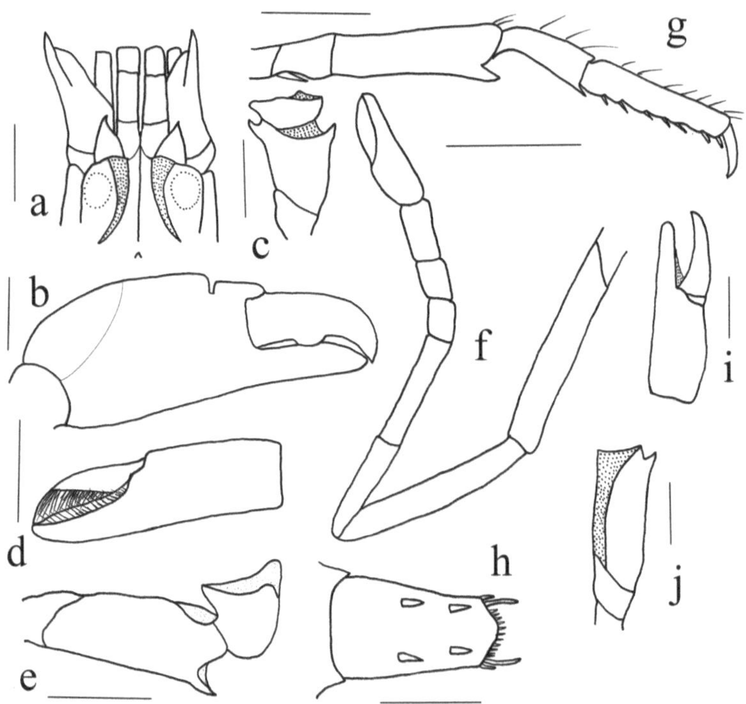


**Fig. 2.36** *Alpheus diadema* Dana, 1852. MBM 158636, male, **a**, anterior part of carapace, dorsal view; **b**, major chela, lateral view; **c**, carpus to ischium of major cheliped, mesial view; **d**, minor chela, mesial view; **e**, carpus to ischium of minor cheliped, mesial view; **f**, second pereiopod; g, third pereiopod; **h**, telson. Female, **i**, minor chela; **j**, merus and ischium of minor cheliped (scale bar = 1 mm)

**Material Examined** CL 5.2 mm, TL 18.1 mm. MBM 159332, 1♂, East Island, Xisha Islands, 28 May 1980; MBM 158099, 1 larva, Xincun, Hainan, Wang Yong-liang Coll., 9 July 1957; MBM 158824, 1 larva, North Island, Yantai, Fan Zhen-gang Coll., 27 April 1958-4-27; MBM 159172, 1 larva, Sanya, Hainan, Li Xin-zheng Coll., 1 December 1997; MBM 158145, 1♀, Sanya, Hainan, 1 December 1997-12-1; MBM 158082, 1 larva, Sanya, Hainan, 20 November 1997; MBM 159187, 1♀, Sanya, Hainan, Li Xin-zheng Coll., 20 November 1997; MBM 159328, 2♂♂, Zhaoshu Island, Xisha Islands, 11 May 1980; MBM 157798, 4♀♀, Shi Island, Xisha Islands, 4 June 1981; MBM 157970, 1♀, North Island, Xisha Islands, Fan Zhen-gang & Xu Jie-shan Coll., 6 May 1958; MBM 158022, 2♂♂1♀, Jinqing Island, Xisha Islands, 30 April 1958; MBM 157975, 3♂♂. Tree Island, Xisha Islands, 1 May 1958; MBM 158218, 3♀♂, Dengqing Island, Xisha Islands, 12 April 1958; MBM 158037, 9♀♀, Bei Reef, Xisha Islands, Fan Zhen-gang & Xu Jie-shan Coll., 23 March 1958; MBM 157958, 1♂, Yongxing Island, Xisha Islands, Wang Yong-liang Coll., 15 May 1957; MBM 158093, 1♀, Luo Island, Guangdong, Ma Xiu-tong Coll., 1 April 1955; MBM 158819, 2♀♀, Yongxing Island, Xisha Islands, 20 March 1958; MBM 159260, 1♀1♂, Nanxun Reef, Nansha Islands, 21 May 1990; MBM 159314–159321, 39♀♂, Xian'e Reef, Nansha Islands, 28 May 1993; MBM 158988, 12♂♂, Yongxing Island, Xisha Islands, Chen Mu Coll., 31 May 1981; MBM 158852, 4♂♂3♀♀, Daai Reef, Nansha Islands, 24 May 1990; MBM 159222, 3♂♂1♀, Zhubi Reed, Nansha Islands, 10 April 1999; MBM 158849, 3♂♂, Yongxing Island, Xisha Islands, Fan Zhen-gang & Xu Jie-shan Coll., 20 March 1958; MBM 158006, 1 larva, Yongxing Island, Xisha Islands, 15 May 1958; MBM 157900, 1♂, Yongshu Reef, Nansha Islands, 11 April 1999; MBM 158993, 2♂♂, Shi Island, Xisha Islands, 4 June 1981;

MBM 158221, 2♂♂1♀, Xincun, Hainan, Wang Yong-liang Coll., 13 July 1957; MBM 159198, 1♂, Xincun, Hainan, 9 July 1957; MBM 158975, 1♀1♂, Shi Island, Xisha Islands, 8 May 1975; MBM 158983, 1♀2♂♂, Yongxing Island, Xisha Islands, 21 March 1980. MBM 159003, 1♂, Shanhu Island, Xisha Islands, Liu Xi-xing Coll., 25 March 1975; MBM 159029, 1♀, Jinqing Island, Xisha Islalnds, Chen Mu Coll., 19 May 1981. MBM 158460, 2♀♀, Yongshu Reef, Nansha Islands, Li Xin-zheng Coll., 18 May 1993; MBM 158355, 1♂, Nanping Reef, Nansha Islands, 2 August 1988; MBM 158636, 1♂, Anda Reef, Nansha Islands, 23 May 1993; MBM 158654, 3♀♀, Yufang Reef, Nansha Islands, 4 May 1990; MBM 157837, 1♀, Shi Island, Xisha Islands, 4 June 1981; MBM 158746, 1♀, Zhangyi Reef, Nansha Islands, 29 May 1990; MBM 159263, 1♂, Nanxun Reef, Nansha Islands, 17 April 1999; MBM 157924, 3♀♀, Yongxing Island, Xisha Islands, 18 May 1981; MBM 157968, 1♀, Yongxing Island, Xisha Islands, Wang Yong-liang Coll., 13 April 1957; MBM 158300, 1♀1♂, Xisha Islands, 13 May 1975; MBM 158578, 1♀1♂, Zhubi Reef, Nansha Islands, 30 May 1993; MBM 158282, 3♂♂5♀♀, Guangjin Island, Xisha Islands, 18 May 1975; MBM 158840, 1♀, Sanya, Hainan, Wang Yong-liang Coll., 11 June 1957; MBM 158625, 1♀, Haikou Reef, Nansha Islands, 3 May 1987; MBM 158272, 2♂♂3♀♀, Jinyin Island, Xisha Islands, 24 May 1975; MBM 157738, 1♀, Xincun, Hainan, Liu Rui-yu Coll., 27 January 1956; MBM 158739, 1♂, Yongshu Reef, Nansha Islands, Li Xin-zheng Coll., 18 May 1993; MBM 158745, 1♀, Nanxun Reef, Nansha Islands, Li Xin-zheng Coll., 21 May 1993; MBM 158768, 3♀♀2♂♂, Shanhu Reef, Nansha Islands, 25 May 1990, MBM 158599, 2♂♂, Yongshu Reef, Nansha Islands, 16 May 1993; MBM 158562, 1♀, Yongshu Reef, Nansha Islands, 16 May 1993; MBM 158320, 1♀, Yongxing Island, Xisha Islands, 31 May 1981; MBM 158693, 3♂♂1♀, East Island, Xisha Islands, 28 May 1980; MBM 158274, 1♂, Yongxing Island, Xisha Islands, 7 June 1975; 04II53AD-22, 1♀, Zhubi Reef, Nansha Islands, 4 May 2004; 04II59AD-30, 1♀1♂, Zhubi Reef, Nansha Islands, 8 May 2004; 04II51AD-19, 1♂Zhubi Reef, Nansha Islands, 4 May 2004; 04II80AD-25, 1♀, Zhubi Reef, Nansha Islands, 13 May 2004; 04II66AD-16, 1♀2♂♂, Zhubi Reef, Nansha Islands, 10 May 2004; 04II52AD-21, 1♀, Zhubi Reef, Nansha Islands, 4 May 2004; MBM 233263, 1♂, Guangjin Island, Xisha Islands, 18 May 1975; MBM 233231, 1♂, East Island, Xisha Islands, 9 June 1975.

**Description** Rostrum triangular, acute, not reaching distal margin of first antennular segment; rostral carina feebly developed, extending to posterior margin of orbital hoods; dorsal surface of carapace bearing one medially protuberance on gastric region; orbitorostral grooves flat; orbital hoods slightly inflated, without tooth.

Telson about 2.1 and 2.5 times as long as anterior and posterior width, respectively; dorsal surface bearing two pairs of moderately large spines, approximately inserted 3/8 and 5/8 length of telson, respectively; posterior margin only slightly convex, not beyond posterolateral slender spines; bearing small spines between posterolateral spines. Uropod with endopod bearing spinules on posterior margin.

Antennular peduncle with distally acute stylocerite, reaching distal margin of first antennular segment; visible part of first antennular segment nearly as long as second segment; third segment shorter than second segment. Antenna with lateral margin of scaphocerite concave, far overreaching distal margin of antennular peduncle; antennal scale with distolateral tooth longer than blade. Carpocerite nearly as long as antennular peduncle.

Major cheliped with chela sub-cylindrical, about 3 times as long as wide. Fingers about 2/5 length of palm, plunger developed. Palm with dorsal surface bearing transverse groove sub-terminally. Merus stout, dorsal margin terminating in acute tooth; ventromesial margin rounded distally. Ischium unarmed. Minor cheliped with chela showing sexual dimorphism: dactylus of male balaeniceps, spatulate, mesial margin of palm bearing short setae; fingers of female tapering. Carpus cup-shaped. Merus slender, dorsal margin bearing sub-terminal tooth; ventromesial margin without tooth distally. Ischium unarmed.

Second pereiopod with carpus 5-segmented, segmented ratio subequal to 6–7: 7–8: 2: 2: 3–4.

Third pereiopod stout, dactylus simple. Propodus with ventral margin bearing 9 spines and lateral margin randomly bearing about 10 spines. Carpus about 1/2 length of merus, ventral margin terminating in acute tooth. Merus with ventral margin bearing sub-terminal tooth, unarmed with spines. Ischium bearing one moderately large spine on ventral margin.

**Remarks** The present specimens have simple dactylus of the third pereiopod, rather than biunguiculate as shown in Banner & Banner (1982, fig. 40i). The variation of the dactylus of the third pereiopod has been discussed by Banner (1959).

**Distribution** The Yellow Sea, The South China Sea, Taiwan; Red Sea; eastern Africa; Indian Ocean; Japan; Philippines; Indonesia; Australia; tropical Pacific to Hawaii.

### 2.2.29 *Alpheus ehlersii* De Man, 1909 (Fig. 2.37)

*Alpheus ehlersii* De Man, 1909c: 663, Pl. 70. Banner AH & Banner DM, 1966: 114b, fig. 40; Banner DM & Banner AH, 1982: 132, fig. 37; Chace, 1988: 25.

*Alpheus macrochirus* –. De Man, 1888b: 519 (Not *A. macrochirus* Richters, 1880)

**Material Examined** CL 4.1–5.9 mm, TL 11.2–16.9 mm. MBM 158199, 1♂, Sanya, Hainan, 1 March 1997; MBM 158892, 1♂, Sanya, Hainan, Li Xin-zheng Coll., 1 March 1997; MBM 158178, 1♂, Sanya, Hainan, 1 March 1997; MBM 158890 , 1♂, Sanya, Hainan, Li Xin-zheng Coll., 1 March 1997; MBM 159153, 1♂, Sanya, Hainan, Li Xin-zheng Coll., 20 November 1997; MBM 159217, 1♂1♀, Sanya, Hainan, Li Xin-zheng Coll., 11 November 1997; MBM 157887, 3♂♂1♀, Dongzhou Island, Zhejiang, 6 April 1958; MBM 159173, 2♂♂, Sanya, Hainan, Li Xin-zheng Coll., 17 November 1997; MBM 159006, 1♀, Sanya, Hainan, 2 April 1958; MBM 157719, 13♀♂, Xizhou Island, Hainan, Liu

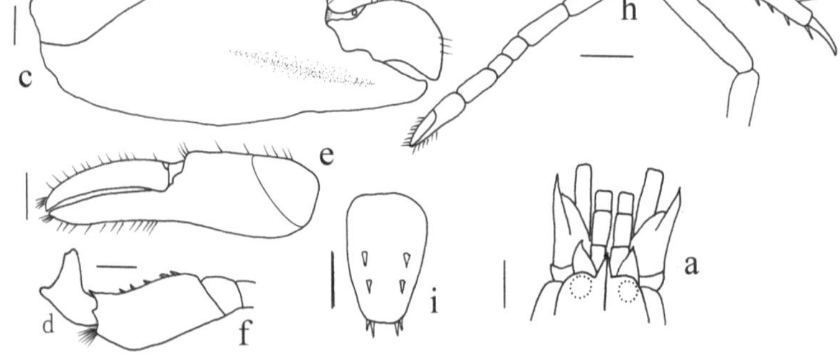


**Fig. 2.37** *Alpheus ehlersii* De Man, 1909. MBM 157722, male, **a**, anterior part of carapace, dorsal view; **b**, major chela, mesial view; **c**, same, lateral view; **d**, carpus to ischium of major cheliped, mesial view; **e**, minor chela, lateral view; **f**, carpus to ischium of minor cheliped, mesial view; **g**, third pereiopod; **h**, second pereiopod; **i**, telson (scale bar = 1 mm)

Rui-yu Coll., 30 December 1955; MBM 157722, 8♀♀6♂♂, Xizhou Island, Hainan, Liu Rui-yu Coll., 30 December 1955; MBM 158740, 1♀1♂, Yongshu Reef, Nansha Island, Li Jin-he Coll., 17 May 1993; MBM 157957, 1♀, Xizhou Island, Hainan, 3 April 1958; MBM 158858, 1♂, Banyue Reef, Nansha Islands, 29 September 1994.

**Description** Rostrum triangular, acute, not reaching distal margin of first antennular segment; rostral carina rounded, extending to posterior margin of orbital hoods; orbitorostral grooves shallow; orbital hoods slightly inflated, without tooth.

Telson about 1.5 and 2.5 times as long as anterior and posterior width, respectively; dorsal surface bearing two pairs of moderately large spines, approximately inserted 1/2 and 2/3 length of telson, respectively; posterior margin nearly straight, posterolateral spines moderately large.

Antennular peduncle with distally acute stylocerite reaching to distal margin of first antennular segment; visible part of first antennular segment nearly as long as second segment; third segment about 2/3 length of second segment. Antenna with lateral margin of scaphocerite sinuous, overreaching distal margin of antennular peduncle; antennal scale with distolateral tooth distinctly longer than blade. Carpocerite beyond antennular peduncle.

Major cheliped with chela slightly compressed, about 2.7 times as long as wide. Fingers about 2/5 length of palm. Dorsal surface of palm bearing transverse feebly developed groove subterminally, which extending obliquely towards mesial face; lateral face bearing depression anteriorly, which extending to middle of lateral face of fixed finger. Carpus cup-shaped. Merus stout, dorsal margin rounded; ventromesial margin bearing 5–9 spines and terminating in subacute tooth or rounded. Ischium unarmed. Minor cheliped with chela not showing sexual dimorphism. Fingers slightly longer than palm, tapering. Palm smooth, without any groove or ridge. Merus similar to that of major cheliped, slender, dorsal margin terminating in rounded tooth. Ischium unarmed.

Second pereiopod with carpus 5-segmented, segmented ratio subequal to 10: 5: 2: 2: 4–5.

Third pereiopod stout, dactylus simple. Propodus with ventral margin bearing 5–6 and distally one pair spines. Carpus about 1/2 1.5 times as long as merus, dorsal margin terminating in rounded tooth. Merus unarmed. Ischium armed with one spine on ventral margin.

**Remarks** *A. ehlersii* is distinguished from other species in Diadema group by the transverse groove on dorsal margin of palm of the major chela obliquely extending to mesial face. The present specimens are generally agreed with previous descriptions, except for the following characters: the lateral tooth of scaphocerite distinctly beyond the antennular peduncle, rather than only reaching to that end as shown in Banner & Banner (1982, fig. 37a); the carpocerite far overreaching distal end of the antennular peduncle, rather than slightly overreaching to that end as shown in Banner & Banner (1982, fig. 37a). The variations of the anterior part have been discussed by Banner & Banner (1982).

**Distribution** The South China Sea; Red Sea; Eastern Africa; Madagascar; Thailand; Philippines; Indonesia; Australia; Caroline, Marshall, Phoenix, Samoa, and Tonga Island groups.

### 2.2.30 *Alpheus gracilipes* Stimpson, 1860 (Fig. 2.38)

*Alpheus gracilipes* Stimpson, 1860: 12; Miers, 1884: 278; Bate, 1888: 561; De Man, 1888b: 500, Pl. 21, fig. 5; 1902: 846; 1911: 380; 1924: 43, fig. 15; Ortmann, 1890: 488; 1894: 15; Borradaile, 1898: 1013; Yokoya, 1933: 22; Banner, 1956: 357; 1957: 204; 1958: 167; Liu & Lan, 1980: 93, fig. 8; Banner DM & Banner AH, 1982: 143, fig. 41; Chace, 1988: 29.

*Crangon gracilipes* Banner, 1953: 115, fig. 41.

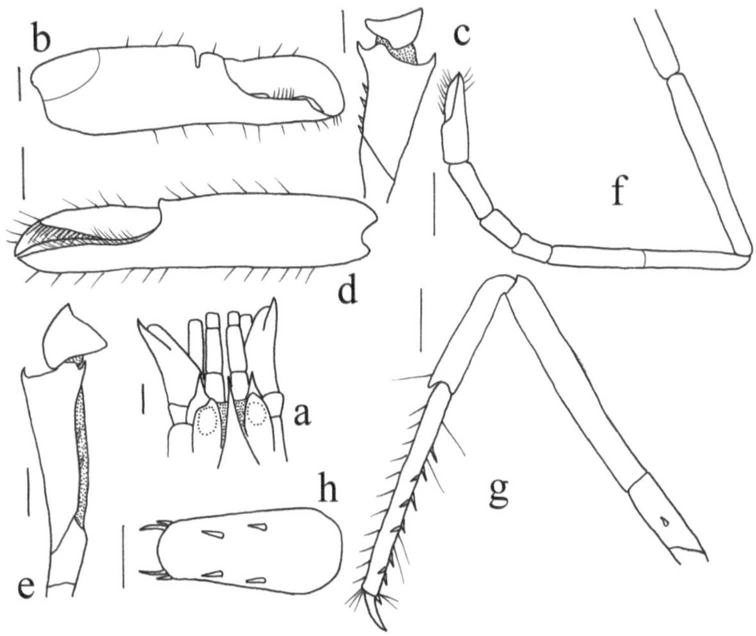


**Fig. 2.38** *Alpheus gracilipes* Stimpson, 1860. MBM 158260, male, **a**, anterior part of carapace, dorsal view; **b**, major chela, lateral view; **c**, carpus to ischium of major cheliped, mesial view; **d**, minor chela, lateral view; **e**, carpus to ischium of minor cheliped, mesial view; **f**, second pereiopod; **g**, third pereiopod; **h**, telson (scale bar = 1 mm)

**Material Examined** CL 6.9–8.1 mm, TL 19.5–20.7 mm. MBM 158868, 1♂, Yongxing Island, Xisha Islands, 11 June 1980-6-11; MBM 159968, 1 larva, Jinqing Island, Xisha Islands, 19 May 1980; MBM 158863, 1♀, Nansha Islands, 23 July 1988-7-23; MBM 158998, 2♂♂, Jinyin Island, Xisha Islands, 27 May 1975; MBM 158996, 6♀♀6♂♂, Guangjin Island, Xisha Islands, 23 May 1981; MBM 158060, 1♀, Jinqing Island, Xisha Islands, Fan Zhen-gang & Xu Jie-shan Coll., 1 May 1958; MBM 158705, 10♀♂, Shanhu Island, Xisha Islands, 19 May 1980; MBM 233256, 1♀, Shanhu Island, Xisha Islands, 19 May 1980; MBM 158227, 6♀♀10♂♂, Yongxing Island, Xisha Islands, 11 June 1980, MBM 159314-5, 3♂♂2♀♀, Nansha Islands; MBM 158988, 1♂, Yongxing Island, Xisha Islands, 31 May 1981; MBM 158852, 1♂2♀♀, Daai Reef, Nansha Islands, 24 May 1990; MBM 159222, 1♀1♂, Zhubi Reef, Nansha Islands, 10 April 1999; MBM 107599, 2♂♂1♀, Qukou, Hainan, Wang Yong-liang Coll., 14 March 1992; MBM 159029, 1♀1♂, Jinqing Island, Xisha Islands, 19 May 1981; MBM 159006, 2♀♀, Sanya, Hainan, 2 April 1958; MBM 138323, 2♀♀2♂♂, Xincun, Hainan, 31 December 1959; MBM 157770, 1♀, Xincun, Hainan, Fan Zhen-gang Coll., 21 December 1955; MBM 158591, 1♀, Yongshu Reef, 16 May 1989; MBM 158654, 3♀♀3♂♂, Yufang Reef, Nansha Island, 4 May 1990; MBM 158448, 1♂, Shalao, Hainan, 4 May 1992; MBM 159298, 1♂, Shalao, Hainan, 4 May 1992; MBM 157857, 2♂♂, Xincun, Hainan, 18 April 1958; MBM 158457, 2♂♂1♀, Xinying, Hainan, 27 March 1992; MBM 157924, 1♀2♂♂, Yongxing Island, Xisha Islands, 18 May 1981; MBM 158301, 1♂, East Island, Xisha Islands, 9 June 1975; MBM 158303, 2♂♂1♀, Shi Island, Xisha Islands, 8 May 1975; MBM 157968, 1♂, Yongxing Island, Xisha Islands, Wang Yong-liang Coll., 13 April 1957; MBM 158296, 1♀1♂, Shi Island, Xisha Islands, 12 June 1975; MBM 158578, 1♀1♂, Zhubi Reef, Nansha Islands, 30 May 1993; MBM 107595, 1♂2♀♀, Yezhu Island, Sanya Bay, 26 March 1992; MBM 158260, 1♀1♂, Guangjin Island, Xisha Islands, 15 May 1975; MBM 157759, 1♀, Shi Island, Xisha Islands, Wang Yong-liang Coll., 13 May 1957;

MBM 159278, 2♀♀1♂, Xincun, Hainan, Wang Yong-liang Coll., 27 March 1992; MBM 158272, 3♂♂1♀, Jinyin Island, Xisha Islands, 24 May 1975; MBM 158510-3, 1♀1♂, Xincun, Hainan, 26 March 1992; MBM 158762, 1♂, Xizhou Island, Hainan, Liu Rui-yu Coll., 29 December 1955; MBM 158734, 1♂, Xinying, Hainan, 14 January 1956; MBM 158590, 2♀♀1♂, Yongshu Reef, Nansha Islands, 18 May 1993; MBM 158732, 2♀♀, Shi Island, Xisha Islands, Fan Zhen-gang & Xu Jie-shan Coll., 19 May 1958; MBM 158582, 4♀♀1♂, Chigua Reef, Nansha Islands, 13 May 1989; MBM 158470, 1♀, Hainan, 20 November 1990; MBM 158703, 3♀♀1♂, Yongxing Island, Xisha Islands, 9 May 1980; MBM 158721, 4♀♀, Sanya, Hainan, 30 December 1959; MBM 158698, 6♀♀4♂♂, East Island, Xisha Islands, 28 May 1980; MBM 158620, 1♂, Meiji Reef, Nansha Islands, 11 May 1989; MBM 158755, 1♀, Yongxing Island, Xisha Islands, 9 May 1980; MBM 270018, 1♂, Yongxing Island, Xisha Islands, 11 June 1980. 04II67AD-4, 3♀♀6♂♂, Zhubi Reef, Nansha Islands, 10 May 2004; 04II56AD-43, 4♀♀2♂♂, Zhubi Reef, Nansha Islands, 6 May 2004; 04II77AD-20,1♂, Zhubi Reef, Nansha Islands, 12 May 2004; 04II59AD-31. 4♀♀2♂♂, Zhubi Reef, Nansha Islands, 8 May 2004; MBM 107621, 1♂, Yongxing Island, Xisha Islands, 11 June 1980; MBM 233253, 1♂, Shanhu Island, Xisha Islands, 20 May 1980.

**Description** Rostrum slender, acute, hardly reaching to distal margin of first antennular segment; rostral carina lacking; orbitorostral grooves deep; orbital hoods inflated, without tooth.

Telson about 2 and 2.2 times as long as anterior and posterior width, respectively; dorsal surface bearing two pairs of moderately large spines, approximately inserted 3/8 and 5/8 length of telson, respectively; posterior margin only slightly convex, not beyond posterolateral slender spines. Uropod with endopod bearing spinules on posterior margin.

Antennular peduncle with distally acute stylocerite reaching to distal margin of first antennular segment; visible part of first antennular segment shorter than second segment; second segment about twice as long as third segment. Antenna with lateral margin of scaphocerite sinuous, overreaching distal margin of antennular peduncle; antennal scale with distolateral tooth longer than narrow blade. Carpocerite reaching about middle length of third antennular peduncle.

Major cheliped with chela slender, slightly compressed, about 4 times as long as wide. Fingers about 1/2 length of palm; dactylus distinctly longer than fixed finger, distally rounded, with plunger feebly developed. Palm with dorsal surface bearing deep transverse groove sub-terminally. Merus stout, dorsal margin terminating in large subacute tooth; ventromesial margin bearing terminal large acute tooth and 2–4 small spines. Ischium unarmed. Minor cheliped with chela not showing sexual dimorphism, both male and female with dactylus balaeniceps, but some female and male juveniles with dactylus balaeniceps inconspicuous; cutting edge of fixed finger bearing setae on proximal half-length. Merus similar to that of major cheliped, but dorsal margin terminating in acute tooth. Ischium unarmed.

Second pereiopod with carpus 5-segmented, segmented ratio subequal to 6: 5–6: 2: 2: 3.

Third pereiopod moderately slender, dactylus simple, about 1/5–1/6 length of propodus. Propodus about 1.5 times as long as carpus, ventral margin bearing 9–16 spines and distally one pair. Merus unarmed. Ischium armed with one spine on ventral margin.

**Remarks** The present specimens are generally agreed with previous descriptions. The length ratio of first and second segment of carpus of second pereiopod shows slightly variability in the present specimens (the first segment slightly longer than or sub-equal to second segment). This variation is also shown in Banner & Banner (1982, 145).

**Distribution** The South China Sea; Red Sea; Eastern Africa to Hong Kong; Ryukyu Archipelago; Philippines; Indonesia; Australia eastward through Pacific islands to Hawaii.

## 2.2.31 *Alpheus mitis* Dana, 1852 (Fig. 2.39)

*Alpheus mitis* Dana, 1852b: 549, Pl. 35, fig. 1; Jeng, 1985: 241–256; Chace, 1988: 42.

**Material Examined** CL 5.1 mm, TL 17.5 mm. MBM 157965, 1♀, Xincun, Hainan, 22 April 1958; MBM 159031, 1♂, Shi Island, Xisha Islands, 2 June 1981; MBM 158514, 1♂, Yezhu Island, Yalong Bay, Neumann Coll., 4 April 1992; MBM 157785-6, 1♂, Shalao, 4 April 1992; MBM 159278, 1♀, Xincun, Hainan, Wang Yong-liang Coll., 27 March 1992; MBM 158457, 1♀, Xinying, Hainan, 27 March 1992; MBM 159306-9, 1♀, Yezhu Island, Yalong Bay, 22 March 1992; MBM 158274, 2♂♂, Yongxing Island, Xisha Islands, Ren Xian-qiu Coll., 7 June 1975; MBM 158566, 1 larva, Yezhu Island, Yalong Bay, 19 November 1990; MBM 158655, 2♀♀, Yezhu Island, Yalong Bay, 25 November 1990; MBM 107613, 1♀1♂, Shalao, Hainan, Wang Yong-liang Coll., 5 April 1992.

**Description** Rostrum triangular, acute, not reaching distal margin of first antennular segment; rostral carina rounded, extending to posterior margin of orbital hoods; orbitorostral grooves shallow; orbital hoods without tooth.

Telson about 1.7 and 3 times as long as anterior and posterior width, respectively; dorsal surface bearing two pairs of moderately large spines, approximately inserted 1/2 and 5/8 length of telson, respectively; posterior margin only slightly convex, not beyond posterolateral slender spines.

Antennular peduncle with distally acute stylocerite reaching to distal margin of first antennular segment; visible part of first antennular segment nearly as long as second segment; third segment 3/5 length of second segment. Antenna with lateral margin of scaphocerite slightly concave, slightly overreaching distal margin of antennular peduncle; antennal scale with distolateral tooth longer than narrow blade. Carpocerite beyond antennular peduncle.

Major cheliped with chela sub-cylindrical, 3 times as long as wide. Fingers about 1/2 length of palm. Palm smooth, without any groove or ridge. Carpus cup-shaped. Merus slender, dorsal margin rounded distally; ventromesial margin terminat-

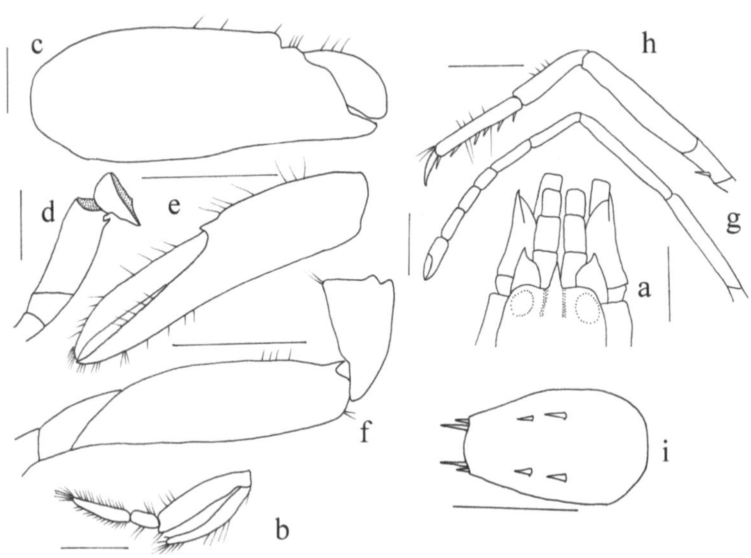


**Fig. 2.39** *Alpheus mitis* Dana, 1852. MBM 157965, female, **a**, anterior part of carapace, dorsal view; **b**, third maxilliped; **c**, major chela, mesial view; **d**, carpus to ischium, mesial view; **e**, minor chela, mesial view; **f**, carpus to ischium, mesial view; **g**, second pereiopod; **h**, third pereiopod; **i**, telson (scale bar = 1 mm)

ing acute tooth. Ischium unarmed. Minor cheliped with chela not showing sexual dimorphism, fingers nearly as long as palm, dactylus conical. Carpus cup-shaped. Merus similar to that of major cheliped, but more slender. Ischium unarmed.

Second pereiopod with carpus 5-segmented, segmented ratio subequal to 7: 7: 4: 4: 5.

Third pereiopod slender, dactylus simple, 1/3 length of propodus. Propodus longer than carpus, ventral margin bearing 5–6 spines and one distally. Carpus with dorsal margin produced as subacute tooth. Merus unarmed. Ischium armed with one spine on ventral margin.

**Remarks** *A. mitis* is very similar to *A. paracrinitus* in Diadema group. The Australian specimens identified as *A. mitis* by Banner & Banner (1982) are *A. paracrinitus* (Banner & Banner, 1983; Chace, 1988). Except for the characters discussed in Chace (1988) used to distinguish *A. paracrinitus* and *A. mitis*, the ventromesial margin of merus of the major and minor cheliped is unarmed with small spines in the present specimens in *A. mitis*. While, the ventromesial margin of merus has small spines in *A. paracrinitus*.

**Distribution** The South China Sea, Taiwan; Balabac Strait.

### 2.2.32 *Alpheus paracrinitus* Miers, 1881 (Fig. 2.40)

*Alpheus paracrinitus* Miers, 1881: 365, Pl. 16, fig. 6; Chace, 1962: 609; Crosnier & Forest, 1966: 253, fig. 15; Banner AH & Banner DM, 1967: 278; Banner DM & Banner AH, 1982: 129, fig. 36; Chace, 1988: 45.
*Alpheus ascensionis* Ortmann, 1893: 45.
*Alpheus paracrinitus bengalensis* Coutière, 1905: 901, Pl. 82, fig. 37.
*Alpheus bengalensis* Holthuis, 1958:25.
*Crangon paracrinita bengalensis* Banner, 1953: 110, fig. 40.
*Crangon togatus* Armstrong, 1940: 2, fig. 1.

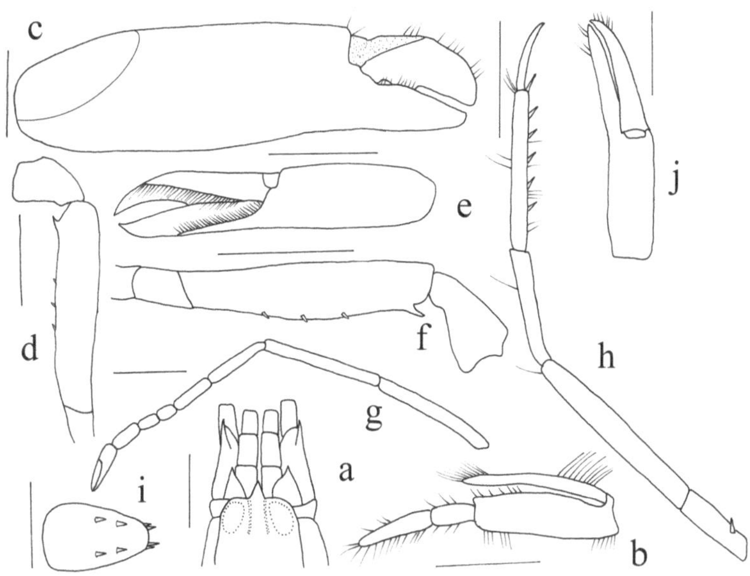


**Fig. 2.40** *Alpheus paracrinitus* Miers, 1881. MBM 159303, male, **a**, anterior part of carapace, dorsal view; **b**, third maxilliped; **c**, major chela, lateral view; **d**, carpus to ischium of major cheliped, mesial view; **e**, minor chela, lateral view; **f**, carpus to ischium of minor cheliped, mesial view; **g**, second pereiopod; **h**, third pereiopod; **i**, telson. Female, **j**, minor chela (scale bar = 1 mm)

*Alpheus paracrinitus*–. Banner DM & Banner AH, 1982: 134, fig. 38 [Not *A. mitis* Dana, 1852]

**Material Examined** CL 4.5 mm, TL 11.5 mm. MBM 159303-5, 1♀1♂, Yalong Bay, Hainan, Wang Yong-liang Coll., 23 March 1992; MBM 158671, 1♀, Yezhu Island, Yalong Bay, Neumann Coll., 25 March 1992; MBM 107611, 1 larva, Yezhu Island, Hainan, Neumann Coll., 27 March 1992.

**Description** Rostrum triangular, slightly longer than wide at base, reaching middle length of first antennular segment; rostral carina rounded, extending to posterior margin of orbital hoods; orbitorostral grooves shallow; orbital hoods not inflated, without tooth.

Telson about 1.6 and 4.2 times as long as anterior and posterior width, respectively; dorsal surface bearing two pairs of moderately large spines, approximately inserted 1/2 and 3/4 length of telson, respectively; posterior margin only slightly convex, not beyond posterolateral spines.

Antennular peduncle with distally acute stylocerite reaching to distal margin of first antennular segment; visible part of first antennular segment nearly as long as second segment; second segment about 1.5 times as long as third segment. Antenna with lateral margin of scaphocerite nearly straight, hardly reaching to distal margin of antennular peduncle; antennal scale with distolateral tooth longer than narrow blade. Carpocerite well beyond antennular peduncle.

Major cheliped with chela sub-cylindrical. Fingers about 1/3 length of palm. Palm smooth, without any groove or ridge. Carpus cup-shaped. Merus slender, dorsal margin rounded distally; ventromesial margin with strong acute tooth subdistally and 3 slender small spines. Ischium unarmed. Minor cheliped with chela showing sexual dimorphism: chela of male occasionally balaeniceps, fingers nearly as long as palm; chela of female slender, not balaeniceps, fingers tapering. Merus similar to that of Major chela. Ischium unarmed.

Second pereiopod with carpus 5-segmented, segmented ratio subequal to 15: 10: 5: 5: 8.

Third pereiopod slender, dactylus simple, elongate, slightly shorter than 1/2 length of palm. Propodus longer than carpus, ventral margin bearing 6–8 spines. Carpus 1/2 length of merus, ventral and margins both rounded distally. Merus unarmed. Ischium armed with one spine on ventral margin.

**Remarks** See "Remarks" under *A. mitis*.

**Distribution** The Shout China Sea, Taiwan; Pan-tropical.

**Edwardsii Group (Fig. 2.41)**

Orbital hoods almost without tooth; Major chela compressed, dorsal surface bearing transverse groove, which extending to mesial and lateral face as triangular or quadrangular depression; ventral surface usually bearing transverse groove nearly base of fixed finger. Minor chela often balaeniceps in male. Third pereiopod usually with simple dactylus; Merus usually bearing tooth on ventral margin (Banner & Banner, 1982; Chace, 1988).

Key to species of Edwardsii Group from China seas

1. Rostrum small, about reaching to anterior margins of orbital hoods ............................................*A. tirmiziae*
-. Rostrum triangular, at least reaching to middle length of first antennular segment ..........................................2
2. Lateral tooth of basicerite about reaching to middle length of second antennular segment..............................3
-. Lateral tooth of basicerite not reaching to end margin of first antennular segment .........................................4
3. Rostral carina absent.........................................................................................................*A. bannerorum*
-. Rostral carina extending to posterior margin of orbital hoods ................................................*A. parvirostris*
4. Carpus of second pereiopod with first segment distinctly shorter than or sub-equal to second segment.............5

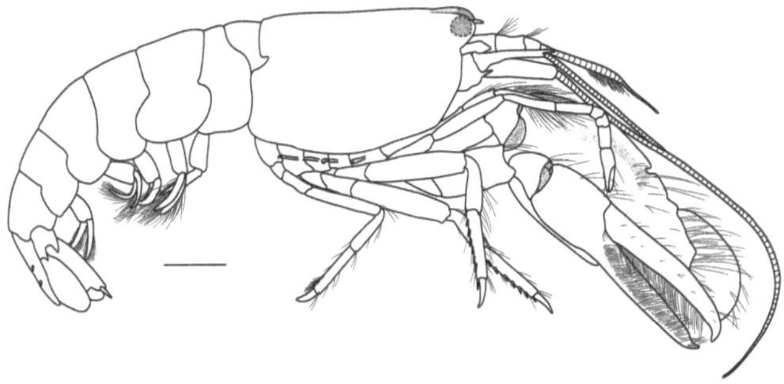

**Fig. 2.41** *Alpheus pacificus* Dana, 1852. entire animal of Edwardsii Group (scale bar = 2 mm)

-. Carpus of second pereiopod with first segment distinctly longer than second segment .................................. 10
5. Minor chela of both male and female balaeniceps or "sub-balaeniceps"........................................................6
-. Minor chela of both male and female not balaeniceps, or only male chela balaeniceps....................................7
6. Major chela with transverse on dorsal margin shallow; dorsal shoulder rounded, gently sloping, not overhanging groove .........................................................................................................*A. balaenodigitus*
-. Major chela with transverse on dorsal margin deep; dorsal shoulder sub-acute, overhanging groove ............................................................................................................ *A. strenuus strenuus*
7. Palm of minor chela with dorsal surface bearing heavy rounded papillae and dense long setae on mesial face ............................................................................................................ *A. funafutensis*
-. Palm of minor chela with dorsal surface not bearing rounded papillae, with or without thin long setae on mesial face ........................................................................................................................8
8. Palm of both major and minor chela with distal margin of mesial and lateral faces bearing large acute teeth, flanking dactylus ..........................................................................................................*A. hoplocheles*
-. Palm of both major and minor chela with distal margin of mesial and lateral faces not bearing large acute teeth ........................................................................................................................ 9
9. Merus of third pereiopod with acute tooth distally on ventral margin.......................................... *A. edamensis*
-. Merus of third pereiopod without acute tooth distally on ventral margin...................................................... ............................................................................................................ *A. pareuchirus*
10. Major chela slender, about 5 times longer than wide..............................................................*A. japonicus*
-. Major chela normal, about 3 times longer than wide .................................................................................. 11
11. Fingers of minor chela more than 3 times length of palm.....................................................*A. malabaricus*
-. Fingers of minor chela sub-equal to or slightly longer than palm.............................................................. 12
12. Major chela with dorsal and ventral shoulder both rounded, gently sloping, not overhanging groove............. 13
-. Major chela with dorsal and ventral shoulder both bearing, or only dorsal shoulder bearing acute or subacute, overhanging transverse grooves.............................................................................................................. 18
13. Antennal scale with blade sub-equal to lateral tooth; rostral carina absent .................................................. ............................................................................................................ *A. euphrosyne*
-. Antennal scale with blade distinctly shorter than lateral tooth; rounded or sharply rostral carina present .......... 14
14. Transverse groove on dorsal margin of major chela extending to mesial face as triangular depression ........... 15
-. Transverse groove on dorsal margin of major chela slightly extending to mesial face as rounded or "U-shaped" depression .................................................................................................................................. 16
15. Fixed finger of male minor chela with mesial face bearing dense long setae; palm of minor chela smooth ............................................................................................................ *A. sudara*
- Fixed finger of male minor chela with mesial face not bearing dense long setae; palm bearing shallow depressions like that of large chela ....................................................................................................*A. lobidens*
16. Minor chela not showing sexual dimorphism: both chela of male and female not balaeniceps......................17

-. Minor chela showing sexual dimorphism: chela of male balaeniceps, chela of female not balaeniceps ........ *A. leviusculus*

17. Merus of major and minor chelipeds bearing acute tooth distally on ventromesial margin ........ *A. ladronis*

-. Merus of major and minor chelipeds not bearing acute tooth distally on ventromesial margin ........ *A. maindroni*

18. Fingers of minor chela more than 1 time as long as palm, cutting edge of dactylus bearing large tooth proximally ........ *A. pacificus*

-. Fingers of minor chela sub-equal or shorter than palm, cutting edge of dactylus not bearing large tooth proximally ........ 19

19. Palm of major chela with only dorsal shoulder subacute, overhanging groove, ventral shoulder general sloping, not overhanging groove ........ 20

-. Palm of major chela with both dorsal and ventral shoulder acute or subacute, overhanging groove ........ 23

20. Merus of third pereiopod bearing large acute tooth distally on ventral margin ........ 21

-. Merus of third pereiopod rounded distally on ventral margin ........ 22

21. Fingers of small chela excavate, dactylus without longitudinal crest ........ *A. hippothoe*

-. Fingers of small chela not excavate, mesial face of dactylus with longitudinal crest, crest at times setiferous ........ *A. serenei*

22. Merus of major chela stout, ventromesial margin bearing small acute tooth distally; third pereiopod stout, with dactylus biunguiculate ........ *A. polyxo*

-. Merus of major chela slender, ventromesial margin bearing large acute tooth sub-distally; third pereiopod slender, with dactylus spatulate ........ *A. spatulatus*

23. Palm of major chela with both dorsal and ventral shoulder acute ........ *A. chiragricus*

-. Palm of major chela with both dorsal and ventral shoulder subacute ........ 24

24. Rostrum flattened in lateral view, demarking from carapace by sharply lateral margins, which overhanging orbitorostral grooves ........ *A. bisincisus*

-. Rostrum rounded in lateral view, lateral margins rounded, not overhanging orbitorostral grooves ........ 25

25. Minor chela bearing dense long setae on distally half-length of mesial face, and chela of male with palm bearing depressions similar to that of major chela ........ *A. inopinatus*

-. Minor chela not bearing dense long setae on mesial face, and chela of male with palm not bearing depressions ........ *A. edwardsii*

### 2.2.33 *Alpheus balaenodigitus* Banner & Banner, 1982 (Fig. 2.42)

*Alpheus balaenodigitus* Banner DM & Banner AH, 1982: 223, fig. 70.

**Material Examined** CL 7.1 mm, TL 18.5 mm. MBM 159193, 1♂, Sanya, Hainan, Li Xin-zheng Coll., 1 December 1997; MBM 157919, 1♂, Cangkou, Qingdao, Zhang Feng-xuan Coll., 1 September 1955; MBM 157972, 1♂, Fujian, 24 May 1982.

**Description** Rostrum triangular, distinctly longer than wide at base, reaching to posterior margin of first antennular segment; rostral carina rounded, extending to posterior margin of orbital hoods; orbitorostral grooves moderately deep; orbital hoods slightly inflated, without tooth.

Telson about 2 and 2.5 times as long as anterior and posterior width, respectively; dorsal surface bearing two pairs of large spines, approximately inserted 1/3 and 2/3 length of telson, respectively; posterior margin convex, but not beyond posterolateral spines.

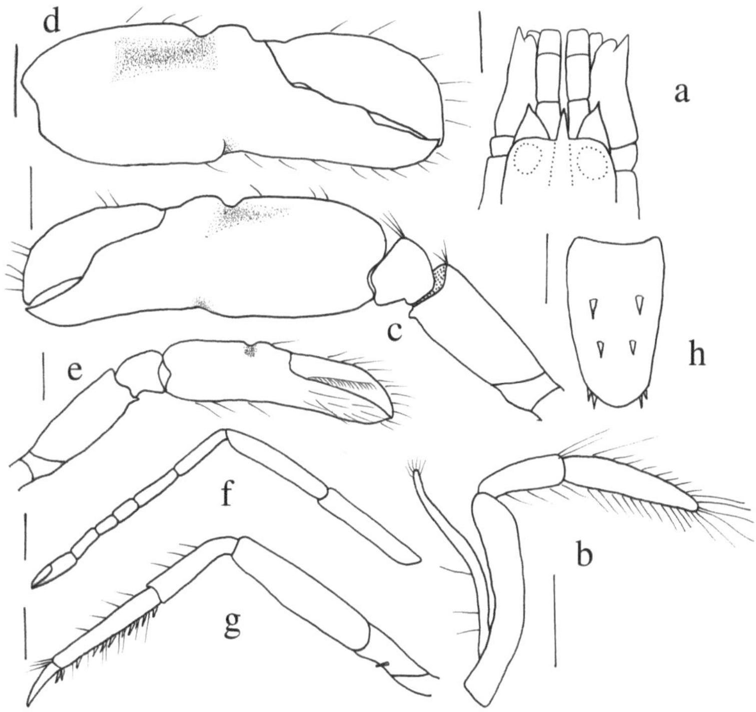


**Fig. 2.42** *Alpheus balaenodigitus* Banner & Banner, 1982. MBM 157972, male, **a**, anterior part of carapace, dorsal view; **b**, third maxilliped; **c**, major cheliped, mesial view; **d**, major chela, lateral view; **e**, minor cheliped, mesial view; **f**, second pereiopod; **g**, third pereiopod; **h**, telson (scale bar = 1 mm)

Antennular peduncle stout, with distally acute stylocerite, slightly overreaching distal margin of first antennular segment; visible part of first antennular segment shorter than second segment; second segment about twice as long as third segment. Antenna with lateral margin of scaphocerite concave, far beyond distal margin of antennular peduncle; antennal scale with distolateral tooth distinctly longer than narrow blade. Carpocerite overreaching antennular peduncle.

Major cheliped with chela compressed, with chela 3.5 times as long as wide. Fingers 2/5 length of palm, slightly curved distally; dactylus as long as fixed finger. Palm with dorsal surface bearing moderately shallow transverse groove, extending to mesial face as triangular depression which extending to about middle length of palm, and to lateral face as quadrangular depression which extending to linea impressa; dorsal shoulder rounded, not overhanging groove; ventral surface bearing shallow transverse groove subterminally; ventral shoulder rounded, not overhanging groove. Merus stout, dorsal margin rounded distally, ventromesial margin bearing subacute tooth distally. Ischium unarmed. Minor cheliped with chela not showing sexual dimorphism, dactylus of both male and female subbalaeniceps, fingers slightly shorter than palm; Palm similar to that of major chela, but depressions on mesial and lateral faces more shallow. Carpus cup-shaped. Merus similar to that of major cheliped.

Second pereiopod with carpus 5-segmented, segmented ratio subequal to 9: 8–9: 3: 3: 5.

Third pereiopod with dactylus simple, about 1/3 length of propodus. Propodus longer than carpus, ventral margin bearing about 9 spines and distally one pair. Carpus 3/5 length of merus, dorsal and ventral margin slightly protruding. Merus unarmed. Ischium armed with one spine on ventral margin.

**Remarks** *A. balaenodigitus* is a newly recorded species from the China seas, and it is originally described from Australia by Banner & Banner (1982). The original description is only based on female specimens, so the character of male is not concluded in that description. The present male specimens are generally agreed with the original description, except for the spines on the ventral margin of propodus of third pereiopod not in pairs as shown in Banner & Banner (1982, fig. 70h).

**Distribution** The Yellow Sea, the South China Sea; Australia.

### 2.2.34 *Alpheus bannerorum* Bruce, 1987 (Fig. 2.43)

*Alpheus bannerorum* Bruce, 1987: 61, Pl. 1, figs. 1–6; 1990: 619, fig. 5.

**Material Examined** This species was reported by Bruce (1990) and Liu (2008) from the South China Sea, but not found in the present specimens.

**Description** Rostrum acute, longer than wide at base, slightly overreaching distal margin of first antennular segment; orbital carina lacking; orbitorostral grooves shallow; orbital hoods rounded, without supraorbital tooth.

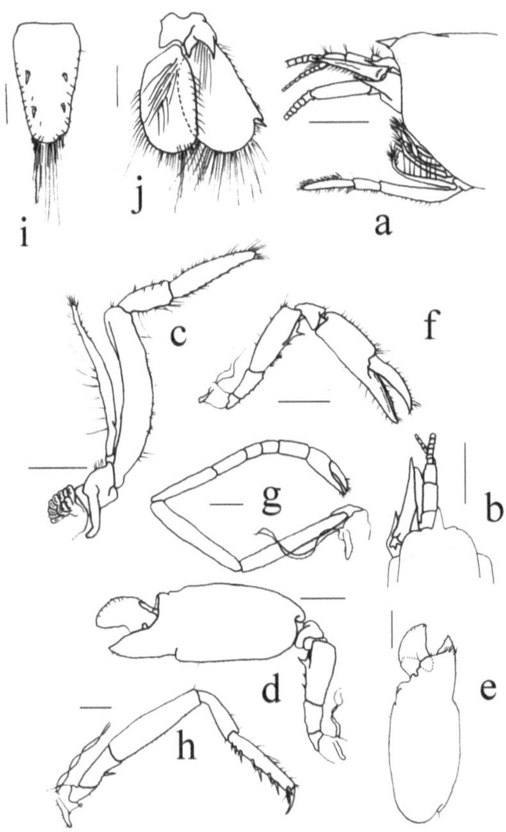


**Fig. 2.43** *Alpheus bannerorum* Bruce, 1987. (Adapted from Bruce, 1987). Female, **a**, anterior part of carapace, lateral view; **b**, same, dorsal view; **c**, third maxilliped; **d**, major cheliped, lateral view; **e**, major chela, mesial view; **f**, minor cheliped, lateral view; **g**, second pereiopod; **h**, third pereiopod; **i**, telson; **j**, right uropod

Telson about 1.75 times as long as anterior width, posterior width 1/2 length of anterior width; dorsal surface bearing two pairs of spines, approximately inserted 2/5 and 4/5 length of telson, respectively; posterior margin only slightly convex, not beyond posterolateral spines.

Antennular peduncle moderately slender, with distally acute stylocerite, distinctly overreaching distal margin of first antennular segment; visible part of first antennular segment nearly as long as third segment; second segment 1.5 times as long as third segment. Antenna with lateral margin scaphocerite slightly concave, beyond distal margin of antennular peduncle; antennal scale with distolateral tooth longer than narrow blade. Basicerite with lateral tooth conspicuous in dorsal and lateral views, acute, reaching to 1/2 length of second antennular segment. Carpocerite slightly beyond antennular peduncle.

Major cheliped with chela slightly compressed. Fingers about 2/5 length of palm, dactylus longer than fixed finger, plunger well developed. Palm almost smooth, dorsal face bearing feebly developed transverse groove subdistally; dorsal shoulder rounded, gently sloping, not overhanging groove; ventral face bearing opposite groove, slightly extending to lateral face; ventral shoulder rounded. Carpus short, cup-shaped, deeply excavate distally. Merus robust, ventromesial margin bearing acute subterminal tooth and two small spines; dorsal margin rounded distally. Ischium unarmed. Minor cheliped with chela subcylindrical not showing sexual dimorphism. Fingers nearly as long as palm, tapering. Palm without groove. Carpus cup-shaped. Merus similar to that of major cheliped, but more slender. Ischium unarmed.

Second pereiopod with carpus 5-segmented, segmented ratio subequal to 1.8: 1: 1: 2: 4.

Third and fourth pereiopod similar in shape, slender. Dactylus simple, compressed, 2/5 length of propodus. Propodus longer than carpus, ventral margin bearing numerous spines (not in single row) and a pair distally. Carpus with dorsal and ventral margins both not projecting distally. Merus unarmed. Ischium armed with one spine on ventral margin (Bruce, 1987).

**Remarks** Bruce (1990) reported this species from Hong Kong.

**Distribution** The South China Sea (Hong Kong); Australia.

### 2.2.35 *Alpheus bisincisus* De Haan, 1849 (Fig. 2.44)

*Alpheus bisincisus* De Haan, 1849 [in Da Haan, 1833–1850]: 179: Pl. 45, fig. 3 (in text as A. avarus, on Plate as A. bisincisus); Pearson, 1911: 182; Tiwari, 1963: 304, fig. 23.

*Alpheus bisincisus malensis* Coutière, 1905: 910, Pl. 86, fig. 48.

*Alpheus bisincisus stylirostris* Coutière, 1905: 911, Pl. 86, fig. 49.

*Alpheus bisincisus variabilis* De Man, 1909a: 109; 1911: 406, fig. 95a–e.

**Material Examined** CL 12.9–15.3 mm, TL 37.5–44.5 mm. MBM 158860, 1♀, Nansha Islands, Wang Shao-wu Coll., 12 May 1993; MBM 158872, 11♀♂, Zhapo, Guangdong, Liu Rui-yu & Wang Yong-liang Coll., 15 November 1954; MBM 157858, 1♀, Fujian, Chen Mu Coll., 26 June 1975; MBM 159034, 1♀, Zhejiang, 4 October 1979; MBM 270025, 2♀♀1♂, Nanji Island, Zhejiang, Ren Xian-qiu Coll., 24 October 1979.

**Description** Rostrum acute, longer than wide at base, hardly reaching to distal margin of first antennular segment; rostrum flattened dorsally with lateral margins overhanging deep orbitorostral groove, extending to posterior margin of orbital hoods; rostral carina lacking; orbital hoods inflated, without tooth.

Telson about 1.5 and 2 times as long as anterior and posterior width, respectively; dorsal surface bearing two pairs of spines, approximately inserted 1/2 and 2/3 length of telson, respectively;

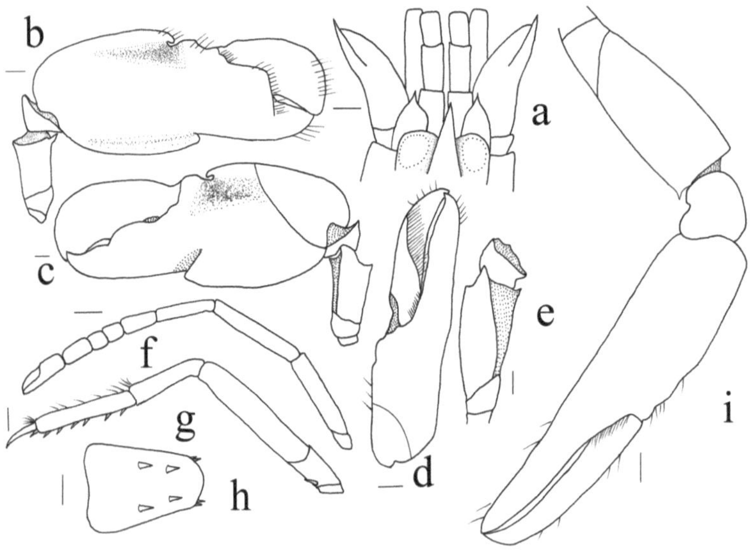


**Fig. 2.44** *Alpheus bisincisus* De Haan, 1849. MBM 270025, male, **a**, anterior part of carapace, dorsal view; **b**, major cheliped, mesial view; **c**, same, lateral view; **d**, minor chela, lateral view; e, carpus to ischium of minor cheliped, lateral view; **f**, second pereiopod; **g**, third pereiopod; **h**, telson . Female, **i**, minor cheliped (scale bar = 1 mm)

posterior margin convex, not beyond posterolateral spines.

Antennular peduncle moderately slender, with distally acute stylocerite reaching to distal margin of first antennular segment; visible part of first antennular segment slightly shorter than second segment; third segment 2/3 length of second segment. Antenna with lateral margin of scaphocerite concave, overreaching distal margin of antennular peduncle; antennal scale with distolateral tooth longer than narrow blade. Carpocerite nearly as long as antennular peduncle.

Major cheliped with chela slightly compressed. Fingers about 2/3 length of palm, dactylus nearly as long as fixed finger, plunger well developed. Palm with dorsal surface bearing transverse groove sub-distally, which extending to mesial and lateral faces as triangular and quadrangular depressions, respectively; depression of mesial face continuing to proximal 1/3 length of palm, depression of lateral face continuing to linea impressa; dorsal shoulder subacute, overhanging groove; ventral face bearing opposite groove, extending to mesial face as a narrow longitudinal groove that about continuing to proximal 1/2 length of palm, and to lateral face as transverse deep groove that continuing to 1/3 width of palm; ventral shoulder rounded, overhanging groove. Carpus cup-shaped. Merus robust, ventromesial margin terminating in acute tooth; dorsal margin rounded distally. Ischium unarmed. Minor cheliped with chela showing sexual dimorphism: chela of male balaeniceps, fingers nearly as long as palm, palm bearing shallow transverse groove sub-terminally; chela of female not balaeniceps, fingers slightly longer than palm, cutting edge of fixed finger bearing short setae proximally, palm almost smooth. Carpus cup-shaped. Merus similar to that of major cheliped. Ischium unarmed.

Second pereiopod with carpus 5-segmented, segmented ratio subequal to 10: 7: 3: 3: 5.

Third and fourth pereiopod similar in shape, slender. Dactylus simple, about 2/5 length of propodus. Propodus longer than carpus, ventral margin bearing 5–8 spines and distally one pair. Carpus with dorsal margin rounded distally; ventral margin variable from rounded to sub-acute. Merus unarmed. Ischium armed with one spine on ventral margin.

**Remarks** *A. bisincisus* is most similar to *A. chiragricus* in Edwardsii group, but can be readily distinguished from *A. chiragricus* by the flattened rostrum overhanging the rostral grooves. The present specimens are generally agreed with previous descriptions. The ventral margin of carpus of the third pereiopod is variable from rounded to sub-acute distally in the present specimens, and not bearing a spine as shown in Banner & Banner (1982, fig. 81f).

**Distribution** The East China Sea, the South China Sea, Taiwan; South Africa; Maldive and Laccadive Archipelagoes; Ceylon; Indonesia; Singapore; Japan.

### 2.2.36 *Alpheus chiragricus* H. Milne Edwards, 1837 (Fig. 2.45)

*Alpheus chiragricus* H. Milne Edwards, 1837: 354; De Man, 1911: 415; Banner DM & Banner AH, 1982: 267, fig. 82; Chace, 1988: 18.

**Material Examined** CL 8.1–15.3 mm, TL 23.5–36.9 mm. MBM 107673, 1♂, Lingao, Hainan, Liu Rui-yu Coll., 2 December 1990; MBM 159141, 1♀1♂, Qukou, Hainan, Wang Yong-liang Coll., 14 March 1992; MBM 138088, 4♂♂2♀♀, Guangzhou, 1 October 1997; MBM 158228, 1♀, Chanwei, Guangzhou, 1 January 1952; MBM 158234, 3♂♂1♀, Zhanjiang, Guangzhou, Liu Rui-yu & Wang Yong-liang Coll., 28 November 1954; MBM 158119, 1♂, Qisha, Guangxi, Shen Jia-rui Coll.,; MBM 158872, 2♀♀2♂♂, Zhakou, Guangzhou, Liu Rui-yu & Wang Yong-liang Coll., 15 November 1954; MBM 157993, 1♀1♂, Zhejiang, Guangzhou, 29 November 1954; MBM 159137, 6♀♀5♂♂, Qukou, Hainan, 13 March 1992; MBM 159404, 1♀, Xinying, Hainan, Li Xin-zheng, 11 April 1992; MBM 158752, 2♀♀2♂♂, Hainan, 11 December 1955; MBM 107675, 1♀, Sanya Bay, Hainan, 27 November 1990; MBM 159136, 1♀1♂, Linchang Reef, Hainan 28 December 1991; MBM 159135, 2♂♂1♀, Qukou, Hainan, Liu Xi-xing Coll., 3 November 1990; MBM 159138, 3♀♀6♂♂, Qukou, Hainan, Liu Xi-xing & Li Xin-zheng Coll., 13 March 1992; MBM 270371, 1♂, Beibu Gulf, Zheng Feng Coll., 18 October 1960; MBM 270372, 1♀, Beibu Gulf, Shen shou-peng Coll., 8 July 1960; MBM 270225, 1♀, Beibu Gulf, 15

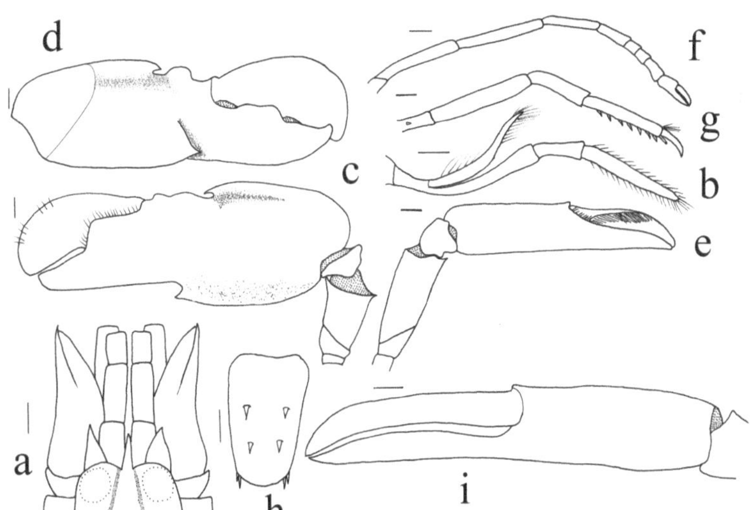


**Fig. 2.45** *Alpheus chiragricus* H. Milne Edwards, 1837. MBM 159138, male, **a**, anterior part of carapace, dorsal view; **b**, third maxilliped; **c**, major cheliped, mesial view; **d**, major chela, lateral view; **e**, minor cheliped, mesial view; **f**, second pereiopod; **g**, third pereiopod; **h**, telson. Female, **i**, minor chela (scale bar = 1 mm)

April 1960; MBM 270224, 1♀, Beibu Gulf, Sun Fu-zeng Coll., 10 July 1960; MBM 270237, 1♀2♂♂, Beibu Gulf, Shen Shou-peng Coll., 18 October 1960; MBM 270327, 1 larva, Beibu Gulf, Sun Ke-zhi Coll., 18 October 1960.

**Description** Rostrum acute, longer than wide at base, hardly reaching to distal margin of first antennular segment; rostral carina extending to posterior margin of orbital hoods; orbitorostral groove moderately deep; orbitorostral margin concave; orbital hoods inflated, without tooth.

Telson about 1.6 and 2 times as long as anterior and posterior width, respectively; dorsal surface bearing two pairs of spines, approximately inserted 2/5 and 2/3 length of telson, respectively; posterior margin slightly convex, not beyond posterolateral spines.

Antennular peduncle slender, with distally acute stylocerite reaching to distal margin of first antennular segment; visible part of first antennular segment slightly shorter than second segment; third segment 1/2 length of second segment. Antenna with lateral margin of scaphocerite concave, far beyond distal margin of antennular peduncle; antennal scale with distolateral tooth slightly longer than narrow blade. Carpocerite slightly longer than antennular peduncle.

Major cheliped with chela slightly compressed. Fingers about 2/3 length of palm, dactylus longer than fixed finger, plunger feebly developed. Palm with dorsal surface bearing transverse groove sub-distally, which extending to mesial and lateral faces as triangular and quadrangular depressions, respectively; depression of mesial face continuing to proximal 1/3 length of palm, depression of lateral face continuing to linea impressa; dorsal shoulder acute, overhanging groove; ventral surface bearing opposite groove, extending to lateral face as longitudinal depression continuing to proximally; ventral shoulder acute projecting, overhanging groove. Carpus cup-shaped. Merus robust, dorsal margin rounded distally; ventromesial margin terminating in acute tooth. Ischium unarmed. Minor cheliped with chela sexual dimorphism: chela of male balaeniceps. Fingers nearly as long as palm. Palm smooth, without groove. Merus similar to that of major cheliped. Ischium unarmed.

Second pereiopod with carpus 5-segmented, segmented ratio subequal to 16: 9: 4: 4: 6.

Third and fourth pereiopod similar in shape, slender. Dactylus simple. Propodus longer nearly as long as merus, ventral margin bearing 8–10 spines. Carpus with dorsal margin terminating in rounded tooth. Merus unarmed. Ischium armed with one spine on ventral margin.

**Remarks** *A. chiragricus* is recognized as a valid species by Banner & Banner (1972). This species is most similar to *A. edwardsii* in Edwardsii group. The differences between this species and *A. edwardsii* have been discussed by Banner & Banner (1982, 268; Chace, 1988: 19). The present specimens are generally agreed with previous descriptions.

**Distribution** The South China Sea; Eastern Africa; Madagascar; Mergui Archipelago; Indonesia; Australia.

### 2.2.37 *Alpheus edamensis* De Man, 1888 (Fig. 2.46)

*Alpheus hippothoe edamensis* De Man, 1888b: 518.

*Alpheus edamensis* –. De Man, 1911:437, fig. 107; Banner AH & Banner DM, 1966b: 157, fig. 61; Banner DM & Banner AH, 1982: 188, fig. 5; Chace, 1988: 24.

*Alpheus acanthomerus* Ortmann, 1890: 474, Pl. 36, fig. 12; Coutière, 1897d: 202.

**Material Examined** CL 8.2–9.5 mm, TL 21.5–23.7 mm. MBM 158712, 1♂, Yongxing Island, Xisha Islands, 8 May 1980; MBM 158682, 5 larva, Shanhu Island, Xisha Islands, 19 May 1980; MBM 157944, 1♂, Yongxing Island, Xisha Islands, 9 May 1980; MBM 158081, 1♀1♂, Xincun, Hainan, 25 December 1955; MBM 159263, 2♀♀1♂, Nanxun Reef, Nansha

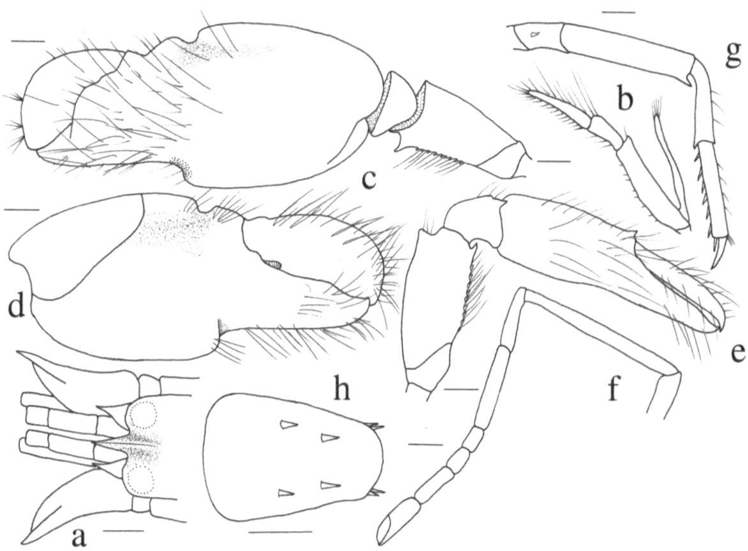


**Fig. 2.46** *Alpheus edamensis* De Man, 1888. MBM 158637, male, **a**, anterior part of carapace, dorsal view; **b**, third maxilliped; **c**, major cheliped, mesial view; **d**, major chela, lateral view; **e**, minor cheliped, mesial view; **f**, second pereiopod; **g**, third pereiopod; **h**, telson (scale bar = 1 mm)

Islands, 17 April 1999; MBM 158189, 1♀, Tree Island, Xisha Islands, 1 May 1958; MBM 158492, 4♀♀4♂♂, Yongxing Island, Xisha Islands, 11 June 1980; MBM 157759, 1♀, Shi Island, Xisha Islands, Wang Yong-liang Coll., 13 May 1957; MBM 157734, 1♂, Xincun, Hainan, 28 January 1956; MBM 157806, 1♀1♂, Shi Island, Xisha Islands, Chen Mu Coll., 2 June 1981; MBM 157779, 1♂, Yongxing Island, Xisha Islands, 9 May 1980; MBM 157760, 1♀, Shi Island, Xisha Islands, 15 May 1957; MBM 158271, 2♀♀, Yongxing Island, Xisha Islands, 5 June 1975; MBM 158272, 1♂, Jinyin Island, Xisha Islands, 24 May 1975; MBM 158306, 1♀1♂, Chenhang Island, Xisha Islands, 22 May 1975; MBM 158261, 1♀1♂, Guangjin Island, Xisha Islands, 18 May 1975; MBM 158637, 1♀1♂, Chigua Reef, Nansha Islands, 19 May 1990; MBM 158609, 1♀, Tree Island, Xisha Islands, Xu Feng-shan Coll., 4 May 1958; MBM 158562, 1♀, Yongshu Reef, Nansha Islands, 16 May 1993; MBM 158679, 1♂, Dongmen Reef, Nansha Islands, Li Xin-zheng Coll., 22 May 1993; MBM 158574, 1♀1♂, Yongshu Reef, Nansha Islands, Li Xin-zheng Coll., 18 May 1993; MBM 233242, 1♀, East Island, Xisha Islands, 3 June 1980.

**Description** Rostrum acute, longer than wide at base reaching to distal margin of first antennular segment; rostral carina high but rounded, extending to posterior margin of orbital hoods; anterior margin between rostrum and orbital hoods slightly concave; orbital hoods inflated, without tooth.

Telson about 1.7 and 2.3 times as long as anterior and posterior width, respectively; dorsal surface bearing two pairs of spines, approximately inserted 2/5 and 3/5 length of telson, respectively; posterior margin slightly convex, not beyond posterolateral spines.

Antennular peduncle slender, with distally acute stylocerite, slightly overreaching distal margin of first antennular segment; visible part of first antennular segment slightly longer than third segment; second segment twice as long as third segment. Antenna with lateral margin of scaphocerite strongly concave, far beyond distal margin of antennular peduncle; antennal scale with distolateral tooth distinctly longer than narrow blade. Carpocerite slightly longer than antennular peduncle.

Major cheliped with chela slightly compressed. Fingers about 2/3 length of palm, dactylus slightly longer than fixed finger, plunger well developed. Palm with dorsal surface bearing transverse groove sub-distally, which extending to mesial and lateral faces as triangular and quadrangular depressions, respectively; depression of mesial face continuing to proximal 1/2 length of palm, depression of lateral face continuing to linea impressa; dorsal shoulder rounded, not overhanging groove; ventral surface bearing opposite groove, slightly extending to lateral face as transverse groove and to mesial face as rounded depression; ventral shoulder rounded, not overhanging groove; distally 1/2 length of palm bearing fine long setae on mesial face. Carpus cup-shaped. Merus robust, dorsal margin rounded distally; ventromesial margin dentate, bearing 10 spinules and terminating in acute large tooth, fringing with long setae. Ischium unarmed. Minor cheliped with chela subcylindrical, showing slightly sexual dimorphism: chela not balaeniceps in both sex, chela of male with finger slightly longer than palm, dorsal margin of palm bearing acute tooth distally, dorsal and ventral shoulder gently slop, mesial face bearing fine setae, cutting edges of fingers with long setae; chela of female more slender, with fingers nearly as long as palm, mesial face bearing less setae than that of male. Merus similar to that of major chela, but more slender. Ischium unarmed.

Second pereiopod with carpus 5-segmented, segmented ratio subequal to 10: 12: 3: 3: 7.

Third and fourth pereiopod similar in shape. Dactylus simple. Propodus slightly shorter than merus, ventral margin bearing 10 spines and distally one pair. Carpus with both dorsal and ventral margin projecting as acute tooth. Merus with ventral margin terminating as acute tooth. Ischium armed with one spine on ventral margin.

**Remarks** The present specimens are generally agreed with the previous descriptions, except for the following characters: the ventromesial margin of merus of the major and minor cheliped dentate, and armed with about 10 spinules which not shown in Banner & Banner (1982, fig. 57c) and Chace (1988, 24); the propodus of the third pereiopod bearing 10 spines in single row on ventral margin, while it bearing 14–16 spines in pairs as shown in Banner & Banner (1982, fig. 57i) and Chace (1988).

**Distribution** The South China Sea; Red Sea; Madagascar; Reunion; Seychelles; Mauritius; Malaya; Thailand; Indonesia; Philippines; Ryukyus; Fiji; Samoa; Society Islands.

### 2.2.38 *Alpheus edwardsii* (Audouin, 1826) (Fig. 2.47)

*Athanas edwardsii* Audouin, 1826: 91.
*Alpheus audouini* Coutière, 1905: 911, Pl. 87, figs. 52
*Alpheus edwardsi*–. Banner AH & Banner DM, 1972: 1141, fig. 1.
*Alpheus edwardsii*–. Banner DM & Banner AH, 1982: 270, fig. 83; Chace, 1988: 25.

**Material Examined** CL 7.9–11.1 mm, TL 22.3–30.5 mm. MBM 159408, 2♀♀2♂♂, Shalao, Wang Yong-liang Coll., 5 April 1992; MBM 158195, 1♀, Sanya, Hainan, 28 February 1997; MBM 157826, 3♂♂1♀, Yongxing Island, Xisha Islands, 8 May 1980; MBM 159219, 2♂♂, Sanya Bay, Li Xin-zheng Coll., 2 December 1997; MBM 158041, 1♀, Sanya, Hainan, 20 November 1997; MBM 158194, 1♀, Sanya, Hainan, Li Xin-zheng Coll., 28 February 1997; MBM 159148, 3♀♀, Sanya, Hainan, 19 November 1997; MBM 158906, 1♂, Sanya, Hainan, 5 March 1997; MBM 159176, 2♀♀, Sanya, Hainan, Li Xin-zheng Coll., 3 December 1997; MBM 159171, 3♀♀, Sanya, Hainan, Li Xin-zheng Coll., 17 November 1997; MBM 158909, 1♀, Sanya, Hainan, 1 March 1997; MBM 159213, 1♀, Sanya, Hainan, 20 November 1997; MBM 158142, 1♀, Sanya, Hainan, 20 November 1997; MBM 159192, 3♀♀, Sanya, Hainan, Li Xin-zheng Coll., 17 November 1997; MBM 159205, 1♂, Sanya, Hainan, 17 November 1997; MBM 158129, 5♀♀4♂♂, Sanya, Hainan, 2 December 1997; MBM 158127, 1♀, Sanya, Hainan, 27 November 1997; MBM 159169, 1♂, Sanya, Hainan, Li Xin-

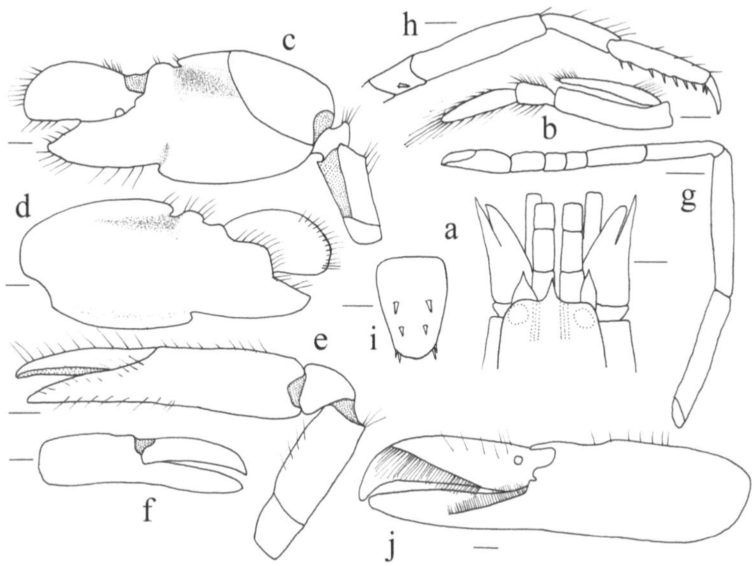


**Fig. 2.47** *Alpheus edwardsii* (Audouin, 1826). MBM 159403, female, **a**, anterior part of carapace, dorsal view; **b**, third maxilliped; **c**, major cheliped, lateral view; **d**, major chela, mesial view; **e**, minor cheliped, mesial view; **f**, minor chela, lateral view; **g**, second pereiopod; **h**, third pereiopod; **i**, telson. Male, **j**, minor chela (scale bar = 1 mm)

zheng Coll., 1 December 1997; MBM 159011, 3♀♀3♂♂, Sanya, Hainan, 7 April 1958; MBM 159006, 1♀2♂♂, Sanya, Hainan, 4 February 1958; MBM 159190, 1♀1♂, Sanya, Hainan, Li Xin-zheng Coll., 20 November 1997; MBM 159155, 2♀♀1♂, Sanya, Hainan, 1 December 1997; MBM 158214, 1♀, Sanya, Hainan, Li Xin-zheng Coll., 1 March 1997; MBM 158198, 1♀, Sanya, Hainan, 1 March 1997; MBM 138323, 8♀♀3♂♂, Xincun, Hainan, 31 December 1959; MBM 233273, 1♀, Dongyang Island, Fujian, 3 May 1984; MBM 158236, 31♀♂, Xizhou Island, 30 December 1955; MBM 158936, 1♀, Guangzhou, Ma Xiu-tong Coll., 26 June 1956; MBM 157878, 1♀, Xincun, Hainan, 19 April 1958; MBM 158254, 23♀♂, Zhanjiang, Guangzhou, Liu Rui-yu & Wang Yong-liang Coll., 1 December 1954; MBM 158948, 58♀♂, Hainan, Wang Yong-laing Coll., 26 June 1957; MBM 158953, 5♀♀5♂♂, Xizhou Island, 29 December 1955; MBM 158146, 2♂♂4♀♀, Weizhou Island, 24 April 1956; MBM 158250, 1♀, Xizhou Island, Liu Rui-yu Coll., 29 February 1956; MBM 158778, 3♀♀, Sanya, Hainan, 3 January 1956; MBM 158143, 7♀♂, Xincun, Hainan, 27 January 1956; MBM 158876, ♀♂, Xinying, Hainan, 11 January 1956; MBM 158247, 10♂♂12♀♀, Xinying, Hainan, Wang Yong-liang Coll., 25 June 1957; MBM 158083, 1♀, Sanya, Hainan, 1 December 1997; MBM 158809, 3♂♂, Xincun, Hainan, 9 July 1957; MBM 158111, 3♀♀, Xincun, Hainan, Wang Yong-liang Coll., 21 December 1955; MBM 157902, 10♀♂, Sanya, Hainan, 24 March 1958; MBM 158018, 1♀, Yongxing Island, Xisha Islands, 19 May 1958; MBM 157993, 2♂♂2♀♀, Zhanjiang, Guangzhou, 9 November 1954; MBM 157880, 1♂, Fujian, 14 April 1957; MBM 158004, 1♀, Sanya, Hainan, 16 March 1992; MBM 157965, 2♀♀1♂, Xincun, Hainan, 22 April 1958; MBM 157957, 1♀, Xizhou Island, 3 April 1958; MBM 158148, 3♀♀4♂♂, Sanya, Hainan, 5 April 1955; MBM 158249, 6♂♂4♀♀, Weizhou Island, Guangzhou, Liu Rui-yu & Wang Yong-liang Coll., 25 December 1954; MBM 159291, 1♀1♂, Yezhu Island, Yalong Bay, Neumann Coll., 27 March 1992; MBM 157785, 11♀♂, Sanya, Hainan, 3 December 1955; MBM 157750, 4♀♀, Xincun, Hainan, 21 December 1955; MBM 158579, 1♀, Linchang Reef, Hainan, 10 April 1992; MBM 158573, 2♀♀1♂, Shalao, Wang Yong-liang Coll., 4 April 1992; MBM 158451,

1♀3♂♂, Linchang Reef, Hainan, 9 April 1992; MBM 159273, 2♀♀, Yezhu Island, Yalong Bay, 27 March 1992; MBM 158510-3, 3♀♀4♂♂, Xincun, Hainan, 26 March 1992; MBM 159303-5, 2♀♀, Sanya Bay, 23 March 1992; MBM 158159-62, 7♀♀2♂♂, 8 April 1992; MBM 159404, 6♀♀6♂♂, Linchang Reef, Hainan, 11 April 1992; MBM 158575, 2♀♀1♂, Linchang Reef, Hainan, Li Jin-he Coll., 10 April 1992; MBM 158385, 1♀, Yalong Bay, 24 November 1990; MBM 158857, 5♂♂8♀♀, 2 December 1990; MBM 159170, 1♀1♂, Fujian, 29 May 1957; MBM 159402, 9♀♂, Shalao, 28 December 1991; MBM 159398, 1♀1♂, Hainan, 22 January 1992; MBM 159390, 1♀1♂, Qukou, Hainan, 27 December 1991; MBM 159400, 1♀, Houshui Bay, Hainan, 1 December 1990; MBM 159403, 3♀♀1♂, 20 December 1991; MBM 159399, 1♀1♂, Sanya, Hainan, 21 November 1990; MBM 159395, 1♂, 2 December 1990; MBM 158666, 60♀♂, Linchang Reef, Hainan, 3 December 1990; MBM 270045, 24♀♂, Weizhou Island, Ma Xiu-tong Coll., 17 April 1978; MBM 159259, 4♂♂2♀♀, Sanya, Hainan, 18 November 1997; MBM 158556, 1♀1♂, Li Xin-zheng Coll., 8 April 1992; MBM 159391, 1♀1♂, Sanya, Hainan, 28 March 1992; MBM 159396, 2♂♂4♀♀, Linchang Reef, Hainan, Li Jin-he Coll., 10 April 1992; MBM 159394, 3♀♀1♂, Linchang Reef, Hainan, 7 April 1992; MBM 107571, 6♀♀4♂♂, Linchang Reef, Hainan, 2 April 1992; MBM 159397, 2♀♀2♂♂, Linchang Reef, Hainan, 7 April 1992.

**Description** Rostrum acute, longer than wide at base, hardly reaching to distal margin of first antennular segment; rostral carina extending to posterior margin of orbital hoods; orbitorostral groove moderately deep; orbital hoods inflated, without tooth.

Telson about 1.4 and 2.6 times as long as anterior and posterior width, respectively; dorsal surface bearing two pairs of spines, approximately inserted 2/5 and 3/5 length of telson, respectively; posterior margin slightly convex, not beyond posterolateral spines.

Antennular peduncle stout, with distally acute stylocerite reaching to distal margin of first antennular segment; visible part of first antennular segment slightly longer than third segment; second segment twice as long as third segment. Antenna with lateral margin of scaphocerite slightly concave, far beyond distal margin of antennular peduncle; antennal scale with distolateral tooth distinctly longer than narrow blade. Carpocerite slightly longer than antennular peduncle.

Major cheliped with chela slightly compressed. Fingers about 2/3 length of palm, dactylus slightly longer than fixed finger, plunger well developed. Palm with dorsal surface bearing transverse groove sub-distally, which extending to mesial and lateral faces as triangular and quadrangular depressions, respectively; depression of mesial face continuing to proximal about 1/2 length of palm, depression of lateral face continuing to linea impressa; dorsal shoulder rounded, overhanging groove; ventral surface bearing opposite groove, extending to mesial face as narrow longitudinal depression that continuing to proximal 1/4 length of palm, and to lateral face as transverse depression that continuing to 1/5 width of palm; ventral shoulder rounded distally, overhanging groove. Merus robust, dorsal margin and ventrolateral margin rounded distally; ventromesial margin terminating in acute tooth. Ischium unarmed. Minor cheliped with chela showing sexual dimorphism: chela of male balaeniceps, fingers shorter than palm, dorsal margin of palm bearing shallow transverse groove that extending slightly to lateral face, ventral groove inconspicuous; chela of female not balaeniceps, fingers nearly as long as palm, grooves on palm inconspicuous. Merus similar to that of major cheliped, but ventromesial margin rounded distally. Ischium unarmed.

Second pereiopod with carpus 5-segmented, segmented ratio subequal to 10: 6: 3: 3: 5.

Third and fourth pereiopod similar in shape. Dactylus simple. Propodus slightly shorter than merus, ventral margin bearing 6–7 spines and distally one pair. Carpus with dorsal margin terminating in rounded tooth. Merus unarmed. Ischium armed with one spine on ventral margin.

**Remarks** Banner & Banner (1972) had established a neotype on specimens from the Suez for *A. edwardsii*, and described the variations of this species. Moreover, the differences between *A. edwardsii* and its similar species were also been discussed by Banner & Banner (1972). The present specimens are generally agreed with the description made by Banner & Banner (1972), except for the merus of the minor cheliped with ventromesial margin rounded distally, not terminating in acute tooth as shown in Banner & Banner (1982, fig. 83d).

**Distribution** The South China Sea, Taiwan; Probably from Red Sea and eastern and South Africa to Thailand; Philippines; Indonesia; Australia; Caroline Islands.

### 2.2.39 *Alpheus euphrosyne* De Man, 1897 (Fig. 2.48)

*Alpheus euphrosyne* De Man, 1897: 745, fig. 64a–d; 1898: 317, Pl. 4, fig. 2; Banner AH & Banner DM, 1966b: 130, fig. 49.

*Alpheus eurydactylus* De Man, 1920: 109; 124: 48, fig. 17.

*Alpheus euphrosyne euphrosyne* –. Banner DM & Banner AH, 1982: 232, fig. 73; Chace, 1988: 27.

**Material Examined** CL 9.1 mm, TL 26.5 mm. MBM 158959, 21♀♂, Ying Gehai, Wang Yong-liang Coll., 26 June 1957; MBM 158928, 1♀, Wotou, Xiamen, Wu Bao-ling Coll., 29 April 1950; MBM 159237, 1♂, Nanji Island, 24 October 1984; MBM 157934, 1♀1♂,, Cangkou, Qingdao, Zhang Feng-xuan Coll., 30 August 1954; MBM 158806, 1♀, Sanya, Hainan, 28 March 1958; MBM 158107, 1♀, Pingyang, Zhejiang, 30 May 1953; MBM 158130, 2♂♂, Shi Island, 8 December 1954; MBM 158845, 2♀♀2♂♂, Zhanjiang, Guangdong, Wu Bao-ling Coll., 16 March 1956; MBM 158873, 1♂, Shi Island, Xisha Islands, 22 June 1955; MBM 158140, 1♀, Nanji Island, Zhejiang, 24 October 1979; MBM 157923, 1♀1♂, Ying Gehai, 6 April 1955; MBM 157899, 1♂, 21 September 1979; MBM 157921, 1♀, Ying Gehai, Wang Yong-liang Coll., 6 May 1955; MBM 159267, 1♀, Pingyang,

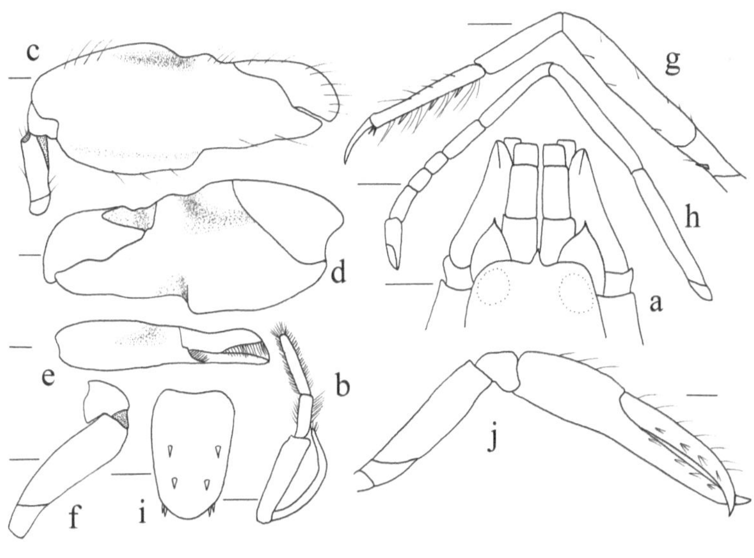


**Fig. 2.48** *Alpheus euphrosyne* De Man, 1897. MBM 159248, male, **a**, anterior part of carapace, dorsal view; **b**, third maxilliped; **c**, major cheliped, mesial view; **d**, major chela, lateral view; **e**, minor chela, lateral view; **f**, carpus to ischium of minor cheliped, mesial view; **g**, third pereiopod; **h**, second pereiopod; **i**, telson. MBM 270007, female, **j**, minor cheliped (scale bar = 1 mm)

Zhejiang, 5 October 1979; MBM 157936, 1♂, Jinchuan, Zhejiang, 19 October 1952; MBM 15923, 1♀1♂, Zhanjiang, Guangzhou, Li Xinzheng Coll., 1958–1959; MBM 158229, 2♀♀5♂♂, Zhanjiang, Guangdong, 28 July 1976; MBM 158497, 1♂, Pingyang, Zhejiang, 4 October 1979; MBM 107598, 2♀♀, Qukou, Hainan, Wang Yong-liang Coll., 14 March 1992; MBM 159248, 1♂, Qiongshan, Hainan, 6 June 1958; MBM 159197, 1♀1♂, Leqing, Zhejiang, 21 September 1979; MBM 157845, 2♂♂4♀♀, Cangkou, Qingdao, 29 September 1954; MBM 157863, 1♀1♂, Sanya, Hainan, 30 March 1958; MBM 157930, 1♀, Cangkou, Qingdao, 7 August 1956; MBM 158519, 1♀1♂, Qinglan, Hainan, 2 April 1992; MBM 270043, 1♂, Fujian, Ren Xianqiu Coll., 1 November 1974; MBM 270505, 15♀♂, Yinggehai, 7 December 1955; MBM 270007, 1♀, Yinggehai, 7 December 1955.

**Description** Rostrum small, nearly as long as wide at base, at most reaching to 1/2 length of first antennular peduncle; rostral carina absent; orbitorostral groove shallow, sometimes absent; orbital hoods without tooth.

Telson about 1.5 and 2 times as long as anterior and posterior width, respectively; dorsal surface bearing two pairs of spines, approximately inserted 2/5 and 7/10 length of telson, respectively; posterior margin convex, slightly beyond posterolateral spines.

Antennular peduncle stout, with distally acute stylocerite reaching to distal margin of first antennular segment, visible part of first antennular 1.5 times as long as third segment, second segment about 2.5 times as long as third segment. Antenna with lateral margin of scaphocerite almost straight, overreaching distal margin of antennular peduncle; antennal scale with distolateral tooth nearly as long as or slightly shorter than broad blade. Carpocerite nearly as long as antennular peduncle.

Major cheliped with chela slightly compressed. Fingers about 2/3 length of palm, dactylus slightly longer than fixed finger, plunger well developed. Palm with dorsal surface bearing transverse groove sub-distally, which extending to mesial and lateral faces as triangular and quadrangular depressions, respectively; depression of mesial face continuing to proximal about 1/2 length of palm, depression of lateral face continuing to linea impressa; dorsal shoulder rounded, gently sloping, not overhanging groove; ventral transverse groove extending to mesial face as narrow longitudinal depression that approximately continuing to base of palm, and to lateral face as transverse triangular depression that continuing to 1/4 width of palm; ventral shoulder, gently sloping, not overreaching groove. Merus without tooth and spines. Ischium unarmed. Minor cheliped with chela showing sexual dimorphism: chela of male balaeniceps, fingers slightly shorter than palm, depressions and grooves similar to that of major chela, but much less pronounced; chela of female not balaeniceps, fingers 1.2–1.3 times as long as palm, palm only bearing slightly dorsal and ventral groove sub-distally. Merus similar to that of major cheliped. Ischium unarmed.

Second pereiopod with carpus 5-segmented, segmented ratio subequal to 10: (4–8) 3: 3: 4.

Third and fourth pereiopod similar in shape. Dactylus simple. Propodus shorter than merus, ventral margin bearing 3–7 spines and distally one pair. Carpus with dorsal margin terminating in rounded tooth. Merus unarmed. Ischium armed with one spine on ventral margin.

**Remarks** Banner & Banner (1966) reported *A. euphrosyne* from Thailand. The present specimens show slight differences form their description as following: the carapace not bearing rostral carina in the present specimens, while it carrying a slightly rostral carina as shown in Banner & Banner (1966, fig. 49a); the lateral face of the major chela bearing quadrangular depression in the present specimens rather than triangular depression as shown in Banner & Banner (1966, fig. 49b). Banner & Banner (1982) has also described this species based on specimens collected from Australia. The present specimens slightly differ from their description on the third pereiopod with the propodus only bearing 3–4 spines on the ventral margin rather than 7 stout spines as shown in Banner & Banner (1982, fig. 73f). Banner & Banner (1966) has discussed the variable characters of this species.

**Distribution** The yellow Sea, the East China Sea, the South China Sea, Taiwan; Kenya, Thailand; Philippines; Indonesia; Queensland, Australia.

### 2.2.40 *Alpheus funafutensis* Borradaile, 1899 (Fig. 2.49)

*Alpheus funafutensis* Borradaile, 1899: 1013, Pl. 15, fig. 10; De Man, 1911: 436; Banner AH & Banner DM, 1964: 96; 1966b: 155, fig. 60; Chace, 1988: 28.

*Alpheus acanthomerus* var. *inermis* Lanchester, 1901: 564.

**Material Examined** CL7.1 mm, TL 17.5 mm. MBM 158704, 2♂♂, Yongxing Island, Xisha Islands, 1 June 1980; MBM 158308, 1♀, Shi Island, Xisha Islands, 8 May 1975; MBM 158317, 1♂, Shi Island, Xisha Islands, Chen Mu Coll., 2 June 1981; MBM 158063, 1♀, Tree Island, Xisha Islands, Fan Zhen-gang Coll., 5 May 1958; MBM 157815, 1♂, East Island, Xisha Islands, 3 June 1980; MBM 158840, 2♀♀, Dazhou Island, Wang Yong-liang Coll., 11 June 1957; MBM 158862, 8♀♀8♂♂, East Island, Xisha Islands, 3 June 1980; MBM 158828, 26♀♀26♂♂, Yongxing Island, Xisha Islands, 11 June 1980; MBM 158068, 1♀, Sanya, Hainan, 18 November 1997; MBM 158074, 6♀♂, Sanya, Hainan 18 November 1997; MBM 159175, 1♀1♂, Sanya, Hainan, Li Xin-zheng Coll., 17 November 1997; MBM 159181, 1♀, Sanya Bay, Li Xin-zheng Coll., 2 December 1997; MBM 159236, 1♀, Sanya Bay, 2 December 1997; MBM 159149, 1♂, Sanya Bay, Li Xin-zheng Coll., 2 December 1997; MBM 159018, 22♂♂23♀♀, Yongxing Island, Xisha Islands, 9 May 1980; MBM 158986, 1♀1♂, East Island, Xisha Islands, 10 June 1975.

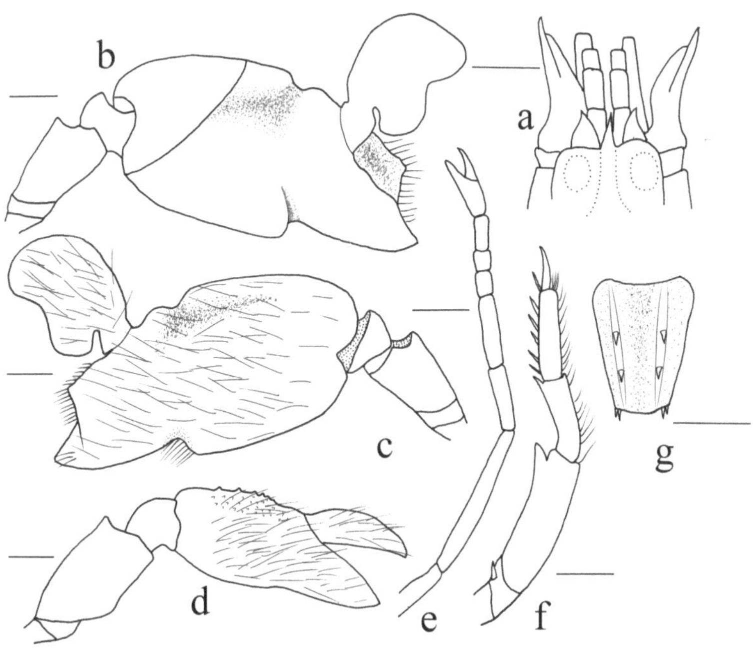


**Fig. 2.49** *Alpheus funafutensis* Borradaile, 1899. MBM 158704, **a**, anterior part of carapace, dorsal view; **b**, major cheliped, lateral view; **c**, same, mesial view; **d**, minor cheliped, mesial view; **e**, second pereiopod; **f**, third pereiopod; **g**, telson (scale bar = 1 mm)

**Description** Rostrum acute, longer than wide at base, variable from not reaching to slightly overreaching distal margin of first antennular segment; rostral carina rounded, extending to posterior margin of orbital hoods; orbitorostral groove moderately deep; orbital hoods without tooth.

Telson about 1.5 and 2 times as long as anterior and posterior width, respectively; dorsal surface with two longitudinal ridges, and on that two pairs of spines approximately inserted 2/5 and 7/10 length of telson, respectively; posterior margin only slightly convex, not beyond posterolateral spines.

Antennular peduncle stout, with distally acute stylocerite reaching to distal margin of first antennular segment; visible part of first antennular segment slightly longer than third segment; second segment about twice as long as third segment. Antenna with lateral margin of scaphocerite strongly concave; antennal scale with blade distinctly shorter than distolateral tooth, which variable from reaching to distal margin of second antennular segment to slightly overreaching distal margin of antennular peduncle. Carpocerite far beyond antennular peduncle.

Major cheliped with chela slightly compressed. Fingers about 2/3 length of palm, dactylus rounded distally, plunger well developed. Palm with mesial face bearing dense long setae; dorsal surface bearing transverse groove sub-distally, which extending to mesial and lateral faces as triangular and quadrangular depressions, respectively; depression of mesial face continuing to proximal about 1/3 length of palm, depression of lateral face continuing to linea impressa; dorsal shoulder rounded, gently sloping, not overhanging groove; ventral surface bearing opposite transverse groove, which extending to lateral face as transverse depression and continuing to 1/4 width of palm; ventral shoulder rounded, gently sloping, not overhanging groove. Merus with dorsal margin terminating in rounded tooth; ventromesial and ventrolateral margin unarmed, rounded distally. Ischium armed. Minor cheliped with chela not showing sexual dimorphism. Fingers shorter than palm, dorsal surface of dactylus bearing dense setae. Palm with dorsal surface bearing heavy rounded papillae; mesial face bearing dense long setae; ventral surface smooth. Merus similar to that of major cheliped. Ischium unarmed.

Second pereiopod with carpus 5-segmented, segmented ratio subequal to 6–7: 7: 2: 2: 4.

Third and fourth pereiopod similar in shape, stout. Dactylus simple. Propodus slightly shorter than merus, ventral surface bearing 8–12 randomly arranged spines and distally one pair. Carpus with ventral margin projecting as spiniform acute tooth. Merus robust, ventral margin bearing large subacute tooth sub-terminally. Ischium armed with one large spine on ventral margin.

**Remarks** Compared with previous descriptions, the present specimens show slight differences as following: the blade of scaphocerite far overreaching distal margin of the antennular peduncle, rather than not reaching to end of the third antennular segment as shown in Banner & Banner (1966, fig. 60a) and Chace (1988, fig. 3a); the second pereiopod with the first carpal segment distinctly shorter than the second segment as shown in Chace (1988, fig. 3d), rather than subequal to the second segment as shown in Banner & Banner (1966, fig. 60d). The variable characters of this species have been discussed by Banner & Banner (1964, 96).

**Type locality** The Ellice Islands

**Distribution** The South China Sea; Kenya; Malaya; Indonesia; Philippines; Central Pacific islands.

### 2.2.41 *Alpheus hippothoe* De Man, 1888 (Fig. 2.50)

*Alpheus hippothoe* De Man, 1888a: 268, Pl. 17, figs. 1–5; Coutière, 1898: 197; Banner DM & Banner AH, 1982: 195, fig. 59.

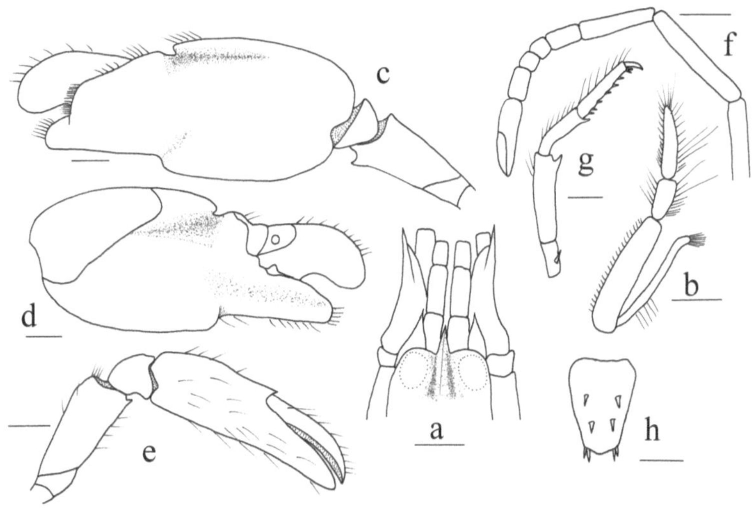


**Fig. 2.50** *Alpheus hippothoe* De Man, 1888. MBM 270436, male, **a**, anterior part of carapace, dorsal view; **b**, third maxilliped; **c**, major cheliped, mesial view; **d**, major chela, lateral view; **e**, minor cheliped, mesial view; **f**, second pereiopod; **g**, third pereiopod; **h**, telson (scale bar = 1 mm)

*Nec Alpheus hippothoe* –. Banner AH & Banner DM, 1966b: 151, fig. 58 (=*A. serenei*).

**Material Examined** CL 6.8–11.5 mm, TL 17.5–28.9 mm. MBM 158190, 1♀, Sanya, Hainan, 26 February 1997; MBM 159212, 1♀, Sanya, 15 November 1997; MBM 157768, 3♀♀, Xincun, Hainan, 21 December 1955; MBM 157857, 1♀1♂, Xincun, Hainan, Ma Xiu-tong Coll., 18 October 1958; MBM 158253, 1♂, Xinying, Hainan, Wang Yong-liang Coll., 25 July 1957; MBM 159022, 1♂, Yongxing Island, Xisha Islands, 12 June 1980; MBM 157911, 2♂♂1♀. Sanya, Hainan, 25 March 1958; MBM 157883, 1♀1♂, Xincun, Hainan, 14 April 1958; MBM 158457, 4♀♀4♂♂, Xincun, Hainan, 27 March 1992; MBM 158475, 2♂♂4♀♀, Xincun, Hainan, Liu Xi-xing Coll., 26 March 1992; MBM 158743, 1♂2♀♀, Xincun, Hainan, 26 March 1992; MBM 159299-302, 1♂, Yezhu Island, Sanya Bay, 23 March 1992; MBM 158544, 1♀1♂, Yezhu Island, Sanya Bay, Neumann Coll., 25 March 1992; MBM 270426, 1♀, Beibu Gulf, Zheng Feng Coll., 18 October 1960; MBM 270437, 1♂, Beibu Gulf, Sun Fu-zheng Coll., 10 December 1959; MBM 270425, 1♀, Beibu Gulf, 15 April 1962; MBM 270427, 1♀, Beibu Gulf, Zheng Feng Coll., 10 July 1960; MBM 270438, 1♀1♂, Beibu Gulf, Fan Zhen-gang Coll., 8 February 1960; MBM 270436, 1♂, Beibu Gulf, Shen Shou-peng Coll., 18 October 1960; MBM 107769, 1♂, Xincun, Hainan, Wang Yong-liang Coll., 26 March 1992.

**Description** Rostrum acute, longer than wide at base reaching to distal margin of first antennular segment; rostral carina rounded, extending to posterior margin of orbital hoods; orbital hoods without tooth.

Telson about 1.3 and 2.6 times as long as anterior and posterior width, respectively; dorsal surface bearing two pairs of spines, approximately inserted 2/5 and 2/3 length of telson, respectively; posterior margin slightly convex, not beyond posterolateral spines.

Antennular peduncle slender, with distally acute stylocerite reaching to distal margin of first antennular segment; visible part of first antennular segment nearly as long as third segment; second segment more than twice as long as third segment. Antenna with lateral margin of scaphocerite strongly concave; antennal scale with narrow blade distinctly shorter than distolateral tooth, which reaching to middle length of third antennular segment. Carpocerite far beyond antennular peduncle.

Major cheliped with chela slightly compressed. Fingers about 1/2 length of palm, dactylus longer than fixed finger, rounded distally, plunger poorly developed, truncate; lateral face of fixed finger bearing a longitudinal depression above ventral margin, extending to middle length of chela. Palm with dorsal surface bearing transverse groove subdistally, which extending to mesial and lateral faces as triangular and quadrangular depressions, respectively; depression of mesial face continuing to proximal about 1/2 length of palm, depression of lateral face continuing to linea impressa; dorsal shoulder subacute, overhanging groove; ventral surface bearing opposite transverse groove, slightly extending to mesial and lateral face; ventral shoulder rounded, gently sloping, not overhanging groove. Merus robust, dorsal margin rounded distally; ventromesial margin bearing fine setae, bearing acute tooth distally. Ischium unarmed. Minor cheliped with chela not showing sexual dimorphism. Fingers slightly shorter than palm. Palm without distinct grooves or depressions; mesial face with distal margin bearing large subacute tooth, flanking dactylus; mesial face bearing thin long setae. Merus similar to that of major cheliped. Ischium unarmed.

Second pereiopod with carpus 5-segmented, segmented ratio subequal to 10: 5: 2: 2: 4.

Third and fourth pereiopod similar in shape, stout. Dactylus simple, 1/3 length of propodus. Propodus slightly shorter than merus, ventral margin bearing 5–7 large spines and 5 accessory small spines and distally one pair. Carpus with dorsal and ventral margin both projecting. Merus robust, ventral margin bearing strong acute tooth sub-terminally. Ischium armed with one large spine on ventral margin.

**Remarks** The present specimens are generally agreed with previous descriptions. However, the character that the merus of both cheliped with ventromesial margin rounded distally shown by De Man (1911) is not found in the present collections. And, the faces of minor chela are smooth without depressions in the present specimens.

**Distribution** The South China Sea; Red Sea; Madagascar; Seychelles; South Africa; Indian Ocean; Malaysia; Indonesia; Philippines; Japan; Fiji; Tonga Islands.

### 2.2.42 *Alpheus hoplocheles* Coutière, 1897 (Fig. 2.51)

*Alpheus hoplocheles* Coutière, 1897d: 197.
*Crangon hoplocheles*–. Yu, 1935: 60.

**Material Examined** CL 13.9–16.5 mm, TL 40.5–45.6 mm. MBM 157882, 1♂, Kulangsu, Xiamen, Chen Mu Coll., 6 April 1982; MBM 159430, 1♀1♂, Shantou, Guangdong, 19 March 1956; MBM 157824, 1♀1♂, Hong Island, Qingdao, Wang Yong-liang & Fan Zhen-gang Coll., 22 September 1956; MBM 158048, 12♀♂, Kulangsu, Xiamen, Fan Zhen-gang & Xu Jie-shan Coll., 30 March 1957; MBM 158228, 4♀♀3♂♂, Shanwei, Guangdong, 1 January 1952; MBM 159005, 1♂, Shenzhen, Guangdong, 10 January 1980; MBM 158113, 2♂♂, Xiangzhou, Guangdong, 4 May 1954; MBM 158000, 1♀1♂, Fujian, 10 April 1982; MBM 157972, 2♂♂, Meihua town, Fuzhou, 24 May 1982; MBM 158120, 2♂♂, Lianyungang, Jiangsu, Huang Zong-guo & Zhang Bao-lin Coll., 29 May 1962; MBM 158875, 5♂♂6♀♀, Kulangsu, Xiamen, 8 April 1982; MBM 157866, 1♂, Kulangsu, Xiamen, 5 April 1982; MBM 157803, 1♀1♂, Lianyungang, Jiangsu, 26 August 1952; MBM 158259, 8♂♂4♀♀, Shantou, Guangdong, 12 December 1980; MBM 157954, 16♀♀19♂♂, Xiamen, Fan Zhen-gang & Xu Jie-shan Coll., 28 March 1957; MBM 158445, 3♂♂, Kulangsu, Xiamen, 11 May 1982; MBM 158480, 2♀♀, Xiamen, 23 November 1984; MBM 158462, 1♂, Kulangsu, Xiamen, 15 October 1985; MBM 233274, 1♂, Kulangsu, Xiamen, Wang Yong-liang Coll., 30 March 1984; MBM 159457, 14♀♂, Hong Island, Qingdao, 10 September 1953.

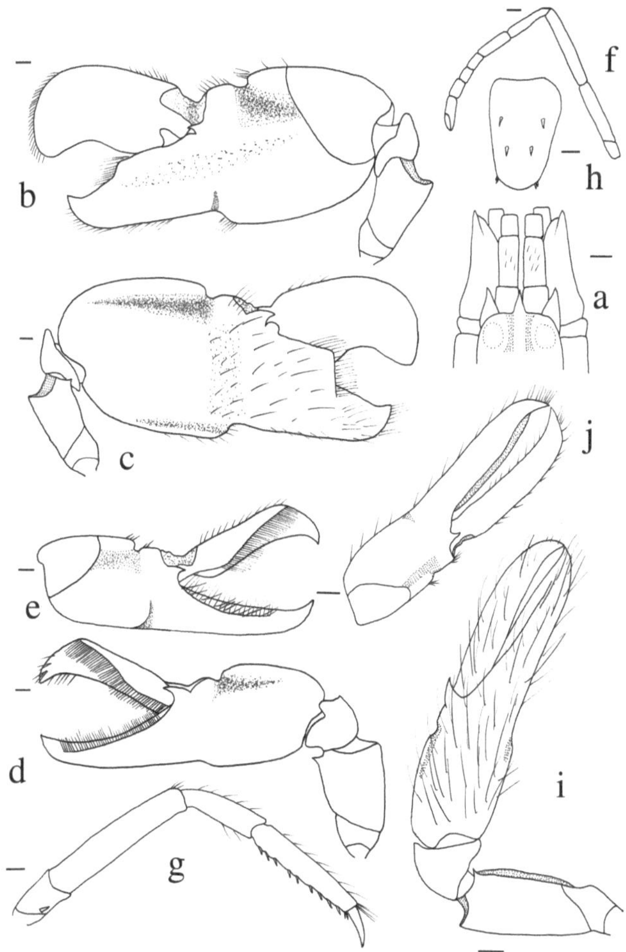


**Fig. 2.51** *Alpheus hoplocheles* Coutière, 1897. MBM 159430, male, **a**, anterior part of carapace, dorsal view; **b**, major cheliped, lateral view; **c**, same, mesial view; **d**, minor cheliped, mesial view; **e**, minor chela, lateral view; **f**, second pereiopod; **g**, third pereiopod; **h**, telson. Female, **i**, minor cheliped, mesial view; **j**, minor chela, lateral view (scale bar = 1 mm)

**Description** Rostrum acute, longer than wide at base reaching to distal margin of first antennular segment; rostral carina extending to posterior margin of orbital hoods; orbitorostral groove moderately deep; orbital hoods without tooth.

Telson about 1.6 times as long as anterior width; dorsal surface bearing two pairs of spines, approximately inserted 1/3 and 2/3 length of telson, respectively; posterior margin slightly convex, not beyond posterolateral spines.

Antennular peduncle slender, with distally acute stylocerite reaching to distal margin of first antennular segment; visible part of first antennular segment slightly longer than third segment; second segment about 3 times as long as third segment. Antenna with lateral margin of scaphocerite concave; antennal scale with narrow blade shorter than distolateral tooth, which slightly overreaching distal margin of antennular peduncle. Carpocerite beyond antennular peduncle.

Major cheliped with chela slightly compressed. Fingers slightly shorter than palm, dactylus longer than fixed finger, rounded distally, plunger poorly developed. Palm with distal margin at base of dactylus bearing two acute teeth on mesial and lateral face, respectively; dorsal surface bearing transverse groove sub-distally, which extending to mesial and lateral faces as triangular and quadrangular depressions, respectively; depression of mesial face continuing to proximal about 1/4 length of palm, depression of lateral face continuing to linea impressa; dorsal shoulder rounded, nearly vertical to palm, not overhanging groove; ventral surface bearing opposite transverse groove, extending to mesial face as longitudinal narrow depression that continuing to proximal 1/3 length of palm, and to lateral face as ill-triangular depression that continuing to middle length of fixed finger; ventral shoulder rounded, gently sloping, not overhanging groove. Merus with distal margin rounded; ventromesial margin bearing acute sub-terminal tooth and terminal tooth. Ischium unarmed. Minor cheliped with chela showing sexual dimorphism: chela of male balaeniceps, fingers nearly as long as palm, palm bearing similar grooves and depressions to major chela, distal margin of palm also bearing two acute teeth at base of dactylus; chela of female not balaeniceps, fingers longer than palm, depressions and grooves on palm less pronounced, distal margin also bearing similar teeth to male. Merus similar to that of major cheliped. Ischium unarmed

Second pereiopod with carpus 5-segmented, segmented ratio subequal to 8: 7: 3: 3: 4.

Third and fourth pereiopod similar in shape, stout. Dactylus simple, 1/3 length of propodus. Propodus slightly shorter than merus, ventral margin bearing 9 spines and distally one pair. Merus unarmed. Ischium armed with one spine on ventral margin.

**Remarks** *A. hoplocheles* is originally described from the South China Sea. Based on present specimens, the characters of this species are re-descripted and illustrated herein.

**Distribution** The Yellow Sea, The East China Sea, the South China Sea; Japan.

### 2.2.43 *Alpheus inopinatus* Holthuis & Gottlieb, 1958 (Fig. 2.52)

*Alpheus inopinatus* Holthuis & Gottlieb, 1958: 42, figs. 8, 9; Tirmizi & Kazmi, 1969: 99, fig.1; Banner DM & Banner AH, 1982: 241, fig. 75.
*Alpheus* sp. Forest & Guinot, 1958: 6, figs. 3–7.

**Material Examined** This species was reported by Banner & Banner (1978) and Liu (2008) from the South China Sea, but not found in the present specimens.

**Description** Rostrum acute, longer than wide at base, slightly overreaching middle length of first antennular peduncle; rostral carina sharp, extending to posterior margin of orbital hoods; orbito-rostral groove moderately deep; orbital hoods without tooth.

Telson 2 times as long as posterior width; dorsal surface bearing two pairs of moderately large spines, approximately inserted 2/5 and 3/5 length of telson, respectively; posterior margin slightly convex, not beyond posterolateral spines, posterior margin between two pairs of posterolateral spines bearing few small spines and fringing long setae.

Antennular peduncle moderately slender, with distally acute stylocerite reaching to distal margin of first antennular segment; visible part of first antennular segment longer than third segment; second segment twice as long as third segment. Antenna with lateral margin of scaphocerite strongly concave at base; antennal scale with narrow blade shorter than distolateral tooth, hardly reaching to distal margin of antennular peduncle. Carpocerite slightly longer than antennular peduncle.

Major cheliped with chela slightly compressed. Fingers slightly shorter than palm, dactylus slightly longer than fixed finger, plunger well developed. Palm with dorsal surface bearing transverse groove sub-distally, which extending to mesial and lateral faces as triangular and quad-

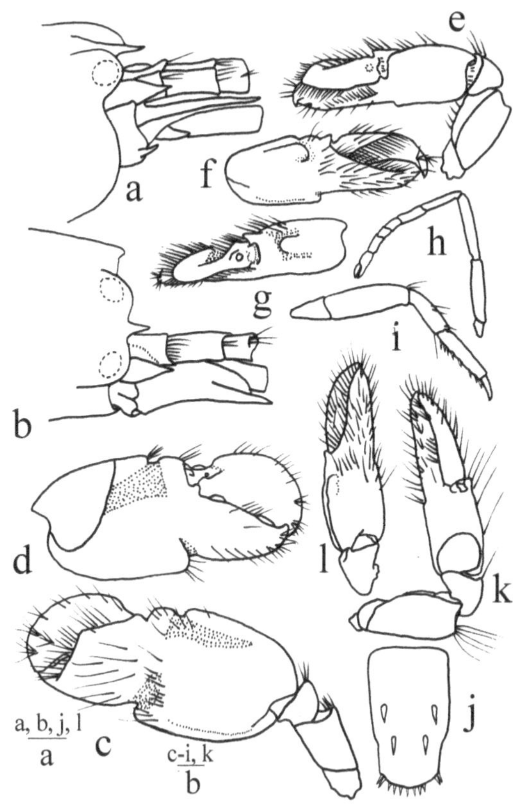


**Fig. 2.52** *Alpheus inopinatus* Holthuis & Gottlieb, 1958. (Adapted from Banner & Banner, 1982). Male, **a**, anterior part of carapace, lateral view; **b**, same, dorsal view; **c**, major cheliped, mesial view; **d**, major chela, lateral view; **e**, minor cheliped, lateral view; **f**, minor chela, mesial view; **g**, same, dorsal view; **h**, second pereiopod; **i**, third pereiopod; **j**, telson. Female, **k**, minor cheliped; **l**, minor chela (scale bar = 1 mm)

rangular depressions, respectively; depression of mesial face continuing to proximal about 1/3 length of palm, depression of lateral face continuing to linea impressa; dorsal shoulder rounded, not overhanging groove; ventral surface bearing opposite transverse groove, which extending to both mesial and lateral faces as small triangular depressions; ventral shoulder rounded, slightly projecting; ventromesial face bearing longitudinal narrow depression extending from base of palm hardly to ventral shoulder. Merus stout, dorsal margin and ventromesial margin rounded. Ischium unarmed. Minor cheliped with chela showing sexual dimorphism: chela of male balaeniceps, fingers nearly as long as palm, palm with dorsal surface bearing transverse groove sub-distally, which extending to both mesial and lateral face as rounded depression, ventral shoulder strong, vertical to palm, cutting edge bearing fine setae, mesial face of palm and fixed finger bearing long setae; chela of female not balaeniceps, fingers nearly as long as palm, dorsal and ventral surface only bearing slightly groove sub-distally. Merus similar to that of major cheliped. Ischium unarmed.

Second pereiopod with carpus 5-segmented, segmented ratio subequal to 10: 7: 3: 3: 5.

Third and fourth pereiopod similar in shape, stout. Dactylus simple, 1/3 length of propodus. Propodus as long as merus, ventral margin bearing 7 spines and distally one pair. Merus and ischium unarmed.

**Remarks** *A. inopinatus* is reported by Banner & Banner (1978) from Hong Kong.

**Distribution** The South China Sea (Hong Kong), Taiwan; Mediterranean coast of Israel; West Pakistan coast; Australia.

### 2.2.44 *Alpheus japonicus* Miers, 1879 (Fig. 2.53)

*Alpheus japonicus* Miers, 1879: 53; Ortmann, 1890: 476, Pl. 36, fig. 14; De Man, 1907: 430, Pl. 33, fig. 53; Balss, 1914: 43; Yokoya, 1930: 527; 1933: 59; Yu, 1935: 59; Liu, 1995: 36; Banner AH & Banner DM, 1984: 39, fig. 1.

*Alpheus longimanus* Spence Bate, 1888: 551, Pl. 98, fig, 4.

**Material Examined** CL 11.5–14.9 mm, TL 35.9–43.5 mm. MBM 157928, 1♀, Nvgukou, Qingdao, 18 September 1936; MBM 159068, 1♂, Yantai, 1 April 1955; MBM 158242-3, 2♀♀3♂♂, the Yellow Sea, 28 April 1976; MBM 158947, 1♀, Guangdong, 26 March 1954; MBM 159975, 5♂♂10♀♀, Shicheng Island, 18 May 1950; MBM 159237, 2♀♀, Nanji Island, Zhejiang, 24 October 1984; MBM 159355; 1♀, Sanya Bay, Hainan, 23 March 1992; MBM 158117, 1♀1♂, Zhejiang, 12 June 1953; MBM 158557, 1♀, Zhejiang, 10 April 1956; MBM 158131, 2♂♂1♀, Chengsi, Zhejiang, 25 April 1956; MBM 158220, 2♂♂, Chanwei, Guangdong, Liu Rui-yu & Wang Yong-liang Coll., 15 April 1954; MBM 158935, 2♀♀2♂♂, Guangdong, 26 March 1954; MBM 157978, 1♀, the East China Sea, 19 September 1976; MBM 158088, 3♀♀, Leqing, Zhejiang, 16 February 1953; MBM 159191, 1♀, Jiaozhou Bay, 9 September 1958; MBM 157961, 1♀, 26 April 1982; MBM 157999, 1♀, the Yellow Sea, 19 May 1992; MBM 157895, 3♀♀1♂, Chengsi, Zhejiang, Chen Mu Coll., 29 April 1975; MBM 157904, 2♀♀, Xu Jie-shan Coll., 10 June 1963; MBM 158257, 2♀♀5♂♂, Hiamen, Jiangsu, 12 March 1956; MBM 157713, 2♀♀, 9 November 1954; MBM 158359, 1♂, the Yellow Sea, Xu Feng-shan Coll., 10 March 1962; MBM 158263, 3♀♀12♂♂, the Yellow Sea, 26 March 1956; MBM 233211, 69♀♂, Qingdao, 23 March 1993; MBM 159983, 26♀♂, Qingdao, Fan Zhen-gang Coll., 9 February

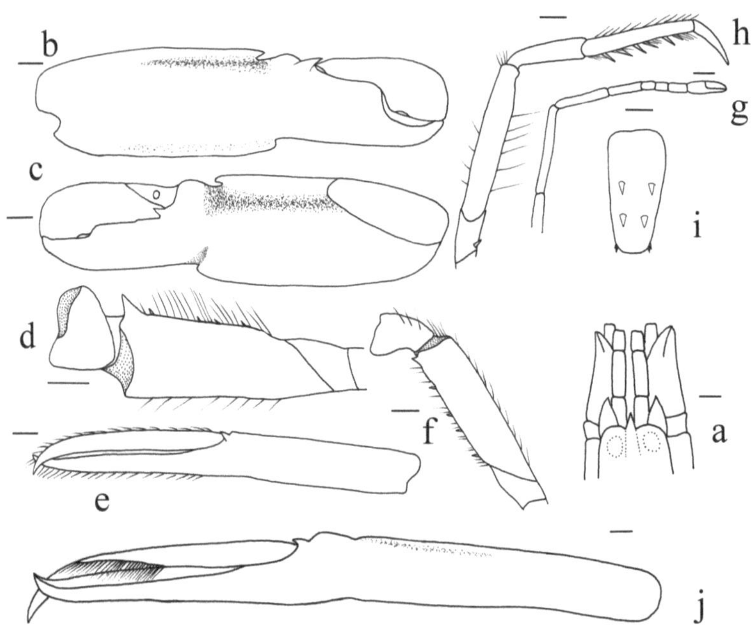


**Fig. 2.53** *Alpheus japonicus* Miers, 1879. MBM 158117, female, **a**, anterior part of carapace, dorsal view; **b**, major chela, mesial view; **c**, same, lateral view; **d**, carpus to ischium of major cheliped, mesial view; **e**, minor chela, mesial view; **f**, carpus to ischium of minor cheliped, mesial view; **g**, second pereiopod; **h**, third pereiopod; **i**, telson. Male, **j**, minor chela (scale bar = 1 mm)

1955; MBM 159987, 32♀♂, Qingdao, 23 February 1955; MBM 158764, 2♀♀, Jiaozhou Bay, Qingdao, 7 December 1985; MBM 158781, 1♀, the Yellow Sea; 14 July 1968; MBM 270704, 24♀♂, the Yellow Sea, 24 June 2007; MBM 270050, 5♀♀1♂, Jiaozhou Bay, Qingdao, 11 February 1993; MBM 270765, 1♀, the Yellow Sea, 16 June 2007; MBM 270797, 13♀♂, the Yellow Sea, 21 June 2007; MBM 270707, 2♂♂4♀♀, the Yellow Sea, 22 June 2007; MBM 270706, 3♀♀3♂♂, the Yellow Sea, 27 June 2007; MBM 270096, 3♀♀1♂, Zhejiang, 7 March 1954; MBM 270708, 1♂, the Yellow Sea, 28 June 2007; MBM 270795, 11♀♂, the Yellow Sea, 19 June 2007; MBM 270793, 201♀♂, the Yellow Sea, 28 June 2007; MBM 270699, 4♀♀, the Yellow Sea, 21 June 2007; MBM 270698, 1♂, the Yellow Sea, 25 June 2007.

**Description** Rostrum acute, longer than wide at base, not reaching to distal margin of first antennular peduncle; anterior margin between rostrum and orbital hoods strongly concave; rostral carina rounded, extending to posterior margin of orbital hoods; orbitorostral grooves shallow; orbital hoods without tooth.

Telson about 2.5 and 3.5 times as long as anterior and posterior width, respectively; dorsal surface bearing two pairs of moderately large spines, approximately inserted 2/5 and 4/5 length of telson, respectively; posterior margin convex, slightly beyond posterolateral small spines.

Antennular peduncle moderately slender, with distally acute stylocerite, slightly overreaching distal margin of first antennular segment; visible part of first antennular segment slightly shorter than second segment; third segment about 1/2 length of second segment. Antenna with lateral margin of scaphocerite slightly concave medially; antennal scale with narrow blade slightly shorter than distolateral tooth, overreaching distal margin of antennular peduncle. Carpocerite reaching to distal margin of antennular peduncle.

Major cheliped with chela elongate, slightly compressed. Fingers slightly shorter than 1/2 length of palm, dactylus as long as fixed finger, plunger poorly developed. Palm with distal margin at base of dactylus bearing two acute teeth on mesial and lateral face, respectively; dorsal surface bearing transverse groove sub-distally, which extending to mesial and lateral faces as triangular and quadrangular depressions, respectively; depression of mesial face continuing to proximal about 1/3 length of palm, depression of lateral face continuing to linea impressa; dorsal shoulder subacute, overhanging groove; ventral surface bearing opposite transverse groove, extending to mesial face as longitudinal narrow depression that posteriorly continuing to proximal 1/3 length of palm, ant to lateral face as small triangular depression; ventral shoulder subacute, overhanging groove. Merus with dorsal margin bearing rounded tooth distally; ventromesial margin bearing 3–7 small spines and acute large tooth distally. Ischium unarmed. Minor cheliped with chela showing sexual dimorphism: chela of male balaeniceps, dactylus distinctly longer than fixed finger, distal margin of palm bearing two acute teeth at base of dactylus, dorsal surface bearing transverse groove sub-distally, which slightly extending to both mesial and lateral face as rounded depression, without dorsal shoulder; chela of female not balaeniceps, distal margin of palm similar to that of male, dorsal groove inconspicuous. Merus similar to that of major cheliped. Ischium unarmed.

Second pereiopod with carpus 5-segmented, segmented ratio subequal to 6: 5: 2: 2: 2.5.

Third and fourth pereiopod similar in shape, moderately slender. Dactylus simple, 1/2 length of propodus. Propodus slightly shorter than merus, ventral margin bearing 6 spines. Carpus with dorsal margin projecting distally. Merus unarmed. Ischium armed with one spine on ventral margin.

**Remarks** The variable characters of *A. japonicus* were discussed by Banner & Banner (1984: 39). The present specimens also show variations in following characters: the stylocerite slightly beyond distal margin of the first antennular segment; the dactylus of minor chela distinctly longer than the fixed finger; the minor chela with dorsal margin of palm without dorsal shoulder; the third pereiopod with ventral margin of propodus bearing 6 spines rather than 3 as shown in Banner & Banner (1984, fig. 1i).

**Distribution** China seas; Russia; Korea; Japan.

## 2.2.45 *Alpheus ladronis* Banner, 1956 (Fig. 2.54)

*Alpheus ladronis* Banner, 1956: 360, fig. 20; Chace, 1988: 31.

**Material Examined** CL 5.1 mm, TL 14.6 mm. MBM 158707, 1♀1♂, Yongxing Island, Xisha Islands, 9 May 1980; MBM 158471, 1♂, Sanya Bay, Hainan, 22 November 1990; MBM 158704, 13♀♂, Yongxing Island, Xisha Islands, 11 June 1980; MBM 158314, 1♀, Shi Island, Xisha Islands, Chen Mu Coll., 2 June 1981; MBM 158680, 1♀, East Island, Xisha Islands, 28 May 1980; MBM 158689, 1♀, Yongxing Island, Xisha Islands, 11 June 1980; MBM 158301, 1♀, Shi Island, Xisha Islands, 9 June 1975; MBM 157759, 1♀, Shi Island, Xisha Islands, Wang Yong-liang Coll., 13 May 1957.

**Description** Rostrum acute, nearly as long as wide at base, reaching to middle length of first antennular segment; rostral carina sharp, extending to posterior margin of orbital hoods; orbito-rostral grooves shallow; orbital hoods without tooth.

Telson about 1.4 and 2.7 times as long as anterior and posterior width, respectively; dorsal surface bearing two pairs of moderately large spines, approximately inserted 1/2 and 3/4 length of telson, respectively; posterior margin slightly convex, not beyond posterolateral small spines.

Antennular peduncle moderately stout, with distally acute stylocerite reaching to distal margin of first antennular segment; visible part of first antennular segment slightly shorter than second segment, second segment about 1.5 times as long as third segment. Antenna with lateral margin of scaphocerite nearly straight; antennal scale with narrow blade distinctly shorter than distolateral tooth, reaching to middle length of third

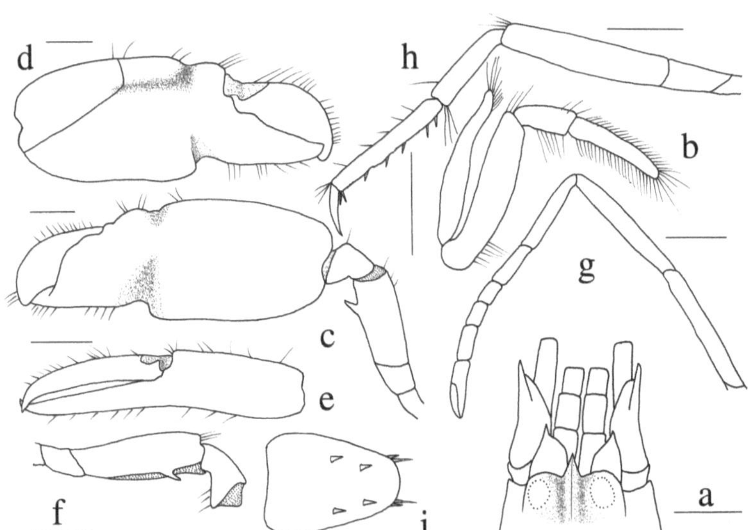


**Fig. 2.54** *Alpheus ladronis* Banner, 1956. MBM 158301, female, **a**, anterior part of carapace, dorsal view; **b**, third maxilliped; **c**, major cheliped, mesial view; **d**, major chela, lateral view; **e**, minor chela, lateral view; **f**, carpus to ischium of minor cheliped, mesial view; **g**, second pereiopod; **h**, third pereiopod; **i**, telson (scale bar = 1 mm)

antennular segment. Carpocerite beyond antennular peduncle.

Major cheliped with chela slightly compressed. Fingers about 1/2 length of palm, tips acute, and dactylus rounded distally. Palm with dorsal surface bearing transverse groove sub-distally, which extending to mesial and lateral faces as rounded and quadrangular depressions, respectively; depression of lateral face continuing to linea impressa; dorsal shoulder U-shaped in outline, not overhanging groove; ventral surface bearing opposite transverse groove, extending to both mesial and lateral face as triangular depressions; ventral shoulder rounded, gently sloping, not overhanging groove. Merus with ventromesial margin bearing strong acute tooth sub-distally. Ischium unarmed. Minor cheliped with chela subcylindrical, not showing sexual dimorphism. Fingers longer than palm. Palm smooth, without grooves. Merus similar to that of major cheliped. Ischium unarmed.

Second pereiopod with carpus 5-segmented, segmented ratio subequal to 14: 9: 3: 3: 5.

Third and fourth pereiopod similar in shape, moderately slender. Dactylus simple, 1/3 length of propodus. Propodus slightly shorter than merus, ventral margin bearing 6–8 spines in single row and distally one pair. Carpus with dorsal margin projecting distally. Merus and ischium unarmed.

**Remarks** *A. ladronis* is a newly recorded species from the China seas, and it is originally described from Saipan. The present specimens show slightly differences from the original description as following: the dactylus of major chela rounded distally rather than bearing hook-like tooth as shown in Banner (1959, fig. 20c, d); the merus of the major and minor cheliped having sub-distal acute tooth on ventromesial margin. Banner (1959) has discussed the variable characters of this species, and the present specimens are included in the variable range of *A. ladronis*.

**Distribution** The South China Sea; Philippines; Marianas; Samoa.

### 2.2.46 *Alpheus leviusculus* Dana, 1852 (Fig. 2.55)

*Alpheus leviusculus* –. Banner AH & Banner DM, 1964: 92, fig. 4; 1966b, 128, fig. 48.

*Alpheus leviusculus leviusculus* –. Banner DM & Banner AH, 1982: 246, fig. 77; Chace, 1988: 34.

*Alpheus edwardsii* var. *leviusculus* Dana, 1852b: 543, Pl. 34, fig. 3a–f.

*Alpheus bouvieri* var. *bastardi* Coutière, 1898a: 133, fig. 1a.

**Material Examined** CL 4.5–5.9 mm, TL 12.1–14.5 mm. MBM 158711, 2♀♀2♂♂, Yongxing Island, Xisha Islands, 8 May 1980; MBM 157998, 1♂, Dengqing Island, Xisha Islands, Fan Zhen-gang & Xu Jie-shan Coll., 15 April 1958. MBM 157756, 8♀♀5♂♂, Yongxing Island, Xisha Islands, Wang Yong-liang Coll., 16 May 1957; MBM 158299, 1♂, Chenhang Island, Xisha Islands, 16 May 1975.

**Description** Rostrum triangular, acute, at most reaching to middle length of first antennular segment; rostral carina rounded, extending to posterior margin of orbital hoods; orbitorostral grooves shallow; orbital hoods slightly inflated, without tooth.

Telson about 1.7 and 2.4 times as long as anterior and posterior width, respectively; dorsal surface bearing two pairs of moderately large spines, approximately inserted 4/9 and 7/9 length of telson, respectively; posterior margin only slightly convex, not beyond posterolateral small spines.

Antennular peduncle stout, with distally acute stylocerite reaching to distal margin of first antennular segment; first antennular segment sub-equal to third segment, second segment about 1.2–1.5 times as long as wide, slightly longer than first segment. Antenna with lateral margin of scaphocerite nearly straight; antennal scale with narrow blade shorter than distolateral tooth, reaching to distal margin of third antennular seg-

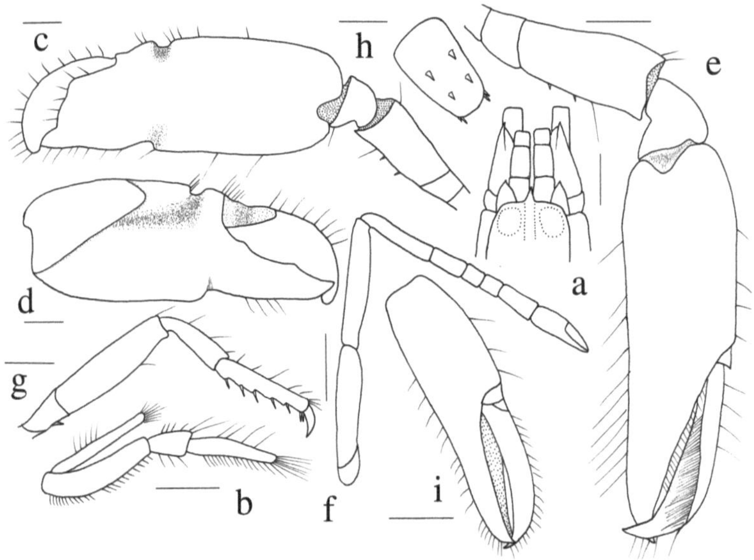


**Fig. 2.55** *Alpheus leviusculus* Dana, 1852. MBM 157756, male, **a**, anterior part of carapace, dorsal view; **b**, third maxilliped; **c**, major cheliped, mesial view; **d**, major chela, lateral view; **e**, minor cheliped, mesial view; **f**, second pereiopod; **g**, third pereiopod; **h**, telson. Female, **i**, minor chela (scale bar = 1 mm)

ment. Carpocerite far beyond antennular peduncle.

Major cheliped with chela slightly compressed. Fingers 1/2 length of palm, dactylus slightly longer than fixed finger, plunger well developed. Palm with dorsal surface bearing transverse groove sub-distally, which extending to mesial and lateral faces as U-shaped and quadrangular depressions, respectively; depression of lateral face continuing to linea impressa; dorsal shoulder U-shaped in outline, not overhanging groove; ventral surface bearing opposite distinct transverse groove, extending to both mesial and lateral face as rounded depressions that continuing to 1/3 width of palm; ventral shoulder rounded, gently sloping, not overhanging groove. Merus about twice as long as wide, ventromesial margin bearing 1–2 spinules and rounded distally; dorsal margin rounded distally. Ischium unnamed. Minor cheliped with chela showing sexual dimorphism: chela of most males balaeniceps, proximal half-length of cutting edges bearing row of setae in fixed finger, balaeniceps poorly developed in some males, fingers nearly as long as palm, palm with dorsal margin terminating in acute tooth overhanging dactylus; chela of female more slender than that of male, not balaeniceps, sometimes bearing slightly fringing with setae on dactylus, fingers nearly as long as palm. Merus similar to that of major cheliped. Ischium unarmed.

Second pereiopod with carpus 5-segmented, segmented ratio subequal to 11: 5: 3: 3: 5.

Third and fourth pereiopod similar in shape, stout. Dactylus simple. Propodus shorter than merus, ventral margin bearing 5 spines and distally one pair. Merus unarmed. Ischium with one spine on ventral margin.

**Remarks** *A. leviusculus* has been well discussed by Banner & Banner (1982). The present specimens are generally agreed with previous descriptions. The dactylus of the third pereiopod is simple, and slightly swelling on flexor margin described by Chace (1988) is not found in the present specimens.

**Distribution** The South China Sea, Taiwan; Red Sea; Eastern Africa; Indian Ocean; Indonesia; Pacific islands at least as far eastward as Wake Island.

### 2.2.47 *Alpheus lobidens* De Haan, 1849 (Fig. 2.56)

*Alpheus lobidens* De Haan, 1849 [in De Haan, 1833–1850]: 179; Ortmann, 1890: 474, Pl. 36, fig. 13; Coutière, 1897d: 199; Banner DM & Banner AH, 1981:29; Chace, 1988: 34.

*Alpheus lobidens polynesica* Banner AH & Banner DM, 1974: 429, fig.3.

*Alpheus crassimanus* Heller, 1865: 107, Pl. 10, fig. 2; Forest & Guinot, 1958: 6, figs. 1, 2; Banner AH & Banner DM, 1966b: 138, fig. 52.

*Alpheus lobidens lobidens* –. Banner DM & Banner AH, 1982: 252, fig. 78.

**Material Examined** CL 7.5–11.1 mm, TL 23.9–32.5 mm. MBM 107575, 32♀♂, Qukou, Hainan, Li Jin-he Coll., 13 March 1992; MBM 107573, 1♀1♂, Qinglan, Hainan, 2 April 1992; MBM 107563, 3♀♀3♂♂, Qinglan, Hainan, Liu Xi-xing Coll., 2 April 1992; MBM 107567, 1♂, Sanya, Hainan, 18 March 1992; MBM 107579, 5♀♀3♂♂, Qinglan, Hainan, 2 April 1992; MBM 107588-9, 6♀♀6♂♂, Luhuitou, Sanya, 20 April 1992; MBM 158062, 1♀1♂, Luhuitou, Sanya, 3 December 1997; MBM 158897, 1♀1♂, Qinglan, Hainan, Li Xin-zheng Coll., 11 March 1997; MBM 158827, 1♀, Qinglan, Hainan, 11 March 1997; MBM 158789, 1♀1♂, Qinglan, Hainan, 11 March 1997; MBM 158201, 1♀1♂, Qinglan, Hainan, 11 March 1997; MBM 159214, 2♂♂, Sanya Bay, Hainan, 1 December 1997; MBM 158185, 1♀1♂, Qinglan, Hainan, 11 March 1997; MBM 159200, 1♀1♂, Sanya Bay, Hainan, 2 December 1997; MBM 158188, 1♀, Sanya Bay, Hainan, 28 February 1997; MBM 159241, 3♂♂, Luhuitou, Sanya, 3 December 1997; MBM 158196, 1♀, Xincun, Hainan, 25 February 1997; MBM 158801, 1♀1♂, Qinglan, Hainan, 11 March 1997; MBM 158787, 1♂, Sanya Bay, Hainan, 5 March 1997; MBM 158941, 1♂, Sanya, Hainan, 19 December 1959; MBM 158938, 1♂, the East China Sea, 25 November 1954; MBM 159011, 3♀♀3♂♂, Sanya Bay, Hainan, Ma Xiu-tong Coll., 7 April 1958; MBM 158972, 1♀1♂, Qinglan, Hainan, 11 March 1997; MBM 159007, 1♀1♂, Qinglan, Hainan, 11 March 1997; MBM 159012, 2♂♂, Wotou, Xiamen, 26 December 1980; MBM 159004, 1♀1♂, Qinglan, Hainan, 11 March 1997; MBM 158940, 3♂♂, Xincun, Hainan, Wang

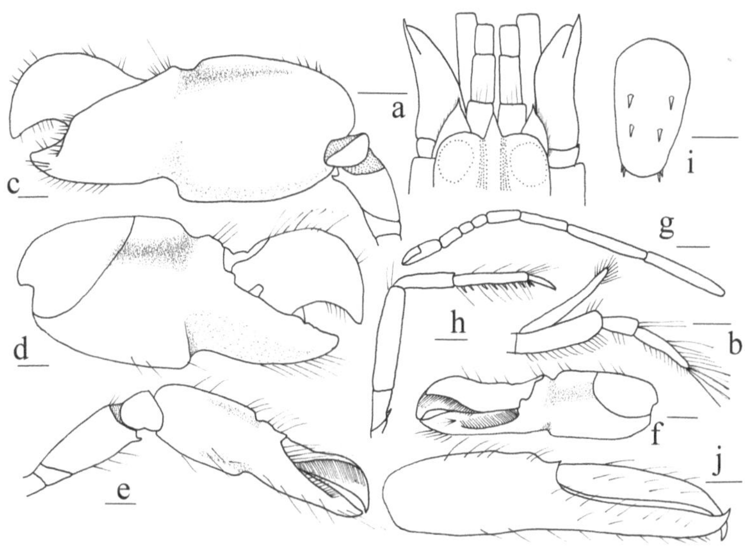


**Fig. 2.56** *Alpheus lobidens* De Haan, 1849. MBM 158730, male, **a**, anterior part of carapace, dorsal view; **b**, third maxilliped; **c**, major cheliped, mesial view; **d**, major chela, lateral view; **e**, minor cheliped, mesial view; **f**, minor chela, lateral view; **g**, second pereiopod; **h**, third pereiopod; **i**, telson. Female, **j**, minor chela (scale bar = 1 mm)

Yong-liang Coll., 9 July 1957; MBM 159028, 1♂, Shi Island, Xisha Islands, 2 June 1981; MBM 158525,10♀♀10♂♂, Qinglan, Hainan, Li Jin-he Coll., 5 April 1992; MBM 159157, 1♀, Sanya Bay, Hainan, 1 December 1997; MBM 158058, 1♀, Sanya Bay, Hainan, 28 November 1997; MBM 157719, 1♀1♂, Xizhou Island, 30 December 1955; MBM 158963, 12♀♂, Xiamen, 12 April 1982; MBM 159013, 1♀1♂, Qukou, Hainan, 25 May 1958; MBM 159237, 1♂, Nanji Island, Zhejiang, 24 October 1984; MBM 158137, 1♀1♂, Weizhou Island, 3 November 1956; MBM 158125, 1♀1♂, Weizhou Island, 11 February 1956; MBM 158226, 3♀♀2♂♂, Zhanjiang, Guangzhou, 29 November 1954; MBM 158255, 5♀♀1♂, Zhapo, Guangdong, 15 November 1954; MBM 158933, 1♀1♂, Xincun, Hainan, 26 January 1956; MBM 157979, 1♀1♂, Jiuzhen Bay, Hainan, 29 April 1982; MBM 158832, Shanwei, Guangdong, 28 February 1956; MBM 158807, 1♂, Qingdao, 1 June 1963; MBM 158929, 1♂, Zhapo, Guangdong, 17 November 1954; MBM 158233, 6♀♀6♂♂, Xincun, Hainan, 9 July 1957; MBM 158836, 1♂, Hong Kong, 30 April 1980; MBM 158754, 1♀1♂, Qinglan, Hainan, 11 March 1997; MBM 158843, 9♀♂, Shangchuan Island, Guangdong, Liu Rui-yu & Wang Yong-liang Coll., 2 November 1954; MBM 158943, 52♀♂, Shi Island, Xisha Islands, 9 December 1954; MBM 158874, 12♀♂, Xincun, Hainan, 7 December 1956; MBM 158795, 1♂2♀♀, Luhuitou, Sanya, 24 February 1997; MBM 157893, 1♂, Sanya, Hainan, 2 April 1958; MBM 158821, 3♀♂, 23 May 1958; MBM 15808, 23♀♂, Luhuitou, Sanya, 23 November 1997; MBM 158902, 1♀1♂, Haikou, Hainan, 28 May 1958; MBM 157939, 1♂, Mai Island, Qingdao, 24 May 1953; MBM 158105, 2♀♀, 20 December 1954; MBM 158237, 56♀♂, Zhapo, Guangdong, Liu Rui-yu & Wang Yong-liang Coll., 15 November 1954; MBM 157937, 1♀, Cangkou, Qingdao, 29 September 1954; MBM 157814, 1♀1♂, 5 November 1981; MBM 157811, 1♀1♂, Cangkou, Qingdao, 4 August 1951; MBM 157950, 1♀1♂, Cangkou, Qingdao, 30 August 1954; MBM 157883, 1♀, Xincun, Hainan, 14 April 1958; MBM 157858, 2♀♀1♂, Fujian, 26 June 1975; MBM 158769, 1♀, Xiamen, 5 February 1963; MBM 158133, 2♂♂, Zhuhai, Guangdong, 3 May 1954; MBM 157949, 9♀♂, Cangkou, Qingdao, 4 August 1955; MBM 157962, 2♀♀3♂♂, Luhuitou, Sanya, 24 March 1958; MBM 158110, 1♀, Wotou, Xiamen, 20 April 1954; MBM 157929, 2♂♂, Cangkou, Qingdao, 7 August 1956; MBM 157919, 13♀♂, Cangkou, Qingdao, 1 September 1955; MBM 157965, 2♂♂1♀, Xincun, Hainan, 22 April 1958; MBM 158148, 2♀♀1♂, Sanya Bay, Wang Yong-liang Coll., 5 April 1955; MBM 158462, 2♂♂1♀, Xiamen, 15 October 1985; MBM 158498, 7♀♀, Cui Yu-xing Coll., 12 January 1960; MBM 157757, 2♀♀, Sanya, 11 June 1957; MBM 157771, 3♂♂, Sanya, Xu Jie-shan Coll., 2 April 1957; MBM 157763, 3♂♂1♀, Weizhou Island, 4 April 1978; MBM 157765, 6♂♂3♀♀, Hainan, 2 May 1980; MBM 157751, 1♀, Luhuitou, Sanya, 1 July 1957; MBM 157789, 1♂, Qinglan, Hainan, 27 April 1980; MBM 158573, 1♀, Shalao, 4 April 1992; MBM 158585, 2♂♂1♀, Qinglan, Hainan, 16 March 1992; MBM 158586, 10♀♂, Houshui Bay, Hainan, 27 March 1992; MBM 157848, 4♂♂14♀♀, Cangkou, Qingdao, Liu Rui-yu Coll., 26 June 1956; MBM 107581, 19♀♂, Luhuitou, Sanya, 17 March 1992; MBM 159274, 4♀♀4♂♂, Qinglan, Hainan, 2 April 1992; MBM 107586, 8♀♀9♂♂, Luhuitou, Sanya, 17 March 1992; MBM 107584, 4♀♀4♂♂, Hailan, Hainan, 2 April 1992; MBM 158667, 2♀♀2♂♂, Hailan, Hainan, 5 November 1990; MBM 107600-3, 6♀♀6♂♂, Qukou, Hainan, 14 March 1992; MBM 107585, 5♂♂10♀♀, Lingao, Hainan, 8 April 1992; MBM 157908, 1♂, Lingao, Hainan, 2 December 1990; MBM 158398, 1♀, Qinglan, Hainan, 5 November 1990; MBM 158626, 1♀, Luhuitou, Hainan, 21 November 1990; MBM 158384, 1♀, Qukou, Hainan, 1 December 1990; MBM 158483, 1♂, Qukou, Hainan, Chen Mu Coll., 14 March 1992; MBM 158857, 2♀♀, Hainan, 2 December 1990; MBM 158613, 2♂♂1♀, Xinying, Hainan, 2 December 1990; MBM 158686, 1♂4♀♀, Xincun, Hainan, Fan Zhen-gang, 25 December 1955; MBM 159401, 1♀1♂, Haikou, Hainan, 3 November 1990; MBM 158632, 2♀♀, Yongxing Island, Xisha Islands, 8 May 1980; MBM 158672, 1♀, Xinying, Hainan, 12 January 1956; MBM 158751, 1♀1♂, Qinglan, Hainan, Li Xin-zheng Coll., 11 March 1997; MBM 158730, 1♀1♂, Qinglan, Hainan, Li Xin-zheng Coll., 11 March 1997; MBM 158741, 3♂♂1♀, Wotou, Xiamen, 21 April 1954; MBM 158704, 3♀♀1♂, Yongxing Island, Xisha Islands, 11 June 1980; MBM 158506, 2♀♀2♂♂, Luhuitou, Sanya, Chen Hui-lian Coll., 18 November 1990; MBM 158433, 8♀♀15♂♂, Luhuitou, Sanya, 1 January 1990; MBM 158124, 1♀, Guangxi, 29 May 1978; MBM 158465, 1♀1♂, , Luhuitou, Sanya, 19 November 1990; MBM 107565, 2♂♂1♀, , Luhuitou, Sanya, 18 November 1990; MBM 107574, 1♀2♂♂, , Luhuitou, Sanya, 24 November 1990; MBM 107572, 2♂♂, Qukou, Hainan, 27 December 1991; MBM 107562, 2♂♂, Qukou, Hainan, Liu Rui-yu Coll., 3 November 1990; MBM 107561, 1♀, , Luhuitou, Sanya, Liu Rui-yu Coll., 19 November 1990; MBM 107582-3, 3♂♂4♀♀, Qinglan, Hainan, 26 November 1990; MBM 158527-30, 4♀♀4♂♂, Qinglan, Hainan, 2 April 1992; MBM 107568, 1♀1♂, Qinglan, Hainan, 27 December 1991; MBM 107570, 7♀♂, Xinying, Hainan, 1 December 1990; MBM 107566, 1♀1♂, Luhuitou, Sanya, 24 November 1990; MBM 107564, 1♀1♂, Qinglan, Hainan, Chen Mu Coll., 2 April 1992.

**Description** Rostrum triangular, longer than wide at base, not reaching to distal margin of first antennular segment; rostral carina sharp, extending to posterior margin of orbital hoods; orbito-rostral grooves shallow; orbital hoods without tooth.

Telson about 1.8 and 3.4 times as long as anterior and posterior width, respectively; dorsal surface bearing two pairs of spines, approximately inserted 2/5 and 7/10 length of telson, respectively; posterior margin only slightly convex, not beyond posterolateral small spines.

Antennular peduncle stout, with distally acute stylocerite reaching to distal margin of first antennular segment; visible part of first antennular segment nearly as long as third segment; third segment 1/2 length of second segment. Antenna with lateral margin of scaphocerite slightly concave; antennal scale with narrow blade shorter than distolateral tooth, reaching to distal margin of third antennular segment. Carpocerite far beyond antennular peduncle.

Major cheliped with chela slightly compressed. Fingers slightly shorter than palm, dactylus longer than fixed finger, tip subacute, with plunger well developed. Palm with dorsal surface bearing transverse groove sub-distally, which extending to mesial and lateral faces as triangular and quadrangular depressions, respectively; depression of mesial face continuing to proximal 1/3 length of palm, depression of lateral face continuing to linea impressa; dorsal shoulder rounded, sloping, not overhanging groove; ventral surface bearing opposite transverse groove, extending to both mesial and lateral face as triangular depressions that continuing to 1/3 width of palm; ventral shoulder rounded, sloping, not overhanging groove. Merus with dorsal margin rounded distally; ventromesial margin with acute tooth sub-distally. Ischium unarmed. Minor cheliped with chela showing sexual dimorphism: chela of male balaeniceps, sculpturing similar to that of major chela but reduced; chela of female not balaeniceps, dorsal and ventral surfaces only bearing inconspicuous transverse groove sub-distally. Fingers nearly as long as palm. Merus similar to that of major cheliped. Ischium unarmed.

Second pereiopod with carpus 5-segmented, segmented ratio subequal to 14: 9: 4: 4: 6.

Third and fourth pereiopod similar in shape, stout. Dactylus simple. Propodus shorter than merus, ventral margin bearing 7 spines and distally one pair. Merus unarmed. Ischium usually with one spine on ventral margin.

**Remarks** Banner & Banner (1974) established one subspecies *A. lobidens polynesica*, and it is synonymized with *A. lobidens* by Chace (1988). *A. lobidens* is a widely distributed species. The present specimens are generally agreed with previous descriptions, except for the following variable characters: the major and minor cheliped with merus always bearing the sub-distal acute tooth on the ventromesial margin; the third pereiopod with propodus armed with 7 spines in single row on the ventral margin.

**Distribution** The Yellow Sea, the South China Sea, Taiwan; Entire Indo-Pacific region.

## 2.2.48 *Alpheus maindroni* Coutière, 1898 (Fig. 2.57)

*Alpheus maindroni* Coutière, 1898a: 133, figs. 2, 2′; Banner DM & Banner AH, 1982: 203, fig. 62; Chace, 1988: 38.

**Material Examined** CL 6.9 mm, TL 19.1 mm. MBM 158175, 1♂, Sanya, Hainan, 1 March 1997; MBM 159255, 1♀, Sanya, Hainan, Liu Rui-yu Coll., 17 November 1997; MBM 158922, 1♂, Sanya, Hainan, Li Xin-zheng Coll., 1 March 1997; MBM 158905, 1♀, Sanya, Hainan, Li Xin-zheng Coll., 1997; MBM 159006, 1♀1♂, Sanya, Hainan, 4 February 1958; MBM 158957, 3♂♂2♀♀, Chenhang Island, Xisha Islands, Chen Mu Coll., 17 May 1981; MBM 158310, 1♀1♂, Yongxing Island, Xisha Islands, Chen Mu Coll., 31 May 1981; MBM 158573, 1♀, Shalao, Hainan, Wang Yong-liang Coll., 4 April 1992-; MBM 158514, 2♀♀, Yezhu Island, Sanya Bay, Neumann Coll., 20 March 1992; MBM 159014, 1♂, Yongxing Island, Xisha Islands, 15 May 1958; MBM 158429, 1♀, Yezhu Island, Sanya Bay, 20 March 1992; MBM 158611, 3♂♂, Yongshu Reef, Nansha Islands, Li Jin-he Coll., 16 May 1993.

**Description** Rostrum small, acute, reaching to 1/2 length of first antennular segment; rostral carina rounded, extending to posterior margin of orbital hoods; orbitorostral grooves shallow; anterior margin between rostrum and orbital hoods concave; orbital hoods inflated, without tooth.

Telson about 2 and 2.5 times as long as anterior and posterior width, respectively; dorsal surface bearing two pairs of spines, approximately inserted 1/2 and 2/3 length of telson, respectively; posterior margin only slightly convex, not beyond posterolateral small spines.

Antennular peduncle stout, with distally acute stylocerite reaching to distal margin of first antennular segment; visible part of first antennular segment and second segment subequal in length; third segment shorter than second segment. Antenna with lateral margin of scaphocerite slightly concave; antennal scale with narrow

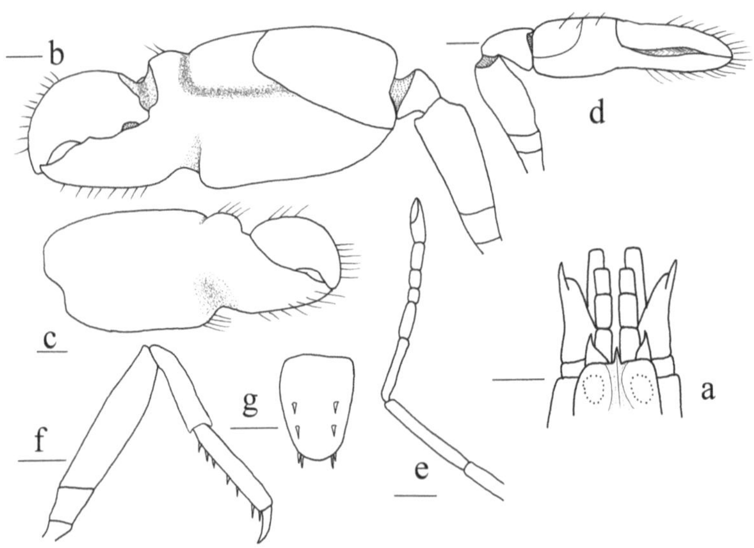


**Fig. 2.57** *Alpheus maindroni* Coutière, 1898. MBM 159006, male, **a**, anterior part of carapace, dorsal view; **b**, major cheliped, lateral view; **c**, major chela, mesial view; **d**, minor cheliped, lateral view; **e**, second pereiopod; **f**, third pereiopod; **g**, telson (scale bar = 1 mm)

blade shorter than distolateral tooth, not reaching to distal margin of third antennular segment. Carpocerite far beyond antennular peduncle.

Major cheliped with chela slightly compressed. Fingers about 1/2 length of palm, dactylus nearly as long as fixed finger, with plunger well developed. Palm with dorsal surface bearing reduced transverse groove sub-distally, which not extending to mesial face and to lateral face as narrow quadrangular depression, respectively; depression of lateral face continuing to linea impressa; dorsal shoulder rounded, gently sloping, not overhanging groove; ventral surface bearing opposite transverse groove, U-shaped in outline, which extending to mesial and lateral faces as rounded and triangular depressions, respectively; depressions of mesial and lateral faces continuing to 1/3 width of palm; ventral shoulder rounded, sloping, not overhanging groove. Merus with dorsal margin rounded distally; ventromesial margin rounded distally. Ischium unarmed. Minor cheliped with chela not showing sexual dimorphism. Fingers slightly longer than palm; cutting edges of fingers fringed with short setae; mesial face bearing numerous patches of short setae; lateral face glabrous. Superior margin of palm bearing inconspicuous groove sub-distally. Merus similar to that of major cheliped. Ischium unarmed.

Second pereiopod with carpus 5-segmented, segmented ratio subequal to 10: 5: 2: 2: 4.

Third and fourth pereiopod similar in shape, stout. Dactylus simple, about 1/3 length of propodus. Propodus shorter than merus, ventral margin bearing 6–8 spines and distally one pair. Carpus shorter than propodus, dorsal margin rounded distally; ventral margin terminating in acute tooth. Merus and ischium unarmed.

**Remarks** *A. maindroni* is a newly recorded species from the China seas, and it is originally described from Djibouti. The present specimens are generally agreed with previous descriptions, excepting the major and minor cheliped with the ventromesial margin of merus rounded distally rather than terminating in acute tooth as shown in Banner & Banner (1982, fig. 62b, e). This character is perhaps variable in *A. maindroni*.

**Distribution** The South China Sea; Gulf of Aden; Gulf of Oman; Mozambique; Madagascar; Philippines; Indonesia; Australia; Caroline and Marshall Islands.

### 2.2.49 *Alpheus malabaricus* (Fabricius, 1775) (Fig. 2.58)

*Astacus malabaricus* Fabricius, 1775: 415.
*Alpheus malabaricus*–. Chace, 1988: 39.
*Alpheus dolichodactylus* Ortmann, 1890: 473, Pl. 36, fig. 11.
*Alpheus dolichodactylus* var. *leptopus* De Man, 1910: 289.
*Alpheus macrodactylus* Ortmann, 1890: 473, Pl. 36, figs. 10, 10e.
*Alpheus malabaricus mackayi* Banner, 1959: 149, fig. 12.
*Alpheus malabaricus songkla* Banner AH & Banner DM, 1966b: 147, fig. 56.
*Alpheus malabaricus trefzae* Banner DM & Banner AH, 1982: 207, fig. 64.

**Material Examined** CL 11.2 mm, TL 32.5 mm. MBM 107609, 1♀1♂, the Yellow Sea, Liu Rui-yu Coll., 26 March 1956; MBM 158138, 2 larva, Songmen, Zhejiang, 25 May 1953; MBM 159035, 4♂♂3♀♀, Pingyang, Zhejiang, 5 October 1979; MBM 107610, 2♂♂1♀, Xiamen, 29 March 1955; MBM 157901, 1♂, Fujian, 21 May 1963; MBM 158794, 1♀1♂, Pingtan, Fujian, 3 May 1984; MBM 158825, 1♂, Pingtan, Fujian, 3 May 1984; MBM 158925, 1♀, Wenling, Zhejiang, 26 June 1955; MBM 107664, 47♀♂, Shantou, Guangdong, Liu Rui-yu Coll., 19 March 1956; MBM 157979, 1♂, Jiuzhen Bay, Fujian, 29 April 1982; MBM 158601, 1♂, Sanya Bay, Hainan, Wang Yong-liang Coll., 23 March 1992; MBM 270028, 1♂, Leqing, Zhejiang, Liu Xi-xing Coll., 20 September 1979; MBM 270459, 2♂♂1♀, Beibu Gulf, Ruan Wen-guang Coll., 14 December 1959; MBM 270580, 1♀, Beibu Gulf, Sun Fu-zeng Coll., 26 August 1962; MBM 270460, 1 larva, Beibu Gulf, Sun Fu-zeng Coll., 4 December 1959.

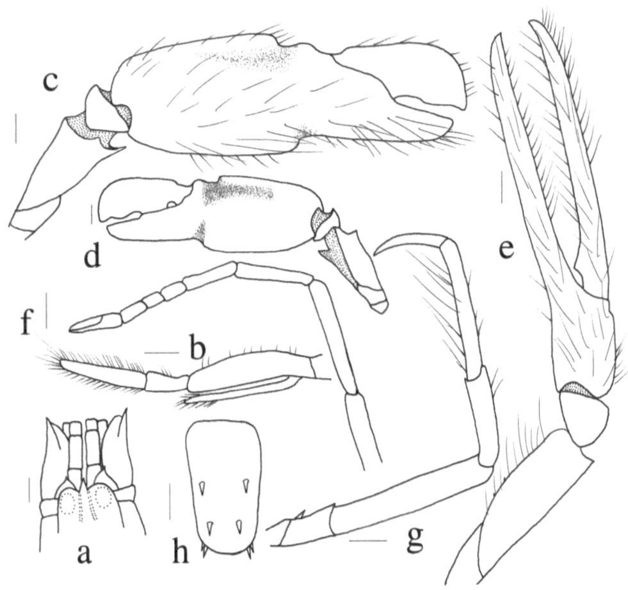


**Fig. 2.58** *Alpheus malabaricus* (Fabricius, 1775). MBM 158925, female, **a**, anterior part of carapace, dorsal view; **b**, third maxilliped; **c**, major cheliped, mesial view; **d**, same, lateral view; **e**, minor cheliped, mesial view; **f**, second pereiopod; **g**, third pereiopod; **h**, telson (scale bar = 1 mm)

**Description** Rostrum acute, rarely reaching to distal margin of first antennular segment; rostral carina variable from rounded to sharp, extending to posterior margin of orbital hoods; anterior margin between rostrum and orbital hoods concave; orbital hoods inflated, without tooth.

Telson about 1.5 and 2 times as long as anterior and posterior width, respectively; dorsal surface bearing two pairs of spines, approximately inserted 1/2 and 5/7 length of telson, respectively; posterior margin strongly convex, beyond posterolateral small spines.

Antennular peduncle stout, with distally acute stylocerite hardly reaching to distal margin of first antennular segment; visible part of first antennular segment slightly longer than third segment; second segment about 3 times as long as third segment. Antenna with lateral margin of scaphocerite nearly straight; antennal scale with narrow blade shorter than distolateral tooth, far overreaching distal margin of antennular peduncle. Carpocerite slightly beyond antennular peduncle.

Major cheliped with chela slightly compressed. Fingers distinctly shorter than palm, dactylus slightly longer than fixed finger, with plunger well developed. Palm with dorsal surface bearing transverse groove sub-distally, which extending to mesial and lateral faces as triangular and quadrangular depressions, respectively; depressions of mesial and lateral face continuing to proximal 1/2 length of palm; dorsal shoulder rounded, sloping, not overhanging groove; ventral surface bearing opposite transverse groove, slightly extending to mesial and lateral face as triangular depression; ventral shoulder rounded, not overhanging groove. Merus with dorsal margin and ventromesial margin terminating in acute tooth, tooth on ventromesial margin stronger. Ischium unarmed. Minor cheliped with chela not showing sexual dimorphism. Fingers 3.5–4 times as long as palm, cutting edges of fingers fringed with long setae and cutting edge of dactylus bearing subacute tooth proximally. Palm nearly smooth; dorsal margin bearing reduced transverse sub-distally; mesial face bearing moderately dense long setae. Merus ventromesial margin bearing 2 small spines and thin long setae and rounded distally. Ischium unarmed.

Second pereiopod with carpus 5-segmented, segmented ratio subequal to 10: 6: 2: 2: 3.

Third and fourth pereiopod similar in shape, stout. Dactylus simple, variable from 1/3 to 1/2 length of propodus. Propodus shorter than merus, ventral margin bearing long setae and several spinules. Merus unarmed. Ischium armed with one spine on ventral margin.

**Remarks** *A. malabaricus* is readily recognized by the minor chela with the fingers more than twice as long as palm. This species is highly variable and widely ranging, so that four subspecies (*A. m. malabaricus*, *A. m. mackayi*, *A. m. songkla*, *A. m. trefzae*) has been established. All the subspecies are synonymized with *A. malabaricus* by Chace (1988).

**Distribution** The Yellow Sea, the East China Sea, the South China Sea; from East Africa to Mexico.

### 2.2.50 *Alpheus pacificus* Dana, 1852 (Fig. 2.59)

*Alpheus pacificus* Dana, 1852b: 544, Pl. 34, fig. 5; Coutière, 1905: 909, fig. 47; Tiwari, 1963: 315, fig. 30; Banner AH & Banner DM, 1966b: 143, fig. 54; Banner DM & Banner AH, 1982: 217, fig. 68; Chace, 1988: 45.
*Alpheus gracilidigitus* Miers, 1884: 287.

**Material Examined** CL 4.1–7.2 mm, TL 11.5–19.3 mm. MBM 158203, 1♂, Sanya, Hainan, Li Xin-zheng Coll., 28 February 1997; MBM 159179, 1♂, Sanya, Hainan, Li Xin-zheng Coll., 1 December 1997; MBM 158129, 1♀1♂, Sanya Bay, Hainan, 2 December 1997; MBM 159006, 1♂, Sanya, Hainan, 2 April 1958; MBM 159409, 1♀, Shalao, Hainan, 4 April 1992; MBM 157839, 2♀♀, Shi Island, Xisha Islands, Chen Mu Coll., 2 June 1981; MBM 158236, 1♀1♂, Xizhou Island, 30 December 1955; MBM 158222, 4♀♀3♂♂, Xizhou Island, Liu Rui-yu Coll., 29 December 1955; MBM 107576, 9♂♂, Yantai, Shandong, 16

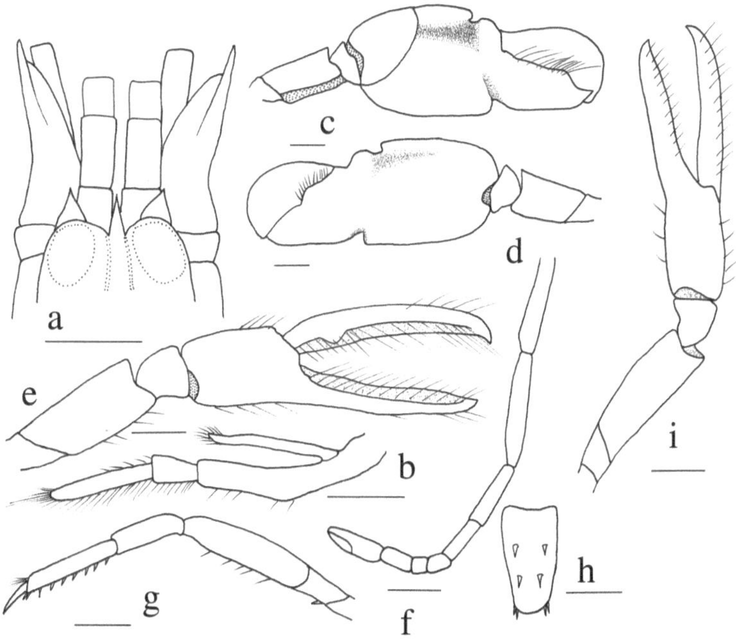


**Fig. 2.59** *Alpheus pacificus* Dana, 1852. MBM 107644, male, **a**, anterior part of carapace, dorsal view; **b**, third maxilliped; **c**, major cheliped, lateral view; **d**, same, mesial view; **e**, minor cheliped, mesial view; **f**, second pereiopod; **g**, third pereiopod; **h**, telson. Female, **i**, minor cheliped (scale bar = 1 mm)

January 1948; MBM 159269, 1♀, Shenhu, Fujian, 5 April 1985; MBM 158948, 6♀♀4♂♂, Hainan, Wang Yong-liang Coll., 26 June 1957; MBM 158951, 1♀, Xincun, Hainan, Fan Zhen-gang Coll., 21 December 1955; MBM 158706, 2♀♀, East Island, Xisha Islands, 28 May 1980; MBM 158802, 1♀, Luhuitou, Sanya, 1 July 1957; MBM 158702, 1♀, Yongxing Island, Xisha Islands, 8 May 1980; MBM 158856, 1♀, Zhong Island, Wang Yong-liang Coll., 28 April 1957; MBM 157912, 7♀♂, Pingtan, Fujian, Chen Mu Coll., 13 May 1975; MBM 157897, 1♂, Xincun, Hainan, 20 April 1958; MBM 158186, 1♀, Dengqing Island, Xisha Islands, Fan Zhen-gang & Xu Jie-shan Coll., 18 April 1958; MBM 157862, 1♀1♂, Haikou, Hainan, 20 May 1958; MBM 157932, 2♂♂, Fujian, 14 April 1957; MBM 157942, 2♂♂1♀, Shilaoren, Qingdao, 10 July 1956; MBM 157956, 2♀♀, Shantou, Guangdong, Fan Zhen-gang & Xu Jie-shan Coll., 20 April 1957; MBM 158606, 2♀♀, Nansha Islands, 19 July 1988; MBM 157938, 2♂♂1♀, Shilaoren, Qingdao, Sun Fu-zeng Coll., 2 November 1956; MBM 157965, 2♀♀1♂, Xincun, Hainan, 22 April 1958; MBM 157957, 1♀, Xizhou Island, 3 April 1958; MBM 107645, 1♀1♂, Xincun, Hainan, 26 March 1992; MBM 157810, 1♂, Huian, Fujian, 15 May 1984; MBM 157818, 1♀, Mai Island, Qingdao, 15 January 1953; MBM 157759, 3♀♀, Shi Island, Wang Yong-liang Coll., 13 May 1957; MBM 157785, 3♀♀1♂, Sanya, Hainan, 3 December 1955; MBM 157760, 5♀♀2♂♂, Shi Island, Xisha Islands, Wang Yong-liang Coll., 15 May 1957; MBM 158464, 2♀♀2♂♂, Linchang Reef, 10 April 1992; MBM 157714, 2♀♀, Hainan, 20 November 1956; MBM 158534-7, 1♀3♂♂, Yezhu Island, Sanya Bay, 26 March 1993; MBM 107597, 2♀♀1♂, Sanya, Hainan, 18 March 1992; MBM 158309, 2♀♀1♂, Jinyin Island, Xisha Islands, 29 May 1975; MBM 158296, 1♀1♂, East Island, Xisha Islands, 12 June 1975; MBM 158300, 1♀, Xisha Islands, 13 May 1975; MBM 158700, 2♂♂, Yongshu Reef, Nansha Islands, 17 May 1993; MBM 158297, 1♀, East Island, Xisha Islands, 8 June 1975; MBM 107644, 6♀♀2♂♂, Sanya, Hainan, 27 December 1991; MBM 107643, 3♂♂1♀, Sanya, Hainan, 14 January 1992; MBM 233230, 1♂, Xisha Islands, 12 May 1975; MBM 159207, 1♀, Sanya, Hainan, Li Xin-zheng Coll., 19 November 1997.

**Description** Rostrum acute, not reaching to distal margin of first antennular segment; rostral carina rounded, extending to posterior margin of orbital hoods; orbitorostral grooves moderately deep; anterior margin between rostrum and orbital hoods strongly concave; orbital hoods slightly inflated, without tooth.

Telson about twice as long as anterior width, posterior margin slightly narrower than anterior margin; dorsal surface bearing two pairs of moderately small spines, approximately inserted 1/3 and 2/3 length of telson, respectively; posterior margin convex, not beyond posterolateral small spines.

Antennular peduncle with distally acute stylocerite reaching to distal margin of first antennular segment; visible part of first antennular segment slightly shorter than second segment; third segment half-length of second segment. Antenna with lateral margin of scaphocerite concave; antennal scale with narrow blade shorter than distolateral tooth, far beyond antennular peduncle. Carpocerite nearly as long as distolateral tooth of antennal scale.

Major cheliped with chela slightly compressed. Fingers slightly shorter than palm, dactylus slightly longer than fixed finger, tip subacute, with plunger well developed. Palm with dorsal surface bearing transverse groove sub-distally, which extending to mesial and lateral face as triangular and quadrangular depression, respectively; depression of mesial face continuing to proximal 1/2 length of palm, depression of lateral face continuing to linea impressa; dorsal shoulder subacute, overhanging groove; ventral surface bearing opposite transverse groove, extending to lateral face as V-shaped depression that continuing to 1/3 width of palm, and to mesial as reduced rounded depression; ventral shoulder subacute, overhanging groove. Merus with dorsal and ventromesial margin rounded distally. Ischium unarmed. Minor cheliped with chela showing sexual dimorphism: chela of male with fingers 1.5–2.2 times as long as palm, cutting edges of fingers fringed with long setae and cutting edge of dactylus bearing large tooth proximally, ventral margin of palm with slight shoulder nearly at base of fixed finger; chela of female with finger 1.2–1.5 times as long as palm, only bearing scattered setae. Merus similar to that of major cheliped. Ischium unarmed.

Second pereiopod with carpus 5-segmented, segmented ratio subequal to 10: 8: 3: 3: 5.

Third and fourth pereiopod similar in shape, stout. Dactylus simple, 1/3 length of propodus. Propodus shorter than merus, ventral margin bearing 7 spines and distally one pair. Merus unarmed. Ischium armed with one spine on ventral margin.

**Remarks** *A. pacificus* is readily recognized by the minor chela with the fingers 1.2–2.0 times as long as palm, and neither the male nor the female has a balaeniceps dactylus. This species has been described and discussed many times (Banner & Banner, 1966, 1982; Chace, 1988). The present species are generally agreed with previous descriptions.

**Distribution** The Yellow Sea, the South China Sea; Red Sea and Eastern Africa to Clipperton Island; Eastern Pacific.

### 2.2.51 *Alpheus pareuchirus* Coutière, 1905 (Fig. 2.60)

*Alpheus pareuchirus* Coutière, 1905: 906, Pl. 84, fig. 43; Bruce, 1990: 621, fig. 6.

*Alpheus pareuchirus pareuchirus*–. Banner DM & Banner AH, 1982: 276, fig. 85a–k; Chace, 1988: 47.

*Alpheus pareuchirus* var. *leucothea* De Man, 1911: 420, fig. 102.

**Material Examined** CL 3.2–5.5 mm, TL 8.7–13.9 mm. MBM 158688, 2♂♂, Linchang Reef, Hainan, 10 April 1992; MBM 158019, 1♂, Hong Kong, 1 August 1999; MBM 270388, 1♂, Beibu Gulf, Sun Fu-zeng Coll., 19 August 1962; MBM 270498, 1♂, Beibu Gulf, Ren Xian-qiu Coll., 10 November 1960; MBM 270464, 1♂, Beibu Gulf, Fan Zhen-gang Coll., 8 February 1960; MBM 270322, 1♀, Beibu Gulf, Fan Zhen-gang Coll., 22 April 1960; MBM 270323, 1♂, Beibu Gulf, 4

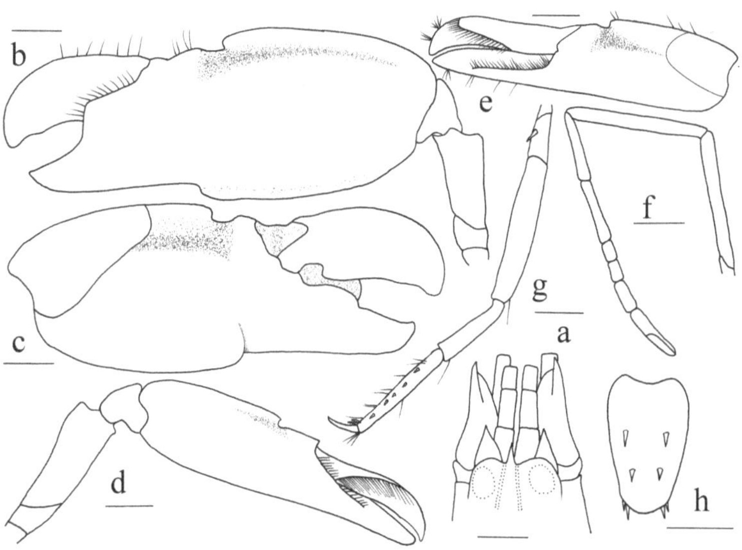


**Fig. 2.60** *Alpheus pareuchirus* Coutière, 1905. MBM 270498, male, **a**, anterior part of carapace, dorsal view; **b**, major cheliped, mesial view; **c**, major chela, lateral view; **d**, minor cheliped, mesial view; **e**, minor chela, lateral view; **f**, second pereiopod; **g**, third pereiopod; **h**, telson (scale bar = 1 mm)

December 1959; MBM 270324, 1♂, Beibu Gulf, Fan Zhen-gang Coll., 10 December 1959; MBM 270313, 1♂, Beibu Gulf, Ruan Wen-guang Coll., 14 January 1962; MBM 270315, 1♂, Beibu Gulf, 20 August 1962; MBM 270156, 1♂, Beibu Gulf, 7 February 1960; MBM 270399, 1♀, Beibu Gulf, 21 August 1962; MBM 270314, 1♂, Beibu Gulf, Sun Fu-zeng Coll., 7 January 1962; MBM 270329, 1♂, Beibu Gulf, Shen Shou-peng Coll., 17 October 1960; MBM 27016, 1♀, Beibu Gulf, Ma Xiu-tong Coll., 9 December 1959.

**Description** Rostrum acute, longer than wide at base, slightly overreaching 1/2 length of first antennular segment; rostral carina blunt, extending to posterior margin of orbital hoods; orbito-rostral grooves shallow; anterior margin between rostrum and orbital hoods slightly concave; orbital hoods without tooth.

Telson about 1.5 and 2.7 times as long as anterior and posterior width, respectively; dorsal surface bearing two pairs of spines, approximately inserted 2/5 and 2/3 length of telson, respectively; posterior margin convex, not beyond posterolateral small spines.

Antennular peduncle slender, with distally acute stylocerite slightly overreaching distal margin of first antennular segment; visible part of first antennular segment nearly as long as second segment; third segment about 1/2 length of second segment. Antenna with lateral margin of scaphocerite slightly concave; antennal scale with narrow blade shorter than distolateral tooth, slightly overreaching distal margin of antennular peduncle. Carpocerite nearly as long as distolateral tooth of antennal scale.

Major cheliped with chela slightly compressed. Fingers slightly longer than 1/2 length of palm, dactylus and fixed finger subequal in length, with plunger reduced. Palm with dorsal surface bearing transverse groove sub-distally, which extending to mesial and lateral face as triangular and quadrangular depression, respectively; depression of mesial face continuing to proximal 1/3 length of palm, depression of lateral face continuing to linea impressa; dorsal shoulder subacute, overhanging groove; ventral surface bearing opposite inconspicuous transverse, not extending to both faces; ventral shoulder rounded, gently sloping, not overhanging groove; mesial face bearing narrow longitudinal depression near ventral margin, extending from base of palm not to ventral shoulder. Merus with ventromesial margin terminating in strong acute tooth. Ischium unarmed. Minor cheliped with chela showing sexual dimorphism: chela of male balaeniceps, fingers shorter than palm, mesial and lateral margin of fixed finger fringed proximal half with row of setae; chela of female not balaeniceps. Palm of male with dorsal surface bearing transverse groove sub-distally, which extending to both mesial and lateral face as triangular depression; dorsal shoulder subacute, overhanging groove; palm of female without sculpturing. Merus similar to that of major cheliped. Ischium unarmed.

Second pereiopod with carpus 5-segmented, segmented ratio subequal to 10: 9: 3: 3: 5.

Third and fourth pereiopod similar in shape, stout. Dactylus simple. Propodus shorter than merus, ventral margin bearing about 12 spines arranging in two rows. Merus unarmed. Ischium armed with one strong spine on ventral margin.

**Remarks** Bruce (1990) reported *A. pareuchirus* from Hong Kong. The present specimens are generally agreed with previous descriptions, except for the minor cheliped with merus terminating in rounded or acute tooth on ventromesial margin.

**Distribution** The South China Sea; Red Sea; Madagascar; Seychelles; Maldives; Thailand; Indonesia; Philippines; Australia; Caroline Islands.

### 2.2.52 *Alpheus parvirostris* Dana, 1852 (Fig. 2.61)

*Alpheus parvirostris* Dana, 1852a: 22; 1852b: 551; Dana, 1855: Pl. 35: fig. 3; Liu & Lan, 1980: 106, fig. 15; Banner DM & AH Banner,

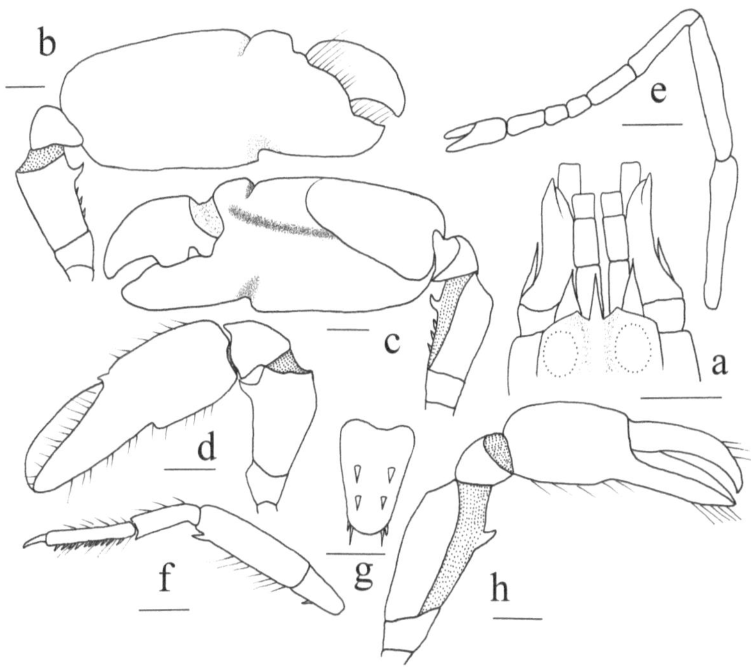


**Fig. 2.61** *Alpheus parvirostris* Dana, 1852. MBM 159006, male, **a**, anterior part of carapace, dorsal view; **b**, major cheliped, mesial view; **c**, same, lateral view; **d**, minor cheliped, mesial view; **e**, second pereiopod; **f**, third pereiopod; **g**, telson. Female, **h**, minor cheliped, lateral view (scale bar = 1 mm)

1982: 185, fig. 56; Banner AH & DM Banner, 1983: 60, fig 9; Chace, 1988: 47.
*Alpheus lineifer* Miers, 1875: 343.
*Alpheus euchiroides* Nobili, 1906a: 257.
*Alpheus braschi* Boone, 1935: 131, Pl. 34, fig. 10.

**Material Examined** CL 6.2 mm, TL 15.3 mm. MBM 107658, 1♀, Jinqing Island, Xisha Islands, 19 May 1980; MBM 107657, 9♀♀, Zhaoshu Island, Xisha Islands, 11 May 1980; MBM 107656, 5♀♀5♂♂, East Island, Xisha Islands, 3 June 1980; MBM 107663, 298♀♂, Shanhu Island, Xisha Islands, 19 May 1980; MBM 107662, 50♀♂, Yongxing Island, Xisha Islands, 9 May 1980; MBM 107661, 45♂♂27♀♀, Yongxing Island, Xisha Islands, 11 June 1980; MBM 107654, 1♀, Yongxing Island, Xisha Islands, 8 May 1980; MBM 107659, 36♂♂18♀♀, East Island, Xisha Islands, 28 May 1980; MBM 107665, 2 larva, Shalao, Hainan, 5 April 1992; MBM 159228, 2♀♀2♂♂, Sanya, Hainan, Li Xin-zheng Coll., 18 November 1997; MBM 159196, 1♀, Sanya, Hainan, 20 November 1997; MBM 159208, 2♀♀, Sanya, Hainan, 18 November 1997; MBM 159166, 1♀1♂, Sanya, Hainan, Li Xin-zheng Coll., 2 December 1997; MBM 159209, 1♀, Sanya, Hainan, Li Xin-zheng Coll., 1 December 1997; MBM 159205, 1♂, Sanya, Hainan, Li Xin-zheng Coll., 17 November 1997; MBM 158191, 1♂, Sanya, Hainan, 28 February 1997; MBM 158983, 3 larva, Yongxing Island, Xisha Islands, 21 March 1980; MBM 159003, 1♀, Shanhu Island, Xisha Islands, 25 March 1975; MBM 158970, 2♀♀, Yongxing Island, Xisha Islands, Fan Zhen-gang & Xu Jie-shan Coll., 15 May 1958; MBM 158971, 1♀1♂ Sanya, Hainan, 1 March 1997; MBM 159028, 1♀, Shi Island, Xisha Islands, 2 June 1981, MBM 159006, 2♀♀2♂♂, Sanya, Hainan, 4 February 1958; MBM 159029, 1♀1♂, Jinqing Island, Xisha Islands, 19 May 1981; MBM 138323, 1♂, Xincun, Hainan, 31 December 1959; MBM 158654, 2♀♀2♂♂, Yufang Reef, Nansha Islands, 4 May 1990; MBM 157837, 2♀♀1♂, Shi Island, Xisha Islands, 4

February 1981; MBM 158540, 1♂, Yezhu Island, Sanya Bay, Neumann Coll., 19 March 1992; MBM 158994, 2♀♀, Chenhang Island, Xisha Islands, 16 May 1981; MBM 158677, 1♂, Sanjiao Reef, Nansha Islands, 24 May 1993; MBM 158253, , 1♀1♂, Xinying, Hainan, 25 July 1957, MBM 158893, 1♂, Sanya, Hainan, Li Xin-zheng Coll., 4 March 1997; MBM 107660, 1♂, East Island, Xisha Islands, 9 June 1975; MBM 159222, 4♀♀1♂, Zhubi Reef, Nansha Islands, 10 April 1999; MBM 159215, 2♀♀1♂, Chigua Reef, Nansha Islands, 16 April 1999; MBM 159268, 1♂, Yongshu Reef, Nansha Islands, 11 April 1999; MBM 158852, 1♀, Daai Reef, Nansha Islands, 24 May 1990; MBM 157909, 5♀♂, Xinyi Reef, Nansha Islands, 29 May 1990; MBM 107590-4, 2♀♀1♂, Linchang Reef, Hainan, 11 April 1992; MBM 158804, 7♀♂, Yongxing Island, Xisha Islands, Fan Zhen-gang & Xu Jie-shan Coll., 30 August 1958; MBM 158923, 3♀♀5♂♂, Xinying , Hainan, 7 May 1958; MBM 158841, 1♀, Dongmen Reef, Nansha Islands, 20 May 1990; MBM 158079, 72♀♂, Yongxing Island, Xisha Islands, 31 May 1981; MBM 158112, 2♀♀1♂, Xisha Islands, 28 April 1958; MBM 159263, 5♀♀, Nanxun Reef, Nansha Islands, 17 April 1999; MBM 157968, 3♂♂1♀, Yongxing Island, Xisha Islands, Wang Yong-liang Coll., 13 July 1957; MBM 158015, 27♀♂, Jinqing Island, Xisha Islands, Fan Zhen-gang & Xu Jie-shan Coll., 25 April 1958; MBM 157943, 23♀♂, Guangjin Island, Xisha Islands, 23 May 1981; MBM 158177, 1♂, Dengqing Island, Xisha Islands, 11 April 1958; MBM 158036, 15♀♂, Shi Island, Xisha Islands, Wang Yong-liang Coll., 13 May 1957; MBM 157762, 6♀♀, Xizhou Island, Liu Rui-yu Coll., 30 December 1955; MBM 158371, 1♀1♂, Xian'e Reef, 28 May 1993; MBM 158469, 3♀♀2♂♂, Xincun, Hainan, 27 March 1992; MBM 107577, 3♂♂4♀♀, Linchang Reef, 10 April 1992; MBM 159278, 4♀♀3♂♂, Xincun, Hainan, 27 March 1992; MBM 159287-90, 2♀♀, Linchang Reef, 11 April 1992; MBM 159310-3, 4♀♀1♂, Xincun, Hainan, 26 March 1992; MBM 158785-6, 1♂, Shalao, Hainan, Wang Yong-liang Coll., 4 April 1992; MBM 158283-8, 42♀♂, Yongxing Island, Xisha Islands, 7 June 1975; MBM 158271, 15♀♂, Yongxing Island, Xisha Islands, Ren Xian-qiu Coll., 5 June 1975; MBM 158272, 2♂♂1♀, Jinyin Island, Xisha Islands, 24 May 1975; MBM 158308, 21♀♂, Shi Island, Xisha Islands, 8 May 1975; MBM 158311, 1♀1♂, Jinqing Island, Xisha Islands, 19 May 1981; MBM 158313, 1♀, Shi Island, Xisha Islands, Chen Mu Coll., 2 June 1981; MBM 158298, 2♀♀1♂, East Island, Xisha Islands, 12 June 1975; MBM 158600, 1♂2♀♀, Xian'e Reef, Nansha Islands, 22 April 1987; MBM 158577, 1♀, Yongshu Reef, Nansha Islands, 16 May 1993; MBM 158645, 1♀1♂, Yongshu Reef, Nansha Islands, 18 May 1993; MBM 158772, 4♀♀, Xinyi Reef, Nansha Islands, 29 May 1990; MBM 158687, 1♀1♂, Yongxing Island, Xisha Islands, Chen Mu Coll., 4 April 1983; MBM 158638, 1♀1♂, Yongshu Reef, Nansha Islands, Li Jin-he Coll., 17 May 1993; MBM 158715, 1♀, Yezhu Island, Yalong Bay, 25 November 1990; MBM 158623, 2♀♀2♂♂, Yongshu Reef, Nansha Islands, 16 May 1993; MBM 158564, 1♂, Zhubi Reef, Nansha Islands, 19 May 1993; MBM 158621, 2♀♀1♂, Yongxing Island, Xisha Islands, 11 June 1980; MBM 158743, 1♂, Xincun, Hainan, 26 March 1992; MBM 158735, 8♀♀7♂♂, Xinying, Hainan, 13 January 1956; MBM 158768, 2♂♂, Shanhu Reef, Nansha Islands, 25 May 1990; MBM 158603, 8♀♀8♂♂, Yongshu Reef, Nansha Islands, Li Jin-he Coll., 18 May 1993; MBM 158578, 3♀♀2♂♂, Zhubi Reef, Nansha Islands, 30 May 1993; MBM 107655, 1♂, Linchang Reef, Hainan, Liu Rui-yu Coll., 20 December 1991; MBM 107653, 1♀1♂, Fujian, Liu Xi-xing Coll., 2 December 1990; 04II61AD-12, 2 larva, Zhubi Reef, Nansha Islands, 8 May 2004; 04II79AD-26, 2♀♀2♂♂, Zhubi Reef, Nansha Islands, 10 May 2004; 04II59AD-31, 2♀♀, Zhubi Reef, Nansha Islands, 8 May 2004; 04II84AD-31, 10 larva, Zhubi Reef, Nansha Islands, 14 May 2004; 04II84AD-32, 1♀, Zhubi Reef, Nansha Islands, 14 May 2004; 04II71AD-28, 2♀♀1♂, Zhubi Reef, Nansha Islands, 11 May 2004; 04II66AD-16, 1 larva, Zhubi Reef, Nansha Islands, 10 May 2004; 04II56AD-43, 17♀♂, Zhubi Reef, Nansha Islands, 6 May 2004; MBM 107752, 1♀, Yongxing Island, Xisha Islands, Fan Zhen-gang & Xu Jie-shan Coll., 28 March 1958.

**Description** Rostrum acute, longer than wide at base, not reaching to distal margin of first antennular segment; rostral carina rounded, extending to posterior margin of orbital hoods; orbitorostral grooves shallow; anterior margin between rostrum and orbital hoods strongly concave; orbital hoods without tooth.

Telson about 1.6 and 2.2 times as long as anterior and posterior width, respectively; dorsal surface bearing two pairs of spines, approximately inserted 1/3 and 2/3 length of telson, respectively; posterior margin convex, not beyond posterolateral small spines.

Antennular peduncle slender, with distally acute stylocerite reaching to distal margin of first antennular segment; visible part of first antennular segment nearly as long as second segment; third segment about half length of second segment. Antenna with lateral margin of scaphocerite strongly concave; antennal scale with narrow blade distinctly shorter than distolateral tooth, reaching to or slightly beyond distal margin of antennular peduncle. Basicerite with lateral tooth conspicuous in dorsal and lateral views, acute, reaching to 1/2 length of second antennular segment. Carpocerite nearly as long as distolateral tooth of antennal scale.

Major cheliped with chela slightly compressed. Fingers about 1/2 length of palm, dactylus slightly longer than fixed finger, with plunger well developed. Palm with dorsal surface bearing transverse groove sub-distally, which not extending to both mesial and lateral face; dorsal shoulder rounded, sloping, not overhanging groove; lateral face bearing narrow longitudinal depression medially, which beginning from linea impressa to below dorsal shoulder, not join with dorsal transverse groove; ventral surface bearing opposite transverse groove, which only slightly extending to both mesial and lateral face as triangular depression; ventral shoulder rounded, gently sloping, not overhanging groove. Merus with dorsal margin rounded distally; ventromesial margin bearing 3 small spines and sub-terminal strong acute tooth. Ischium unarmed. Minor cheliped with chela showing slightly sexual dimorphism: chela of female more slender, dorsal and ventral shoulder lacking in female. Merus with ventromesial margin bearing sub-terminal strong acute tooth. Ischium unarmed.

Second pereiopod with carpus 5-segmented, segmented ratio subequal to 10: 6: 3: 3: 4.

Third and fourth pereiopod similar in shape, stout. Dactylus simple. Propodus shorter than merus, ventral margin bearing about 10 spines arranging in two rows. Merus with ventral margin bearing strong acute terminal tooth. Ischium armed with one strong spine on ventral margin.

**Remarks** *A. parvirostris* is readily recognized by the basicerite bearing unusually long lateral tooth in the Edwardsii group. This species is very common thorough the Pacific. The present specimens are generally agreed with previous descriptions, except for the third pereiopod with the propodus bearing 10 spines in single row rather than in two rows as shown in Banner & Banner (1982, fig. 56i).

**Distribution** The South China Sea, Taiwan; Red Sea; Eastern and South Africa to Japan; Philippines; Indonesia; Australia; to the Tuamotu Archipelago but not Hawaii.

### 2.2.53 *Alpheus polyxo* De Man, 1909 (Fig. 2.62)

*Alpheus polyxo* De Man, 1909a: 108; 1911: 423, fig. 104; Liu & Lan, 1980: 108, fig. 16; Banner DM & Banner AH, 1982: 274, fig. 84; Chace, 1988: 48.

**Material Examined** CL 6.9–7.5 mm, TL 19.5–20.1 mm. MBM 270463, 1♂, Beibu Gulf, Sun Fu-zeng Coll., 21 January 1962; MBM 159279, 1♀1♂, Yezhu Island, Neumann Coll., 3 February 1992.

**Description** Rostrum acute, longer than wide at base, reaching to distal margin of first antennular segment; rostral carina rounded, extending to posterior margin of orbital hoods; anterior mar-

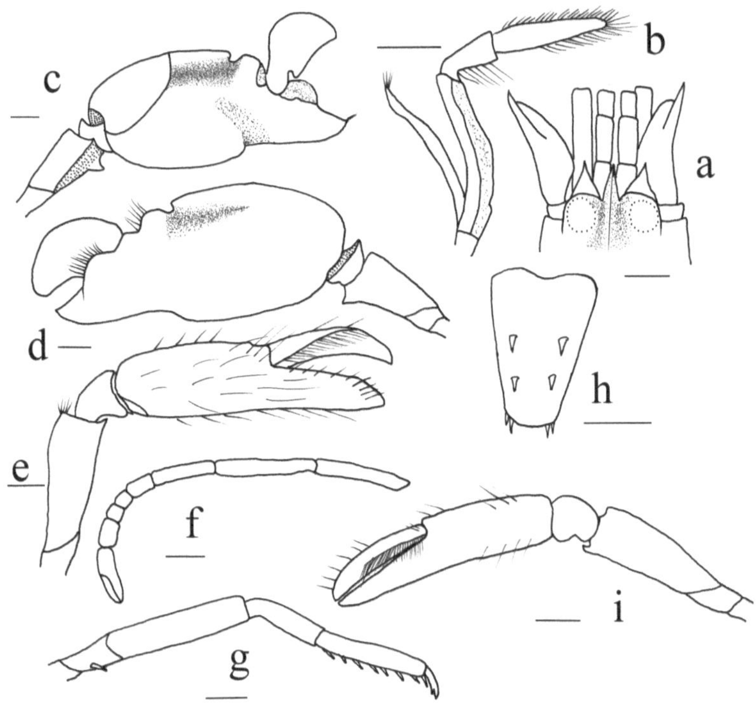


**Fig. 2.62** *Alpheus polyxo* De Man, 1909. MBM 159279, male, **a**, anterior part of carapace, dorsal view; **b**, third maxilliped; **c**, major cheliped, lateral view; **d**, same, mesial view; **e**, minor cheliped, mesial view; **f**, second pereiopod; **g**, third pereiopod; **h**, telson. Female, **i**, minor cheliped, mesial view (scale bar = 1 mm)

gin between rostrum and orbital hoods concave; orbital hoods without tooth.

Telson about 1.5 and 2.5 times as long as anterior and posterior width, respectively; dorsal surface bearing two pairs of spines, approximately inserted 1/3 and 2/3 length of telson, respectively; posterior margin only slightly convex, not beyond posterolateral small spines.

Antennular peduncle slender, with distally acute stylocerite reaching to distal margin of first antennular segment; visible part of first antennular segment distinctly shorter than second segment, but longer than third segment; second segment about twice as long as third segment. Antenna with lateral margin of scaphocerite concave; antennal scale with narrow blade distinctly shorter than distolateral tooth, not overreaching distal margin of antennular peduncle. Carpocerite nearly as long as antennular peduncle.

Major cheliped with chela slightly compressed. Fingers about 1/2 length of palm, dactylus slightly longer than fixed finger, with plunger reduced. Palm with dorsal surface bearing transverse groove sub-distally, which extending to mesial and lateral face as triangular and quadrangular depression, respectively; depression of mesial face continuing to proximal half-length of palm, depression of lateral face continuing to linea impressa; dorsal shoulder subacute, overhanging groove; ventral surface bearing opposite transverse groove, only extending to lateral face as rounded depression that continuing about 2/3 width of palm; ventral shoulder rounded, gently sloping, not overhanging groove. Merus with dorsal margin slightly projecting distally; ventromesial margin terminating in acute tooth. Ischium unarmed. Minor cheliped with chela showing sexual dimorphism: chela of male balaeniceps, fingers slightly shorter than palm, dorsal surface of palm bearing slightly transverse groove sub-distally; chela of female not balaeniceps, fingers tapering, as long as palm, mesial margin and lateral margin of dactylus fringing with setae only on proximal half-length. Merus similar to that of major cheliped. Ischium unarmed.

Second pereiopod with carpus 5-segmented, segmented ratio subequal to 10: 5: 2: 2: 4.

Third and fourth pereiopod similar in shape, stout. Dactylus biunguiculate, unguis on flexor margin acute, smaller. Propodus shorter than merus, ventral margin bearing about 8 and 7 accessory spines and distally one pair. Merus unarmed. Ischium armed with one small spine on ventral margin.

**Remarks** The present specimens are generally agreed with the previous descriptions, except for the dactylus of the major chela rounded distally rather than subacute as shown in Banner & Banner (1982, fig. 84e); the third pereiopod with propodus bearing 7 accessory small spines, while Banner & Banner (1982, fig. 84i) not showing that spines.

**Distribution** The South China Sea; Madagascar; Philippines; Indonesia; Australia.

### 2.2.54 *Alpheus serenei* Tiwari, 1964 (Fig. 2.63)

*Alpheus serenei* Tiwari, 1964: 314; 1963: 310, figs. 27, 28; Banner DM & Banner AH, 1982: 196, fig. 60; Chace, 1988: 51.

**Material Examined** MBM 159152, 1♂, Sanya, Hainan, Li Xin-zheng Coll., 1 December 1997; MBM 159134, 1♀1♂, Xinying, Hainan, 2 January 1992; MBM 159139, 4♀♀4♂♂, Xincun, Hainan, 26 March 1992; MBM 159140, 1♀1♂, Yezhu Island, Hainan, Neumann Coll., 20 March 1992; MBM 159310-3, 2♂♂3♀♀, Xincun, Hainan, 26 March 1992; MBM 158657-9, Yezhu Island, Hainan, Neumann Coll., 19 March 1992; MBM 159306-9, 1♀1♂, Yezhu Island, Hainan, Neumann

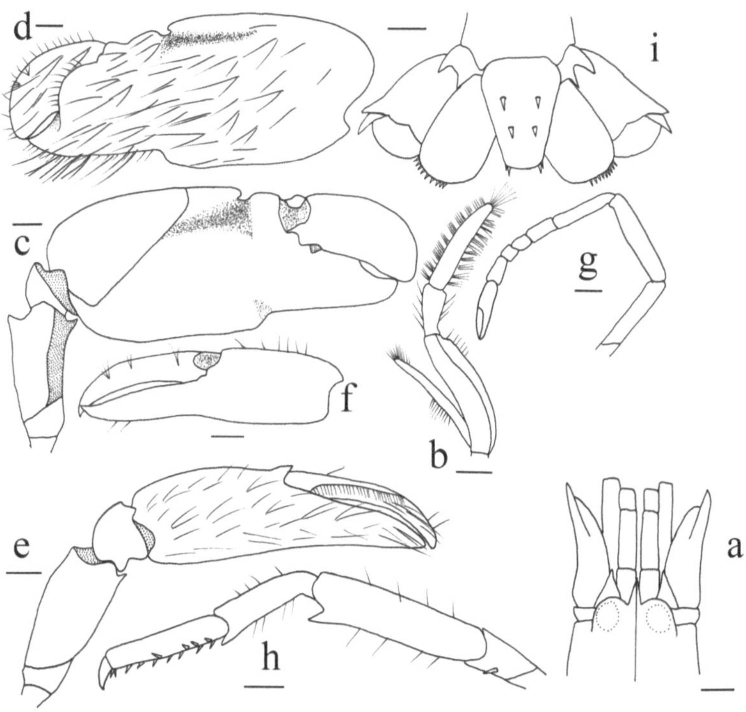


**Fig. 2.63** *Alpheus serenei* Tiwari, 1964. MBM 159404, male, **a**, anterior part of carapace, dorsal view; **b**, third maxilliped; **c**, major cheliped, lateral view; **d**, major chela, lateral view; **e**, minor cheliped, mesial view; **f**, minor chela, lateral view; **g**, second pereiopod; **h**, third pereiopod; **i**, uropods and telson (scale bar = 1 mm)

Coll., 22 March 1992; MBM 159404, 2♂♂1♀, Linchang Reef, Hainan, 11 April 1992; MBM 158420-1, 1♀, Xincun, Hainan, 26 March 1992; MBM 107794-9, 4♀♀4♂♂, Xincun, Hainan, 27 March 1992; MBM 158485, 18♀♂, Xincun, Hainan, 27 March 1992; MBM 107671, 1♀1♂, Xincun, Hainan, 27 March 1992.

**Description** Rostrum acute, longer than wide at base, not reaching to distal margin of first antennular segment; rostral carina strong, extending to middle length of carapace; anterior margin between rostrum and orbital hoods concave; orbital hoods slightly inflated, without tooth.

Telson about 1.3 and 2.2 times as long as anterior and posterior width, respectively; dorsal surface bearing two pairs of spines, approximately inserted 2/5 and 3/5 length of telson, respectively; posterior margin only slightly convex, not beyond posterolateral small spines. Uropod with posterior margin and distal lateral margin of endopod bearing numerous small spines.

Antennular peduncle slender, with distally acute stylocerite hardly reaching to distal margin of first antennular segment; visible part of first antennular segment longer than third segment; second segment longer than twice length of third segment. Antenna with lateral margin of scaphocerite strongly concave; antennal scale with narrow blade distinctly shorter than distolateral tooth, slightly overreaching middle length of third antennular segment. Carpocerite nearly as long as distolateral tooth of antennal scale.

Major cheliped with chela slightly compressed. Fingers slightly shorter than palm, dactylus slightly longer than fixed finger, with plunger well developed. Palm with dorsal surface bearing transverse groove sub-distally, which extending to mesial and lateral face as triangular and quadrangular depression, respectively; depression of mesial face continuing to proximal half-length of palm, depression of lateral face continuing to linea impressa; dorsal shoulder subacute, overhanging groove; ventral surface bearing opposite transverse groove, slightly extending and lateral face as shallow rounded depression; ventral shoulder rounded, gently sloping. Merus with dorsal margin terminating subacute tooth; ventromesial margin beating strong acute tooth distally. Ischium unarmed. Minor cheliped with chela showing sexual dimorphism: dactylus of male with mesial face bearing an oblique crest of setae that almost reaching to dorsal margin, lateral face nearly face almost glabrous; dactylus of female both mesial and lateral face without crest of setae. Fingers and palm subequal in length, palm without grooves. Merus similar to that of major cheliped. Ischium unarmed.

Second pereiopod with carpus 5-segmented, segmented ratio subequal to 10: 5: 2: 2: 4.

Third and fourth pereiopod similar in shape, stout. Dactylus usually obscurely biunguiculate, unguis on flexor margin smaller. Propodus shorter than merus, ventral margin bearing about 16 spines arranging in two rows. Carpus with dorsal margin terminating in rounded tooth; ventral margin bearing acute large tooth distally. Merus with ventral margin terminating in strong acute tooth. Ischium armed with one spine on ventral margin.

**Remarks** The differences between *A. serenei* and its similar species were discussed by Banner & Banner (1982, 197). The present specimens are generally agreed with previous descriptions.

**Distribution** The South China Sea; Red Sea, Eastern Africa; Madagascar; Singapore; Gulf of Thailand; Vietnam; Philippines; Indonesia; Australia.

### 2.2.55 *Alpheus spatulatus* Banner & Banner, 1968 (Fig. 2.64)

*Alpheus spatulatus* Banner AH & Banner DM, 1968a: 146, fig. 3; 1981: 233; Banner DM & Banner AH, 1978: 229; Chace, 1988: 52.

**Material Examined** This species was reported by Banner & Banner (1978) and Liu (2008) from

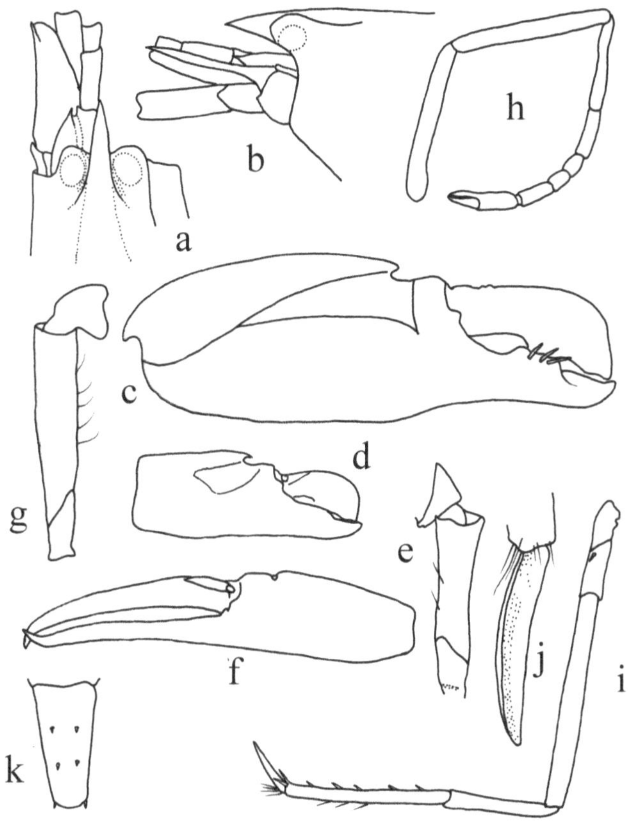


**Fig. 2.64** *Alpheus spatulatus* Banner & Banner, 1968. (Adapted from Banner & Banner, 1968). Female, **a**, anterior part of carapace, dorsal view; **b**, same, lateral view; **c**, major chela, mesial view; **d**, major chela (another specimen); **e**, carpus to ischium of major cheliped; **f**, minor chela; **g**, carpus to ischium of minor cheliped ; **h**, second pereiopod; **i**, third pereiopod; **j**, dactylus of third pereiopod; **k**, telson

the South China Sea, but not found in the present specimens.

**Description** Rostrum acute, longer than wide at base, reaching to 1/4 length of second antennular segment; rostral carina rounded, extending to posterior margin of orbital hoods; anterior margin between rostrum and orbital hoods concave; orbitorostral grooves narrow and deep; orbital hoods slightly inflated, without tooth.

Telson slender, about 2 and 4.3 times as long as anterior and posterior width, respectively; dorsal surface bearing two pairs of moderately small spines, approximately inserted 1/3 and 2/3 length of telson, respectively; posterior margin convex, not beyond posterolateral small spines.

Antennular peduncle slender, with distally acute stylocerite reaching to distal margin of first antennular segment; visible part of first antennular nearly as long as second segment; third segment about 1/3 length of second segment. Antenna with lateral margin of scaphocerite nearly straight; antennal scale with broad blade slightly shorter than distolateral tooth, well beyond distal margin of antennular peduncle. Carpocerite slightly overreaching distolateral tooth of antennal scale.

Major cheliped with chela compressed. Fingers about1/2 length of palm, dactylus slightly shorter than fixed finger, with plunger well developed. Palm with dorsal surface bearing transverse groove sub-distally, which extending to both mesial and lateral face as elongate triangular depression; depression of mesial face continuing to proximal 1/2 length of palm, depression of lateral face continuing to proximal 1/4 length of palm; dorsal shoulder subacute, overhanging groove; ventral surface without opposite transverse groove. Merus with dorsal margin rounded distally; ventromesial margin terminating in small acute tooth. Ischium unarmed. Minor cheliped with fingers slightly longer than palm, tapering. Palm with dorsal margin bearing reduced transverse groove sub-distally; dorsal shoulder rounded, not overhanging groove. Merus slender, ventromesial margin only bearing setae. Ischium unarmed.

Second pereiopod with carpus 5-segmented, segmented ratio subequal to 10: 5: 2: 2: 3.

Third and fourth pereiopod similar in shape, slender. Dactylus spatulate, about 1/3 length of propodus. Propodus shorter than merus, ventral margin bearing about 5 slender spines and distally one pair. Merus about 7 times as long as wide. Ischium armed with one small spine on ventral margin.

**Remarks** *A. spatulatus* is originally described from the South China Sea. Banner & Banner (1978) added the characters of male minor chela.

**Distribution** The South China Sea, Taiwan; (?) Philippines.

## 2.2.56 *Alpheus strenuus strenuus* Dana, 1852 (Fig. 2.65)

*Alpheus strenuus* Dana, 1852b: 543, Pl. 34, fig. 4; Coutière, 1905: 913, fig. 53; Pearson, 1911: 185, Pl. 7, fig. 6; Gravely, 1930: 79, Pl. 1, fig. 6a, b; Banner AH & Banner DM, 1966a: 191, fig. 20; 1966b: 140, fig. 53.

*Alpheus strenuus strenuus*–. Banner DM & Banner AH, 1982: 225, fig. 71; 1985: 31; Chace, 1988: 56.

*Alpheus alope* White, 1847: 74. [nomen nudum]

*Alpheus rhode* White, 1847: 74. [nomen nudum]

*Alpheus doris* White, 1847: 75. [nomen nudum]

*Alpheus strenuus* var. *angulatus* Coutière, 1905: 914.

**Material Examined** CL10.1 mm, TL 27.5 mm. MBM 159195, 1♂, Sanya, Hainan, Li Xin-zheng Coll., 20 November 1997; MBM 158796, 1♀1♂, Sanya, Hainan, Li Xin-zheng Coll., 28 February 1997; MBM 159186, 3♀♀1♂, Sanya Bay, Hainan, Li Xin-zheng Coll., 2 December 1997; MBM 159165, 2♂♂1♀, Sanya, Hainan, Li Xin-zheng Coll., 18 November 1997; MBM 159245, 2♂♂, Sanya, Hainan, Li Xin-zheng Coll., 19 November 1997; MBM 158059, 1♀1♂, Luhuitou, Sanya, Hainan, Li Xin-zheng Coll., 27 November 1997; MBM 159189, 2♂♂, Sanya, Hainan, Li Xin-zheng Coll., 1 December 1997; MBM 159218, 1♀, Sanya, Hainan, Li Xin-zheng Coll., 1 December 1997; MBM 159159, 1♀1♂, Sanya, Hainan, Li Xin-zheng Coll., 3 December 1997; MBM 158911, 1♀1♂, Sanya, Hainan, Li Xin-zheng Coll., 4 March 1997; MBM 158976, 1♀, Jinqing Island, Xisha Islands, 24 April 1978; MBM 158966, 1♀, East Island, Xisha Islands, 10 June 1975; MBM 159029, 1♀, Jinqing Island, Xisha Islands, 19 May 1981; MBM 158253, 6♀♀, Xinying, Hainan, Wang Yong-liang Coll., 25–28 July 1957; MBM 158930, 1♀, Yongxing Island, Xisha Islands, Fan Zhen-gang & Xu Jie-shan Coll., 12 June 1958; MBM 158032, 1♀, Xizhou Island, Sanya, 2 April 1958; MBM 158128, 1♂, Xincun, Hainan, 9 July 1957; MBM 157952, 6 larva, Dengqing Island, Xisha Islands, 11 April 1958; MBM 157924, 1♂, Yongxing Island, Xisha Islands, 31 May 1981; MBM 158244, 5♀♀7♂♂, Xinying, Hainan, 25 July 1957; MBM 158044, 2♀♀, Tree Island, Xisha Islands, 5 May 1958; MBM 157968, 1♂, Yongxing Island, Xisha Islands, Wang Yong-liang Coll., 13 April 1957; MBM 159261, 2♂♂, Sanya, Hainan, Li Xin-zheng Coll., 1 December 1997; MBM 157917, 1♂, Linchang Reef, Hainan, 8 May 1958; MBM 157879, 1♀1♂, Xinying, Hainan, 10 May 1958; MBM 159229, 3♂♂2♀♀, Sanya, Hainan, 1 December 1997; MBM 158176, 1♀, Yongxing Island, Xisha Islands, Fan Zhen-gang & Xu Jie-shan Coll., 20 March 1958; MBM 157963, 1♂, Xincun, Hainan, 18 April 1958;

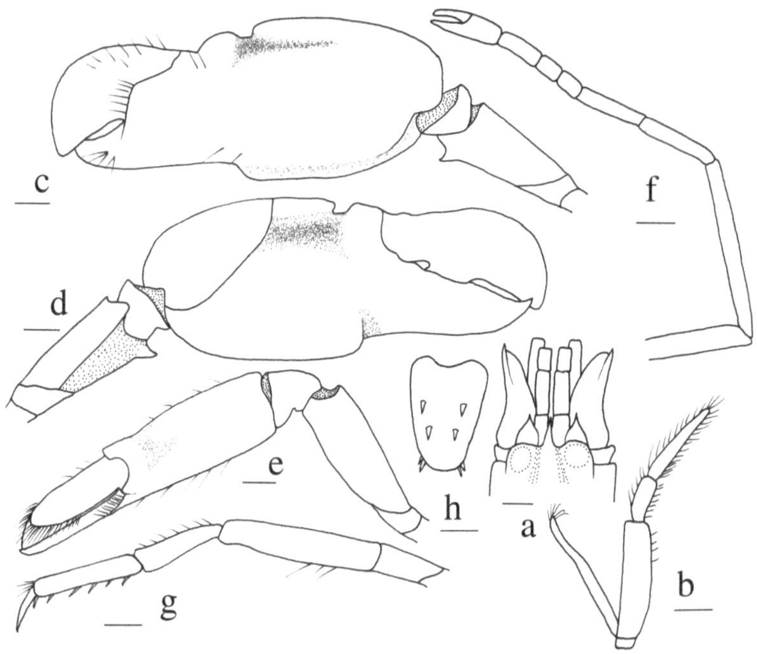


**Fig. 2.65** *Alpheus strenuus strenuus* Dana, 1852. MBM 158307, male, **a**, anterior part of carapace, dorsal view; **b**, third maxilliped; **c**, major cheliped, mesial view; **d**, same, lateral view; **e**, minor cheliped, mesial view; **f**, second pereiopod; **g**, third pereiopod; **h**, telson (scale bar = 1 mm)

MBM 157948, 2♂♂1♀, Yongxing Island, Xisha Islands, 9 May 1980; MBM 157960, 3♂♂, Sanya, Hainan, 21 March 1958; MBM 157957, 8♀♂, Xizhou Island, 3 April 1958; MBM 158148, 1♂, Sanya, Hainan, Wang Yong-liang Coll., 5 April 1955; MBM 157725, 2♂♂5♀♀, East Island, Xisha Islands, 3 June 1980; MBM 157715, 1♀, Yongxing Island, Xisha Islands, 7 May 1980; MBM 157835, 2♂♂, Jinqing Island, Xisha Islands, 19 May 1981; MBM 157779, 2♂♂5♀♀, Yongxing Island, Xisha Islands, 9 May 1980; MBM 158568, 1♂, Linchang Reef, Hainan, 11 April 1992; MBM 157760, 2♂♂, Rongcheng, Shandong, 15 May 1957; MBM 157774, 1♀, Yongxing Island, Xisha Islands, 8 May 1980; MBM 157842, 3♂♂1♀, Xizhou Island, 4 April 1958; MBM 107578, 3♀♀1♂, Linchang Reef, Hainan, 3 December 1990; MBM 158784, 7♂♂23♀♀, Yongxing Island, Xisha Islands, 11 June 1980; MBM 107599, 1♀, Xizhou Island, Wang Yong-liang Coll., 19 March 1992; MBM 158271, 3♀♀1♂, Yongxing Island, Xisha Islands, 5 June 1975; MBM 158270, 1♂2♀♀, Xisha Islands, Ren Xian-qiu Coll., 12 May 1975; MBM 158305, 1♂, Guangjin Island, Xisha Islands, 18 May 1975; MBM 158307, 1♂, Shi Island, Xisha Islands, 9 May 1975; MBM 158275, 2♀♀, Yongxing Island, Xisha Islands, 7 June 1975; MBM 158289, 3♀♀, Sanya, Hainan, 25 April 1975; MBM 158290, 3♀♀1♂, Sanya, Hainan, 30 April 1975; MBM 158297, 2♀♀1♂, East Island, Xisha Islands, 8 June 1975; MBM 158298, 2♀♀, East Island, Xisha Islands, 12 June 1975; MBM 158302, 4♀♀2♂♂, Yongxing Island, Xisha Islands, 5 June 1975; MBM 157576, 1♂, Luhuitou, Sanya, 19 November 1990; MBM 158574, 2♂♂, Yongshu Reef, Hainan, Li Xin-zheng Coll., 18 May 1993; MBM 158576, 1♀1♂, Wufang Reef, Nansha Islands, 7 May 1989; MBM 158704, 4♀♀, Yongxing Island, Xisha Islands, 11 June 1980; MBM 159268, 1 larva, Yongshu Reef, Nansha Islands, 11 April 1999; MBM 157903, 1♀, Sanya, Hainan, 19 April 1975; MBM 233255, 1 larva, Yongxing Island, Xisha Islands, 9 May 1980; MBM 233251, 1♂, Yongxing Island, Xisha Islands, 8 May 1980; MBM 233240, 2♀♀, East Island, Xisha Islands, Ren Xian-qiu Coll., 9 June 1975.

**Description** Rostrum acute, longer than wide at base, not reaching to distal margin of first antennular segment; rostral carina rounded, extending to posterior margin of orbital hoods; anterior margin between rostrum and orbital hoods slightly concave; orbitorostral grooves shallow; orbital hoods slightly inflated, without tooth.

Telson slender, about 1.4 and 4.3 times as long as anterior and posterior width, respectively; dorsal surface bearing two pairs of spines, approximately inserted 1/3 and 2/3 length of telson, respectively; posterior margin convex, not beyond posterolateral spines.

Antennular peduncle slender, with distally acute stylocerite reaching to distal margin of first antennular segment; visible part of first antennular about 2/3 length of second segment; third segment shorter than first segment. Antenna with lateral margin of scaphocerite concave; antennal scale with blade shorter than distolateral tooth, slightly overreaching distal margin of antennular peduncle. Carpocerite overreaching antennular peduncle.

Major cheliped with chela slightly compressed. Fingers nearly as long as palm, dactylus slightly longer than fixed finger, with plunger well developed. Palm with dorsal surface bearing transverse groove sub-distally, which extending to mesial and lateral face as triangular and quadrangular depression, respectively; depression of mesial face continuing to proximal 1/2 length of palm, depression of lateral face continuing to linea impressa; dorsal shoulder subacute, overhanging groove; ventral surface bearing opposite transverse groove, extending to lateral face as triangular depression that continuing about 1/4 width of palm and to mesial face as narrow quadrangular depression that continuing to nearly base of palm; ventral shoulder rounded, not overhanging groove. Merus with dorsal margin rounded distally; ventromesial margin terminating in strong acute tooth. Ischium unarmed. Minor cheliped with chela not showing sexual: both male and female with chela balaeniceps. Fingers slightly shorter than palm, tips narrow and hooked. Palm with dorsal surface bearing transverse groove, that extending to lateral face as poorly defined in large specimens; dorsal shoulder rounded, gently sloping, not overhanging groove; ventral surface no bearing opposite transverse groove. Merus similar to that of major cheliped. Ischium unarmed.

Second pereiopod with carpus 5-segmented, segmented ratio subequal to 10: 10: 3: 3: 6.

Third and fourth pereiopod similar in shape, stout. Dactylus simple, about 2/5 length of propodus. Propodus shorter than merus, ventral margin bearing 6 and 4 accessory spines and distally one pair. Carpus about 1/2 length of merus, dorsal margin terminating in rounded tooth. Merus about 2.3 times as long as wide, unarmed. Ischium armed with one spine on ventral margin.

**Remarks** *Alpheus strenuus strenuus* is readily recognized by both the male and the female with the dactylus of minor chela balaeniceps. Banner & Banner (1966a, b) has discussed the variable characters and the differences between this species and its similar species. The present specimens are generally agreed with previous descriptions.

**Distribution** The Yellow sea, The East China Sea and The South China Sea, Taiwan; Red Sea to Society Islands, not including Hawaii.

### 2.2.57 *Alpheus sudara* Banner & Banner, 1966 (Fig. 2.66)

*Alpheus sudara* Banner AH & Banner DM, 1966b: 153, fig. 59; Banner DM & Banner AH, 1978: 233; 1982: 243, fig. 76.

*Alpheus crassimanus*–. Tiwari, 1963: 307, figs. 25, 26 (part).

**Material Examined** This species was reported by Banner & Banner (1978) and Liu (2008) from the South China Sea, but not found in the present specimens.

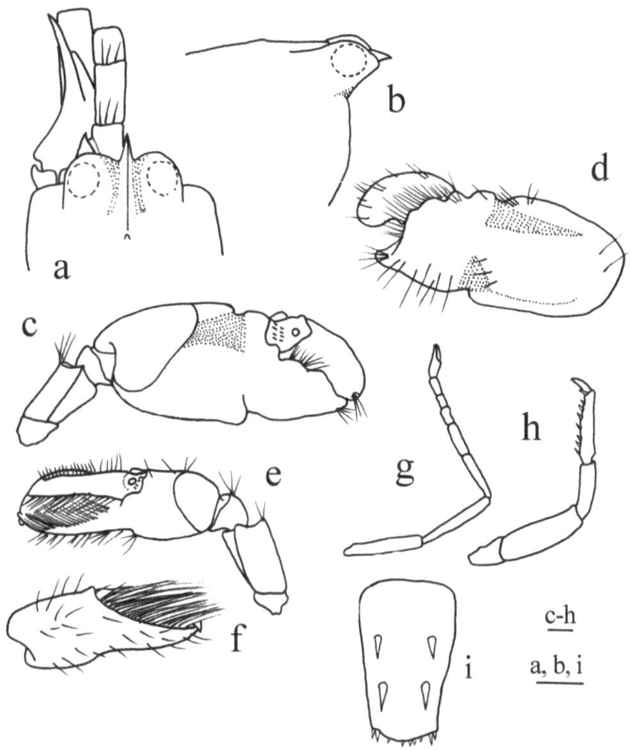


**Fig. 2.66** *Alpheus sudara* Banner & Banner, 1966. (Adapted from Banner & Banner, 1982). Male, **a**, anterior part of carapace, dorsal view; **b**, same, lateral view; **c**, major cheliped, lateral view; **d**, major chela, mesial view; **e**, minor cheliped, lateral view; **f**, minor chela, mesial view; **g**, second pereiopod; **h**, third pereiopod; **i**, telson (scale bar = 1 mm)

**Description** Rostrum triangular, not reaching to distal margin of first antennular segment; rostral carina sharp, extending to posterior margin of orbital hoods, and terminating by slightly protrusion in male (lacking in female); anterior margin between rostrum and orbital hoods slightly concave; orbitorostral grooves moderately deep; orbital hoods slightly inflated, without tooth.

Telson about 2.2 times as long as posterior width; dorsal surface bearing two pairs of moderately large spines, approximately inserted 1/3 and 2/3 length of telson, respectively; posterior margin convex, not beyond posterolateral spines; posterior margin between posterolateral spines bearing several spinules.

Antennular peduncle moderately stout, with distally acute stylocerite distinctly not reaching to distal margin of first antennular segment; visible part of first antennular segment nearly as long as third segment; second segment about twice as long as third segment. Antenna with lateral margin of scaphocerite strongly concave; antennal scale with reduced narrow blade distinctly shorter than distolateral tooth, hardly reaching to 1/2 length of third antennular segment. Carpocerite nearly as long as distolateral tooth of antennal scale, far beyond antennal peduncle.

Major cheliped with chela slightly compressed. Fingers about 1/2 length of palm, dactylus rounded distally, slightly longer than fixed finger, with plunger well developed; mesial margin of fixed finger fringed with long setae. Palm with dorsal surface bearing transverse groove sub-distally, which extending to mesial and lateral face as elongate triangular and quadrangular depression, respectively; depression of mesial face continuing to proximal 1/4 length of palm, depression of lateral face continuing to linea impressa; dorsal shoulder rounded, sloping, not overhanging groove; ventral surface bearing opposite transverse groove, extending to mesial face as triangular depression that continuing to 1/4 width of palm, and to lateral face as narrow U-shaped depression; ventral shoulder rounded, gently sloping, not overhanging groove. Merus with margins rounded distally, without tooth. Ischium unarmed. Minor cheliped with chela showing sexual dimorphism: chela of male with fingers 1.6 times as long as palm, balaeniceps, medial side of dactylus bearing moderate tooth, fixed finger fringed with rows of dense setae, that longer than setae on dactylus, palm with dorsal surface slightly flattened, without transverse groove sub-distally; chela of female not balaeniceps, more slender, only bearing sparse setae. Merus similar to that of major cheliped, but more slender. Ischium unarmed.

Second pereiopod with carpus 5-segmented, segmented ratio subequal to 10: 7: 3: 3: 4.

Third and fourth pereiopod similar in shape, stout. Dactylus simple, about 1/3 length of propodus. Propodus shorter than merus, ventral margin bearing 7 spines and distally one pair. Carpus with both dorsal and ventral margin rounded, not projecting. Merus 3 times as long as wide, unarmed. Ischium without spine on ventral margin.

**Remarks** *A. sudara* is originally described from Thailand, and Banner & Banner (1978) reported this species from Hong Kong.

**Distribution** The South China Sea (Hong Kong); Thailand; probably Vietnam; Australian.

### 2.2.58 *Alpheus tirmiziae* Kazmi, 1974 (Fig. 2.67)

*Alpheus tirmiziae* Kazmi, 1974: 170, fig. 1–4; Banner DM & Banner AH, 1978: 233.

**Material Examined** This species was reported by Banner & Banner (1978) and Liu (2008) from the South China Sea, but not found in the present specimens.

**Description** Rostrum very small, only slightly overreaching anterior margin of orbital hoods; rostral carina low, extending to posterior margin of orbital hoods; orbital hoods beyond anterior margin of carapace, without tooth, and anterior margin of it strongly convex.

Telson about 1.3 times as long as anterior width; dorsal surface bearing two pairs of moderately small spines, approximately inserted 1/3 and 2/3 length of telson, respectively; posterior margin strongly convex, rounded, bearing small posterolateral spines.

Antennular peduncle slender, with distally acute stylocerite not reaching to distal margin of first antennular segment; visible part of first antennular segment longer than third segment, inner margins bearing one small curved spine; second segment about twice as long as third segment. Antenna with lateral margin of scaphocerite nearly straight; antennal scale with broad blade beyond distolateral tooth.

Major cheliped with chela slightly compressed, surfaces covered with minimum acute granules. Fingers longer than 1/2 length of palm, dactylus and fixed finger subequal in length. Palm with dorsal surface bearing transverse groove sub-distally, which extending to both mesial and lateral face as depression; dorsal shoulder subacute, overhanging groove; ventral surface bearing opposite transverse groove, slightly extending to faces as depression; ventral shoulder rounded, sloping, not overhanging groove. Minor chela with fingers about twice as long as palm, lateral face of palm bearing few granules.

Second pereiopod with carpus 5-segmented, segmented ratio subequal to 9: 6: 2: 2: 4.

Third and fourth pereiopod similar in shape, stout. Dactylus simple. Propodus shorter than merus, ventral margin bearing 6 spines and distally one pair. Merus unarmed. Ischium with one spine on ventral margin.

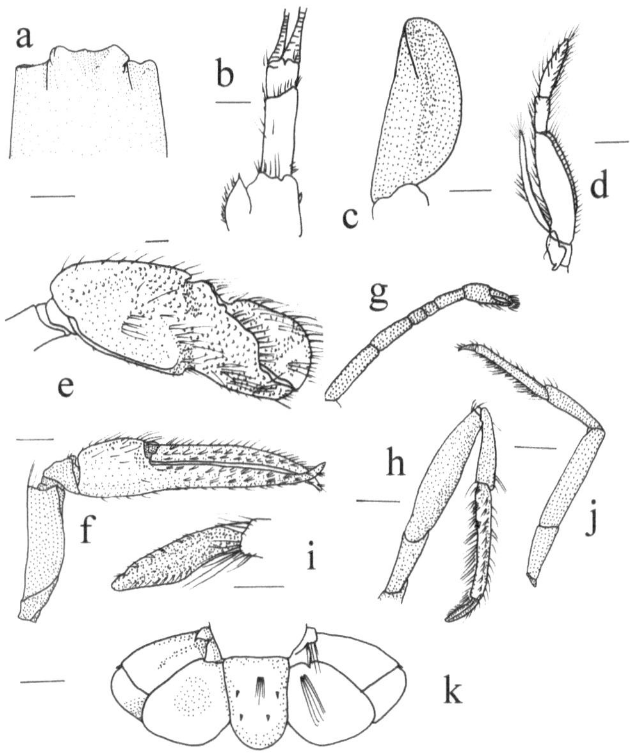


**Fig. 2.67** *Alpheus tirmiziae* Kazmi, 1974. (Adapted from Kazmi, 1974). Female, **a**, anterior part of carapace, dorsal view; **b**, antennular peduncle; **c**, antennal scale; **d**, third maxilliped; **e**, major chela; **f**, minor cheliped; **g**, chela and carpus of second pereiopod; **h**, third pereiopod; **i**, dactylus of third pereiopod; **j**, fourth pereiopod; **k**, uropods and telson (scale bar = 1 mm)

**Remarks** Banner & Banner (1978) checked 2 specimens collected from Hong Kong, and listed them under the name *A. tirmiziae*. The differences between the specimens collected from Hong Kong and the type specimens were discussed by Banner & Banner (1978).

**Distribution** The South China Sea (Hong Kong); the Bay of Bengal.

**Macrocheles Group (Fig. 2.68)**

Orbital hoods always with tooth. Major chela with dactylus usually high and thin; palm compressed, somewhat twisted, with 3 longitudinal ridges and grooves, which sometimes interrupted, terminating distally in strong tooth overhanging dactylus, the adhesive plaque, and in a strong tooth below dactylus. Third pereiopod with dactylus simple or biunguiculate; merus with or without tooth on ventral margin (Banner & Banner, 1982).

Key to species of Macrocheles Group from China seas

1. Major chela with crest on dorsal margin of palm not interrupted by transverse groove ........................*A. astrinx*
-. Major chela with crest on dorsal margin of palm interrupted by transverse groove ........................................2
2. Mesial faces of major and minor chela bearing dense long setae................................................*A. deuteropus*
-. Mesial faces of major and minor chela only bearing thin short setae ........................................ *A. collumianus*

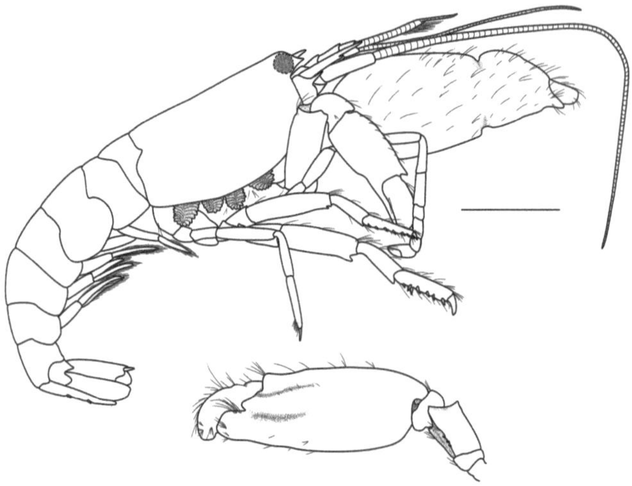

**Fig. 2.68** *Alpheus collumianus* Stimpson, 1860. entire animal of Macrocheles Group (scale bar = 2 mm)

### 2.2.59 *Alpheus astrinx* Banner & Banner, 1982 (Fig. 2.69)

*Alpheus astrinx* Banner DM & Banner AH, 1982: 35, fig. 5; Banner AH & Banner DM, 1984: 37.

**Material Examined** CL 4.9–5.3 mm, TL 12.5–14.8 mm. MBM 158534-7, 2♀♀1♂, Yezhu Island, Sanya Bay, Neumann Coll., 26 March 1992.

**Description** Rostrum acute, slightly longer than wide at base, hardly reaching to distal margin of first antennular segment; margins between rostrum and orbital hoods concave; rostral carina absent; orbital teeth acute, about reaching to 1/3 length of rostrum.

Telson slender, about 2 and 2.6 times as long as anterior and posterior width, respectively; dorsal surface bearing two pairs of spines, approximately inserted 1/3 and 2/3 length of telson, respectively; posterior margin slightly convex, not beyond posterolateral spines.

Antennular peduncle stout, with distally acute stylocerite reaching to distal margin of first antennular segment; visible part of first antennular segment slightly shorter than second segment; third segment about 3/5 length of second segment. Antenna with lateral margin of scaphocerite slightly concave; antennal scale with slightly reduced narrow blade shorter than distolateral tooth, only reaching to middle length of third antennular segment. Carpocerite slightly longer than distolateral tooth of antennal scale, nearly as long as antennular peduncle.

Major cheliped with chela compressed, somewhat twist. Fingers 1/3 length of palm, dactylus bulbous, with plunger slightly reduced. Palm with dorsal surface curving laterally to form rounded ridge, distal portion of ridge demarked superiorly by flattened area, and inferiorly by rounded groove; superior flattened area with distal ridge on medial margin, terminating in strong acute tooth, that flanking dactylus; distal portion of lateral face with third longitudinal rounded ridge, that terminating in rounded tooth flanking dactylus, below ridge a rounded but deep depression present. Merus with dorsal margin slightly projecting; ventromesial margin bearing 7 small spines and terminating in an acute tooth. Ischium with ventral margin bearing 2 small spines; dorsal margin bearing 1 spines distally. Minor cheliped with chela of male not showing sexual dimorphism, fingers nearly as long as palm; cutting edges bearing short stiff setae. Palm with distal margin of mesial and lateral face both bearing acute tooth medially that flanking dactylus.

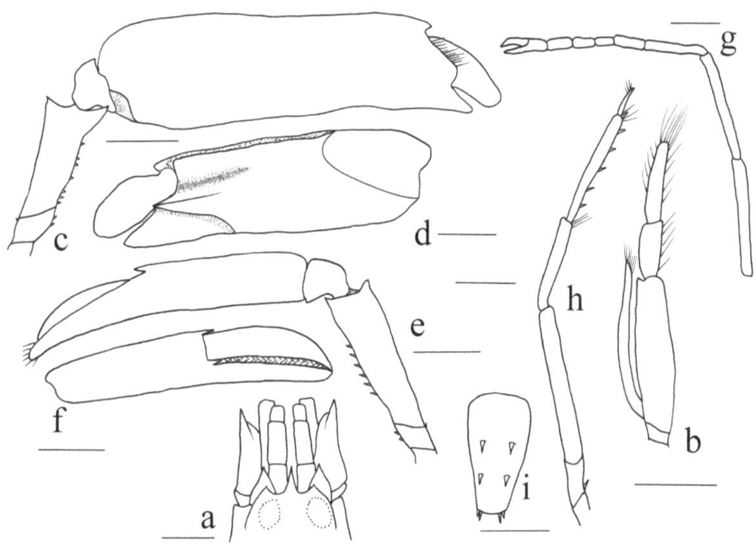


**Fig. 2.69** *Alpheus astrinx* Banner & Banner, 1982. MBM 158534, female, **a**, anterior part of carapace, dorsal view; **b**, third maxilliped; **c**, major cheliped, mesial view; **d**, major chela, lateral view; **e**, minor cheliped, mesial view; **f**, minor chela, lateral view; **g**, second pereiopod; **h**, third pereiopod; **i**, telson (scale bar = 1 mm)

Merus and ischium similar to that of major cheliped.

Second pereiopod with carpus 5-segmented, segmented ratio subequal to 10: 5: 4: 4: 5.

Third and fourth pereiopod similar in shape, slender. Dactylus simple, dorsal margin bearing setae sub-distally. Propodus shorter than merus, ventral margin bearing 9–11 spines and distally one pair. Merus about 7 times as long as wide, unarmed. Ischium with one spine on ventral margin.

**Remarks** *A. astrinx* is a newly recorded species from the China seas, and it is originally described from Australia. The type specimens are only female, so that the characters of the male are missed in the original description. One male specimen is checked herein, and the character of the male is added. The minor chela not shows sexual dimorphism: neither the male nor the female has a balaeniceps dactylus. Because of only one male specimen is checked herein, so the character of the male is needed to be confirmed based more male specimens in the future. Moreover, the second pleopod of the male with the endopod has the appendix interna slightly shorter than the appendix masculine, and the appendix masculine has several long setae on the tip. The present specimens are generally agreed with the original description.

**Distribution** The South China Sea; Australia.

### 2.2.60 *Alpheus collumianus* Stimpson, 1860 (Fig. 2.70)

*Alpheus collumianus* Stimpson, 1860: 30; De Man, 1911: 334, fig. 65; Banner, 1953: 67, fig. 21; Banner DM & Banner AH, 1982: 45, fig. 9; Chace, 1988: 19.

*Alpheus collumianus probabilis* Banner 1956: 338, fig. 10.

*Alpheus collumianus medius* Banner, 1956: 340, fig. 11.

*Alpheus collumianus inermis* Banner, 1956: 342, fig. 12.

*Alpheus malhaensis* Coutière, 1908: 205; 1921: 419, Pl. 62, fig. 14.

*Alpheus seurati* –.Coutière, 1921: 419.

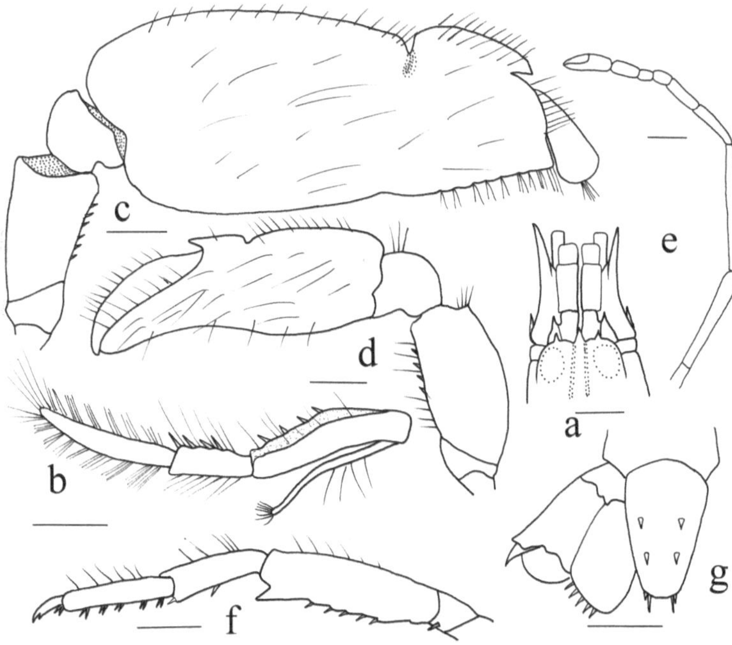


**Fig. 2.70** *Alpheus collumianus* Stimpson, 1860. MBM 158578, male, **a**, anterior part of carapace, dorsal view; **b**, third maxilliped; **c**, major cheliped, mesial view; **d**, minor cheliped, mesial view; **e**, second pereiopod; **f**, third pereiopod; **g**, left uropod and telson (scale bar = 1 mm)

**Material Examined** CL 3.7 mm, TL 9.5 mm. MBM 159142, 1♀, Zhaoshu Island, Xisha Islands, 11 May 1980; MBM 159145, 1♂, East Island, Xisha Islands, 28 May 1980; MBM 159143, 1♀, Sanya, Hainan, 12 January 1992; MBM 159147, 1♀1♂, Shalao, Hainan, Wang Yong-liang Coll., 4 April 1992; MBM 158990, 1♂1♀, Yongxing Island, Xisha Islands, Chen Mu Coll., 31 May 1981; MBM 158084, 1 larva, Yongxing Island, Xisha Islands, Fan Zhen-gang & Xu Jie-shan Coll., 12 June 1958; MBM 158675, 1♂, Yongshu Reef, Nansha Islands, Li Jin-he Coll., 16 May 1993; MBM 158676, 1♂, Yongshu Reef, Nansha Islands, 29 May 1993; MBM 158456, 1♂, Zhubi Reef, Nansha Islands, 19 May 1993; MBM 159306-9, 1♂, Yezhu Island, Sanya Bay, 22 March 1992; MBM 158578, 2♂♂2♀♀, Zhubi Reef, Nansha Islands, 30 May 1993; MBM 158726, 1♂, Jinqing Island, Xisha Islands, Fan Zhen-gang & Xu Jie-shan Coll., 25 April 1958; MBM 158628, 1♂1♀, Yongshu Reef, Nansha Islands, Li Jin-he Coll., 17 May 1993; 04II86AD-15, 1 larva, Zhubi Reef, Nansha Islands, 14 May 2004; 04II54AD-22, 6♂♀, Zhubi Reef, Nansha Islands, 6 May 2004; 04II51AD-19, 1♂1♀, Zhubi Reef, Nansha Islands,4 May 2004; 04II73AD-9, 4♂♀, Zhubi Reef, Nansha Islands, 12 May 2004; 04II82AD-12, 4♂♀, Zhubi Reef, Nansha Islands, 14 May 2004; 04II59AD-31, 1 larva, Zhubi Reef, Nansha Islands, 8 May 2004; 04II83AD-27, 2♂♀, Zhubi Reef, Nansha Islands, 4 May 2004; 04II50AD-6, 2 larva, Zhubi Reef, Nansha Islands, 4 May 2004; 04II62AD-26, 9♂♀, Zhubi Reef, Nansha Islands, 8 May 2004; 04II69AD-32, 3 larva, Zhubi Reef, Nansha Islands, 11 May 2004; 04II65AD-19, 2♀♀1♂, Zhubi Reef, Nansha Islands, 10 May 2004; 04II61AD-12, 1♂1♀, Zhubi Reef, Nansha Islands, 2004; 04II64AD-19, 6♂♀, Zhubi Reef, Nansha Islands, 10 May 2004; 04II68AD-13, 1♂1♀, Zhubi Reef, Nansha Islands, 11 May 2004; MBM 233247, 1♀, Jinyin Island, Xisha Islands, 26 May 1975; MBM 233248, 1 larva, Jinyin Island, Xisha Islands, 26 May 1975; MBM 107679, 14♂♂♂7♀♀, Yongxing Island, Xisha Islands, 11 June 1980; MBM 107670, 1♀, Jinyin Island, Xisha Islands, 26 May 1975; MBM 107667, 1♀, Jinyin Island, Xisha Islands, 15 May

1980; MBM 107669, 8♂♂8♀♀, East Island, Xisha Islands, 28 May 1980; MBM 158852, 2♂♂1♀, Daai Reef, Nansha Islands, 24 May 1990; MBM 158583, 1 larva, Zhubi Reef, Nansha Islands, 23 May 1990; MBM 158885, 1♂, Dengqing Island, Xisha Islands, 11 April 1958; MBM 107668, 4 larva, Jinyin Island, Xisha Islands, 14 May 1980; MBM 158309, 1♂1♀, Jinyin Island, Xisha Islands, 29 May 1975; MBM 158301, 5♀♀3♂♂, East Island, Xisha Islands, 9 June 1975; MBM 159146, 1♀, Jinyin Island, Xisha Islands, 15 May 1980; MBM 157992, 1♂, Yongxing Island, Xisha Islands, Fan Zhen-gang & Xu Jie-shan Coll., 20 May 1958; MBM 15914, 1♀, Yongxing Island, Xisha Islands, 11 June 1980; MBM 158499, 1 larva, Meiji Reef, Nansha Islands, 27 May 1993; MBM 158412, 2♂♂2♀♀, Banyue Reef, Nansha Islands, 28 September 1994; MBM 158383, 2♂♂1♀, Yongshu Reef, Nansha Islands, 16 May 1993; MBM 158728, 1♀, Jinqing Island, Xisha Islands, 25 April 1958; MBM 158570, 6♀♀9♂♂, Nanxun Reef, Nansha Islands, Li Xin-zheng Coll., 21 May 1993; MBM 233249, 6♂♂3♀♀, Jinyin Island, Xisha Islands, 26 May 1975.

**Description** Rostrum acute, slightly longer than wide at base, about reaching to middle length of first antennular segment; rostral carina strong, rounded, extending slightly beyond posterior margin of orbital hoods; orbitorostral groove deep; orbital teeth acute, slightly shorter than or nearly as long as rostrum, or absent.

Telson slender, about 1.6–2 and 2.5–3 times as long as anterior and posterior width, respectively; dorsal surface bearing two pairs of spines, approximately inserted 1/3 and 2/3 length of telson, respectively; posterior margin slightly convex, not beyond posterolateral long spines. Uropod with lateral distal margin of endopod bearing several spines.

Antennular peduncle slender, with distally acute stylocerite not reaching to distal margin of first antennular segment; visible part of first antennular segment slightly longer than third segment; second segment about 2–3.5 times as long as wide, about 3 times as long as third segment. Antenna with lateral margin of scaphocerite strongly concave; antennal scale with reduced narrow blade far shorter than distolateral tooth, variable from 2/3 length of second antennular segment to slightly overreaching distal margin of that. Carpocerite slightly beyond antennular peduncle

Major cheliped with chela slightly compressed, somewhat twist. Fingers about 1/3 length of palm, dactylus twist laterally, truncate, tip overhanging fixed finger. Palm with dorsal margin bearing deep transverse groove sub-distally; superior ridge strong, arising from proximal of transverse groove, continuing to 2/5 width of lateral face, and terminating in acute tooth that overhanging dorsal margin of dactylus; plaque rounded ridge merging with dorsal margin of palm, demarking from superior and lateral ridge by deep and rounded grooves; lateral ridge flattened, terminating in strong lateral tooth flanking dactylus. Merus with ventromesial margin bearing 3–4 spines and strong acute or rounded tooth distally. Ischium unarmed. Minor cheliped with fingers and palm subequal in length. Palm with dorsal margin bearing slight transverse groove sub-distally; dorsal margin terminating in strong acute tooth that overhanging dactylus; distal margin of lateral face bearing acute tooth medially that flanking dactylus. Merus similar to that of major cheliped, but more slender.

Second pereiopod with carpus 5-segmented, segmented ratio subequal to 7: 5–6: 2: 2: 3.

Third and fourth pereiopod similar in shape, moderately stout. Dactylus biunguiculate, inferior unguis variable from slight angle to strong acute unguis, about 1/3 length of propodus. Propodus shorter than merus, ventral margin bearing 5 pairs of spines and distally one pair. Carpus about half-length of merus, ventral margin bearing 0–4 spines. Merus with ventral margin bearing 0–5 spines and strong acute or rounded tooth distally. Ischium with one spine on ventral margin.

**Remarks** *A. collumianus* is an extremely variable species in Macrocheles group. The present specimens are generally agreed with the previous descriptions.

**Distribution** The South China Sea; Red Sea; Madagascar; South Africa to Japan, Australia and Pacific islands to Hawaii and Societies Islands.

### 2.2.61 *Alpheus deuteropus* Hilgendorf, 1879 (Fig. 2.71)

*Alpheus deuteropus* Hilgendorf, 1879: 834, Pl. 4, figs. 8–10; Coutière, 1899: 215, figs. 254, 255; Banner AH & Banner DM, 1966b: 80, fig. 26; Tiwari, 1963: 281, fig. 6; Banner DM & Banner AH, 1982: 42, fig. 8; 1985: 15; Chace, 1988: 22.

**Material Examined** MBM 158858, 1♂, CL 7.1 mm, TL 21.2 mm, Banyue Reef, Nansha Island, 29 September1994.

**Description** Rostrum acute, tip curved upward, about reaching to middle length of first antennular segment; rostral carina rounded, extending to posterior margin of orbital hoods; orbitorostral groove moderately deep; orbital hoods inflated, with acute teeth nearly as long as or slightly shorter than rostrum.

Telson slightly longer than anterior width and about twice as long as posterior width; dorsal surface bearing two pairs of small spines, approximately inserted 1/2 and 3/4 length of telson, respectively; posterior margin nearly straight, distolateral angles bearing two pairs of moderately long spines. Uropod with distal margin of endopod bearing about 5 slender spines laterally.

Antennular peduncle slender, with distally acute stylocerite not reaching to distal margin of first antennular segment; visible part of first antennular segment slightly longer than third segment; second segment about 3 times as long as wide, twice as long as third segment. Antenna with lateral margin of scaphocerite strongly concave; antennal scale with reduced, narrow blade

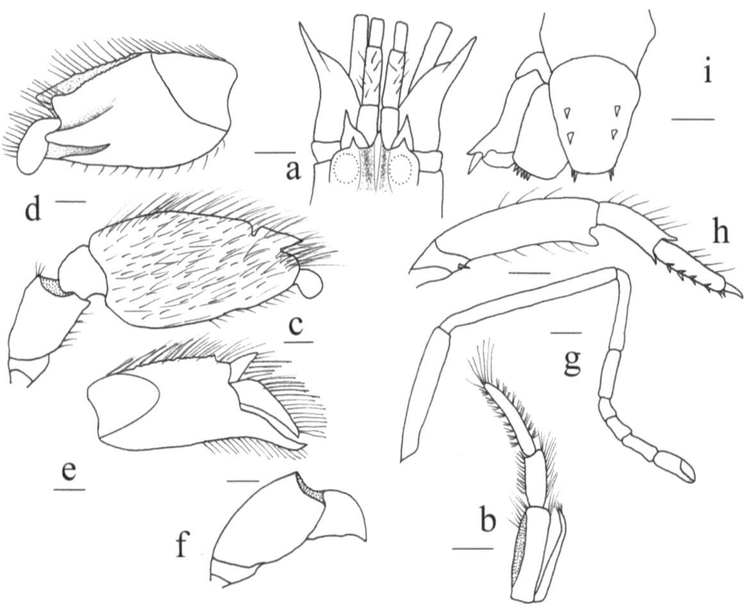


**Fig. 2.71** *Alpheus deuteropus* Hilgendorf, 1879. MBM 158858, male, **a**, anterior part of carapace, dorsal view; **b**, third maxilliped; **c**, major cheliped, mesial view; **d**, major chela, lateral view; **e**, minor chela, lateral view; **f**, carpus to ischium of minor cheliped, lateral view; **g**, second pereiopod; **h**, third pereiopod; **i**, uropods and telson (scale bar = 1 mm)

far shorter than distolateral tooth, not reaching to distal margin of second antennular segment. Basicerite bearing strong lateral tooth, reaching to distal margin of first antennular peduncle. Carpocerite nearly as long as antennular peduncle, slightly longer than distolateral tooth of antennal scale.

Major cheliped with chela somewhat twist. Fingers about 1/5 length of palm, dactylus with bulbous tip overhanging fixed finger, plunger poorly developed. Palm with distal margin of mesial and lateral face bearing strong acute tooth medially, overhanging dactylus; dorsal margin bearing transverse groove sub-distally; superior ridge rounded demarking by grooves; plaque and lateral ridge rounded, demarking by grooves; mesial and dorsal face bearing dense setae. Merus with dorsal margin bearing triangular projection distally; ventromesial margin terminating in small acute tooth. Ischium unarmed. Minor cheliped with chela compressed. Fingers longer than 1/2 length of palm. Palm with distal margin of mesial face bearing exceeding large acute tooth, flanking dactylus, and distal margin of lateral face bearing small acute tooth medially; dorsal surface bearing transverse groove sub-distally, with proximal shoulder slightly overhanging groove. Merus similar to that of large cheliped. Ischium unarmed.

Second pereiopod with carpus 5-segmented, segmented ratio subequal to 10: 6: 3: 3: 5.

Third and fourth pereiopod similar in shape, moderately stout. Dactylus simple. Propodus shorter than merus, ventral margin bearing 4 pairs of spines and distally one pair. Carpus with both margins terminating in acute tooth. Merus with ventral margin terminating in strong large acute tooth. Ischium with one spine on ventral margin.

**Remarks** The variations of *A. deuteropus* had been discussed by Banner (1953). The present one male specimen is agreed well with the previous descriptions.

**Distribution** The South China Sea; Red Sea; Eastern and South Africa through the Indian Ocean to Japan; Philippines; Australia, and eastward across the Pacific to Hawaii.

### Obesomanus Group (Fig. 2.72)

Rostrum reduced or absent. Antennular peduncle usually elongate, with reduced or absent stylocerite. Major cheliped with chela rounded proximally, tapering; dactylus single- or double-headed hammer-shaped; palm bearing slight to moderate deep longitudinal grooves. Minor cheliped with chela never balaeniceps. Second pereiopod at times exceedingly elongate, and asymmetrical development. Uropod with exopod may lacking diaeresis. Telson variable. Sometimes showing strong sexual dimorphism.

Key to species of Obesomanus Group from China seas

1. Rostrum absent ........ *A. perplexus*
-. Rostrum present ........ 2
2. Second pereiopod with first carpal segment sub-equal to second segment ........ 3
-. Second pereiopod with first carpal segment much shorter than second segment ........ 4
3. Major chela with dorsal margin of palm bearing shallow transverse depression sub-distally ........ *A. malleodigitus*
-. Major chela with dorsal margin of palm not bearing transverse depression sub-distally ........ *A. idiocheles*
4. Antennal scale with narrow blade reaching to middle length of third antennular segment ........ *A. microstylus*
-. Antennal scale with blade reduced, reaching to middle length of second antennular segment ........ *A. obesomanus*

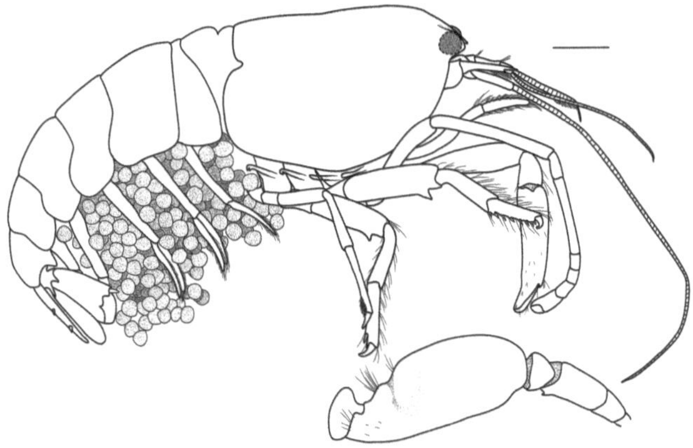

**Fig. 2.72** *Alpheus microstylus* (Bate, 1888). entire animal of Obesomanus Group (scale bar = 2 mm)

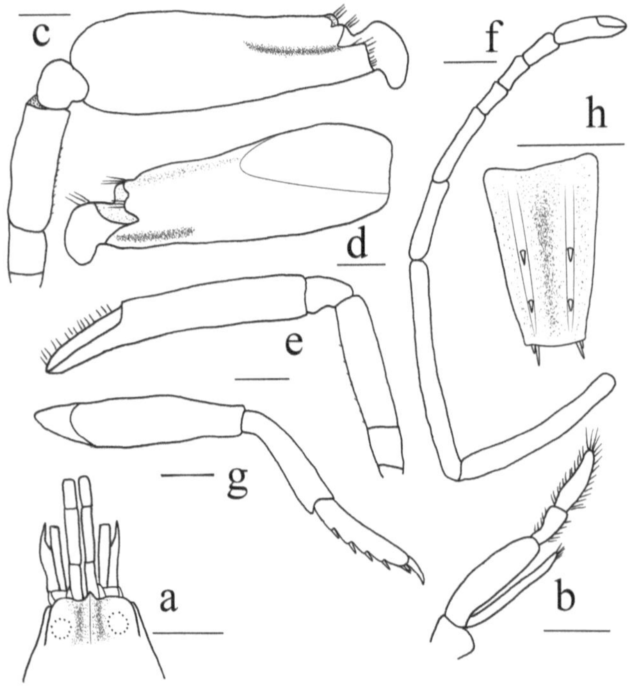


**Fig. 2.73** *Alpheus idiocheles* Coutière, 1905. 04II57AD-16, **a**, anterior part of carapace, dorsal view; **b**, third maxilliped; **c**, major cheliped, mesial view; **d**, major chela, lateral view; **e**, minor cheliped, mesial view; **f**, second pereiopod; **g**, third pereiopod; **h**, telson (scale bar = 1 mm)

## 2.2.62 *Alpheus idiocheles* Coutière, 1905 (Fig. 2.73)

*Alpheus idiocheles* Coutière, 1905: 883, Pl. 75, fig. 21; Pearson, 1905: 84; Banner AH & Banner DM, 1967: 271.

*Alpheus baculifer* Coutière, 1908: 16; 1921: 422, Pl. 62, fig. 22.

*Thunor idiocheles* –. Banner, 1956: 367, fig. 22.

**Material Examined** CL 5.2 mm, TL 10.3 mm. MBM 159969, 1♂, East Island, Xisha Islands, 28 May 1980; MBM 1599701♂, Yongxing Island, Xisha Islands, 11 June 1980; MBM 158013, 1♂, Shi Island, Xisha Islands, Fan Zhen-gang & Xu

Jie-shan Coll., 6 April 1958; 04II83AD-27, 1♂, Zhubi Reef, Nansha Islands, 4 May 2004; 04II57AD-16, 3♂, Zhubi Reef, Nansha Islands, 6 May 2004; 04II53AD-21, 1♂, Zhubi Reef, Nansha Islands, 4 May 2004.

**Description** Rostrum small, not reaching to distal margin of first antennular segment; rostral carina sharp, extending slightly beyond posterior margin of orbital hoods; orbitorostral groove moderately shallow; orbital hoods inflated, without teeth.

Telson slender about 2 and 2.2 times as long as anterior and posterior width, respectively; dorsal surface bearing two pairs of spines, approximately inserted 3/8 and 5/8 length of telson, respectively; posterior margin nearly straight, distolateral angles bearing two pairs of moderately long spines.

Antennular peduncle slender, with distally rounded stylocerite reaching about 1/3 length of first antennular segment; visible part of first antennular segment slightly longer than third segment; second segment about twice as long as third segment. Antenna with lateral margin of scaphocerite concave; antennal scale with blade distinctly shorter than lateral tooth, reaching middle length of second antennular segment. Carpocerite slightly shorter than distolateral tooth of antennal scale.

Major cheliped with chela elongate. Fingers about 1/4 length of palm, dactylus hammer-shaped, distinctly longer than fixed finger. Palm with distal 1/3 length of mesial face bearing longitudinal depression; longitudinal depression at dorsal margin of lateral face extending to linea impressa; lateral face of fixed finger bearing moderately deep depression continuing to distal 1/4 length of palm. Carpus cup-shaped. Merus with dorsal and ventromesial margin rounded distally; ventromesial margin bearing several spinules. Ischium unarmed. Minor chela elongate, with fingers 1/2 length of palm. Palm smooth. Carpus cup-shaped. Merus similar to that of major cheliped. Ischium unarmed.

Second pereiopod asymmetrically. Carpus 5-segmented, proportions of segments not showing apparent difference, segmented ratio subequal to 10: 12: 4: 4: 5.

Third and fourth pereiopod similar in shape, moderately stout. Dactylus simple. Propodus slightly shorter than carpus, ventral margin bearing 4 spines and distally one pair. Carpus distinctly shorter than merus. Merus and Ischium unarmed.

**Remarks** *A. idiocheles* is originally described only based on one male specimen, and then reported by Pearson (1905), Banner (1956) and Banner & Banner (1967). The variable characters had been discussed by Banner (1956), and Banner & Banner (1967) had discussed the sexual dimorphism of this species. The present specimens show similar variability.

**Distribution** The Shout China Sea; Chagos Islands; Saipan Harbor; Society Islands.

### 2.2.63 *Alpheus malleodigitus* (Spence Bate, 1888) (Fig. 2.74)

*Betaeus malleodigitus* Spence Bate, 1888: 565, Pl. 101, fig. 5.

*Alpheus malleodigitus*–. De Man, 1902: 866; Banner AH & Banner DM, 1966a: 174, figs. 8c, d; 1966b: 103, fig. 35i, j; Miya, 1974: 130, Pl. 22, fig. f, g; Liu & Lan, 1980: 90, fig. 6: 10–14, 16; Banner DM & Banner AH, 1982: 92, fig. 22o–m; Chace, 1988: 40.

*Alpheus malleodigitus* var. *gracilicarpus* De Man, 1911: 345, fig. 68; Banner AH & Banner DM, 1966b: 171.

*Alpheus obesomanus* var. *japonicus* Ortmann, 1890: 487.

*Alpheus persicus* Nobili, 1906a: 33.

*Alpheus phrygianus* Coutière, 1905: 886, fig. 25.

**Material Examined** CL 5.1–5.5 mm, TL 11.7–14.9 mm. MBM 158758, 1 larva, Ganquan Island, Xisha Islands, 26 April 1983; MBM 107607, 1♂,

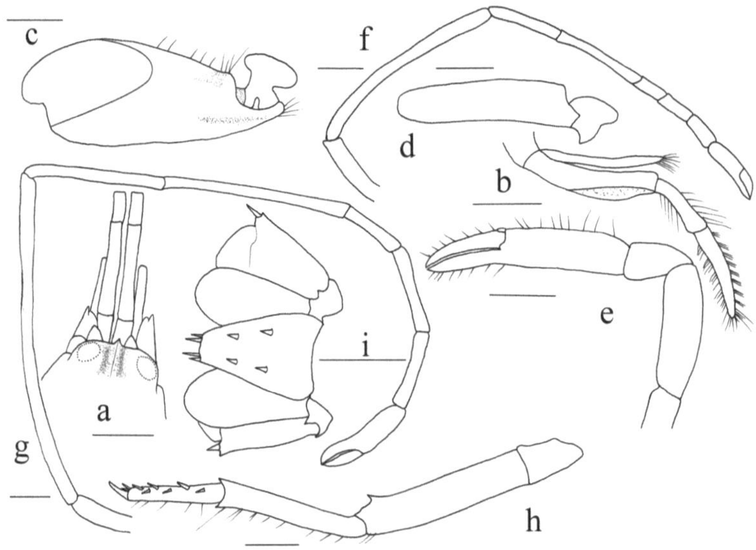


**Fig. 2.74** *Alpheus malleodigitus* (Bate, 1888). MBM 158651, male, **a**, anterior part of carapace, dorsal view; **b**, third maxilliped; **c**, major chela, lateral view; **d**, carpus to ischium of major cheliped, lateral view; **e**, minor cheliped, lateral view; **f**, left second pereiopod; g, right second pereiopod; **h**, third pereiopod; **i**, telson (scale bar = 1 mm)

Yongxing Islands, Xisha Islands, 11 June 1980; MBM 107606, 1♂, Jinyin Island, Xisha Islands, 15 May 1980; MBM 158883, 1♂, Xisha Islands, 28 April 1958; MBM 158392, 1♂, Yongshu Reef, Nansha Islands, 16 May 1993; MBM 158651, 3♀♀2♂♂, Yongshu Reef, Nansha Islands, Li Xin-zheng Coll.; MBM 158578, 5♂♂, Zhubi Reef, Nansha Islands, Li Xin-zheng Coll., 30 May 1993; 04II73AD-9, 2♂♂, Zhubi Reef, Nansha Islands, 12 May 2004; 04II68AD-13, 1♀1♂, Zhubi Reef, Nansha Islands, 11 May 2004; 04II51AD-19, 1♀, Zhubi Reef, Nansha Islands, 4 May 2004; 04II87AD-11, 3♀♀3♂♂, Zhubi Reef, Nansha Islands, 14 May 2004; 04II56AD-43, 2♂♂, Zhubi Reef, Nansha Islands, 6 May 2004; 04II86AD-14, 4♀♀, Zhubi Reef, Nansha Islands, 14 May 2004; 04II64AD-19, 1♂, Zhubi Reef, Nansha Islands, 10 May 2004.

**Description** Rostrum exceeding small, far not reaching to distal margin of first antennular segment; rostral carina sharp, extending to posterior margin of orbital hoods; orbitorostral grooves shallow; orbital hoods slightly inflated, without tooth.

Telson slender, about 3.5 times as long as posterior width; dorsal surface bearing two pairs of spines, approximately inserted 1/3 and 2/3 length of telson, respectively; posterior margin nearly straight, distolateral angles bearing two pairs of moderately long spines.

Antennular peduncle slender, with distally rounded stylocerite about reaching to middle length of first antennular segment; visible part of first antennular segment nearly long as third segment; second segment at least 3 times as long as wide. Antenna with lateral margin of scaphocerite concave; antennal scale with reduced blade far shorter than distolateral tooth, not reaching to middle length of second antennular segment; distolateral tooth of antennal not reaching to distal margin of second antennular segment. Carpocerite variable from not reaching to distal margin of second antennular segment to hardly reaching to distal margin of antennular peduncle.

Major cheliped with chela subcylindrical. Dactylus double-headed hammer-shaped, overhanging fixed finger, with plunger well developed. Palm with dorsal surface bearing transverse groove sub-distally, which slightly extending to both mesial and lateral face as longitudinal depression; lateral face of fixed finger bearing longitudinal shallow depression. Merus with dorsal and ventromesial margins both rounded distally, without tooth. Ischium unarmed. Minor cheliped with chela cylindrical, not balaeniceps in both sexes. Fingers about 2/3 length of palm. Palm without grooves or ridge. Merus similar to that of major cheliped. Ischium unarmed.

Second pereiopod extremely asymmetrical in length, but similar in shape; carpus 5-segmented, segmented ratio subequal to 10: 10–22: 4: 4: 5 in both right and left pereiopod.

Third and fourth pereiopod similar in shape, moderately stout. Dactylus simple. Propodus shorter than carpus, ventral margin bearing about 6 spines not arranging in single row and distally one pair. Carpus nearly as long as merus, both dorsal and ventral margin terminating in acute tooth. Merus with ventral margin bearing strong acute tooth sub-distally. Ischium unarmed with spin on ventral margin.

**Remarks** Some of the present specimens show slightly differences with previous descriptions on the anterior part of carapace: the blade of the antennal scale only reaching to distal margin of first antennular peduncle, while it reaching to middle length of the second antennular peduncle shown in Banner & Banner (1982, fig. 22m); the carpocerite only reaching to middle length of the second antennular peduncle, while it reaching to middle length of third antennular peduncle shown in Banner & Banner (1982, fig. 22m). However, other specimens are well agreed with the illustration made by Banner & Banner (1982). Additionally, fingers of the minor chela is about 2/3 length of palm, rather than 1/2 length of palm as described by previous descriptions.

**Distribution** The South China Sea; Red Sea; Eastern Africa to Japan; Philippines; Indonesia; Australia; eastward to the Society Islands but not Hawaii.

### 2.2.64 *Alpheus microstylus* (Spence Bate, 1888) (Fig. 2.75)

*Betaeus microstylus* Spence Bate, 1888: 566, Pl. 101, fig. 6.

*Alpheus microstylus* –. Banner DM & Banner AH, 1982: 92, fig. 22p–r; Banner AH & Banner DM, 1966b: 105, fig. 35a–f; 1983: 45, fig. 6a–f.; Liu & Lan, 1980: 89, fig. 6: 7–9; Chace, 1988: 41.

**Material Examined** CL 7.9–8.2 mm, TL 18.1–21.5 mm. MBM 159024, 50♀♂, Shi Island, Xisha Islands, Chen Mu Coll., 2 June 1981; MBM 158999, 1♂, Shanhu Island, Xisha Islands, 19 May 1980; MBM 158812, 1♂, Shanhu Island, Xisha Islands, 22 April 1983; MBM 158216, 1♂, Dengqing Island, Xisha Islands, Fan Zhen-gang & Xu Jie-shan Coll., 11 April 1958; MBM 158002, 1♂, Shi Island, Xisha Islands, Wang Yong-liang Coll.,14 May 1957; MBM 158995, 1♂, Guangjin Island, Nansha Islands, 20 May 1981; MBM 158994, 1♂, Chenhang Island, Xisha Islands, Chen Mu Coll., 16 May 1981; MBM 158969, 1♂, Guangjin Island, Xisha Islands, Chen Mu Coll., 23 May 1981; MBM 159029, 2♀♀, Jinqing Island, Xisha Islands, 19 May 1981; MBM 158392, 2♀♀2♂♂, Yongshu Reef, Nansha Islands, 16 May 1993; MBM 158649, 1♀1♂, Yongshu Reef, Nansha Islands, Li Xin-zheng Coll., 18 May 1993; MBM 107615, 4♀♀4♂♂, Shi Island, Xisha Islands, 10 May 1980; MBM 107624, 22♀♂, Yongxing Island, Xisha Islands, 9 May 1980; MBM 107617, 7♀♂, East Island, Xisha Islands, 28 May 1980; MBM 107619, 1♀, East Island, Xisha Islands, 28 May 1980; MBM 107620, 1♀, Yongxing Island, Xisha Islands 9 May 1980-5-9; MBM 107623, 58♀♂, Yongxing Island, Xisha Islands, 11 June 1980; MBM 159028, 1♀1♂, Shi Island, Xisha Islands, 2 June 1981; MBM 157973, 1♂, Jinqing Island, Xisha Islands, 25 April 1958; MBM 158271, 1♀, Yongxing Island, Xisha Islands, Ren Xian-qiu Coll., 5 June 1975; MBM 159263, 1♀1♂, Nanxun Reef, Nansha Islands, 17 April 1999; MBM 158287, 1♂, Chenhang Island, Xisha Islands, 18 May 1975; MBM 233276, 9♀♂, Yongxing Island,

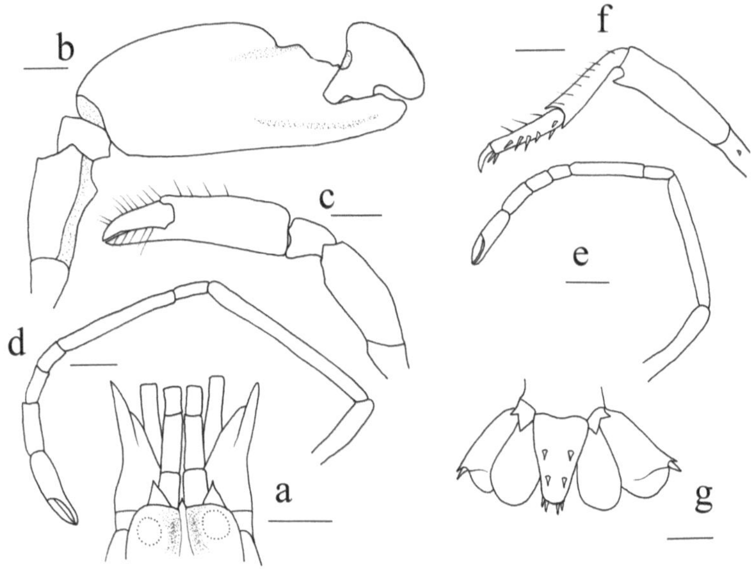


**Fig. 2.75** *Alpheus microstylus* (Bate, 1888). MBM 158287, male, **a**, anterior part of carapace, dorsal view; **b**, major cheliped, lateral view; **c**, minor cheliped, lateral view; **d**, left second pereiopod; **e**, right second pereiopod; **f**, third pereiopod; **g**, uropods and telson (scale bar = 1 mm)

Xisha Islands, Chen Mu Coll., 31 May 1981; MBM 270842, 4♂♂3♀♀, Dengqing Island, Xisha Islands, 11 April 1958; 04II84AD-31, 1♂, Zhubi Reef, Nansha Islands, 14 May 2004; MBM 107616, 9♀♀7♂♂, East Island, Xisha Islands, 3 June 1980.

**Description** Rostrum exceeding small, far not reaching to distal margin of first antennular segment; rostral carina sharp, extending to posterior margin of orbital hoods; orbitorostral grooves shallow; orbital hoods slightly inflated, without tooth.

Telson slender, about 3.5 times as long as posterior width; dorsal surface bearing two pairs of spines, approximately inserted 1/3 and 2/3 length of telson, respectively; posterior margin narrow, nearly straight, distolateral angles bearing two pairs of moderately long spines. Uropod with diaeresis on exopod absent.

Antennular peduncle slender, with distally acute stylocerite not reaching to distal margin of first antennular segment; visible part of first antennular segment nearly as long as third segment; second segment about twice as long as third segment. Antenna with lateral margin of scaphocerite slightly concave; antennal scale with narrow blade distinctly shorter than distolateral tooth, reaching to middle length of third antennular segment; distolateral tooth of antennal scale overreaching distal margin of antennular peduncle. Carpocerite nearly as long as distolateral tooth of antennal scale.

Major cheliped with chela subcylindrical. Dactylus double-headed hammer-shaped, overhanging fixed finger, with plunger well developed. Palm with dorsal surface bearing transverse groove sub-distally, which slightly extending to both mesial and lateral face as longitudinal depression, the depression on mesial face shallow than that of lateral face; lateral face of fixed finger bearing longitudinal shallow depression, extending distal 1/3 length of palm. Merus with dorsal margin rounded distally; ventromesial margin bearing sub-terminal strong acute tooth. Ischium unarmed. Minor cheliped with chela subcylindrical. Fingers about 1/2 length of palm. Merus with ventromesial face not bearing large acute tooth. Ischium unarmed.

Second pereiopod extremely asymmetrical in length, but similar in shape; carpus 5-segmented, segmented ratio subequal to 4: 12: 3: 3: 5 in both right and left pereiopod.

Third and fourth pereiopod similar in shape, moderately stout. Dactylus simple. Propodus nearly as long as carpus, ventral margin bearing 7 spines not arranging in single row and distally one pair. Carpus shorter than merus, both dorsal and ventral margin terminating in acute tooth. Merus with ventral margin terminating in strong acute tooth. Ischium with one inconspicuous small spine on ventral margin.

**Remarks** *A. microstylus* is very similar to *A. obesomanus* by the merus of the third pereiopod bearing large acute tooth on ventral margin distally, but can be distinguished from *A. obesomanus* by the well-developed blade of antennal scale overreaching distal margin of second antennular segment, while the reduced blade at most reaching to middle length of second antennular peduncle in *A. obesomanus*. The present specimens are well agreed with previous descriptions of *A. microstylus*.

**Distribution** The South China Sea; Red Sea; Eastern Africa to Indonesia; Vietnam; Philippines; Indonesia; Australia; Caroline, Mariana, and Samoan Islands.

### 2.2.65 *Alpheus obesomanus* Dana, 1852 (Fig. 2.76)

*Alpheus obesomanus* Dana, 1852b: 547, Pl. 34, fig. 7; Boone, 1935: 135, Pl. 35; Banner AH & Banner DM, 1966a: 101, fig. 35h, i; 1966b: 174, fig. 8a, b; 1983: 53, fig. 6g, h; Miya, 1974: 127, Pl. 22, fig. a–e; Liu & Lan, 1980: 87, fig. 6: 1–6, 15; Banner DM & Banner AH, 1982, fig. 22a–l; Chace, 1988: 44.

*Alpheus lutini* Coutière, 1905: 885, Pl. 76, fig. 24.

**Material Examined** CL 4.9–6.5 mm, TL 13.5–18.7 mm. MBM 159222, 1♂, Zhubi Reef, Nansha Islands, 10 April 1999; MBM 158412, 2♀♀1♂, Banyue Reef, Nansha Islands, 28 September 1994; MBM 158534, 2♀♀, Sanya Bay, Hainan,

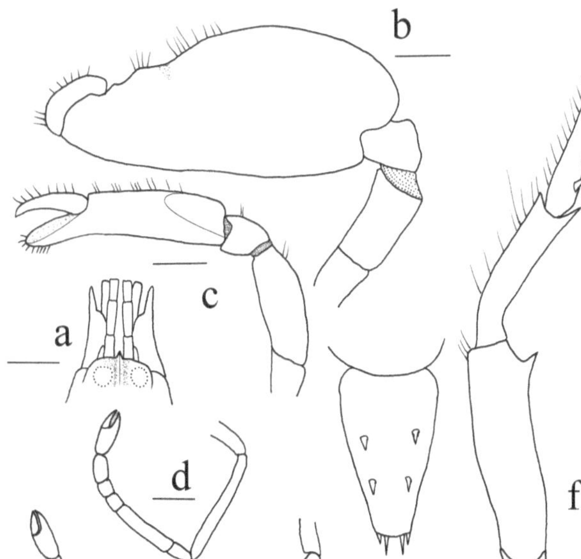


**Fig. 2.76** *Alpheus obesomanus* Dana, 1852. MBM 107642, male, **a**, anterior part of carapace, dorsal view; **b**, major cheliped, mesial view; **c**, minor cheliped, mesial view; **d**, left second pereiopod; **e**, right pereiopod; **f**, third pereiopod; **g**, telson (scale bar = 1 mm)

26 March 1992; MBM 158656, 1♂, Yezhu Island, Sanya Bay, Neumann Coll., 19 March 1992; MBM 107636, 2 larva, Chenhang Island, Xisha Islands, 18 May 1980; MBM 157837, 3♀♀1♂, Shi Island, Xisha Islands, Chen Mu Coll., 2 June 1981; MBM 107640, 48♀♂, Shi Island, Xisha Islands, 10 May 1980; MBM 107639, 1♀1♂, Yongxing Island, Xisha Islands, 9 May 1980; MBM 107638, 4♀♀4♂♂, Yongxing Island, Xisha Islands, 9 May 1980; MBM 107642, 1♂, Yongxing Island, Xisha Islands, 11 June 1980; MBM 10763, 1♂, East Island, Xisha Islands, 28 May 1980; MBM 107628, 2 larva, Shi Island, Xisha Islands, 10 May 1980; MBM 107627, 6♀♀ Yongxing Island, Xisha Islands, 11 June 1980; MBM 107632, 15♂♂6♀♀, East Island, Xisha Islands, 28 May 1980; MBM 107634, 26♂♂3♀♀, Shanhu Island, Xisha Islands, 19 May 1980; MBM 107635, 61♀♂, Yongxing Island, Xisha Islands, 11 June 1980; MBM 158541, 1 larva, Yezhu Island, Sanya Bay, 20 March 1992; MBM 159028, 1♀1♂, Shi Island, Xisha Islands, 2 June 1981; MBM 158618, 2♂♂1♀, Yongshu Reef, Nansha Islands, Li Xin-zheng Coll., 18 May 1993; MBM 158651, 1♂, Yongshu Reef, Nansha Islands, 16 May 1993; MBM 158592, 1 larva, Yongshu Reef, Nansha Islands, 18 May 1993; MBM 158717, 1♂, Xisha Islands, 28 April 1958; MBM 158570, 1♂, Nanxun Reef, Nansha Islands, 21 May 1993; MBM 159263, 3♂♂4♀♀, Nanxun Reef, Nansha Islands, 17 April 1999; MBM 158309, 1♀1♂, Guangjin Island, Xisha Islands, 18 May 1975; MBM 158274, 3♀♀1♂, Xisha Islands, 12 May 1975; MBM 158627, 1♀1♂, Xinyi Reef, Nansha Islands, 29 May 1990; MBM 158768, 1♂, Shanhu Reef, Nansha Islands, 25 May 1990; MBM 158322, 1♀, Guangjin Island, Xisha Islands, Chen Mu Coll., 20 May 1981; MBM 158578, 3♀♀3♂♂, Zhubi Reef, Nansha Islands, 30 May 1993; MBM 107629, 1♀, Yalong Bay, Sanya, Liu Rui-yu Coll., 19 November 1990; MBM 233268, 1♂, Quanfu Island, Xisha Islands, Liu Xi-xing Coll., 27 March 1975; MBM 158728, 1♂, Jinqing Island, Xisha Islands, 25 April 1958; 04II68AD-13, 2♂♂1♀, Zhubi Reef, Nansha Islands, 11 May 2004; 04II86AD-14, 1♀, Zhubi Reef, Nansha Islands, 14 May 2004; 04II71AD-28, 2♀♀, Zhubi Reef, Nansha Islands, 11 May 2004; 04II86AD-14, 1♂3♀♀, Zhubi Reef, Nansha Islands, 11 May 2004; 04II56AD-43, 2 larva, Zhubi Reef, Nansha Islands, 2 April 2004; 04II66AD-16, 7 larva, Zhubi Reef, Nansha Islands, 10 May 2004; 04II62AD-26, 3 larva, Zhubi Reef, Nansha Islands, 8 May 2004; 04II86AD-15, 8♂♂, Zhubi Reef, Nansha Islands, 14 May 2004; 04II57AD-16, 1♂, Zhubi Reef, Nansha Islands, 6 May 2004; 04II61AD-12, 3♂♂, Zhubi Reef, Nansha Islands, 8 May 2004; 04II50AD-6, 1♂2♀♀, Zhubi Reef, Nansha Islands, 4 May 2004; 04II78AD-39, 3♂♂1♀, Zhubi Reef, Nansha Islands, 13 May 2004; 04II83AD-12, 15♂♂, Zhubi Reef, Nansha Islands, 14 May 2004; 04II52AD-20, 1♀, Zhubi Reef, Nansha Islands, 4 May 2004; 04II54AD-22, 9♂♂, Zhubi Reef, Nansha Islands, 6 May 2004; MBM 107633, 1♂, Shi Island, Xisha Islands, 10 May 1980; MBM 107637, 1♂, Yongxing Island, Xisha Islands, 11 June 1980; MBM 233232, 1♂, Jinyin Island, Xisha Islands, 27 May 1975; MBM 107784, 1♀, Shalao, Wang Yong-liang Coll., 4 April 1992; MBM 107824, 1♂, Shalao, 5 April 1992; MBM 233263, 1♀, Guangjin Island, Xisha Islands, 18 May 1975; MBM 233265, 1♂, East Island, Xisha Islands, 10 June 1975; MBM 159350, 2♀♀2♂♂, Shalao, 4 April 1992.

**Description** Rostrum exceeding small, far not reaching to distal margin of first antennular segment; rostral carina sharp, extending to posterior margin of orbital hoods; orbitorostral grooves shallow; orbital hoods slightly inflated, without tooth.

Telson slender, about 4.5 times as long as posterior width; dorsal surface bearing two pairs of spines, approximately inserted 1/3 and 2/3 length of telson, respectively; posterior margin narrow, nearly straight, distolateral angles bearing two pairs of moderately long spines. Uropod with diaeresis on exopod absent.

Antennular peduncle slender, with distally rounded stylocerite not reaching to distal margin of first antennular segment; visible part of first antennular segment slightly longer than third segment; second segment about twice as long as third segment. Antenna with lateral margin of scaphocerite slightly concave; antennal scale with reduced narrow blade, only slightly protruded, reaching to middle length of second segment; distolateral tooth of antennal scale hardly reaching to distal margin of antennular peduncle. Carpocerite nearly as long as distolateral of antennal scale.

Major cheliped with chela subcylindrical. Dactylus double-headed hammer-shaped, overhanging fixed finger, with plunger well developed. Palm with dorsal surface bearing transverse shallow depression sub-distally, extending to lateral face as moderately deep longitudinal depression; lateral face of fixed finger bearing shallow longitudinal depression. Merus with dorsal and ventromesial margin both rounded distally. Ischium unarmed. Minor cheliped with chela subcylindrical. Fingers slightly shorter than 1/2 length of palm. Merus similar to that of major cheliped. Ischium unarmed.

Second pereiopod extremely asymmetrical in length, but similar in shape; carpus 5-segmented, segmented ratio subequal to 4: 12: 3: 3: 5 in both right and left pereiopod.

Third and fourth pereiopod similar in shape, moderately stout. Dactylus simple. Propodus nearly as long as carpus, ventral margin bearing 6 spines not arranging in single row and distally one pair. Carpus shorter than merus, both dorsal and ventral margin terminating in acute tooth. Merus with ventral margin terminating in strong acute tooth. Ischium unarmed on ventral margin.

**Remarks** The variability of *O. obesomanus* had been discussed by Banner & Banner (1982, 87). The present specimens are generally agreed with previous descriptions.

**Distribution** The South China Sea; Red Sea; Eastern Africa, and Madagascar to the Society Islands, including Japan; Philippines; Australia, but not from Hawaii.

### 2.2.66 *Alpheus perplexus* Banner, 1956 (Fig. 2.77)

*Alpheus perplexus* Banner, 1956: 347, fig. 13; Banner AH and DM Banner, 1967: 270; Banner DM & Banner AH, 1985: 27, fig. 3.

**Material Examined** 04II61AD-12, 1ovgerious♀1♂, CL 6.5–7.1 mm, TL 12.9–15.9 mm, Zhubi Reef, Nansha Islands, 8 May 2004.

**Description** Rostrum absent; rostral carina rounded, extending to posterior margin of orbital hoods; orbitorostral grooves shallow; orbital hoods slightly inflated, without tooth.

Telson about 1.4 and 3.3 times as long as anterior and posterior width; dorsal surface free of spines; posterior margin narrow, nearly straight, distolateral angles bearing two pairs of extremely small spines.

Antennular peduncle slender, with distally rounded stylocerite slightly shorter than antennal scale; visible part of first antennular segment longer than third segment; second segment longer than twice length of third segment. Antennal scale reduced, only reaching to middle length of first antennular segment. Carpocerite slightly beyond distal margin of first antennular segment.

Major cheliped with chela subcylindrical. Dactylus hammer-shaped, overhanging fixed finger, with plunger reduced, truncate. Palm with distal 1/3 length of dorsal surface bearing triangular depression; lateral face bearing longitudinal depression arising from base of fixed finger to distal 1/3 length of palm. Merus with dorsal and ventromesial margin both rounded distally. Ischium unarmed. Minor cheliped with chela subcylindrical. Fingers 1/2 length of palm. Merus similar to that of major cheliped. Ischium unarmed.

Second pereiopod extremely asymmetrical in length, but similar in shape; carpus 5-segmented, segmented ratio subequal to 16: 19: 8: 8: 9.

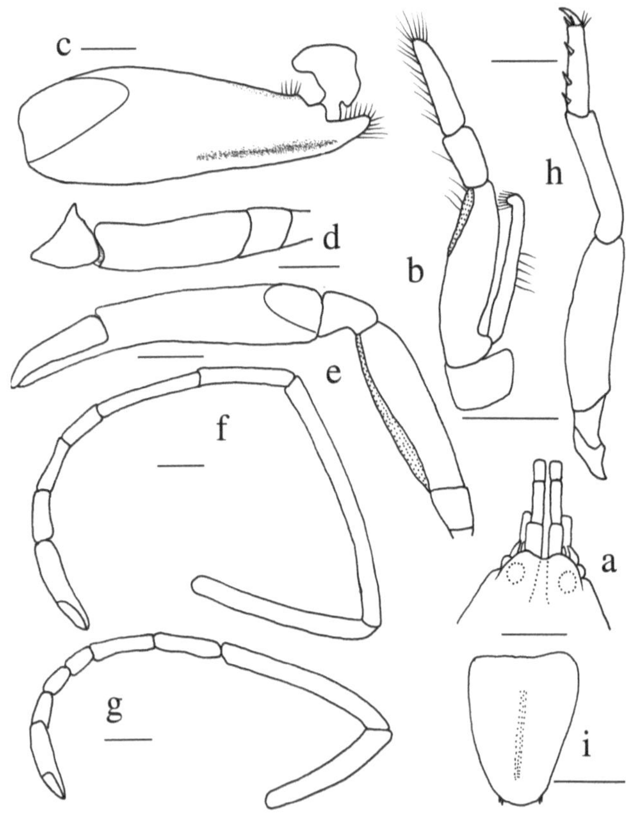


**Fig. 2.77** *Alpheus perplexus* Banner, 1956. 04II61AD-12, female, **a**, anterior part of carapace, dorsal view; **b**, third maxilliped; **c**, major chela, lateral view; **d**, carpus to ischium of major cheliped, lateral view; **e**, minor cheliped, lateral view; **f**, right second pereiopod; **g**, left second pereiopod; **h**, third pereiopod; **i**, telson (scale bar = 1 mm)

Third and fourth pereiopod similar in shape, moderately stout. Dactylus simple. Propodus slightly shorter than carpus, ventral margin bearing 3 spines and distally one pair. Carpus shorter than merus, both dorsal and ventral margin rounded distally. Merus and ischium unarmed.

**Remarks** *A. perplexus* is a newly recorded species from China seas, and it is originally described from Mariana Archipelago. The present specimens are identified as *A. perplexus* for the absence of rostrum, the shape of minor cheliped and the second pereiopod. However, the present specimens differs from the original description of *A. perplexus* by the following characters: the antennal scale reduced not reaching to distal margin of first antennular segment, rather than slightly overreaching distal margin of second antennular segment as shown in Banner (1956, fig. 13a); the carpocerite not reaching to middle length of the second antennular segment, rather than reaching to middle length of the third antennular segment as shown in Banner (1956, fig. 13a); the major chela not twist as shown in Banner (1956, fig. 13d); the propodus of third pereiopod with ventral margin bearing 3 spines rather than 5 pairs of spines; the merus of third pereiopod with ventral margin rounded distally, rather than terminating in subacute tooth; the dorsal surface of telson free of spines, rather than bearing two pairs of spinules.

**Distribution** The South China Sea; Mariana Archipelago; Society Islands.

**Sulcatus Group (Fig. 2.78)**

Orbital teeth usually present. Major chela never unusually compressed, bearing light to heavy longitudinal grooves, usually without transverse grooves. Minor chela never balaeniceps. Third pereiopod with merus with or without tooth on ventral margin distally; dactylus simple or biunguiculate (Banner & Banner, 1982).

Key to species of Sulcatus Group from China seas

1. Third pereiopod with ventral margin of merus armed with one large tooth distally ........ 2
-. Third pereiopod with ventral margin of merus rounded distally........ 4
2. Major chela with dorsal margin bearing shallow transverse depression........ 3
-. Major chela with dorsal margin not bearing shallow transverse depression ........ *A. canaliculatus*
3. Minor chela with palm more than twice as long as fingers........ *A. acutofemoratus*
-. Minor chela with palm nearly as long as fingers........ *A. amirantei*
4. Orbital hoods each armed with large acute tooth on anterior dorsal margin........ *A. splendidus*
-. Orbital hoods with or without tooth ........ 5
5. Orbital hoods without tooth ........ *A. sulcatus*
-. Orbital hoods with tooth ........ 6
6. Third maxilliped with ventral margin of antepenultimate and penultimate segment bearing several spines; third pereiopod with ventral face of dactylus bearing two longitudinal ridges ........ *A. lottini*
-. Third maxilliped with ventral margin of antepenultimate and penultimate segment only bearing long setae; third pereiopod with ventral face of dactylus not bearing longitudinal ridge ........ 7
7. Dactylus of third pereiopod simple........ *A. facetus*
-. Dactylus of third pereiopod biunguiculate........ 8
8. Mesial face of major chela bearing groups of stiff setae ........ *A. architectus*
-. Mesial face of major chela at most bearing thin setae........ *A. gracilis*

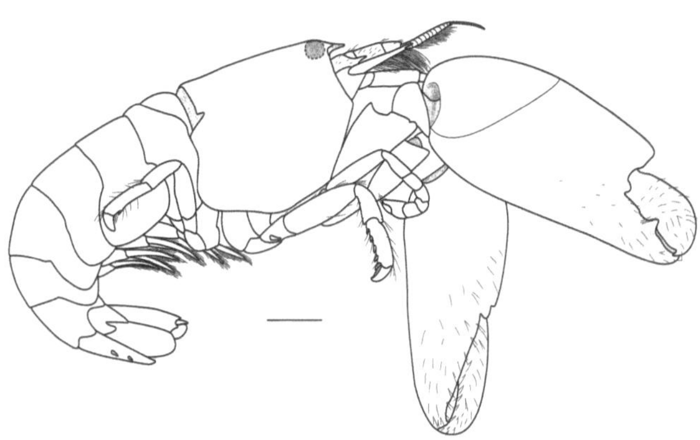

**Fig. 2.78** *Alpheus lottini* Guérin, 1829. entire animal of Sulcatus Group (scale bar = 2 mm)

### 2.2.67 *Alpheus acutofemoratus* Dana, 1852 (Fig. 2.79)

*Alpheus acutofemoratus* Dana, 1852b: 550, Pl. 35, fig. 2; De Man, 1902: 888, Pl. 27, fig. 63; Banner AH & Banner DM, 1966b: 87, fig. 29; Banner DM & Banner AH, 1982: 77, fig. 19; Chace, 1988: 15.
*Alpheus parabrevipes* Coutière, 1898b: 151.

**Material Examined** 04II84AD-31, 2♂♂, CL 7.1 mm, TL 18.2 mm, Zhubi Reef, Nansha Islands, 14 May 2004.

**Description** Rostrum small, about reaching to middle length of first antennular segment; rostral carina sharp, extending slightly beyond to posterior margin of orbital hoods; anterior margin of carapace between rostrum and orbital hoods strongly concave; orbital hoods without tooth, anterior margin convex.

Telson about 1.5 and 2.5 times as long as anterior and posterior width; dorsal surface bearing two pairs of spines, approximately inserted 2/5 and 2/3 length of telson, respectively; posterior margin slightly convex, not beyond two pairs of posterolateral spines.

Antennular peduncle with distally acute stylocerite not reaching to distal margin of first antennular segment; visible part of first antennular segment about 1/2 length of second segment; second segment twice as long as wide; third segment slightly shorter than first segment. Antenna with lateral margin of scaphocerite concave; antennal scale with blade distinctly shorter than distolateral tooth, beyond distal margin of antennular peduncle. Carpocerite nearly as long as distolateral tooth of antennal scale.

Major cheliped with chela subcylindrical, slightly twist. Fingers about 1/2 length of palm, dactylus usually truncate at tip, slightly longer than fixed finger, with plunger well developed. Palm with dorsal and upper lateral faces bearing extremely long setae that arising from a small pustule; dorsal surface bearing narrow transverse groove sub-distally, that slightly continuing to

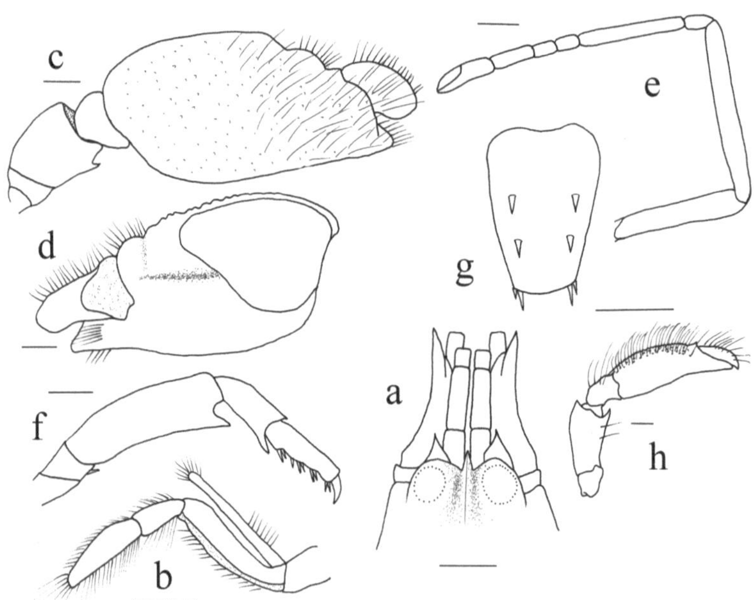


**Fig. 2.79** *Alpheus acutofemoratus* Dana, 1852. 04II84AD-31, male, **a**, anterior part of carapace, dorsal view; **b**, third maxilliped; **c**, major cheliped, mesial view; **d**, major chela, lateral view; **e**, second pereiopod; **f**, third pereiopod; **g**, telson. Adapted from Banner & Banner (1982), **h**, minor cheliped (scale bar = 1 mm)

lateral face; lateral face bearing narrow longitudinal groove, that arising from below transverse groove on dorsal surface continuing to linea impressa. Merus robust, about 1.5 times as long as wide; ventromesial margin bearing acute tooth sub-distally. Minor cheliped with chela subcylindrical. Fingers shorter than 1/2 length of palm. Palm with dorsal and upper mesial faces bearing long setae, similar to that of major chela. Merus similar to that of major cheliped, but more slender. Ischium unarmed.

Second pereiopod with carpus 5-segmented, segmented ratio subequal to 3: 15: 3: 3: 5.

Third and fourth pereiopod similar in shape, moderately stout. Dactylus simple. Propodus slightly longer than carpus; ventral margin bearing 5 and accessory 3 spines and distally one pair. Carpus nearly 1/2 length of merus; dorsal and ventral margins terminating in rounded and large acute tooth, respectively. Merus with ventral margin bearing large acute tooth distally. Ischium armed with one spine on ventral margin.

**Remarks** *A. acutofemoratus* is readily recognized by the mesial face of major and minor chela bearing long setae. The present specimens are generally agreed with previous descriptions, except for the second pereiopod with first carpal segment about 1/5 length of the second carpal segment rather than 1/3 length of that segment as shown in Banner & Banner (1982, fig. 19g).

**Distribution** The South China Sea; Andaman Sea; Gulf of Thailand; Philippines; Indonesia; Queensland; Australia; Marshall; Fiji; Samoa, and Tonga Islands.

### 2.2.68 *Alpheus amirantei* Coutière, 1908 (Fig. 2.80)

*Alpheus amirantei* Coutière, 1908: 205; 1921, 421, Pl. 63, fig. 16.

**Material Examined** This species was reported by Liu (2008) from the South China Sea, but not found in the present specimens.

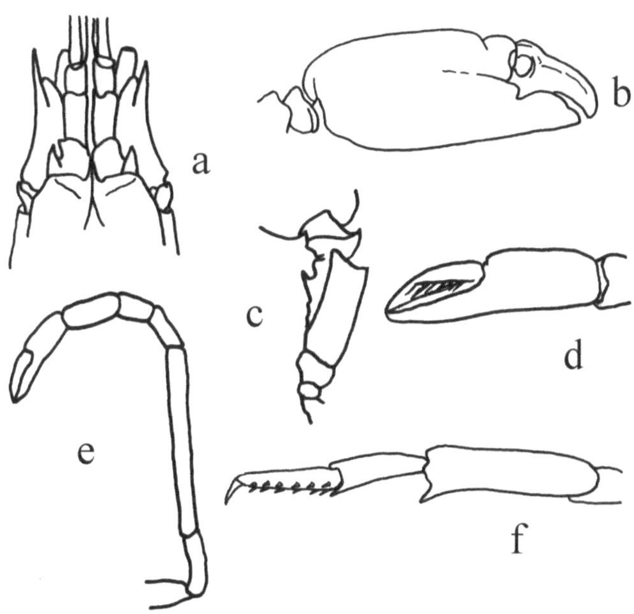


**Fig. 2.80** *Alpheus amirantei* Coutière, 1908. (Adapted from Coutière, 1921). Male, **a**, anterior part of carapace, dorsal view; **b**, major chela, mesial view; **c**, carpus to ischium of major cheliped, lateral view; **d**, minor chela, mesial view; **e**, chela and carpus of second pereiopod; **f**, third pereiopod

**Description** Rostrum slender, not reaching to distal margin of first antennular segment; rostral carina extending to posterior margin of orbital hoods; anterior margin between rostrum and orbital hoods concave; orbital hoods without tooth.

Antennular peduncle with stylocerite not reaching to distal margin of first antennular segment; visible part of first antennular segment slightly longer than third segment; second segment longest, about twice as long as wide. Antenna with lateral margin of scaphocerite concave; antennal scale with narrow blade distinctly shorter than distolateral tooth, reaching to middle length of third segment. Carpocerite beyond antennular peduncle.

Major cheliped with chela subcylindrical. Fingers about 1/2 length of palm, dactylus slightly longer than fixed finger. Palm with dorsal surface bearing narrow transverse groove subdistally; lateral face bearing narrow longitudinal groove medially, that arising from distal margin continuing to about middle length of palm. Merus robust, about 1.5 times as long as wide; ventromesial margin bearing acute tooth sub-distally. Minor cheliped with chela subcylindrical. Fingers slightly shorter than palm.

Second pereiopod with carpus 5-segmented, segmented ratio subequal to 4: 13: 3: 3: 4.

Third pereiopod moderately stout. Dactylus simple. Propodus slightly longer than carpus; ventral margin bearing about 6 pair's spines and distally one pair. Carpus nearly 1/2 length of merus; dorsal and ventral margins terminating in rounded and large acute tooth, respectively. Merus with ventral margin bearing large acute tooth distally. Ischium unarmed.

**Distribution** Taiwan; Amirante Island.

### 2.2.69 *Alpheus architectus* De Man, 1897 (Fig. 2.81)

*Alpheus architectus* De Man, 1897: 726, fig. 60; Banner DM & Banner AH, 1982: 55, fig. 11.
*Alpheus bullatus* Barnard, 1955: 45, fig. 22.

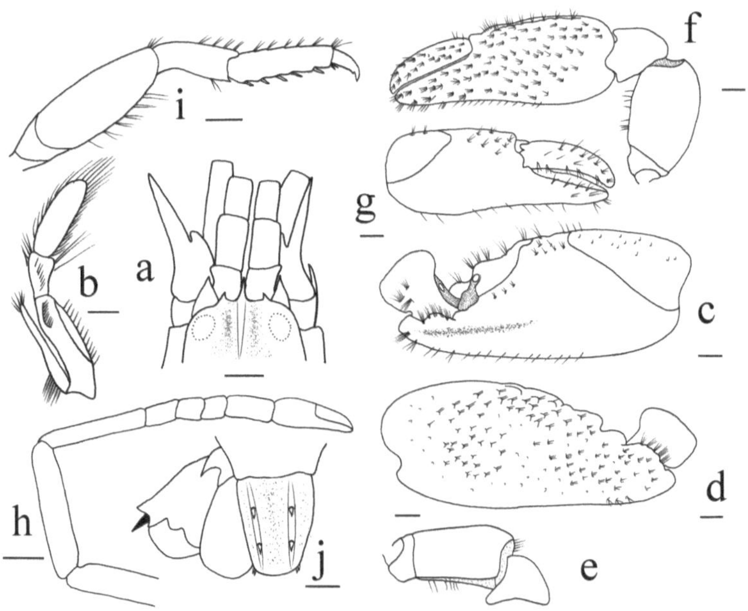


**Fig. 2.81** *Alpheus architectus* De Man, 1897. MBM 270033, female, **a**, anterior part of carapace, dorsal view; **b**, third maxilliped; **c**, major chela, lateral view; **d**, same, mesial view; **e**, carpus to ischium of major cheliped, mesial view; **f**, minor cheliped, mesial view; **g**, minor chela, lateral view; **h**, second pereiopod; **i**, third pereiopod; **j**, left uropod and telson (scale bar = 1 mm)

**Material Examined** CL 10.5–10.8 mm, TL 25.3–29.5 mm. MBM 158121, 1♀1♂, Sanya, Hainan, 10 June 1958; MBM 158023, 1♀, Shanhu Reef, Cui Yu-heng Coll., 16 June 1959; MBM 270033, 2♀♀4♂♂, Xizhou Island, Liu Rui-yu Coll., 30 December 1955.

**Description** Rostrum triangular, longer than wide at base, reaching to middle length of first antennular segment; rostral carina rounded, extending to middle of carapace; orbitorostral grooves deep; orbital hoods slightly inflated, with teeth not reaching distal end of rostrum.

Telson about 1.5 and 2 times as long as anterior and posterior width; dorsal surface bearing two longitudinal ridges, with two pairs of spines inserted on that, anterior and posterior pair inserted 1/3 and 2/3 length of telson, respectively; posterior margin slightly convex, not beyond two pairs of posterolateral spines.

Antennular peduncle with distally acute stylocerite not reaching to distal margin of first antennular segment; visible part of first antennular segment nearly as long as third segment; second segment about 1.5 times as long as wide, slightly longer than third segment. Antenna with lateral margin of scaphocerite slightly concave; antennal scale with blade reduced, reaching to middle length of second segment, distolateral tooth slightly overreaching distal margin of antennular peduncle. Carpocerite nearly as long as distolateral tooth of antennal scale. Lateral tooth of basicerite overreaching distal margin of antennular peduncle.

Major cheliped with chela subcylindrical. Fingers about 1/4 length of palm, bearing tufts of short stiff setae on faces; dactylus longer than fixed finger. Palm with dorsal surface developed as flattened shelf distally, to accommodate end of dactylus when flexed; mesial and dorsal face bearing groups of stiff setae. Merus robust, distal margin rounded, without tooth. Minor cheliped with chela cylindrical. Fingers slightly shorter than palm. Palm with mesial and dorsal face bearing groups of stiff setae. Merus similar to that of major cheliped.

Second pereiopod with carpus 5-segmented, segmented ratio subequal to 12: 4: 3: 3: 5.

Third and fourth pereiopod similar in shape, moderately stout. Dactylus biunguiculate, unguis on flexor margin small. Propodus slightly longer than carpus, ventral margin bearing 5 spines and distally one pair. Carpus with dorsal margin terminating in subacute tooth. Merus longer than propodus, margins rounded distally. Ischium unarmed with spine on ventral margin.

**Remarks** *A. architectus* is a newly recorded species from China seas, and it is originally described from Atjeh. Banner & Banner (1982) reported this species from Australia. The present species are generally agreed with previous descriptions, except for the following variable characters: the blade of antennal scale reaching to middle length of the second antennular peduncle; the lateral tooth of basicerite slightly overreaching distal margin of the first antennular peduncle; the second pereiopod with the first carpal segment about 3 times as long as the second segment. Banner & Banner (1982, fig. 11j) has discussed the appendix masculina of the second pleopod distinctly long than the appendix interna, which is also found in the present specimens.

**Distribution** The South China Sea; Atjeh; South Africa; Madagascar; Australia.

### 2.2.70 *Alpheus canaliculatus* Banner & Banner, 1968 (Fig. 2.82)

*Alpheus canaliculatus* Banner AH & Banner DM, 1968a: 141, fig. 1; Chace, 1988: 18.

**Material Examined** This species was reported by Banner & Banner (1968) and Liu (2008) from the South China Sea, but not found in the present specimens.

**Description** Rostrum triangular, longer than wide at base, reaching to middle length of first

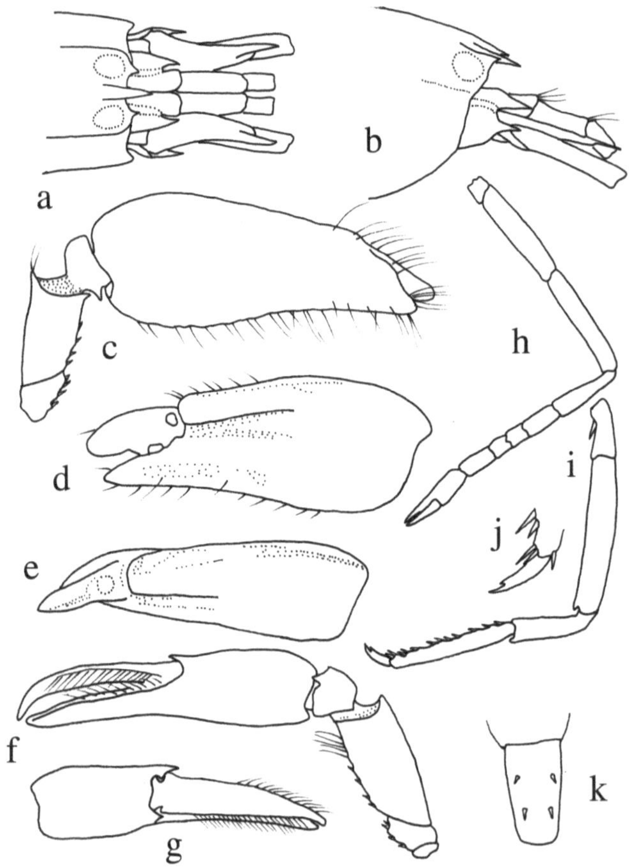


**Fig. 2.82** *Alpheus canaliculatus* Banner & Banner, 1968. (Adapted from Banner & Banner, 1968a). Female, **a**, anterior part of carapace, dorsal view; **b**, same, lateral view; **c**, major cheliped, mesial view; **d**, major chela, lateral view; **e**, same, dorsal view; **f**, minor cheliped, mesial view; **g**, minor chela, dorsal view; **h**, second pereiopod; **i**, third pereiopod; **j**, dactylus of third pereiopod; **k**, telson

antennular segment; rostral carina rounded, extending slightly posterior to middle of eyes; orbital hoods slightly inflated, anterior margins bearing acute teeth, not reaching to end of rostrum.

Telson about 2.5 times as long as posterior width; dorsal surface bearing two pairs of moderately large spines, approximately inserted 1/3 and 2/3 length of telson, respectively; posterior margin truncate, only slightly convex.

Antennular peduncle slender, with distally acute stylocerite reaching to distal margin of first antennular segment; visible part of first antennular segment longer than third segment; second segment about 3 times as long as wide, about twice as long as third segment. Antenna with lateral margin of scaphocerite concave; antennal scale with narrow blade distinctly shorter than distolateral tooth, reaching to middle length of third antennular segment. Carpocerite far beyond distal margin of antennular peduncle.

Major cheliped with chela subcylindrical. Fingers about 1/3 length of palm. Palm with dorsal surface bearing shallow longitudinal groove, that arising sub-distally slightly towards inners face and continuing proximally gradually becoming narrow deep groove to base of palm; outer face bearing longitudinal groove medially arising distally to middle length of palm; outer face bearing another short, narrow longitudinal groove nearly ventral margin, that arising from middle of fixed finger to middle length of palm; mesial face smooth, ventral margin fringed with setae. Merus with dorsal margin bearing rounded tooth distally; ventromesial margin bearing 4 spines and terminating in acute tooth. Ischium with ventromesial margin bearing 5 spines; ventrolateral margin bearing large acute tooth. Minor cheliped with chela "sub-balaeniceps". Fingers slightly longer than palm. Palm with distal margin bearing two acute teeth flanking dactylus. Merus and ischium similar to that of major cheliped.

Second pereiopod with carpus 5-segmented, segmented ratio subequal to 10: 8: 3: 3: 4.

Third pereiopod moderately slender. Dactylus biunguiculate, unguis on flexor margin small. Propodus slightly shorter than merus, ventral margin bearing 14 irregularly spaced spines and distally one pair. Carpus with dorsal margin terminating in subacute tooth. Merus 5.4 times as long as wide, ventral margin terminating in acute tooth. Ischium armed with one strong spine on ventral margin.

**Remarks** *A. canaliculatus* is originally described Hong Kong.

**Distribution** The South China Sea (Hong Kong); Philippines.

### 2.2.71 *Alpheus facetus* De Man, 1908 (Fig. 2.83)

*Alpheus facetus* De Man, 1908: 100; 1911: 340, fig. 67; Tiwari, 1963: 288, fig. 9; Banner AH & Banner DM, 1966b: 96, fig. 33; Banner DM & Banner AH, 1982: 62, fig. 14; Chace, 1988: 27.

**Material Examined** CL 5.1 mm, TL 12.6 mm. MBM 159408, 3♀♀1♂, Shalao, 5 April 1992; MBM 158835, 1♀1♂, Xiamen, Fujian, 27 April 1956; MBM 159409, 4♀♀3♂♂, Shalao, 4 April 1992; MBM 159410, 1♀, Shalao, 31 December 1991.

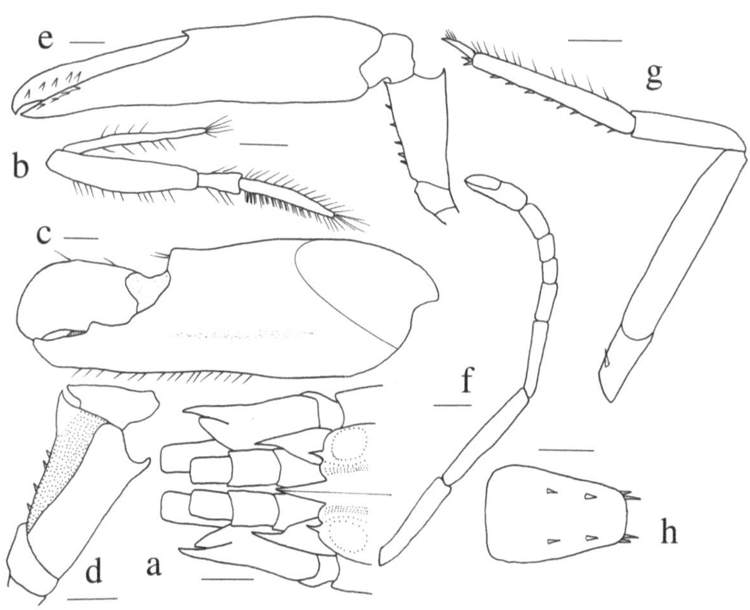


**Fig. 2.83** *Alpheus facetus* De Man, 1908. MBM 159409, male, **a**, anterior part of carapace, dorsal view; **b**, third maxilliped; **c**, major chela, lateral view; **d**, carpus to ischium of major cheliped, lateral view; **e**, minor cheliped, mesial view; **f**, second pereiopod; **g**, third pereiopod; **h**, telson (scale bar = 1 mm)

**Description** Rostrum acute, twice as long as wide at base, reaching to distal margin of first antennular segment; rostral carina blunt, nearly extending to middle length of carapace; orbito-rostral grooves shallow; orbital hoods slightly inflated, anterior margins bearing acute teeth, about 1/3 length of rostrum.

Telson about 2 and 2.5 times as long as anterior and posterior width, respectively; dorsal surface bearing two pairs of spines, approximately inserted 3/7 and 5/7 length of telson, respectively; posterior margin slightly convex, posterolateral angles bearing two pairs of spines.

Antennular peduncle stout, with distally acute stylocerite reaching middle length of second antennular segment; three antennular segments nearly equal in length, second segment about 1.4 times as long as wide. Antenna with lateral margin of scaphocerite nearly straight; antennal scale with blade shorter than distolateral tooth, hardly reaching to distal margin of antennular peduncle. Carpocerite far beyond distal antennular peduncle.

Major cheliped with chela slightly compressed. Fingers about 1/3 length of palm. Palm with dorsal surface bearing two longitudinal ridges, arising from proximal 1/5 length of palm and continuing distal 1/5 length of palm, with a depression between two ridges; lateral face bearing moderately shallow longitudinal depression,, arising from base of fixed finger to proximally 1/3 length of palm. Merus with dorsal margin bearing acute tooth distally; ventromesial margin bearing 3–6 small spines and terminating in large acute tooth. Ischium unarmed. Minor cheliped with chela subcylindrical. Fingers and palm nearly equal in length. Palm with dorsal margin of mesial face terminating in acute tooth, overhanging dactylus. Merus similar to that of major cheliped. Ischium unarmed.

Second pereiopod with carpus 5-segmented, segmented ratio subequal to 10: 5: 3: 3: 5.

Third and fourth pereiopod similar in shape, moderately slender. Dactylus simple. Propodus slightly shorter than merus, ventral margin bearing 7–9 spines and distally one pair. Carpus slightly 1/2 length of merus, dorsal and ventral margin only slightly projecting. Merus unarmed. Ischium armed with one spine on ventral margin.

**Remarks** Banner & Banner (1982) had discussed the variable characters of *A. facetus*. The present specimens are generally agreed with previous descriptions and show similar variations.

**Distribution** The South China Sea; Western Indian Ocean; Thailand; Vietnam; Philippines; Indonesia; Japan; Australia; Caroline Islands.

### 2.2.72 *Alpheus gracilis* Heller, 1861 (Fig. 2.84)

*Alpheus gracilis* Heller, 1861: 271, Pl. 3, figs. 19, 20; De Man, 1897: 733, fig. 60g, h; Tiwari, 1963: 283, fig. 7; Banner DM & Banner AH, 1982: 60, fig. 13; Chace, 1988: 30.

*Alpheus gracilis* var. *alluaudi* Coutière, 1905: 882.

*Alpheus gracilis* var. *luciparensis* de Man, 1911: 338.

*Crangon gracilis* var. *simplex* Banner, 1953: 75, fig. 25.

**Material Examined** This species was reported by Liu (2008) from the South China Sea, but not found in the present specimens.

**Description** Rostrum acute, rounded dorsally, not reaching to distal margin of first antennular segment; rostral carina absent; orbital hoods slightly inflated, anterior margins bearing narrow acute teeth, not reaching to distal margin of rostrum.

Telson about 2.5 times as long as posterior width; dorsal surface bearing two pairs of spines, approximately inserted 1/2 and 3/4 length of telson, respectively; posterior margin only slightly

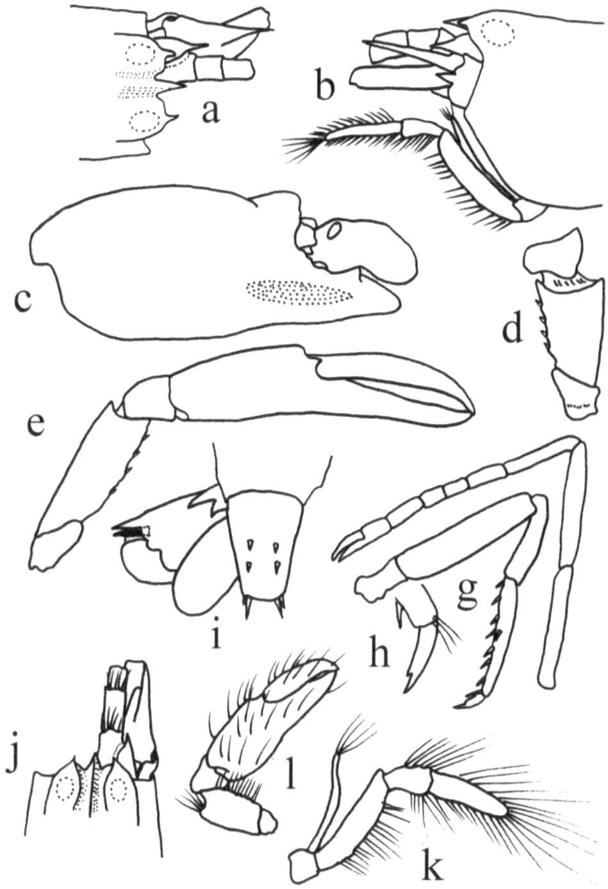


**Fig. 2.84** *Alpheus gracilis* Heller, 1851. (Adapted from Banner & Banner, 1982). Female, **a**, anterior part of carapace, dorsal view; **b**, same, lateral view; **c**, major chela, lateral view; **d**, carpus to ischium of major cheliped, mesial view; **e**, minor cheliped, mesial view; **f**, second pereiopod; **g**, third pereiopod; **h**, dactylus of third pereiopod; **i**, left uropod and telson. Male, **j**, anterior part of carapace, dorsal view; **k**, third maxilliped; **l**, minor cheliped, mesial view

convex, posterolateral angles bearing two pairs of spines.

Antennular peduncle moderately stout, with distally acute stylocerite reaching to distal margin of first antennular segment; visible part of first antennular segment slightly longer than second segment; second segment slightly longer than wide, longer than third segment. Antenna with lateral margin of scaphocerite only slightly concave; antennal scale with narrow blade distinctly shorter than distolateral tooth, reaching to distal margin of antennular peduncle. Carpocerite nearly as long as distolateral tooth of antennal scale.

Major cheliped with chela compressed. Fingers about 1/2 length of palm, dactylus rounded at tip, with plunger moderately developed. Palm with dorsal margin bearing shallow transverse depression sub-distally; lateral face bearing longitudinal depression, that arising from distally 1/4 length of palm and continuing to middle length of fixed finger. Merus robust, dorsal margin bearing rounded tooth distally; ventromesial margin bearing 4 spines and terminating in small acute tooth. Ischium unarmed. Minor cheliped with chela subcylindrical. Fingers slightly longer than palm. Palm with dorsal margin terminating in acute tooth, overhanging dactylus.

Merus similar to that of major cheliped. Ischium unarmed.

Second pereiopod with carpus 5-segmented, segmented ratio subequal to 10: 4: 3: 3: 4.

Third and fourth pereiopod similar in shape. Dactylus biunguiculate, flexor margin only swelling or bearing small acute unguis. Propodus slightly shorter than merus, ventral margin bearing 8 spines and distally one pair. Carpus with dorsal margin terminating in rounded tooth. Merus and ischium unarmed.

**Distribution** The South China Sea, Taiwan; Red Sea, Eastern and possibly South Africa to Thailand; Philippines; Japan; Indonesia; Queensland; Australia and Pacific islands, including Hawaii and Society Islands.

### 2.2.73 *Alpheus lottini* Guérin-Méneville, 1838 (Fig. 2.85)

*Alpheus lottini* Cuérin-Méneville, 1838 [in Guérin-Méneville, 1829–1838]: Pl. 3, fig. 3; Holthuis, 1958: 22; 1961: 168; Tiwari, 1963: 285, fig. 8; Banner AH & Banner DM, 1966b: 91, fig. 31; Liu & Lan, 1980: 84, fig. 4; Banner DM & Banner AH, 1982: 65, fi. 15; Chace, 1988: 35.

*Alpheus ventrosus* H. Milne Edwards, 1837: 352.

*Alpheus laevis* Randall, 1839: 141.

*Alpheus thetis* White, 1847: 5, Pl. 4, fig. 7.

*Crangon latipes* Banner, 1953: 82, fig. 27.

**Material Examined** CL 7.1–7.5 mm, TL 19.8–23.2 mm. MBM 158997, 3♂♂2♀♀, Shalao, Liu Xi-xing Coll., 7 November 1990; MBM 159045, 1♀1♂, Hainan, Liu Rui-yu Coll., 29 December 1955; MBM 159995, 2♂♂, Yongxing Island, Xisha Islands, 11 June 1980; MBM 107556, 1♂, Jinyin Island, Xisha Islands, 26 May 1975; MBM

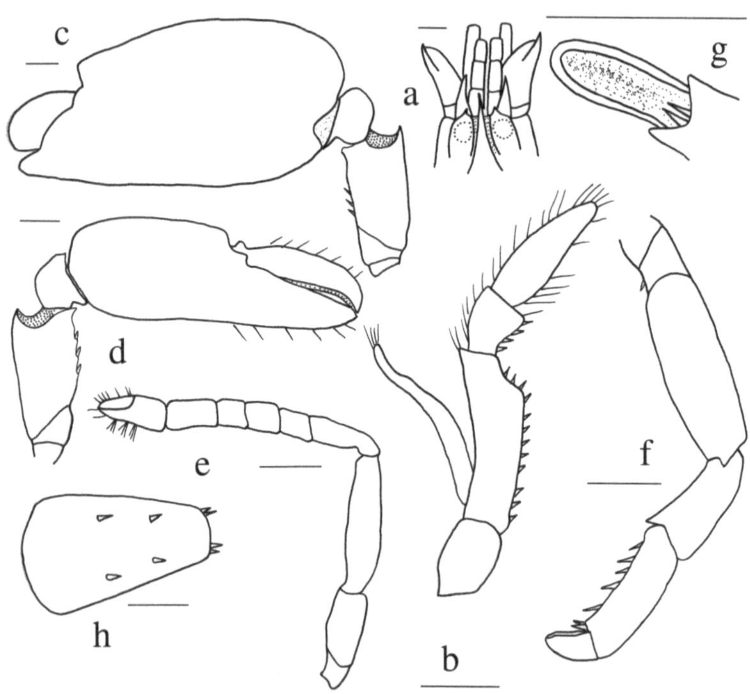


**Fig. 2.85** *Alpheus lottini* Guérin, 1829. MBM 158608, male, **a**, anterior part of carapace, dorsal view; **b**, third maxilliped; **c**, major cheliped, mesial view; **d**, minor cheliped, mesial view; **e**, second pereiopod; **f**, third pereiopod; **g**, dactylus of third pereiopod; **h**, telson (scale bar = 1 mm)

15999, 1♂, Jinyin Island, Xisha Islands, 25 May 1975; MBM 159992, 2♀♀1♂, Shanhu Island, Xisha Islands, 19 May 1980; MBM 159996, 1♂, Jinyin Island, Xisha Islands, 14 May 1980; MBM 159997, 4♀♀6♂♂, Jinyin Island, Xisha Islands, 14 May 1980; MBM 159993, 4♀♀4♂♂, East Island, Xisha Islands, 28 May 1980; MBM 107559, 19♀♀20♂♂, Xizhou Island, Fan Zhen-gang Coll., 29 December 1955; MBM 107555, 21♀♀22♂♂, Xizhou Island, 29 December 1955; MBM 107560, 11♀♀11♂♂, Zhaoshu Island, Xisha Islands, 11 May 1980; MBM 159163, 1♀1♂, Sanya, Hainan, Li Xin-zheng Coll., 2 December 1997; MBM 159204, 1♀1♂, Xincun, Hainan; MBM 157966, 2♂♂, Yongxing Island, Xisha Islands, Wang Yong-liang Coll., 13 April 1957; MBM 107557, 1♂2♀♀, Yezhu Island, 22 March 1992; MBM 158817, 2♂♂1♀, Yongxing Island, Xisha Islands, 15 May 1958; MBM 159177, 4♂♂, Sanya, Hainan, 3 December 1997; MBM 158852, 1♂, Daai Reef, Nansha Islands, 24 May 1990; MBM 158800, 1♀1♂, Sanya, Hainan, 9 April 1958; MBM 158147, 1♂, Sanya, Hainan, 19 November 1997; MBM 157913, 1♀, Huayang Reef, Nansha Islands, 15 April 1999; MBM 157969, 3♀♀4♂♂, Yongxing Island, Xisha Islands, 13 April 1957; MBM 107558, 5♂♂7♀♀, Shi Island, Xisha Islands, Wang Yong-liang Coll., 17 May 1957; MBM 159253, 1♂4♀♀, Sanya, Hainan, Fan Zhen-gang Coll., 21 March 1962; MBM 158024, 1♂3♀♀, Shalao, Wang Yong-liang Coll., 29 March 1955; MBM 157867, 1♂, 6 April 1957; MBM 159021, 1♀1♂, Sanya, Hainan, Li Xin-zheng Coll., 4 March 1997; MBM 158676, 1♂, Yongshu Reef, Nansha Islands, Li Xin-zheng Coll., 29 May 1993; MBM 158086, 1♂, Sanya, Hainan, 27 November 1997; MBM 158014, 2♀♀1♂, Nansha Islands, Fan Zhen-gang & Xu Jie-shan Coll., 28 April 1958; MBM 159154, 1♂, Sanya, Hainan, 2 December 1997; MBM 158232, 2♀♀5♂♂, Dazhou Island, Wang Yong-liang Coll., 12 June 1957; MBM 233260, 5♂♂1♀, East Island, Xisha Islands, 12 June 1975; MBM 159268, 4♀♀5♂♂, Yongshu Reef, Nansha Islands, 11 April 1999; MBM 158964, 1♂, Shanhu Reef, 25 March 1955; MBM 158246, 40♂33♀♀, Yongxing Island, Xisha Islands, Wang Yong-liang Coll., 13 April 1957; MBM 158847, 3♂♂, Nanping Reef, Nansha Islands, 2 August 1988; MBM 159192, 1♀, Sanya, Hainan, 17 November 1997; MBM 157754, 6♀♀3♂♂, Yongxing Island, Xisha Islands, 13 April 1957; MBM 158472, 1♀1♂, Hainan. 18 December 1992; MBM 157721, 1♀, Dengqing Island, Xisha Islands, 11 April 1958; MBM 157745, 2♀♀2♂♂, North Island, Xisha Islands, Wang Yong-liang Coll., 29 April 1957; MBM 158459, 2♀♀, Anda Reef, Li Xin-zheng Coll., 23 May 1993; MBM 157747, 18♀♂, Yongxing Island, Xisha Islands, Fan Zhen-gang & Xu Jie-shan Coll., 20 May 1958; MBM 158291, 1♀1♂, Sanya, Hainan, Ren Xian-qiu Coll., 25 April 1975; MBM 158292, 1♂7♀♀, Sanya, Hainan, 12 April 1975; MBM 158293, 5♀♀4♂♂, Luhuitou, Sanya, 13 April 1975; MBM 158304, 1♀1♂, Guangjin Island, Xisha Islands, 18 May 1975; MBM 158301, 1♀, East Island, Xisha Islands, 9 June 1975; MBM 158300, 1♂, Xisha Islands, 13 May 1975; MBM 158309, 1♀1♂, Jinyin Island, Xisha Islands, 29 May 1975; MBM 158650, 1♀1♂, Banyue Reef, Nansha Islands, 5 May 1987; MBM 158768, 3♀♀1♂, Shanhu Reef, Nansha Islands, 25 May 1990; MBM 158608, 3♀♀3♂♂, Shalao, Wang Yong-liang Coll., 3 April 1992; MBM 158746, 2♀♀1♂, Nansha Islands, Xinyi Reef, 29 May 1990; MBM 107554, 1♂, Yezhu Island, Sanya Bay, Neumann Coll., 16 January 1992; MBM 158434, 1♀, Yezhu Island, Sanya Bay, Neumann Coll., 21 March 1992; MBM 158435, 1♀, Yezhu Island, Sanya Bay, Neumann Coll., 22 March 1992; 04II65AD-18, 2♀♀, Zhubi Reef, Nansha Islands, 10 May 2004; 04II50AD-6, 1♀, Zhubi Reef, Nansha Islands, 4 May 2004; MBM 233220, 1♂, East Island, Xisha Islands, 10 June 1975; MBM 158486, 9♀♂, Shalao, 4 April 1992; MBM 158767, 1♀1♂, Yalong Bay, 26 November 1990; MBM 159998, 1♀1♂, Xincun, Hainan, Li Xin-zheng Coll., 26 March 1992.

**Description** Rostrum slender, acute, reaching to distal margin of first antennular segment; lateral margins of rostrum demarked from carapace by narrow and deep grooves, that extending to posterior margin of orbital hoods; rostral carina absent; orbital hoods slightly inflated, anterior margins bearing acute teeth, reaching to middle length of rostrum.

Telson about 3.5 times as long as posterior width; dorsal surface bearing two pairs of spines, approximately inserted 3/8 and 5/8 length of telson, respectively; posterior margin slightly convex, posterolateral angles bearing two pairs of spines.

Antennular peduncle with distally acute stylocerite reaching to middle length of second antennular segment; visible part of first antennular segment nearly as long as second segment; second segment about twice as long as wide; third segment slightly shorter than third segment. Antenna with lateral margin of scaphocerite nearly straight, antennal scale with narrow blade distinctly shorter than distolateral tooth, beyond distal margin of antennular peduncle. Carpocerite slightly longer than distolateral tooth of antennal scale.

Major cheliped with chela compressed. Fingers about 1/3 length of palm in male, and 1/5 length of palm in female. Palm almost smooth, without grooves. Merus with dorsal margin rounded distally; ventromesial margin bearing 5 spines and terminating in subacute tooth. Ischium unarmed. Minor cheliped almost as long as major cheliped. Fingers and palm subequal in length. Merus similar to that of major cheliped. Ischium unarmed.

Second pereiopod with carpus 5-segmented, segmented ratio subequal to 5: 3: 2: 2: 4.

Third and fourth pereiopod similar in shape, moderately stout. Dactylus blunt, compressed laterally, ventral margin bearing longitudinal ridge on entire length, surrounded by ridge of soft flexible chitin. Propodus with ventral margin bearing 5–7 spines and distally one pair. Carpus about 2/3 length of merus, both dorsal and ventral margins terminating rounded tooth. Merus unarmed. Ischium armed with one small spine on ventral margin.

**Remarks** *A. lottini* is readily recognized in *Alpheus* by the shape of dactylus of the last three pereiopods. This species has been described many times, and the present specimens are generally agreed with previous descriptions.

**Distribution** The South China Sea, Taiwan; Red Sea to South Africa, eastward to the Philippines; Indonesia; Australia, to Hawaii, the Galapagos Islands, and the American mainland from the Gulf of California to Colombia.

### 2.2.74 *Alpheus splendidus* Coutière, 1897 (Fig. 2.86)

*Alpheus splendidus* Coutière, 1897c: 235; De Man, 1924: 41, fig. 14; 1929: 23, Pl. 3, fig. 8; Banner DM & Banner AH, 1982: 56, fig.12; Chace, 1988: 54.

*Alpheus pomatoceros* Banner & Banner, 1966b: 93, fig. 32.

**Material Examined** CL 3.9–5.1 mm, TL 9.5–13.3 mm. MBM 158889, 2♂♂, Xinying, Hainan, Wang Yong-liang Coll., 26 July 1957; MBM 158855, 2♂♂, Xinying, Hainan, Wang Yong-liang Coll., 26 July 1957; MBM 107676, 1♂, Linchang Reef, Hainan, 4 December 1990; MBM 158772, 2♂♂, Xinyi Reef, Nansha Islands, 29 May 1990; MBM 158157, 3♂♂2♀♀, Nansha Islands, Li Xin-zheng Coll., 9 April 1992.

**Description** Rostrum acute, triangular, about twice as long as wide at base, reaching to distal margin of first antennular segment; rostral carina rounded, extending to posterior margin of orbital hoods; anterior margin between rostrum and orbital hoods concave; orbitorostral grooves moderately deep; orbital hoods slightly inflated, acute teeth arising from upper surfaces not from anterior margins.

Telson about 2.3 times as long as posterior width; dorsal surface bearing two pairs of spines, approximately inserted 1/3 and 2/3 length of telson, respectively; posterior margin slightly convex, posterolateral angles bearing two pairs of

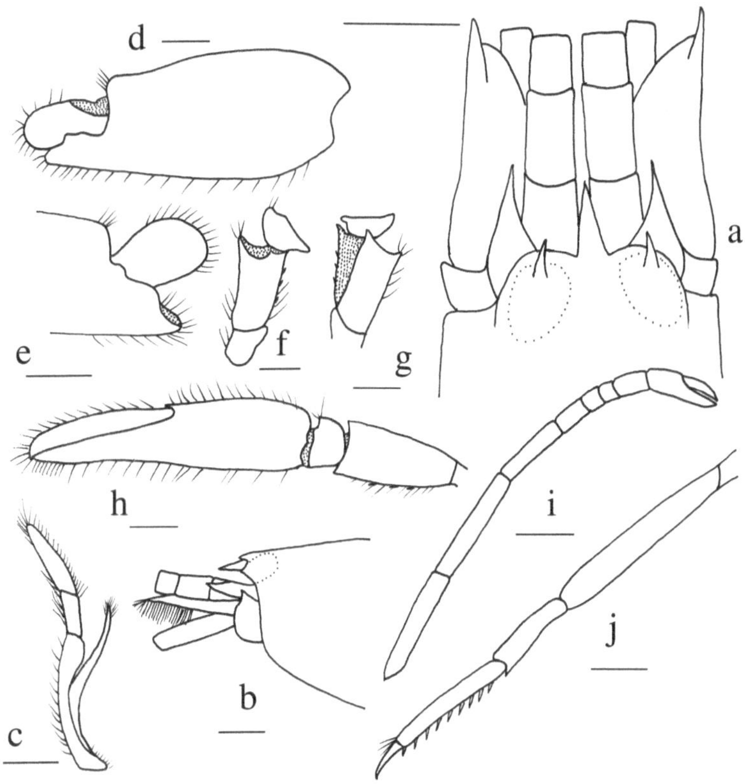


**Fig. 2.86** *Alpheus splendidus* Coutière, 1897. MBM 158157, female, **a**, anterior part of carapace, dorsal view; **b**, same, lateral view; **c**, third maxilliped; **d**, major chela, lateral view; **e**, same, fingers, mesial view; **f**, carpus to ischium of major cheliped, mesial view; **g**, same, lateral view; **h**, minor cheliped, mesial view; **i**, second pereiopod; **j**, third pereiopod (scale bar = 1 mm)

spines. Uropod with posterolateral margin of endopod bearing several spinules.

Antennular peduncle stout, with distally acute stylocerite reaching to middle length of second antennular segment; three antennular segments subequal in length, second segment about 1.2 times as long as wide. Antenna with lateral margin of scaphocerite slightly concave; antennal scale with narrow blade distinctly shorter than distolateral tooth, reaching to distal margin of antennular peduncle. Carpocerite nearly as long as distolateral tooth of antennal scale.

Major cheliped with chela moderately compressed. Fingers about 1/3 length of palm; dactylus with tip truncate. Palm with dorsal margin bearing a narrow, shallow longitudinal groove, extending from proximal 1/3 length of palm to sub-distally. Merus robust, dorsal margin bearing strong rounded tooth distally; ventromesial margin terminating in acute tooth. Ischium unarmed. Minor cheliped with chela subcylindrical. Fingers slightly longer than palm. Palm with dorsal margin terminating in acute tooth, overhanging dactylus. Merus similar to that of major cheliped. Ischium unarmed.

Second pereiopod with carpus 5-segmented, segmented ratio subequal to 10: 4: 3: 3: 4.

Third and fourth pereiopod similar in shape, moderately slender. Dactylus simple, 1/4 length of propodus. Propodus shorter than merus, ventral margin bearing 8 spines and distally one pair. Merus and ischium unarmed.

**Remarks** *A. splendidus* is readily recognized in Sulcatus group by the teeth on the orbital hoods. The present specimens are generally agreed with previous descriptions.

**Distribution** The South China Sea, Taiwan; Red Sea; Eastern Africa; Seychelles; Malaysia; Thailand; Philippines; Indonesia; Australia.

### 2.2.75 *Alpheus sulcatus* Kingsley, 1878 (Fig. 2.87)

*Alpheus sulcatus* Kingsley, 1878: 193; Crosnier & Forest, 1966: 237, fig. 9a–d; Banner DM & Banner AH, 1982: 79, fig. 20; Chace, 1988: 56.
*Alpheus luciae* Barnard, 1947: 389; 1950: 755, fig. 143j–m.
*Alpheus macrochirus* Richters, 1880: 164, Pl. 17, figs. 31–33.

**Material Examined** MBM 157839, 1♀, CL 11.7 mm, TL 33.5 mm, Shi Island, Xisha Islands, Chen Mu Coll., 2 June 1981.

**Description** Rostrum triangular, not reaching to distal margin of first antennular segment; lateral margins of rostrum demarked from carapace by deep grooves; rostral carina absent; orbital hoods slightly inflated, without teeth.

Telson about 1.4 and 2.5 times as long as anterior and posterior width; dorsal surface bearing two pairs of spines, approximately inserted 2/5 and 3/5 length of telson, respectively; posterior margin convex, posterolateral angers bearing two pairs of spines.

Antennular peduncle stout, with distally acute stylocerite teaching to distal margin of first antennular segment; visible part of first antennular segment slightly longer than third segment; second segment about twice as long as wide, 1.5 times as long as third segment. Antenna with lateral margin of scaphocerite slightly concave; antennal scale with narrow blade shorter than distolateral tooth, slightly overreaching distal margin of antennular peduncle. Carpocerite nearly as long as distolateral tooth of antennal scale.

Major cheliped with chela compressed. Fingers 1/3–1/2 length of palm; dactylus rounded, with plunger well developed. Palm with lateral face glabrous, medially bearing longitudinal groove that arising from middle length of palm and continuing to distally; dorsal surface bearing rounded ridge on entire length of palm; mesial face covering with patches of setae, bearing similar longitudinal groove like that on lateral face but less deep;

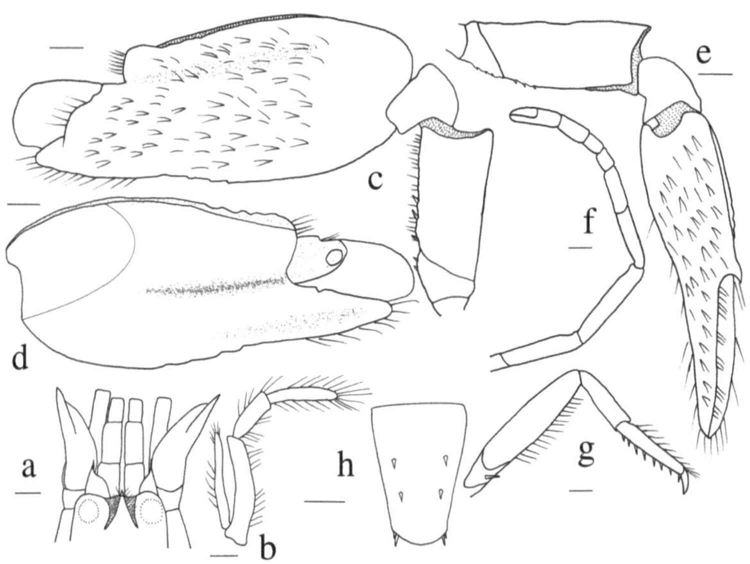


**Fig. 2.87** *Alpheus sulcatus* Kingsley, 1878. MBM 157839, female, **a**, anterior part of carapace, dorsal view; **b**, third maxilliped; **c**, major cheliped, mesial view; **d**, major chela, lateral view; **e**, minor cheliped, mesial view; **f**, second pereiopod; **g**, third pereiopod; **h**, telson (scale bar = 1 mm)

lateral face bearing moderately deep longitudinal depression medially and moderately shallow longitudinal depression near ventrally. Merus robust, both dorsal and ventromesial margin rounded distally; ventromesial margin dentate, bearing 3 small spines. Ischium with ventral margin dentate, bearing several small spines. Minor cheliped with chela subcylindrical. Fingers slightly longer than palm. Mesial face of palm with distal margin bearing acute tooth medially, flanking dactylus; dorsal margin bearing ridge on entire length; without longitudinal groove on both faces. Merus and ischium similar to that of major cheliped.

Second pereiopod with carpus 5-segmented, segmented ratio subequal to 22: 17: 7: 7: 12.

Third and fourth pereiopod similar in shape, slender. Dactylus biunguiculate with unguis on flexor margin small, or simple flexor margin only swelling. Propodus shorter than merus, ventral margin bearing 6 spines and distally one pair. Merus unarmed. Ischium armed with one spine on ventral margin.

**Remarks** *A. sulcatus* is common in the Indo-Pacific and has been described many times. The present specimens are generally agreed with the previous description.

**Distribution** The South China Sea; Southeastern Africa to Philippines; Indonesia; Japan; Australia to Society and Galapagos islands and Pacific American mainland from Gulf of California to Peru; western Africa from equator to southern Angola.

## 2.3 Genus *Arete* Stimpson, 1860 (Fig. 2.88)

**Diagnose** Body relatively stout. Carapace smooth, not setose. Rostrum bearing acute tip. Extra-corneal teeth acute, not protruding beyond anterior margin of eye; supra-corneal teeth absent or feebly developed; infra-corneal teeth absent. Orbital hoods absent. Eyes exposed in dorsal and lateral views. Pterygostomial angle at most subacute. Sixth abdominal segment with articulated plate at posteroventral angle. Antenna with normal, not particularly stout basicerite, distolateral tooth always present; scaphocerite broadly oval or rounded. Mandible with palp, incisor process slightly expanded, distally with no more than 12 small teeth. Maxillule with bilobed palp, both ventral and dorsal lobes with setae. Third maxilliped with distally projecting dorsal margin of antepenultimate segment; penultimate segment rectangular, about as long as or slightly longer than wide; ultimate segment with short spines on tip. Chelipeds feebly sexually dimorphic, usually sub-symmetrical, sometimes asymmetrical, stout, carried extended with dactylus in lateral or ventrolateral position; chela smooth. Second pereiopod with carpus 4-segmented. Third pereiopod with dactylus stout, biunguiculate. Second pleopod with appendix interna and appendix masculina in males and sometimes in females. Telson with two pairs of dorsal spines and two pairs of posterolateral spines (Anker & Jeng, 2007).

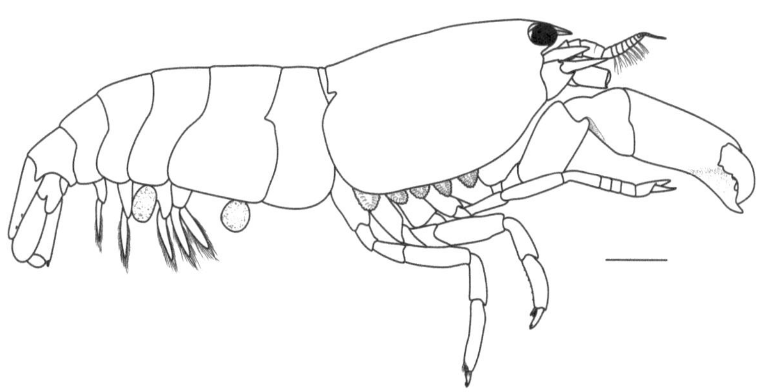

**Fig. 2.88** *Arete dorsalis* Stimpson, 1860. Ovigerous female, CL 3.3 mm, entire animal (scale bar = 1 mm)

Key of species of genus of *Arete* from China seas

-. Third pereiopod with ventromesial margin terminating in acute tooth; rostrum usually reaching near distal margin of third antennular segment ........................................................................ *A. indicus*

-. Third pereiopod with ventromesial margin terminating in feeble tooth; rostrum usually reaching near distal margin of second antennular segment ........................................................................ *A. dorsalis*

### 2.3.1 *Arete dorsalis* Stimpson, 1860 (Fig. 2.89)

*Arete dorsalis* Stimpson, 1860: 32; Coutière, 1905: 866, figs. 136, 137; Hayashi, 1995: 1854, figs. 277c, d, 278b, c; Sha & Liu, 2008: 8.

*Athanas mascarenicus* Richters, 1880: 164; Coutière, 1899: 33.

*Arete dorsalis* var. *pacificus* Coutière, 1903: 87, fig. 30.

*Arete maruteensis* Coutière, 1905: 864; Nobili, 1907: 353; Banner AH & Banner DM, 1960: 151.

*Arete maruteensis* var. *salibabuensis* De Man, 1910: 313; 1911: 169, fig. 13a–c; Banner AH & Banner DM, 1960: 151.

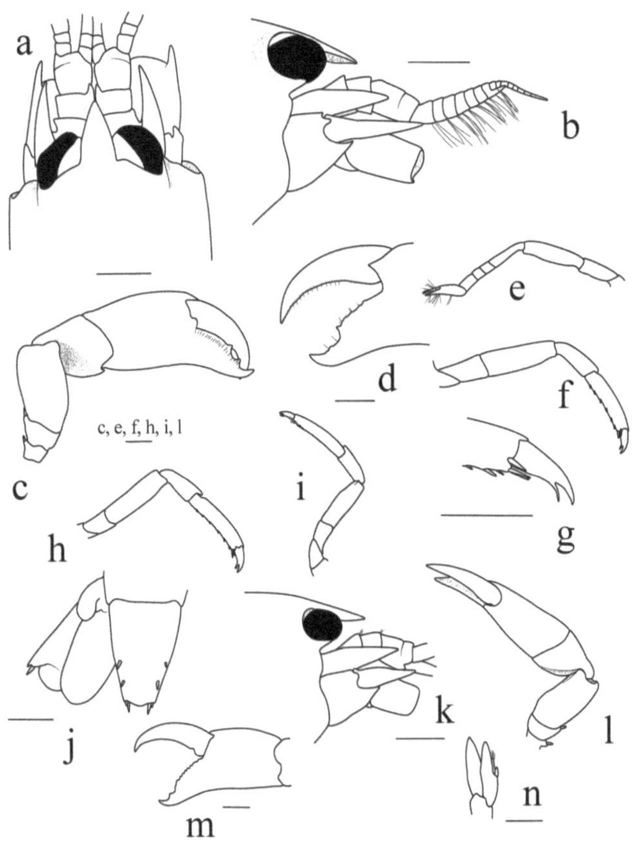


**Fig. 2.89** *Arete dorsalis* Stimpson, 1860. Ovigerous female, CL 3.3 mm, **a**, anterior part of carapace, dorsal view; **b**, same, lateral view; **c**, right major cheliped, lateral view; **d**, same, fingers, mesial view; **e**, right second pereiopod; **f**, right third pereiopod; g, same, dactylus; **h**, right fourth pereiopod; **i**, right fifth pereiopod; **j**, left uropod and telson. Male, CL 3.5 mm, **k**, anterior part of carapace, lateral view; **l**, minor cheliped; **m**, same, chela; **n**, second pleopod (scale bar = 0.5 mm)

**Material Examined** 1♂, CL 2.74 mm, TL 9.59 mm, 1♀, CL 3.42 mm, TL 10.8 mm, 90C-261, Xiaodonghai, Hainan, 14 November 1990, associated with sea urchin; 2♀♀, CL 1.51–1.62 mm, TL 5.51–5.92 mm, 57-P076, Bei Island, Xisha Islands, 30 April 1957, associated with starfish; 1♂, CL 5.57 mm, TL 18.62 mm, 1♀, CL 6.64 mm, TL 20.62 mm, CNS93-053, Yongshu Reef, Nansha Islands, 18 June 1993, associated with *Anthocidaris crassispina*.

**Description** Rostrum triangular, distally acute, variable from reaching to distal margin of first antennular to mid-length of second antennular segment. Extra-corneal well-developed; supra-corneal and infra-corneal teeth absent.

Antennular peduncle stout, with distally acute stylocerite hardly reaching to distal margin of third segment; third segment slightly longer than first and second segment. Antenna with broad blade nearly as long as lateral tooth, reaching to distal margin of antennular peduncle; carpocerite stout, hardly reaching to distal margin of antennal scale.

Major cheliped robust, with chela variable, showing slightly sexual dimorphism: in female, cutting edge of fingers armed with irregular teeth proximally; in male, cutting edge of fixed finger armed with large tooth against which the dactylus when closed; in immature specimens, cutting edges of finger dentate on 2/3 length proximally. Carpus cup-shaped. Merus stout, margins unarmed. Ischium unarmed. Minor cheliped with fingers nearly as long as palm; cutting edges of fingers dentate, bearing irregular teeth or smooth. Carpus cup-shaped. Merus and ischium similar to that of major cheliped.

Second pereiopod stout. Fingers shorter than palm. Carpus 4-segmented, ratios sub-equal to 3.7: 1: 1: 2.1. Merus about 2.5–3.7 times as long as wide. Ischium unarmed with spine on ventral margin.

Third pereiopod stout. Dactylus biunguiculate, inferior unguis shorter than superior unguis. Propodus nearly as long as merus, ventral margin bearing numerous spinules and distally one pair slender spines. Carpus about 2/3 length of merus, dorsal margin projecting as sub-acute tooth. Merus about 2.5–3.4 times as long as wide, with ventral margin terminating as small acute tooth. Ischium unarmed with spine on ventral margin.

Uropod much longer than telson. Telson 3.2–4.0 times as long as posterior width; dorsal surface armed with two pairs of spines, situated approximately on 3/5 and 4/5 length of telson; posterior margin slightly convex, bearing two pairs of small posterolateral spines.

**Remarks** *A. dorsalis* is always associated with echinoderms, and the present specimens are associated with sea urchins and starfish.

**Distribution** The South China Sea; Red Sea; Djibouti; Southern Africa; Arabian Sea; Laccadive Islands; Bay of Bengal; Japan; Indonesia; Australia; Marshall Islands; New Caledonia; Fiji; Samoa; Kermadec Island.

### 2.3.2 *Arete indicus* Coutière, 1903 (Fig. 2.90)

*Arete dorsalis indicus* Coutière, 1903: 84, figs 25–30.

*Arete indicus*.– Coutière, 1905: 863; Banner DM & Banner AH, 1973: 327; Hayashi, 1995: 186, figs. 277e, f, 278d, e; Sha & Liu, 2008: 8.

*Arete iphianassa* De Man, 1910: 302; 1911: 164, Pls. 3, 4, fig. 11; 1922: 22, Pl. 3, fig. 11; Banner AH & Banner DM, 1960: 149.

*Arete intermedius* Yu, 1931: 513, fig. 1; Banner AH & Banner DM, 1960: 149.

*Arete kominatoensis* Kubo, 1942: 82, figs. 1, 2; Miya & Miyake, 1968: 152

**Material Examined** 2 ♂♂, CL 4.49–4.77 mm, TL 14.2–16.17 mm, 87C-200, Aotou, Guangdong, Ma Xiu-tong Coll., 16 April 1987.

**Description** Rostrum triangular, slightly over-reaching distal margin of second antennular seg-

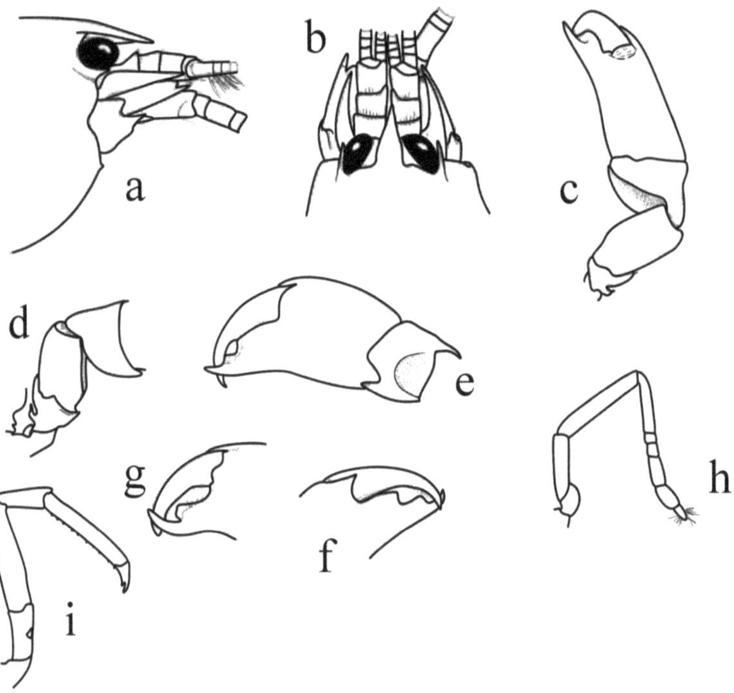


**Fig. 2.90** *Arete indicus* Coutière, 1903. (Adapted from Anker & Jeng. 2007). *Arete* cf. *indicus* Coutière, 1903, 3 males, **a**, frontal region, lateral view; **b**, same, dorsal view; **c**, left cheliped, lateral view; **d**, same, coxa to carpus, mesial view; **e**, same, chela and carpus, lateral view; **f**, fingers of chela of a different specimen, lateral view; **g**, chela fingers of a larger specimen, mesial view; **h**, second pereiopod; **i**, third pereiopod

ment. Extra-corneal teeth well-developed, reaching to anterior margin of cornea; supra- and infra-corneal teeth absent.

Antennular peduncle stout, with distally acute stylocerite not reaching to distal margin of third segment; second segment shortest, wider than long. Antennal scale with broad blade distinctly shorter than lateral teeth, reaching distal margin of antennular peduncle; carpocerite stout, slightly shorter than antennular peduncle.

Chelipeds slightly asymmetrical, carried forward. Fingers nearly as long as palm; cutting edges bearing irregular teeth. Carpus cup-shaped. Merus about twice as long as wide. Ischium with dorsal margin bearing one spine.

Second pereiopod stout. Carpus 4-segmented, ratios sub-equal to 10: 2.5: 2.5: 5.

Third pereiopod stout. Dactylus biunguiculate. Propodus distinctly longer than merus, with ventral margin bearing 9 small spines. Carpus about 1/2 length of propodus, with dorsal margin terminating in sub-acute tooth. Merus with ventral margin terminating in large acute tooth. Ischium unarmed with spine on ventral margin.

**Remarks** *A. indicus* is most similar to *A. dorsalis*, but can be distinguished from the latter by the rostrum reaching near distal margin of the second antennular segment, and the third pereiopod with merus bearing large acute tooth on ventromesial margin distally.

**Distribution** Guangdong, China; Western Somal Basin, Africa; Persian Gulf; Red Sea; Madagascar; Bay of Bengal; Japan; Indonesia; Australian; across the Central Pacific from the Marshall Islands to the Tuamotu Archipelago.

## 2.4 Genus *Aretopsis* De Man, 1910

**Diagnose** Body not unusually compressed. Carapace without high carina throughout length of dorsal midline. Rostrum present. Cornea of eye usually largely exposed in dorsal view. Mandible typically with incisor process but without palp. Chelipeds with major chela lacking socket on fixed finger to receive plunger on movable finger. second pereiopod with carpus 5-segmented. Abdomen with triangular articulated plate at posterolateral angle of sixth segment. Telson not terminating posteriorly in triangular tooth (Chace, 1988).

### 2.4.1 *Aretopsis amabilis* De Man, 1910 (Figs. 2.91, 2.92, and 2.93)

*Aretopsis amabilis* De Man, 1910: 311; 1911: 171, fig. 14: Miyake & Miya, 1967: 267, fig. 1, 2. Banner AH & Banner DM, 1968b: 272. Bruce, 1969: 175, fig. 1–4; Liu, 2008, 714.

*Aretopsis aegyptiaca*–. Ramadan, 1936: 2, fig. 9–17.

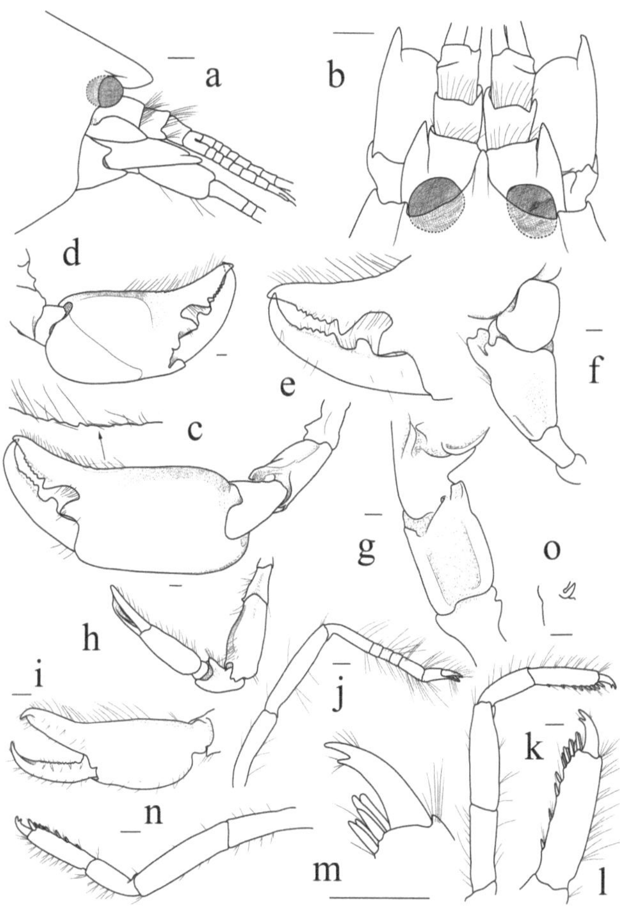


**Fig. 2.91** *Aretopsis amabilis* De Man, 1910. MBM 156248, female, CL 8.5 mm, **a**, anterior part of carapace, lateral view; **b**, same, dorsal view; **c**, left major cheliped, mesial view; **d**, major chela, lateral view; **e**, same, fingers; **f** & **g**, carpus to ischium of major cheliped; **h**, right minor cheliped; **i**, minor chela, lateral view; **j**, right second pereiopod; **k**, right third pereiopod; **l**, dactylus and propodus of right third pereiopod; **m**, same, dactylus; **n**, right fourth pereiopod. MBM 107727, female, CL 6.1 mm, **o**, basicerite of major cheliped (scale bar = 0.5 mm, **c** & **h** share same scale bar)

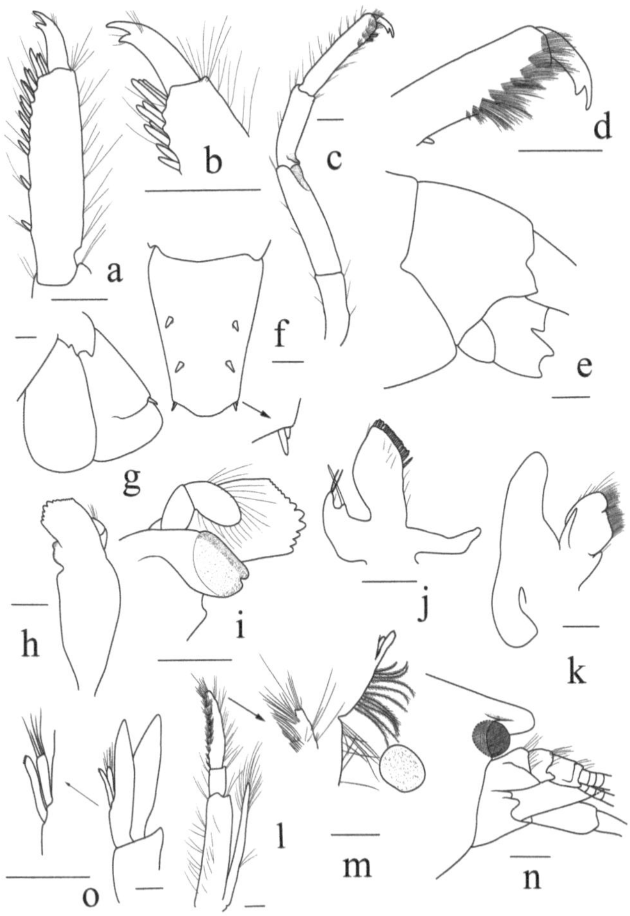


**Fig. 2.92** *Aretopsis amabilis* De Man, 1910. MBM 156248, female, CL 8.5 mm, **a**, dactylus and propodus of right fourth pereiopod; **b**, same, dactylus; **c**, right fifth pereiopod; **d**, same, dactylus; **e**, sixth abdominal segment; **f**, telson; **g**, right uropod; **h** & **i**, left mandible; **j**, left maxillule; **k**, left maxilla; **l**, left third maxilliped. MBM107727, female, CL 6.1 mm, **m**, endopod of second pleopod; **n**, anterior part of carapace, lateral view. MBM 233275, male, CL 6.9 mm, **o**, second pleopod (scale bar = 0.5 mm)

**Material Examined** MBM 158248, 1♀, CL 8.5 mm, TL 24.6 mm, Yongxing Island, Xisha Islands, 13 May 1958; MBM 107727, 1♀, CL 6.1 mm, TL 15.2 mm, Dong Island, Xisha Islands, Fang Xiu-bin Coll., 1 June 1980; MBM 233257, 1♀, CL 7.6 mm, TL 19.2 mm, 1♂, CL 6.9 mm, TL 17.1 mm, Dengqing Island, Xisha Islands, 11–13 April 1958; MBM 158733, 1♀, CL 7.2 mm, TL 18.3 mm, Shi Island, Xisha Islands, Fan Zhen-gang & Xu Jie-shan Coll., 19 May 1958; MBM 107723, 1♀, CL 6.1 mm, TL 14.8 mm, 1♂, CL 4.2 mm, TL 10.9 mm, Bei Island, Xisha Islands, Wang Yong-liang Coll., 30 April, 1957; MBM 107726, 1♂, CL 4.9 mm, TL 12.3 mm, Jinyin Island, Xisha Islands, Fang Xiu-bin Coll., 15 May 1980; MBM 284953, 1♀, CL 3.2 mm, TL 9.4 mm, Dong Island, Xisha Islands, Fang Xiu-bin Coll., 1 June 1980; MBM 107724, 1♀, CL 3.2 mm, TL, 8.3 mm, Zhaoshu Island, Xisha Islands, Fang Xiu-bin Coll., 11 May 1980; MBM 284952, 1♂, CL 5.1 mm, TL 13.2 mm, Yalong Bay, Yezhu Island, Neumann Coll., 19 March 1992; MBM 107725, 1♀, CL 5.0 mm, TL 13.2 mm, Zhaoshu Island, Xisha Islands, Fang Xiu-bin Coll., 11 May 1980; MBM 270479, 1♂, CL 9.5 mm, TL 24.1 mm, Dengqing Island, Xisha Islands, Fan Zhen-gang & Xu Jie-shan Coll., 5 May 1958; MBM 270479, 1♀, CL

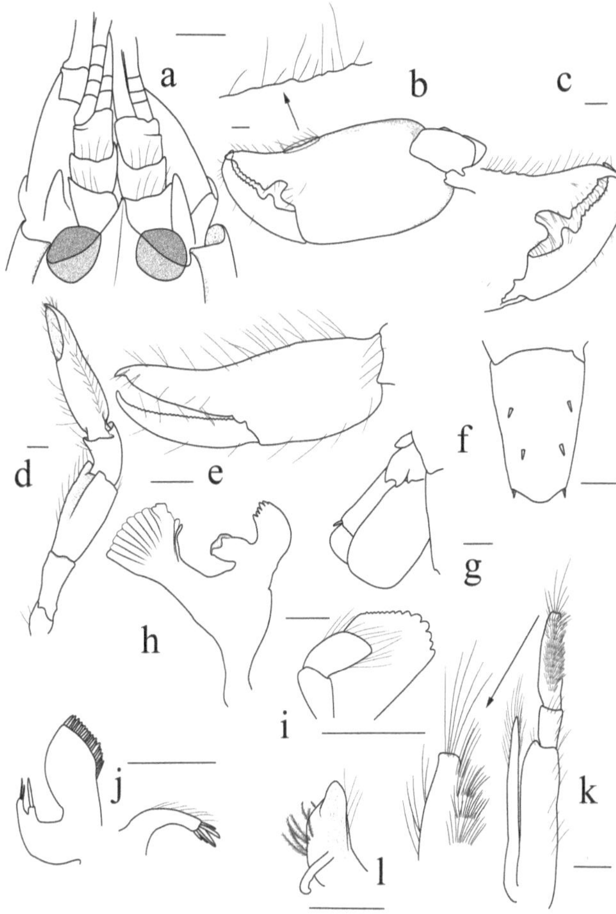


**Fig. 2.93** *Aretopsis amabilis* De Man, 1910. MBM 1077274, female, CL 6.1 mm, **a**, anterior part of carapace, dorsal view; **b**, major chela; **c**, same, fingers; **d**, minor cheliped; **e**, minor chela; **f**, telson; **g**, left uropod; **h** & **i**, left mandible; **j**, left maxillule; **k**, left third maxilliped; **l**, epipod of third maxilliped (scale bar = 0.5 mm)

14.9 mm, TL 32.5 mm, Dengqing Island, Xisha Islands, Fan Zhen-gang & Xu Jie-shan Coll., 5 May 1958.

**Description** Carapace smooth. Infra-corneal teeth regressive, tip acute. Rostrum varied from slightly overreaching end of first antennular segment to nearly distal margin of second antennular segment, horizontal, triangular; upper and lower margins well carinate. Eyes visible both in lateral and dorsal view. Pterygostomian angle not produced, obtuse. Cardiac notch well developed. Posterior about 1/2 length of carapace with thin plumose setae in lower margins.

Antennular peduncle moderately stout; visible part of first segment slightly longer than second; second segment about as long as wide at base; third segment slightly longer than first; distally acute stylocerite not reaching middle of second segment; lateral flagellum biramous, base composed of about eight segments, bearing tufts of aesthetascs. Antenna with basicerite bearing strong ventrolateral tooth and dorsolateral tooth at distal margin, the former longer than latter; scaphocerite oval, anterior margin of blade rounded, not overreaching end of third antennular segment; distolateral tooth strong, far overreaching blade, slightly curved inward; carpocerite moderately slender, far overreaching end of third segment of antennular peduncle.

Mandible with palp 2-segmented; incisor process bearing distally about 7 rounded, strong, irregular teeth and about 6 serrated, regular smaller teeth; molar process well developed. Maxillule with two endites, distal endite with numerous large spine-like setae on inner margin, basal endite with several setae on distal margins; palp bilobed, lower lobe with one robust seta, upper lobe rounded, with two moderately slender setae. Maxilla with slender tapering palp; distal endite bilobed, but not deeply, densely setose medially; proximal endite with single lobe, lanceolate, distally with setae; scaphognathite well developed. First and second maxilliped normal. Third maxilliped slender; exopod slightly overreaching end of basal segment; antepenultimate segment longest, superior margin with moderately long setae; penultimate segment about twice as long as wide at base; ventral margin of ultimate segment setose, distally with long spine-like setae; basipodite with epipod hook-shaped.

First cheliped asymmetrical in shape and size, with dactylus in ventral position. Major chela laterally compressed, with fingers about 5/6 length of palm; dactylus strongly curved to fixed finger at tip, cutting edge of dactylus and fixed finger both bearing serrate, smaller teeth distally 1/2 length, mesh when fingers closed, serrate teeth on dactylus continues to hooked, acute tip, anterior 1/2 part of dactylus armed with one larger, stronger teeth medially and a shallow concavity in proximal half, anterior part of fixed finger bearing a massive rounded tooth separated by a distinct deep interval from the crest. Palm rectangular in lateral view, compressed, lower margin bearing strong carina, concave approximate to fixed finger, bearing irregular small teeth and long setae, setae extending to about end of fixed finger. Carpus cup-shaped about 1/3 length of palm. Merus stout, about twice as long as wide, distally bearing about four irregular teeth. Ischium bearing a small triangular process approximate to distal. Minor chela compressed, with fingers about 2/3 length of palm, tips strongly hooked, crossing together when closed; cutting edge of dactylus bearing serrated small teeth almost full length, cutting edge of fixed finger flattened. Lower margin of palm slightly carina, and bearing long setae. Carpus, merus and ischium similar to that of major cheliped, but more slender.

Second pereiopod moderately slender; chela simple, fingers slightly longer than palm; carpus 5-segmented, fringed with long setae, length ratio subequal to 1.2: 0.2: 0.2: 0.2: 0.6.

Third and fourth pereiopod stout, with dactylus biunguiculate, upper unguis much longer than lower one. Propodus stout, ventral margin bearing spines (anterior part bearing rows of spines, every row consisted with about 2–3 spines), dorsal margin bearing long setae, slight shorter than merus, about four times as long as wide at base. Carpus about 1/2 length of merus, distal ventral margin without spine. Merus about 3.6 times as long as wide at base, longer than ischium, unarmed. Ischium unarmed. Fifth pereiopod more slender than third and fourth pereiopod, and distal ventral margin of propodus bearing rows of moderately long setae.

Endopod of second pleopod with appendix interna and appendix masculina in ovigerous female. Sixth abdominal segment with articulate pleura. Uropod rectangular, far overreaching telson; diaeresis of exopod ending at middle, distal corner bearing a moderately large spine. Telson about 1.3 times as long as wide anterior margin, lateral margin slightly sinuous; dorsal surface bearing two pairs of spines, situated at some distance from lateral margin, anterior and posterior pair situated about 0.41 and 0.68 length of telson, respectively; posterior margin of telson broadly rounded, bearing two pairs of spines, mesial about three times as long as lateral.

**Remarks** *Aretopsis amabilis* is originally described from Indonesia. Miyake & Bruce (1969) re-describe it from the Seychelle Islands, with describing the mouthparts of it and discussing the variations of the major first cheliped. In addition, Banner & Banner (1973) reported it from Australia, and they (1981, 1985) describe the color in life of it. Marin (2010, 2014) reports this species from Vietnam, and also provides some colorful photos of the species. Liu (2008) ever listed it from Taiwan based on Jeng (1997) from Kenting. Ramadan (1936) described another species of this genus *A. aegyptiaca*. Then, it is synonymized with *A. amabilis* (Banner and Banner, 1973; Chace, 1988; De Grave and Fransen, 2011), for the degree of the covered eyes by carapace was not enough to name a species. The present materials are plainly belonging to *A. amabilis*. The present specimens slightly differ with the previous descriptions. The posterior margin of the telson is broadly rounded in the present materials and Banner & Banner (1973, fig. 12l), while it is truncate in De Man (1910, 1911, fig. 14c) and Miyake & Miya (1967, fig. 2f); the scaphocerite is slightly overreaching the end of third segment of the antennular peduncle in present materials and Banner & Banner (1973, fig. 12a), while it is far overreaching that end in De Man (1911, fig. 14) and Miyake & Miya (1967, fig. 2a); the covered eyes described by Ramadan (1936) and Banner & Banner (1981) is not found in the present materials. Apparently, the present materials are more similar to the description made by Banner & Banner (1973) from Australia waters. Those above differences are taken as interspecific difference by Banner & Banner (1973). Moreover, the teeth on the cutting edges of major and minor chela show variable in the present specimens.

Additionally, the first chelipeds and the second pereiopod present a well-developed rudimentary exopod on the Basis in all the present materials, while Bruce (1969) states the first and second pereiopods lacked the exopods. Moreover, Miyake & Miya (1967) also show that exopod (pl. 270, fig. 2d). The presence or absence of the exopod is an important character to distinguish different species in the Caridea, so it is necessary to re-examine those specimens from the Seychelle Islands.

The ovigerous female specimens simultaneously bearing the appendix interna and appendix masculina in this species were possible hermaphrodite. Additionally, Miyake & Miya (1967) and Banner & Banner (1973) both ever reported same female specimens when described this species. The possible hermaphrodite is also found in other genus in Alpheidae: e.g. *Athanas*, and *Salmoneus* (Suzuki, 1971; Oliveira, et al., 2015). Therefore, *A. amabilis* is a possible hermaphrodite.

**Distribution** Xisha Islands, Taiwan; Gulf of Aqaba; Eylath, Israel; Seychelles Is.; Indian Ocean; E. Borneo; Okinawa and Marshall Islands; Vietnam; southern portion of the Great Barrier Reef, Australia.

## 2.5 Genus *Athanas* Leach, 1814 (Fig. 2.94)

**Diagnose** Body not unusually compressed. Rostrum distinct, acute in lateral aspect. Carapace without high carina throughout length of dorsal midline. Cornea of eye largely exposed in dorsal aspect. Abdomen with triangular flap articulated diagonally at posterolateral angle of 6th segment. Telson not terminating posteriorly in triangular tooth. Mandible with palp and molar process; third maxilliped not unusually broadened to form partial operculum over other mouthparts. Major chela without molar-like tooth on cutting edges of fingers. Second pereiopod with carpus 5-segmented (Chace, 1988).

Key of species of genus of *Athanas* from China seas

1. Supra-corneal tooth present ........ 2
-. Supra-corneal tooth absent ........ 3
2. Infra-corneal tooth more prominent than extra-corneal tooth; chelipeds carried straight forward .... *A. areteformis*
-. Extra-corneal tooth more prominent than infra-corneal; cheliped carried folded ........
........ *A. djiboutensis*
3. Dactylus of last three pereiopods biunguiculate ........ 4
-. Dactylus of last three pereiopods simple ........ 5

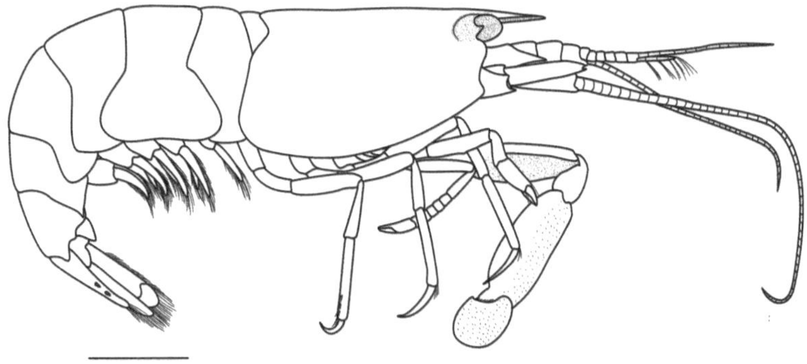

**Fig. 2.94** *Athanas ohsima* Yokoya, 1936. Male, CL 3.9 mm, entire animal (scale bar = 2 mm)

4. Palm of major chela bearing pronounced lobe on ventral margin; fingers of minor chela dorsoventrally flattened ........ *A. ornithorhynchus*

-. Palm of major chela smooth, without lobes; fingers of minor chela as usual, not dorsoventrally flattened ........ *A. parvus*

5. Chela of major and minor cheliped furnished with long and fine setae ........ *A. squillophilus*

-. Chela of major and minor cheliped not furnished with setae ........ 6

6. Dactylus of male major chela strongly curved, semilunar ........ *A. oshimai*

-. Dactylus of male major chela only slightly curved, not semilunar ........ 7

7. Pterygostomian angle with acute tooth ........ *A. polymorphus*

-. Pterygostomian angle rounded ........ 8

8. Infra-corneal tooth acute ........ 9

-. Infra-corneal tooth projecting and blunt ........ *A. dimorphus*

9. Second pereiopod slender, with second and third carpal segments 3–4 times as long as wide ........ *A. gracilipes*

-. Second pereiopod robust, with second and third carpal segments nearly as long as wide ........ 10

10. Rostral carina absent ........ *A. hongkongensis*

-. Rostral carina present, extending to posterior margin of orbital hoods ........ *A. japonicus*

### 2.5.1 *Athanas areteformis* Coutière, 1903 (Fig. 2.95)

*Athanas areteformis* Coutière, 1903: 79, fig. 17, 18; 1905: 860; Banner AH & Banner DM, 1960: 138; Banner DM & Banner AH, 1973: 304, fig. 2; 1978: 234; Chace, 1988: 61; Sha & Liu, 2007: 750.

*Athanas naifaroensis* Coutière, 1903: 77, figs. 14–16.

*Athanas erythraeus* Ramadan, 1936: 13, fig. 1.

*Athanas dubius* Banner, 1956: 322, fig. 2.

**Material Examined** TL 9.0–9.5 mm. CJ97C-207, 1♂, Dadonghai, Hainan, Li Xin-zheng Coll., 4 March 1997; 97C-235, 2♀♀, Xiaodonghai, Hainan, Li Xin-zheng Coll., 5 March 1997; 80X-229-2, 1♀, Yongxing Island, Xisha Islands, 11–13 June 1980.

**Description** Rostrum triangular, usually over-reaching distal margin of second antennular segment to middle length of third segment; orbit with obscure supra-corneal tooth, and extra-corneal tooth much stronger than infra-corneal tooth, beyond distal margin of cornea. Pterygostomian angle rounded.

Antennular peduncle robust; visible part of first segment longer than second segment; second

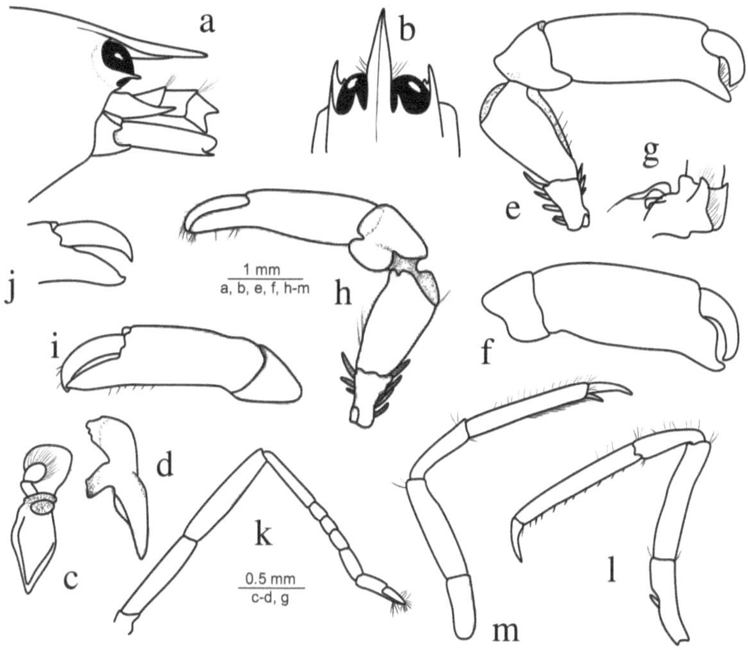


**Fig. 2.95** *Athanas areteformis* Coutière, 1903 (adapted from Anker & Jeng, 2007). Male, **a**, anterior part of carapace, lateral view; **b**, anterior margin of carapace, dorsal view; **c** & **d**, mandible from 2 different aspects; **e**, left major cheliped, mesial view; **f**, same, chela and carpus, mesial view; **g**, same, coxa and basis with exopod; **h**, right minor cheliped, mesial view; **i**, same, chela and carpus, mesial view; **j**, same, chela fingers, lateral view; **k**, second pereiopod; **l**, third pereiopod; **m**, fifth pereiopod

segment longer than third segment; distally acute stylocerite slightly overreaching mid-length of second segment. Antenna with basicerite bearing strong ventro-lateral tooth; antennal scale with distally convex blade slightly longer than disto-lateral tooth; carpocerite reaching middle length of third antennular segment.

Chelipeds nearly symmetrical in size. Major cheliped robust, with chela sub-cylindrical. Fingers slightly longer than 1/3 length of palm; tip of dactylus curved. Palm smooth, without depressions or ridges. Carpus cup-shaped. Merus with dorsal and ventromesial margin rounded distally, distally widening, unarmed. Ischium with dorsal margin bearing 4 slender spines and distally one; ventral margin armed with two slender spines and distally one. Minor cheliped slender. Fingers about half-length of palm; cutting edges of fingers bearing irregular small teeth. Palm smooth. Carpus cup-shaped. Merus more slender than that of major cheliped. Ischium similar to that of major cheliped.

Second pereiopod robust, with chela simple, fingers slightly shorter than palm. Carpus 5-segmented, with ratio of segments from proximal to distal approximately sub-equal to 3.5: 1: 1: 1: 1.6.

Third and fourth pereiopod moderately stout, with dactylus simple, shorter than 1/3 length of propodus, slightly curved. Propodus longer than merus, ventral margin bearing about 10 spines and distal one pair. Carpus shorter than merus, ventral margin bearing one spine distally. Merus stout, unarmed. Ischium armed with one spine on ventral margin. Fifth pereiopod similar to third and fourth, but ventral margin bearing rows of transverse setae; carpus elongate, only slightly shorter than merus.

**Remarks** *A. areteformis* is readily distinguished from *A. djiboutensis* by the well-developed extra-corneal tooth and the forward directed major and minor cheliped. The present specimens are generally agreed with previous descriptions, except for the second antennular segment longer than third segment rather than shorter than the third segment as shown in Banner & Banner (1973, 305, fig. 2a).

**Distribution** The South China Sea; Red sea; South Africa; Maldives; Philippines; Indonesia; Mariana Archipelago; Australia; Marshall; Samoa; Fiji; Tonga; Society Island.

### 2.5.2 *Athanas dimorphus* Ortmann, 1894 (Fig. 2.96)

*Athanas dimorphus* Ortmann, 1894: 12; Coutière, 1903: 77; 1905: 858; Tattersall, 1921: 371; Banner AH & Banner DM, 1966b: 28; 1983: 76; Banner DM & Banner AH, 1973: 313; Chace, 1988: 61; Hayashi, 1995: 2, figs. 268b, 279b, 270c, d; Sha & Liu, 2007: 751.

*Athanas setoensis* Kubo, 1951: 265; Banner DM & Banner AH, 1973: 313.

*Athanas solenomerus* Coutière, 1897b: 381; 1897: 233.

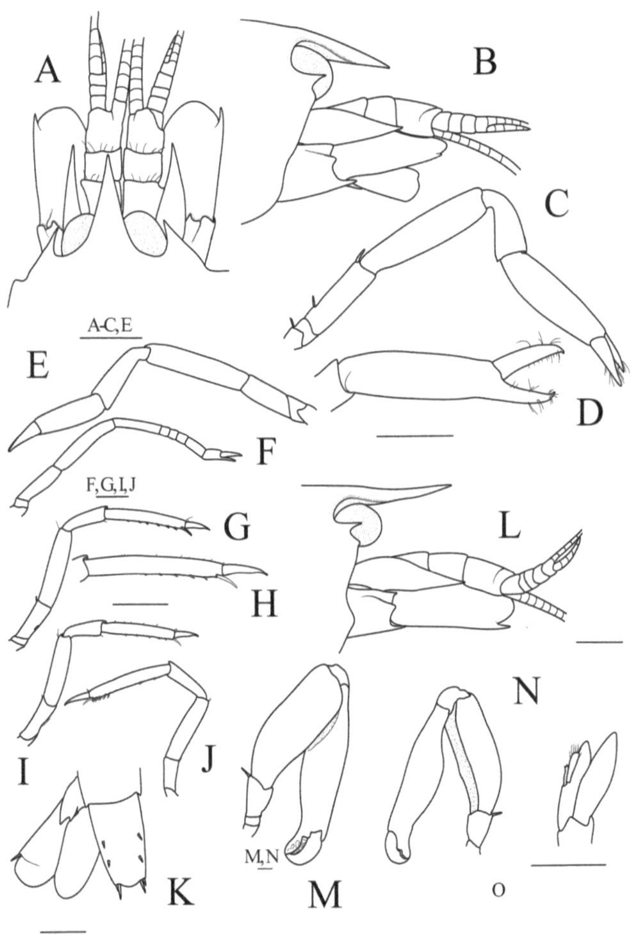


**Fig. 2.96** *Athanas dimorphus* Ortmann, 1894. ovigerous female, CL 2.9 mm, **a**, anterior part of carapace, dorsal view; **b**, same, lateral view; **c**, right major cheliped; **d**, major chela; **e**, left minor cheliped; **f**, right second pereiopod; **g**, right third pereiopod; **h**, same, propodus and dactylus; **i**, right fourth pereiopod; **j**, right fifth pereiopod; **k**, telson and left uropod. Male, CL 5.1 mm. **l**, anterior part of carapace, lateral view; **m**, right first pereiopod, lateral view; **n**, same, mesial view. Male, CL 2.8 mm. o, right second pleopod (scale bar = 0.5 mm)

*Athanas leptocheles* Coutière, 1897b: 381; 1897: 233.
*Athanas dispar* Coutière, 1897c: 233; 1897: 301.
*Athanas dimorphus seeding* Banner AH & Banner DM, 1966b: 28, fig4.

**Material Examined** TL 5.5–7.8 mm. 3♀♀, Cangkou, Jiaozhou Bay, Shandong, 22 September 1952; 2♀♀, Cangkou, Jiaozhou Bay, Shandong, Zhang Xiu-ji Coll., 21 November 1953; 3♀♀, Cangkou, Jiaozhou Bay, Shandong, Zhang Feng-Xuan Coll., 5 October 1955; 4♀♀, 56-638, Cangkou, Jiaozhou Bay, Shandong, Zhang Feng-Xuan Coll., 30 July 1956; 1♀, Cangkou, Jiaozhou Bay, Shandong, Liu Yi-qing Coll., 2 July 1957; 3♂♂3♀♀, 54-270, Weizhou Island, Guangxi, 23 December 1954; 3♂♂4♀♀, 54-154, Zhapo, Guangdong, 15 November 1954; 1♂1♀, 92C-1467, Linchang Reef, Hainan, Turkey Coll., 10 April 1992; 1♂1♀, C597C-335, Dadonghai, Hainan, Li Xin-zheng Coll., 10 November 1997; 1♂1♀, C597C-103, Dadonghai, Hainan, 1 March 1997; 1♂3♀, C597C-093, Dadonghai, Hainan, 1 March 1997; 2♂♂13♀♀, 56-K006, Xinying, Hainan, Liu Rui-yu Coll., 11–12 January 1956; 1♂2♀♀, 92C-1124, Lanmai, Hainan, Turkey Coll., 10 September 1992; 1♂, 90C-352, Meixia, Hainan, Wang Yong-liang Coll., 2 November 1990.

**Description** Rostrum reaching to or slightly overreaching distal margin of second antennular segment. Supra-corneal teeth absent; distally acute extra-corneal teeth stronger than distally rounded infra-corneal. Pterygostomial angle subacute.

Antennular peduncle stout, with distally acute stylocerite variable from reaching to distal margin of second segment to mid-length of third segment; visible part of first segment about 1.6 times as long as second segment; third segment nearly as long as second segment. Antenna with basicerite with strong ventrolateral tooth; antennal scale with broad blade distinctly shorter than distolateral tooth, overreaching distal margin of antennular peduncle.

Major and minor cheliped sexual dimorphism, symmetrical in both male and female, carried flexed at meral-carpal articulation. Chelipeds of female smaller and more slender than that of male; fingers longer than 1/2 length of palm; carpus elongate, nearly as long as merus; merus variable from 6.5 to 8.8 times as long as wide; ischium unarmed. Cheliped of male robust, inflated, with chela cylindrical in section; dactylus curved distally, 2/5 length of palm, cutting edges of fingers bearing usually bearing rounded teeth, and cutting edge of dactylus fringed with setae; carpus cup-shaped; merus slightly shorter than chela, about 3 times as long as wide, ventral face strongly excavated to accommodate chela when flexed; ischium slightly longer than wide, dorsal margin bearing 1–2 spines.

Second pereiopod with carpus 5-segmented, with ratio of segments from proximal to distal approximately sub-equal to 10: 1: 1: 1: 4.

Third and fourth pereiopod with dactylus simple. Propodus nearly as long as merus, ventral margin bearing several spines and distally one pair. Carpus half-length of merus. Merus about 4.6 times as long as wide.

**Remarks** *A. dimorphus* is often found in pools under rocks, coral heads at low tide.

**Distribution** The Yellow Sea, the South China Sea; Red sea; East Africa; Bay of Bengal (India); Thailand; Vietnam; Southern Philippines; Japan; Australian; New Caledonia.

### 2.5.3 *Athanas djiboutensis* Coutière, 1897 (Fig. 2.97)

*Athanas djiboutensis* Coutière, 1897c: 233; 1905: 856; Boone, 1935: 119; Banner AH & Banner, 1960: 140; Miya & Miyake, 1968: 131; Banner DM & Banner AH, 1973: 306; 1978: 234; Chace, 1988: 62; Hayashi, 1995: 4, figs. 267a, 268c, 269c, 270e, f; Sha & Liu, 2007: 751.

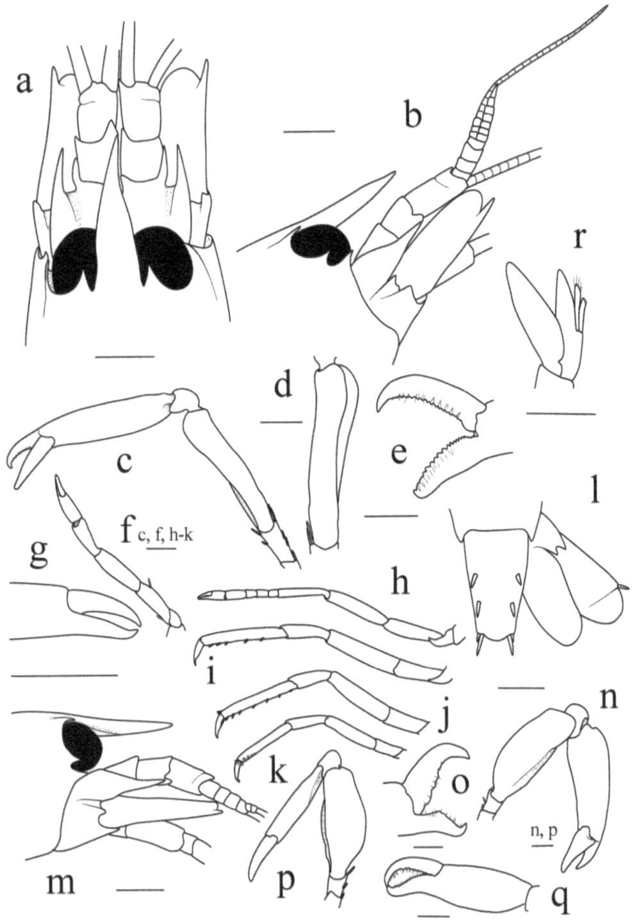


**Fig. 2.97** *Athanas djiboutensis* Coutière, 1897. Ovigerous female, CL 3.2 mm, **a**, anterior part of carapace, dorsal view; **b**, same, lateral view; **c**, left major cheliped; **d**, same, merus; **e**, fingers of major chela; **f**, right minor cheliped; **g**, same, fingers; **h**, right second pereiopod; **i**, right third pereiopod; **j**, left fourth pereiopod; **k**, left fifth pereiopod; **l**, right uropod and telson. Male, CL 3.9 mm, **m**, anterior part of carapace, dorsal view; **n**, right major cheliped; **o**, same, fingers; **p**, left minor cheliped; **q**, same, chela; **r**, second pleopod (scale bar = 0.5 mm)

*Athanas sulcatipes* Borradaile, 1898: 1011, fig. 9; Coutière, 1905: 856.

**Material Examined** TL 5.81–8.80 mm. 1♂7♀♀, 57-P104, Shi Island, Xisha Islands, Wang Yong-liang Coll., 13–15 May1957; 3♂♂5♀♀, 58C-150, Shi Island, Xisha Islands, Fan Zhen-gang & Xu Jie-shan Coll., 6 April 1958; 1♂, 58C-005, Wude Island, Xisha Islands, Fan Zhen-gang & Xu Jie-shan Coll., 20 March 1958; 1♀, 58C-086, Wude Island, Xisha Islands, Fan Zhen-gang & XU Jie-Shan Coll., 31 March 1958; 1♂1♀, 58C-104, Wude Island, Xisha Islands, Fan Zhen-gang & Xu Jie-shan Coll., 2 April 1958; 1♀, Wude Island, Xisha Islands, Fan Zhen-gang & Xu Jie-shan Coll., 15 May 1958; 1♀, 57-P033, Wude Island, Xisha Islands, 13–17 April 1957; 2♂♂7♀♀, 80X-216-8, Yongxing Island, Xisha Islands, Wang Yong-liang & Ren Xian-qiu Coll., 11–13 June 1980; 1♀, 80X-35-9, Yongxing Island, Xisha Islands, Wang Yong-liang & Ren Xian-qiu Coll., 9 June 1980; 3♀♀, 80X-223-4,Yongxing Island, Xisha Islands, Wang Yong-liang & Ren Xian-qiu Coll., 11–13 June 1980; 1♂, 80X-229-2, Yongxing Island, Xisha Islands, Wang Yong-liang & Ren Xian-qiu Coll., 11–13 June 1980; 3♂♂23♀♀, 80X-224, Yongxing Island, Xisha Islands, Wang Yong-liang & Ren Xian-qiu Coll.,

11–13 June 1980; 3♀♀, Jinqing Island, Xisha Islands, Fan Zhen-gang & Xu Jie-shan Coll., 30 April 1958; 14♀♀, 58C-534, Jinqing Island, Xisha Island, Fan Zhen-gang & Xu Jie-shan Coll., 24 April 1958; 1♂12♀♀, 80-191-13, Dong Island, Xisha Islands, Wang Yong-liang & Ren Xian-qiu Coll., 28–31 May 1980; 1♂1♀, 80x-189-4, Dong Island, Xisha Islands, Wang Yong-liang & Ren Xian-qiu Coll., 28–31 May 1980; 3♀♀, 80X-151-7, Shanhu Reef, Xisha Islands, Wang Yong-liang & Ren Xian-qiu Coll., 19–21 May 1980; 1♀, 80X-140-7, Shanhu Reef, Xisha Islands, Wang Yong-liang & Ren Xian-qiu Coll., 19–21 May 1980; 1♀, Dong Island, Xisha Islands, Wang Yong-liang & Ren Xian-qiu Coll., 3 June 1980; 1♀, 80X-47-5, Shi Island, Xisha Islands, Wang Yong-liang & Ren Xian-qiu Coll., 28 May 1980; 1♀, 80X-62-10, Zhaoshu Island, Xisha Islands, 11 May 1980; 1♂, 93-076B, Anda Island, Nansha Islands, 23 May 1993; 1♂, 93-063B, Zhubi Reef, Nansha Islands, 20 May 1993.

**Description** Rostrum slender, slightly overreaching distal margin of second antennular segment; rostral carina sharp, extending to posterior margin of orbit. Supra-corneal tooth present distally acute; distally acute infra-corneal tooth stronger than extra-corneal tooth. Pterygostomial angle rounded.

Antennular peduncle slender, with distally acute stylocerite slightly overreaching mid-length of second segment; visible part of first segment about 1.5 times as long as second segment; third segment sub-equal to second segment. Antenna with basicerite bearing ventrolateral tooth; antennal scale with distally convex blade slightly overreaching distolateral tooth.

Major and minor cheliped asymmetrical, folded beneath. Major cheliped with chela sub-cylindrical; showing slightly sexual dimorphism, with that of female more slender than male. Fingers about 1/2 length of palm; tips curved; cutting edges crenate. Palm smooth, without depressions or ridges. Carpus cup-shaped. Merus sub-equal to palm, slightly swollen, with ventral face deeply excavated. Ischium slightly shorter than half-length of merus, with dorsal and ventral margins armed with 3 and 2 spines, respectively. Minor cheliped slender, cylindrical. Fingers slightly shorter than palm. Carpus elongate, about 1.5 times as long as palm. Merus longer than carpus, with ventral margin flattened. Ischium sub-equal to carpus, with dorsal margin armed with 3 spines.

Second pereiopod with carpus 5-segmented, ratio of segments from proximal to distal approximately sub-equal to 10: 3: 3: 3: 4.

Third and fourth pereiopod slender. Dactylus simple, about 1/3 length of propodus. Propodus with ventral margin bearing several spines and distally slender one pair. Carpus about 1/2 length of propodus. Merus nearly as long as propodus, unarmed. Ischium armed with 2 spines on ventral margin.

Telson about 3 times as long as posterior margin; dorsal surface bearing two pairs of spines, situate about 2/5 and 7/10 length of telson, respectively; posterior margin slightly concave, bearing two pairs of slender posterolateral spines.

**Remarks** The present specimens are generally agreed with the previous descriptions, but show slightly difference in the second antennular longer than the third segment, rather than as long as third segment as described by Banner & Banner (1973).

**Distribution** The South China Sea; Israel; Red Sea; Djibouti; Maldives and Laccadive Arch.; Indonesia; Japan; Australia; Marshall; Funafuti; Samoa; Fiji; Phoenix; Tonga; Society and Marquesas Islands.

### 2.5.4 *Athanas gracilipes* Banner & Banner, 1978 (Fig. 2.98)

*Athanas gracilipes* Banner DM & Banner AH, 1978: 234, fig. 3a–g.

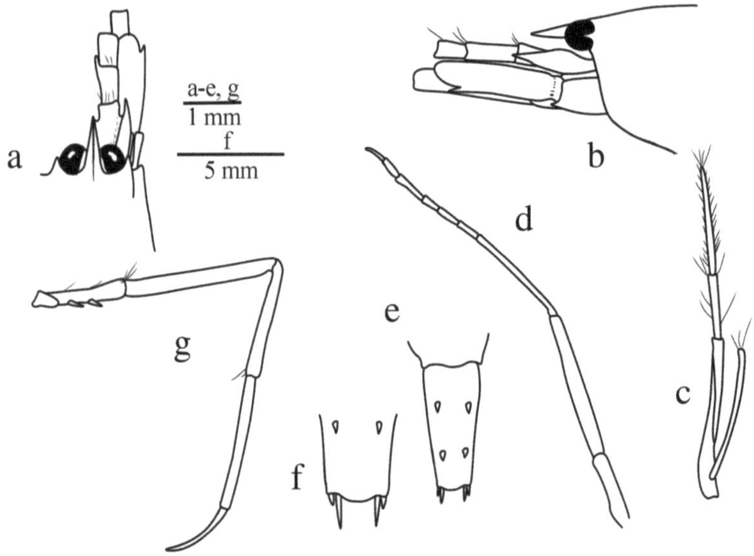


**Fig. 2.98** *Athanas gracilipes* Banner & Banner, 1978. (Adapted from Banner & Banner, 1978). Female, **a**, anterior part of carapace, dorsal view; **b**, same, lateral view; **c**, third maxilliped, lateral face; **d**, second pereiopod; **e**, **f**, telson with tip enlarged. Female, **g**, third pereiopod

**Material Examined** This species was reported by Banner & Banner (1978) and Liu (2008) from Hong Kong, but not found in the present specimens.

**Description** Rostrum acute, 2.5 times as long as wide at base, hardly reaching to distal margin of first antennular segment; rostral carina sharp, extending to posterior margin of eyes. Extra-corneal tooth distally acute, reaching to mid-length of eyes; infra-corneal tooth distally rounded, not greatly projecting; supra-corneal tooth absent. Pterygostomial angle rounded.

Antennular peduncle slender, with distally acute stylocerite slightly overreaching distal margin of first segment; visible part of first segment much longer than second segment; second segment 1.7 times as long as wide distally; third segment slightly shorter than second segment. Antenna with basicerite small acute lateral tooth; antennal scale with distally convex blade beyond distolateral tooth, reaching to distal margin of antennular peduncle; carpocerite far overreaching distal margin of antennular peduncle.

Major and minor cheliped missing.

Second pereiopod much slender, with carpus 5-segmented, ratio of segments from proximal to distal approximately sub-equal to 10: 3: 3: 3: 4, first carpal segment 12.5 times as long as wide.

Third and fourth pereiopod slender, with dactylus simple, long and strongly curved, 0.45 times as long as merus. Propodus 0.8 times as long as merus, bearing tuft of distodorsal setae, unarmed with spines on ventral margin. Carpus half-length of merus. Merus 6 times as long as wide, unarmed. Ischium nearly as long as carpus, ventral margin armed with 2 spines.

Telson 3.5 times as long as posterior margin; dorsal margin bearing two pairs of spines, situate about 1/3 and 2/3 length of telson, respectively; posterior margin slightly convex, bearing two slender posterolateral spines.

**Remarks** *A. gracilipes* is originally described from Hong Kong. Unfortunately, the major and minor cheliped lost in the type specimens.

**Distribution** The South China Sea (Hong Kong).

### 2.5.5 *Athanas hongkongensis* Bruce, 1990 (Fig. 2.99)

*Athanas hongkongensis* Bruce, 1990: 624, figs. 7, 8.

**Material Examined** This species was reported by Bruce (1990) and Liu (2008) from Hong Kong, but not found in the present specimens.

**Description** Rostrum distally acute, slender, reaching to mid-length of second antennular segment; rostral carina absent. Supra-corneal tooth absent; extra-corneal tooth well-developed, dis-

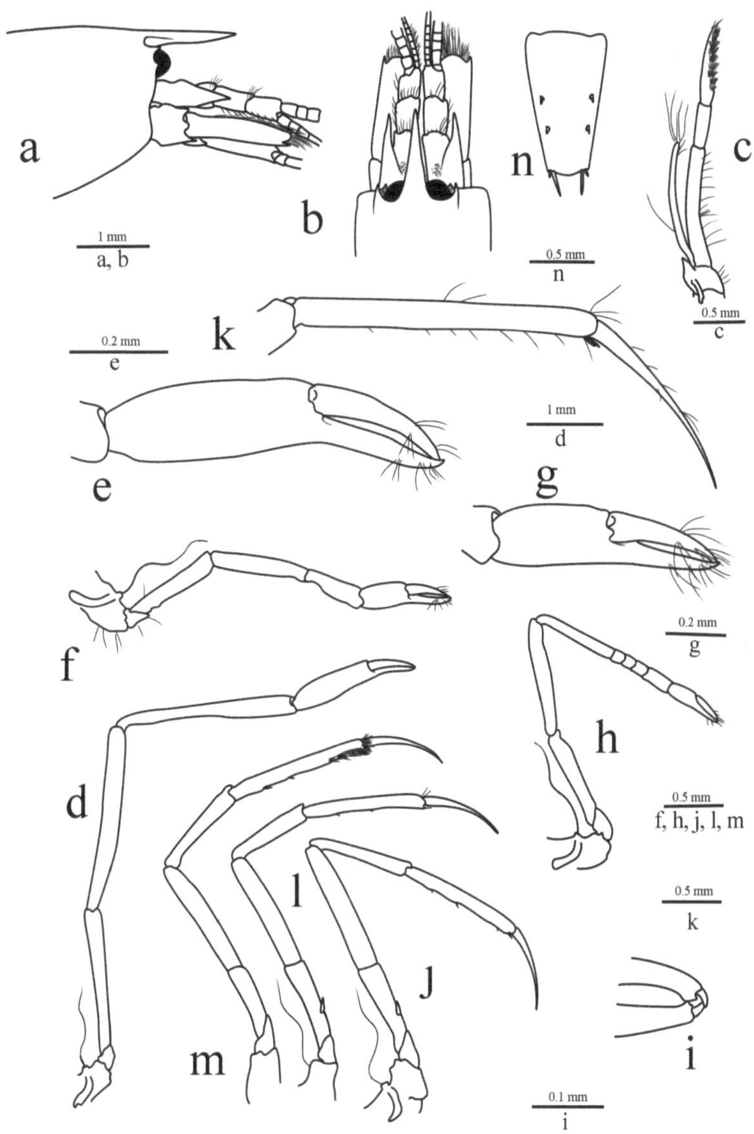


**Fig. 2.99** *Athanas hongkongensis* Bruce, 1990. (Adapted from Bruce, 1990). **a**, anterior part of carapace, lateral view; **b**, same, dorsal view; **c**, third maxilliped; **d**, first pereiopod; **e**, same, chela; **f**, first pereiopod; **g**, same, chela; **h**, second pereiopod; **i**, same, tips of fingers; **j**, third pereiopod; **k**, same, dactylus and propodus; **l**, fourth pereiopod; **m**, fifth pereiopod; **n**, telson

tally acute; infra-corneal tooth distally acute, smaller than extra-corneal tooth. Pterygostomial angle broadly rounded.

Antennular peduncle slender, with distally acute stylocerite, reaching to mid-length of second segment; first segment about twice as long as second segment, second segment nearly as long as third segment. Antenna with stout basicerite bearing acute ventrolateral tooth; antennal with distally convex blade not beyond distolateral tooth, overreaching distal margin of antennular peduncle; carpocerite slightly shorter than antennal scale.

Major and minor cheliped symmetrical in shape and size, short or elongate, variable from overreaching carpocerite by about 0.3 length of palm to by 0.5 length of merus. Fingers sub-equal to or slightly longer than palm, with tips feebly bidentate. Palm smooth, without depression or ridge. Carpus about 0.7–1.5 times as long as chela. Merus 1–1.6 times as long as carpus, 6.0–1.0 times as long as wide. Ischium 1–0.77 times as long as merus, unarmed.

Second pereiopod slender, with fingers sub-equal to palm. Carpus 5-segmented, ratio of segments from proximal to distal approximately sub-equal to 27: 4: 4: 5: 9.

Third and fourth pereiopod slender. Dactylus slender, simple, feebly curved, 0.66–1 times as long as propodus. Propodus with ventral margin bearing 2 spines and distally one slender. Carpus about 0.66 times as long as propodus, unarmed. Merus slightly shorter than propodus, about 6.3 times as long as wide, unarmed. Ischium slightly shorter than carpus, about 4.2 times as long as wide, ventral margin armed with one strong spine.

Telson about 1.6 times as long as proximal width; dorsal surface bearing two pairs of spines, situate about 1/2 and 3/5 length of telson, respectively; posterior margin slightly convex, bearing two small posterolateral spines.

**Remarks** *A. hongkongensis* is originally described from Hong Kong.

**Distribution** The South China Sea (Hong Kong).

### 2.5.6 *Athanas japonicus* Kubo, 1936 (Fig. 2.100)

*Athanas japonicus* Kubo, 1936a: 43; Miya & Miyake, 1968: 139, figs. 4–6; Banner DM & Banner AH, 1973: 308, fig. 4; Hayashi, 1995: 4, figs. 268e, 270h, i; Anker, 2003a: 301, figs. 14–19.

*Athanas lamellifer* Kubo, 1940: 102, fig. 22; Miya & Miyake, 1968: 139; Sha & Liu, 2007: 752.

**Material Examined** TL 7.1–12.4 mm. 2♀♀, 54-0127, Cangkou, Shandong, 30 August 1954; 1♂, 54-0172, Cangkou, Shandong, Zhang Feng-xuan Coll., 29 September 1954; 1♀, Cangkou, Shandong, March 1955; 2♂♂1♀, Cangkou, Shandong, Zhang Feng-xuan Coll., 8 July 1955; 2♀♀, Cangkou, Shandong, Zhang Feng-xuan Coll., 5 October 1955; 1♀, 57-409, Xuejiadao, Shandong, 28 May 1957; 1♂41♀♀, 63-59, Xuejiadao, Shandong, Fan Zhen-gang Coll., 20 July 1963; 4♀♀, 54-157, Zhapo, Guangdong, 16 November 1954; 2♂♂14♀♀, 54-302, Leizhou, Guangdong, Wang Yong-liang & Liu Rui-yu Coll., 9 December 1954; 1♂1♀, 90C-282, Luhuitou, Hainan, Turkey Coll., 19 November 1990; 1♂2♀♀, 92Cc-021, Luhuitou, Hainan, Liu Rui-yu & Li Jing-he Coll., 17 March 1992; 1♂, 92C-1468, Linchang, Hainan, Turkey Coll., 14 April 1992.

**Description** Carapace sparsely setose. Rostrum acute slightly overreaching distal margin of first antennular segment. Supra-corneal teeth absent; infra-corneal teeth slightly longer than extra-corneal teeth, distally acute. Eyestalks visible in dorsal and lateral view. Pterygostomial angle rounded.

Antennular peduncle moderately stout, with distally acute almost reaching mid-length of sec-

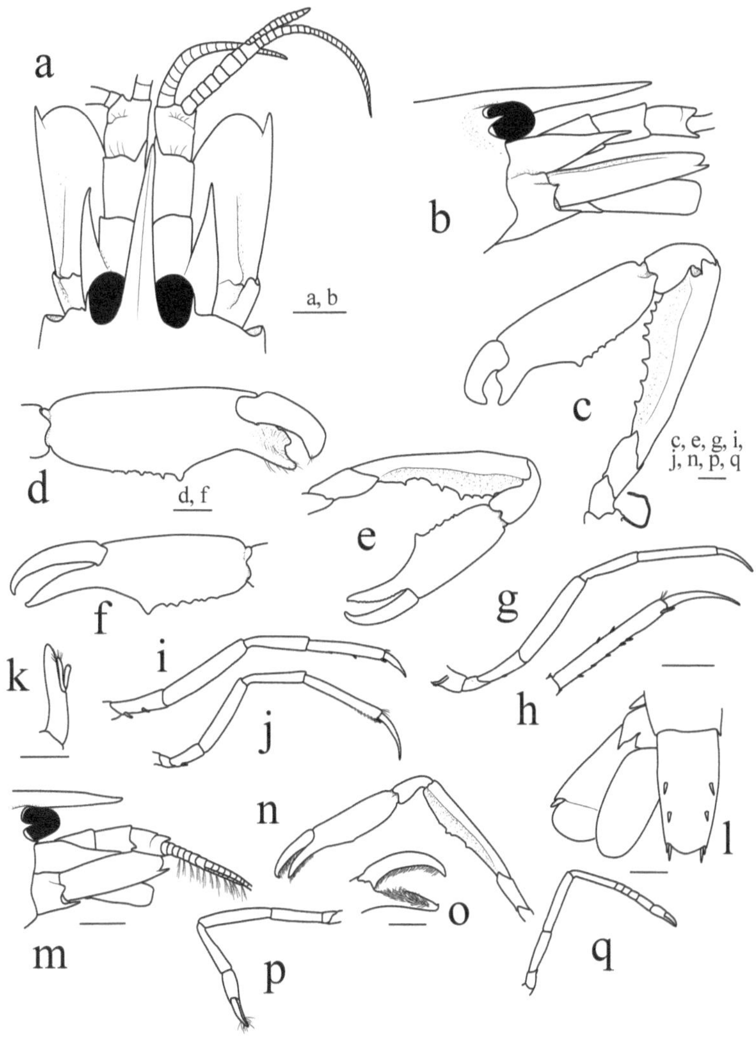


**Fig. 2.100** *Athanas japonicus* Kubo, 1936. Male, CL 3.7 mm, **a**, anterior part of carapace, dorsal view; **b**, same, lateral view; **c**, major cheliped, mesial view; **d**, major chela, lateral view; **e**, minor cheliped, mesial view; **f**, minor chela, lateral view; **g**, third pereiopod; **h**, same, dactylus and propodus; **i**, fourth pereiopod; **j**, fifth pereiopod; **k**, endopod of second pleopod; **l**, left uropod and telson. Ovigerous female, CL 4.1 mm, **m**, anterior part of carapace, lateral view; **n**, major cheliped, mesial view; **o**, same, fingers, lateral view; **p**, minor cheliped; **q**, second pereiopod (scale bar = 0.5 mm)

ond segment; visible part of first segment longer than second segment. Antenna with basicerite bearing acute ventrolateral tooth; antennal scale with distally convex blade shorter than distolateral tooth, hardly reaching to distal margin of antennular peduncle; carpocerite slightly shorter than antennal scale.

Major and minor cheliped showing sexual dimorphism. In male sub-equal in length, but very asymmetrical in shape. Major cheliped robust, with fingers distally curved, about 1/3 length of palm; cutting edge of dactylus slightly serrated, and cutting edge of fixed finger armed with one large rectangular tooth; lateral face of dactylus and fixed finger covered with fine setae. Palm sub-cylindrical, mesial face flattened and forming an angle with ventral face; lateral margin of ventral face bearing 3 tubercles situated approximately at mid-length. Carpus cup-shaped, distal margin with lobes. Merus elongate, with ventral face shallowly excavated; ventrolateral margin armed with 8 teeth; ventromesial margin straight. Ischium unarmed. Minor cheliped with fingers about 0.7 times as long as palm, curved distally; cutting edges irregularly serrated. Palm with ventral margin bearing tubercles. Carpus and merus similar to that of major cheliped. Ischium unarmed. In female, major and minor cheliped very asymmetrical in shape and unequal in size. Major cheliped robust, with chela subcylindrical. Fingers slightly less than half-length of palm, distally curved; cutting edges serrated; lateral faces bearing dense row of fine setae. Palm smooth, ventral margin unarmed. Carpus cup-shaped. Merus slender, with ventral face shallowly excavated; ventrolateral margin bearing one large triangular tooth situated approximately medially and several small teeth situated on distal half. Ischium elongate, almost twice as long as that of major cheliped. Minor cheliped slender, much smaller and less robust than major cheliped. Fingers about 4/5 length of palm, distally curved; cutting edges unarmed, with some tufts of setae. Carpus about 3/5 length of merus, elongate, cylindrical. Ischium nearly as long as merus.

Second pereiopod with carpus 5-segmented, ratio of carpal segments sub-equal to 5: 1: 1: 1: 2.

Third and fourth pereiopod slender, with dactylus slender, simple, gradually curved, about 1/2 length of propodus. Propodus slightly shorter than merus, armed with 2 spines on ventral margin. Carpus and merus unarmed. Ischium armed with 2 spines on ventral margin.

Telson broadly oval-rectangular; about 4 times as long as wide; dorsal surface bearing two pairs of spines, situate about 1/2 and 3/5 length of telson, respectively; posterior margin slightly convex, bearing two slender posterolateral spines.

**Remarks** The variations of chelipeds of *A. japonicus* had been discussed by Anker (2003), and the same variations are also found in the present specimens. This species is commonly found in muddy shore interspersed with rock and mangrove roots, and in dead corals.

**Distribution** The Yellow Sea, the South China Sea; Japan; Singapore; Australia.

### 2.5.7 *Athanas ohsimai* Yokoya, 1936 (Fig. 2.101)

*Athanas ohsimai* Yokoya, 1936: 129, fig. 1; Hayashi, 1995: 108, figs. 272d, 273f, g; Sha & Liu, 2007: 752.

**Material Examined** 1♂, TL 9.2 mm, Cangkou, Jiaozhou Bay, Shandong, 16 November 1953; 1♂, TL 12.01 mm, Cangkou, Jiaozhou Bay, Shandong, 30 July 1954; 10♂♂, TL 6.52–11.36 mm, Cangkou, Jiaozhou Bay, Shandong, 14 August 1954; 6♀♀, TL 8.25–10.66 mm, 54-0172, Cangkou, Jiaozhou Bay, Shandong, Zhang Feng-xuan Coll., 29 September 1954; 1♂, TL 11.05 mm, 54-0192, Cangkou, Jiaozhou Bay, Shandong, 13 October 1954; 12♂♂, TL 8. 93–13.43 mm, 54-0198, Cangkou, Jiaozhou Bay, Shandong, 28 October 1954; 1♂, TL 8.37 mm, 56-757, Cangkou, Jiaozhou Bay, Shandong, Fan Zhen-gang Coll., 5 October 1955; 1♂, TL 8.37 mm, Cangkou, Jiaozhou Bay, Shandong, 15 October 1955; 1♂, TL 13.73 mm, 56-K155, Pingtai, Guangdong, Liu Rui-yu Coll., 16 February 1956; 10♂♂11♀♀, TL 8.03–11.47 mm, Cangkou, Jiaozhou Bay, Shandong, Fan Zhen-gang Coll., 20 September 1956; 4♂♂8♀♀, TL 6. 17–10.85 mm. Cangkou, Jiaozhou Bay, Shandong, Fan Zhen-gang Coll., 20 September 1956; 2♂♂3♀♀, TL 6.69–11.00 mm, 56-931, Cangkou, Jiaozhou Bay, Shandong, Fan Zhen-gang & Xu Jie-shan Coll., 20 November 1956; 1♂, TL 6·93 mm, Cangkou, Jiaozhou Bay, Shandong, Liu Yi-qing Coll., 2 July 1957; 2♂♂, TL 5.94–8.75 mm, 63-59, Xuejiadao, Shandong, Fan Zhen-gang Coll., 20 August 1963; 1♂, TL 10.43 mm, Hongshiya, Shandong, 16 September 1981; 1♂, TL 10.47 mm, 92C-1048, Qinglan, Hainan, Li Jin-he Coll., 5 April 1992.

**Description** Rostrum triangular, not reaching to distal margin of second antennular segment.

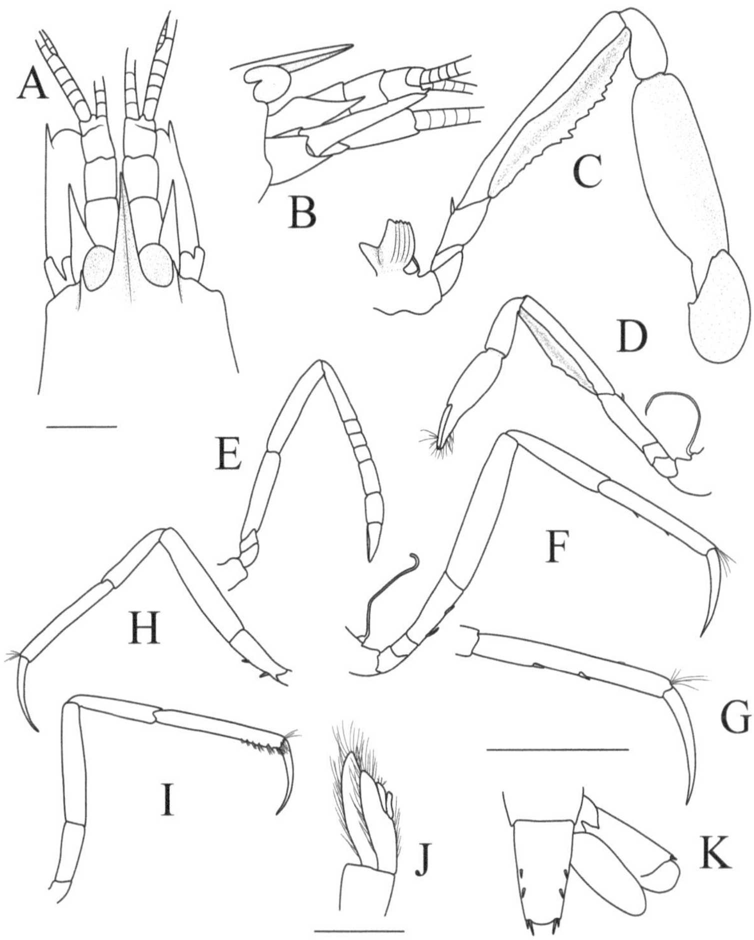


**Fig. 2.101** *Athanas ohsima* Yokoya, 1936. Male, CL 3.9 mm, **a**, anterior part of carapace, dorsal view; **b**, anterior part of carapace, lateral view; **c**, left major cheliped, mesial view; **d**, right minor cheliped, mesial view; **e**, right second pereiopod; **f**, right third pereiopod; **g**, right third pereiopod, propodus and dactylus; **h**, right fourth pereiopod; **i**, right third pereiopod; **j**, right second pleopod; **k**, telson and right uropod (scale bar = 0.5 mm)

Supra-corneal tooth absent; distally acute extra-corneal tooth stronger than infra-corneal tooth. Pterygostomial angle sub-acute.

Antennular peduncle slender, with distally acute stylocerite hardly reaching to distal margin of second segment; visible part of first segment much longer than second segment; third segment shorter than second segment. Antenna with basicerite bearing acute lateral tooth; antennal scale with blade shorter than distolateral tooth, hardly reaching to distal margin of antennular peduncle; carpocerite stout, reaching to distal margin of second antennular segment.

Major and minor cheliped asymmetrical in size and shape, showing sexual dimorphism. In male: the major cheliped elongate, stout; dactylus half-moon shaped, much longer than fixed finger; palm about twice as long as dactylus, smooth; carpus cup-shaped; merus elongate, with ventromesial margin bearing 3–4 irregular small teeth. In female: the major cheliped slender; fingers nearly as long as palm; carpus elongate, longer than palm, a little shorter than 2/3 length of merus; merus nearly as long as ischium.

Second pereiopod slender, with carpus 5-segmented, ratio of carpal segments sub-equal to 4: 1: 1: 1: 2.3.

Third and fourth pereiopod slender. Dactylus simple, slender, about 1/2 length of propodus. Propodus nearly as long as merus, ventral margin bearing 2–3 spines. Carpus about 2/3 length of propodus. Merus armed with one spine on ventral margin. Ischium nearly as long as carpus, ventral margin armed with 2 spines. Merus of fourth and fifth pereiopod unarmed, and ischium of fifth pereiopod only armed with one spine on ventral margin.

Telson about twice as long as wide at base; dorsal margin bearing two pairs of spines, situate about 1/2 and 3/4 length of telson, respectively; posterior margin convex, with two pairs of slender posterolateral spines.

**Remarks** The present specimens slightly differ from previous descriptions by the following characters: the stylocerite reaching to mid-length of second antennular segment rather than the distal margin of that segment; the rudimentary epipod found in first to third pereiopods rather than on first to fourth pereiopods (Yokoya, 1936). Moreover, the shape and symmetry of chelipeds show variable in the present specimens as described in the part of description.

**Distribution** From the southern Yellow Sea to the South China Sea; Japan.

### 2.5.8 *Athanas ornithorhynchus* Banner & Banner, 1973 (Fig. 2.102)

*Athanas ornithorhynchus* Banner DM & Banner AH, 1973: 319, fig.8; Bruce, 1990: 623.

**Material Examined** This species was reported by Bruce (1990) and Liu (2008) from Hong Kong, but not found in the present specimens.

**Description** Rostrum distally acute, triangular, overreaching distal margin of antennular peduncle; rostral carina sharp, extending to posterior margin of eyes. Supra-corneal tooth absent; extra-corneal tooth distally acute, almost reaching to distal margin of first antennular segment; infra-corneal tooth only slightly projecting. Pterygostomial angle rounded.

Antennular peduncle stout, with distally acute stylocerite reaching to mid-length of third segment; visible part of first segment about twice as long as second segment; third segment slightly shorter than first segment. Antenna with basicerite bearing strong ventrolateral tooth, almost reaching to distal margin of first antennular segment; antennal scale with blade hardly reaching to distal margin of antennular peduncle, distinctly shorter than distolateral tooth; carpocerite shorter than antennal scale.

Major and minor cheliped asymmetrical in size and shape. Major cheliped elongate. Fingers slightly less than 1/3 length of palm; tips rounded and slightly crossing; cutting edge of fixed finger bearing two large teeth proximally and distally. Palm with 2/3 length of ventral margin armed with one thin lobe. Carpus cup-shaped. Merus excavated to accommodate chela; 6 times as long as wide at middle; ventromesial margin bearing expanded into one thin lobe, lies over part of carpus and carpal-propodal articulation when appendage flexing. Ischium armed with 5 spines on dorsal margin. Minor cheliped about 1/3 length of major cheliped. Fingers 0.7 times as long as palm, flattened and broadened; tip of dactylus broadly rounded, forming a closure like the bill of a duck with opposing surface excavated or spoon-shaped. Carpus 0.3 times as long as merus, cup-shaped. Merus 2.7 times as long as wide, not excavated, unarmed. Ischium armed with 2 spines on dorsal margin.

Second pereiopod slender, carpus 5-segmented, ratio of carpal segments sub-equal to 10: 1: 1: 1: 2. Ischium 3/5 length of merus, dorsal margin armed with 3 spines; ventral margin bearing several long setae.

Third pereiopod with dactylus about 1/3 length of propodus, biunguiculate, inferior unguis only 1/5 length of superior. Propodus slightly longer than merus, ventral margin bearing numerous spines. Carpus 3/5 length of merus, distodor-

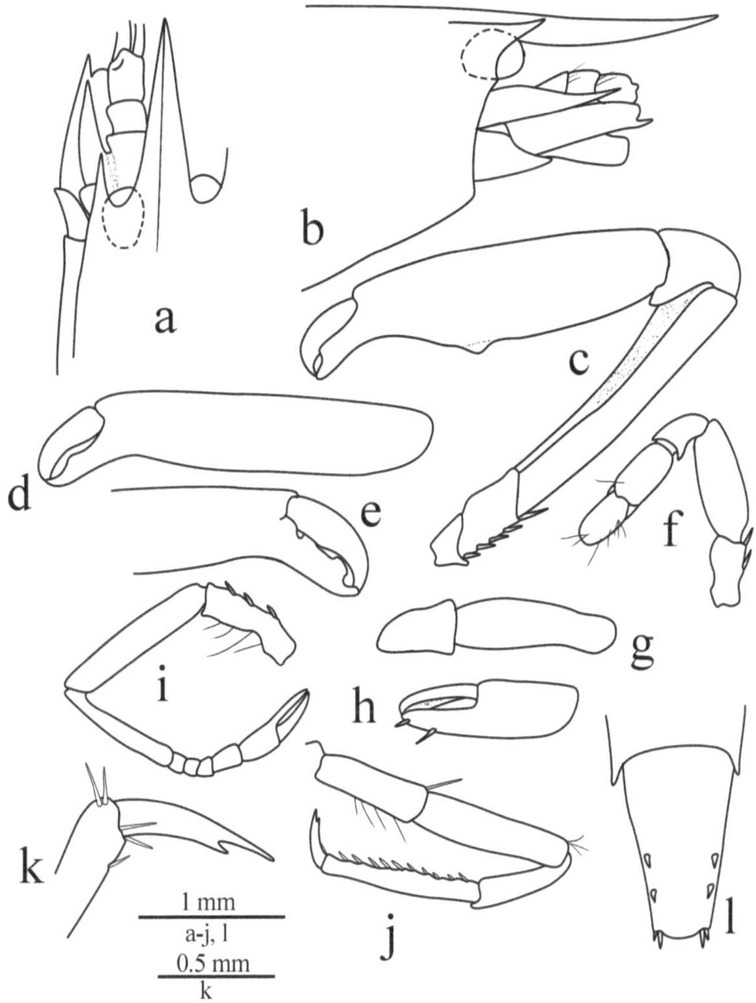


**Fig. 2.102** *Athanas ornithorhynchus* Banner & Banner, 1973 (Adapted from Banner & Banner, 1973). **a**, anterior part of carapace, dorsal view; **b**, same, lateral view; **c**, major cheliped, lateral view; **d**, major chela, mesial view; **e**, same, fingers; **f**, minor cheliped, lateral view; **g**, same, medial view; **h**, minor chela, lateral view; **i**, second pereiopod; **j**, third pereiopod; **k**, same, dactylus; **l**, telson

sal projected. Merus unarmed. Ischium 0.7 times as long as merus; dorsal margin bearing one spine distally.

Telson about 3.2 times as long as posterior margin wide; dorsal surface bearing two pairs of spines, situate about 1/2 and 3/4 length of telson, respectively; posterior margin convex, with two pairs of slender posterolateral spines.

**Remarks** *A. ornithorhynchus* is originally described from Australia and reported by Bruce (1990) from Hong Kong.

**Distribution** The South China Sea (Hong Kong); Australia.

### 2.5.9 *Athanas parvus* De Man, 1910 (Fig. 2.103)

*Athanas sibogae* De Man, 1910: 314; 1911: 151; Sha & Liu, 2007: 753.

*Athanas parvus* De Man, 1910: 315; 1911: 148; Miya & Miyake, 1968: 134.

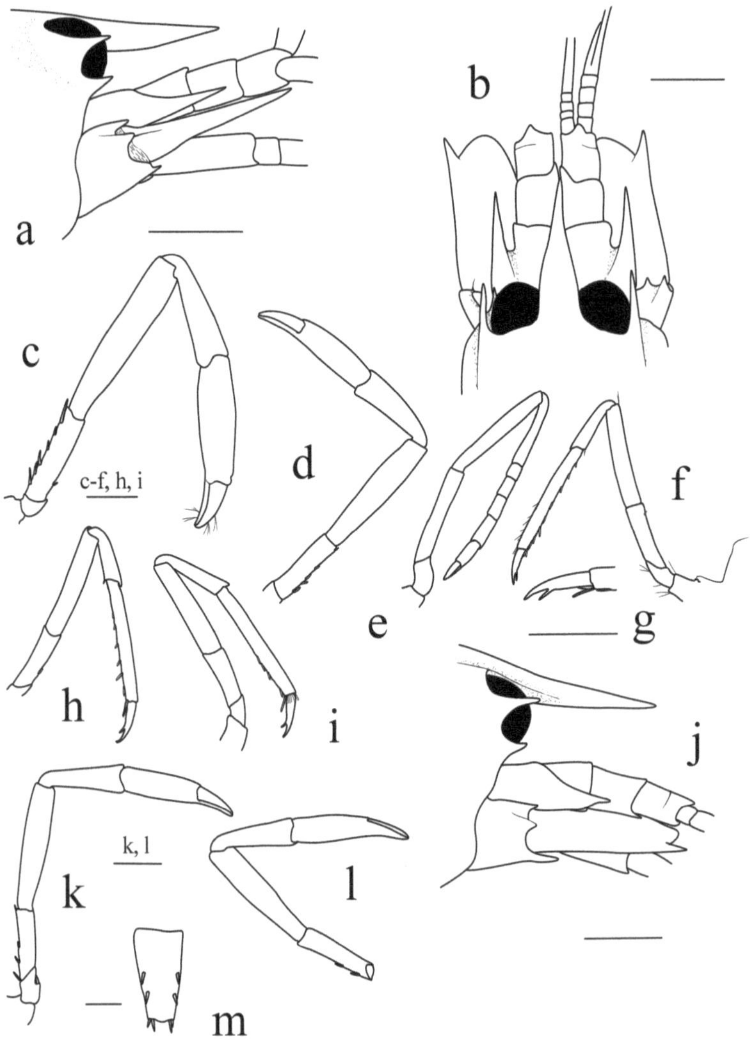


**Fig. 2.103** *Athanas parvus* de Man, 1910. Ovigerous female, CL 2.5 mm, **a**, anterior part of carapace, lateral view; **b**, same, dorsal view; **c**, right major cheliped, lateral view; **d**, left minor cheliped, lateral view; **e**, right second pereiopod; **f**, right third pereiopod; **g**, same, dactylus; **h**, right fourth pereiopod; **i**, right fifth pereiopod. Male, CL 2.9 mm, **j**, anterior part of carapace, lateral view; **k**, right first pereiopod; **l**, left first pereiopod; m, telson (scale bar = 0.5 mm)

**Material Examined** TL 6.2–9.0 mm. 1♀, 56-K094, Xincun harbor, Hainan, 26 January 1956; 1♀, 90C-381c,Yezhu Island, Hainan, Newmann Coll., 25 November 1990; 1♂, 92C-067, Yezhu Island, Hainan, Newmann & Li Jin-he Coll., 18 March 1992; 3♀♀, 92C-063, Yezhu Island, Hainan, Li Jin-he Coll., 19 March 1992; 3♀♀, 92C-147, Yezhu Island, Hainan, 25 March 1992; 1♂8♀♀, 92C-1416, Linchang reef, Hainan, 10 April 1992; 1♂, 92C-1430A, Linchang reef, Hainan, Li Jin-he Coll., 11 April 1992; 3♀♀, 92C-1855, Yezhu Island, Hainan, 11 April 1992; 1♂, C597C-207, Dadonghai, Hainan, Li Jin-he Coll., 4 March 1997.

**Description** Rostrum reaching to distal margin of second antennular segment; rostral carina sharp, extending to posterior margin of eyes. Supra-corneal lacking; extra-corneal tooth acute, extending far beyond cornea; infra-corneal tooth acute, extending slightly beyond cornea. Pterygostomial angle rounded.

Major and minor cheliped almost symmetrical, sexual dimorphism. In male cheliped stout, with chela cylindrical, 3.5 times as long as wide, folding back into expanded and excavated merus; dactylus crescent with cutting edge variable in bearing either a row of small irregular teeth or 1–2 large rounded teeth; Ischium 1/5 length of merus, dorsal margin bearing several strong spines. In female, chelipeds feeble, with fingers slender, almost straight, tapering; merus excavated distally, sub-equal to chela, unarmed; ischium bearing about 6 strong spines on dorsal margin.

Second pereiopod with carpus 5-segmented, ratio of carpal segments sub-equal to 10: 3: 3: 3: 5.

Third pereiopod with dactylus biunguiculate, inferior unguis much shorter and more slender than superior. Propodus armed with 9 spinules on ventral margin. Carpus distally with rounded tooth on dorsal margin; ventral margin terminating in spinules. Merus 4.5 times as long as wide. Ischium armed with one spinule on dorsal margin.

**Remarks** *A. parvus* is commonly found under rocks on sandy beach, and in dead coral.

**Distribution** The South China Sea; Red Sea; Japan; Singapore; Indonesia; Malay; Philippine; Australia; Samoa Island; Tonga.

### 2.5.10 *Athanas polymorphus* Kemp, 1915 (Fig. 2.104)

*Athanas polymorphus* Kemp, 1915: 289, figs. 31–32; Tattersall, 1921: 370; Bruce & Coombes, 1997: 325; Anker, 2003a: 294; Sha & Liu, 2007: 753.

**Material Examined** 1♂, TL 13.3 mm, 3♀♀, TL 6.9–7.7 mm, Beigang, Guangxi, 23 May 1958; 2♂♂, TL 12.9–14.2 mm, 4♀♀, TL 11.34–12.3 mm, 92C-144b, Yezhu Island, Hainan, Ren Xian-qiu Coll., 16 March 1992; 5♀♀, 7.1–7.8 mm, 56-K094, Xincun, Hainan, 26 January 1956.

**Description** Carapace smooth. Rostrum reaching middle of second antennular segment, with apex sharply pointed. Supra-corneal tooth absent; extra-corneal tooth reaching middle of cornea; infra-corneal tooth present and acute, slightly longer than extra-corneal. Pterygostomian angle with a large sharp tooth.

Antennular peduncle with stylocerite acute, reaching beyond end of second segment; second segment long, almost twice as long as wide; third segment shorter than second. Scaphocerite reaching end of antennular peduncle; carpocerite reaching almost middle of third antennular article; basicerite with developed spine on ventral side of distal margin.

First cheliped with chelae carried flexed ventrally, variable in shape and symmetry. In male first chelipeds heavy, symmetrical or sub-symmetrical; tip of fingers curved, about 1/2 length of palm, cutting edge with small serrated teeth and short setae; palm with 2–4 proximal blunt teeth; carpus cup-shaped, external margin with 2–3 teeth or without tooth; merus excavated, ventrolateral margin serrated with 9–18 teeth; ischium as long as fingers, armed with a small spine on dorsal margin. In female variable as follows: type 1, small, slender and symmetrical; fingers slightly longer than palm; carpus shorter than merus; ischium armed with a small spine on dorsal margin; type 2, cheliped asymmetrical; large cheliped with tips of finger slightly crossing, merus excavated, ventrolateral margin serrated with 1–2 teeth, ischium without or with a small spine on dorsal margin; small cheliped small and slender, as first chelipeds in type 1.

Second pereiopod slender; fingers slightly longer than palm; carpus five-segmented, ratio of segments sub-equal to 6.3: 1.0: 1.0: 1.0: 2.2; merus as long as ischium; ischium unarmed.

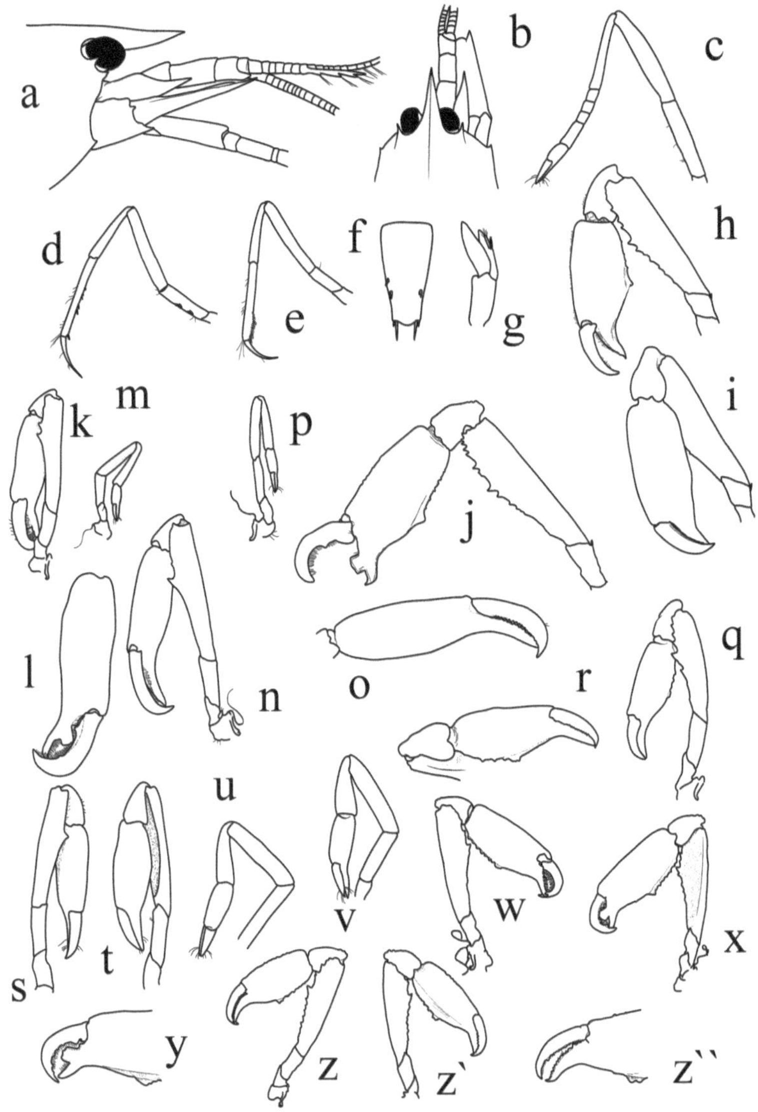


**Fig. 2.104** *Athanas polymorphus* Kemp, 1915. (Adapted from Anker, 2003). Male, **a**, anterior part of carapace, lateral view; **b**, same, dorsal view; **c**, second pereiopod; **d**, third pereiopod; **e**, fifth pereiopod; **f**, telson; **g**, second pleopod; **h**, left first pereiopod, lateral view; **i**, right first pereiopod, lateral view. Male, **j**, left first pereiopod, lateral view. Female, **k**, major cheliped, lateral view; **l**, same, chela, mesial view; **m**, minor chela. Female, **n**, major cheliped, lateral view; **o**, same, chela, mesial view; **p**, minor cheliped. Male, **q**, major cheliped; **r**, same, chela. Male, **s**, major cheliped, lateral view; **t**, same, mesial view; u, minor cheliped. Female, **v**, first pereiopod (chelipeds symmetrical). Male, **w**, major cheliped, lateral view; **x**, same, mesial view; **y**, same, fingers; **z**, minor cheliped, lateral view; **z**, same, mesial view; '**z'**', same, fingers

Third pereiopod slender, with dactylus simple, 0·64 times as long as propodus. Propodus with ventral margin bearing 2–3 small spines and 1 distal spine. Merus 6 times as long as broad, unarmed. Ischium armed with 2 spines on dorsal margin. Fourth pereiopod similar to third pereiopod, but propodus armed with 3–4 small spines on ventral margin. Propodus of fifth pereiopod armed with 2 small spines on ventral margin, and 8 rows of short setae near distally, ischium bearing 1 spine on dorsal margin.

**Remarks** The present specimens are generally agreed with previous descriptions, and show similar variations.

**Distribution** The South China Sea; India; Northern Gulf of Thailand; Singapore; Australia.

### 2.5.11 *Athanas squillophilus* Hayashi, 2002 (Fig. 2.105)

*Athanas squillophilus* Hayashi, 2002: 396, figs. 1–3; Sha & Liu, 2007: 754.

**Material Examined** 5♂♂, TL 13.1–20.4 mm, 5♀♀, TL 7·2–14.9 mm, 54-0192, Cangkou, Jiaozhou Bay, Shandong, 13 October 1954; 1♂, TL 13. 9 mm, 6♀♀, TL 7.2–14.9 mm, 54C-0198, Cangkou, Jiaozhou Bay, Shandong, 28 October 1954; 1♀, TL 14.5 mm, 58-019, Cangkou,

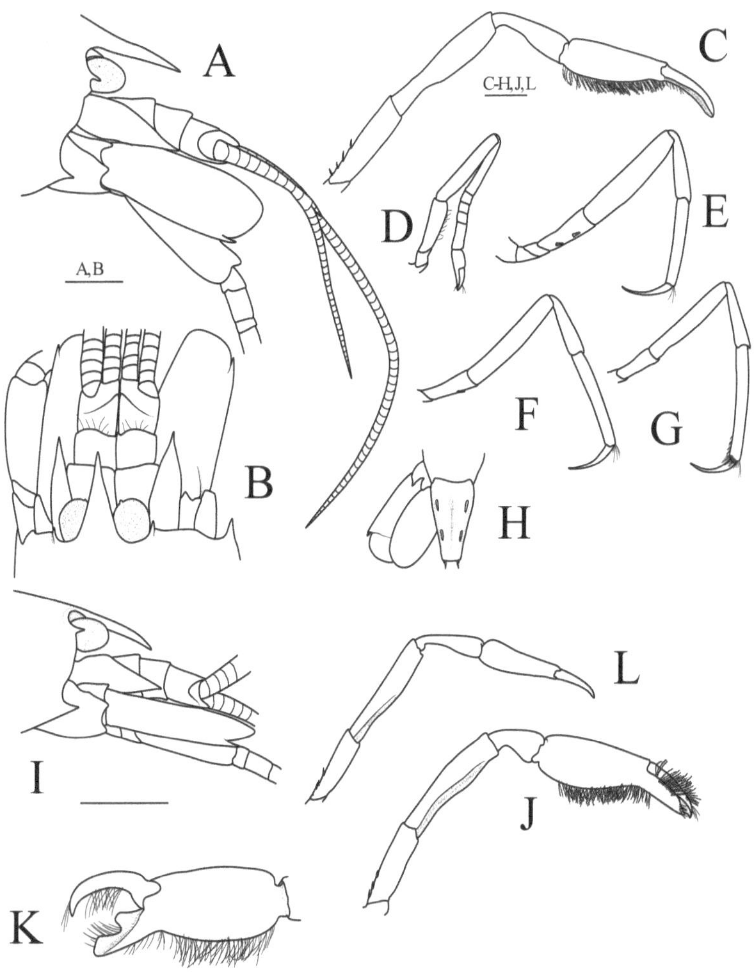


**Fig. 2.105** *Athanas squillophus* Hayashi, 2002, ovigerous female, CL 7.8 mm, **a**, anterior part of carapace, lateral view; **b**, same, dorsal view; **c**, right cheliped; **d**, right second pereiopod; **e**, right third pereiopod; **f**, right fourth pereiopod; **g**, right fifth pereiopod; **h**, telson and left uropod. Female, CL 6.1 mm, **i**, anterior part of carapace, lateral view; **j**, left major cheliped; **k**, same, chela; **l**, right minor cheliped (scale bar 1 mm)

Jiaozhou Bay, Shandong, Zhang Feng-xuan & Xu Jie-shan Coll., 6 January 1958; 1♂4♀♀, 10.3–15.7 mm, 54-0217, Cangkou, Jiaozhou Bay, Shandong, Fan Zhen-gang Coll., 25 November 1954; 1♂2♀♀, TL 11.1–19.4 mm, 55-1240, Cangkou, Jiaozhou Bay, Shandong, 3 November 1955; 1♂, TL 15.2 mm, 56-695, Cangkou, Jiaozhou Bay, Shandong, Zhang Feng-xuan & Xu Jie-shan Coll., 6 September 1956; 2♀♀, TL 10.1–11.5 mm, Cangkou, Jiaozhou Bay, Shandong, Zhang Feng-xuan Coll., 23 September 1956; 2♂♂, TL 12.2–14.19 mm, 56-796, Cangkou, Jiaozhou Bay, Shandong, Fan Zhen-gang & Xu Jie-shan Coll., 19 October 1956; 1♀, TL 17.1 mm, 56-929, Cangkou, Jiaozhou Bay, Shandong, Fan Zhen-gang & Xu Jie-shan Coll., 20 November 1956; 1♀, TL 14.79 mm, 57-274, Cangkou, Jiaozhou Bay, Shandong, Zhang Feng-xuan Coll., 15 April 1957; 1♀, TL 21.8 mm, Xuejiadao, Shandong, Fan Zhen-gang & Fang Xiu-bin Coll., 7 September 1981.

**Description** Carapace smooth. Rostrum variable reaching mid-length to distal margin of second antennular segment, apex sharply pointed. Supra-corneal and infra-corneal teeth absent; extra-corneal tooth acute. Pterygostomian angle with stout tooth.

Antennular peduncle stout, with distally acute stylocerite reaching to or slightly overreaching distal margin of second segment; second segment wider than long, shorter than third. Antenna with basicerite bearing acute ventrolateral tooth; scaphocerite reaching beyond end of third antennular segment; carpocerite reaching to distal margin of scaphocerite.

Major and minor cheliped with chelae carried flexed ventrally, high variation of shape and symmetry appearing in Chinese specimens. In male: first cheliped varied as follows: type 1, cheliped heavy, sub-symmetrical; in major cheliped, most part or whole ventral margin of chela with dense and long setae; finger 0.6 times as long as palm, with tips crossed over, cutting edges with long setae and strong or small blunt teeth; distal half of flexor margin of palm with 2–9 blunt teeth; carpus shorter than fingers; merus and ischium slightly excavated to hold part of flexed carpus and chela, mesial part of ventral margin with 2–5 feeble teeth; ischium almost as long as fingers, armed with 4 teeth on basal half of external margin; fingers of minor cheliped 0.7 times as long as palm, cutting edges with serration; distal half of ventral margin of palm with 0–5 blunt teeth; ventral margin of chela unarmed or fringed with dense, long setae; carpus longer than fingers; merus and ischium slightly excavated to hold part of flexed carpus and chela, distal part of ventral margin of merus with 3 feeble teeth; ischium 0.8 times as long as merus, armed with 4 teeth on basal half of external margin; type 2, chelipeds heavy and symmetrical; ventral margin of chela fringed with dense, long setae; fingers 0.6 times as long as palm, cutting edges with small teeth; ventral margin of palm with 2 rows of blunt teeth; carpus shorter than palm, flexor margin with three round processes; merus and ischium slightly excavated to hold part of flexed carpus and chela, mesial part of ventral margin with 8 feebly teeth; ischium almost as long as palm, armed with 5 teeth on basal half of external margin; type 3, chelipeds asymmetrical, all ventral margin of chela fringed with dense, long setae in major cheliped, but with or without in minor cheliped; shape of major cheliped similar to type 1, palm with7–9 teeth, merus with 5–10 teeth; palm of minor cheliped with 2 rows of teeth; ischium armed with 4 teeth on basal half of external margin; the other part similar to type 2. In female, first cheliped symmetrical or sub-symmetrical, shape similar to type 1 in male except the number of teeth on palm and fixed finger of major chela without large teeth.

Second pereiopod slender, with fingers slightly longer than palm. Carpus 5-segmented, ratio of segments sub-equal to 6.3: 1: 1: 1.3: 2.5. Merus longer than ischium, basal of merus and all ischium fringed with sparse, long setae on ventral margin.

Third and fourth pereiopod slender, with dactylus simple, 0.6 times as long as propodus. Propodus unarmed with spines on ventral margin. Merus 4 times longer than wide. Ischium armed with 2 spines on ventral margin. Fifth pereiopod similar to third pereiopod, but propodus bearing 3 rows of short setae on ventral margin near distally.

**Remarks** Based on the original description, the chelipeds are similar and slightly unequal in length. However, in present specimens, strong variations have been found in shape of chelipeds: the palm of some individuals armed with two rows of teeth, the cutting edge of fingers armed with large, round tooth; the chela of minor cheliped not fringed with long setae. Additionally, the extra-corneal tooth is not found in the present specimens.

**Distribution** The Yellow Sea; Japan.

## 2.6 Genus *Automate* De Man, 1888

**Diagnose** Body laterally compressed. Rostrum absent or only slightly developed. Orbital hoods absent. Eyes not covered by carapace, free with reduced corneas. Antennular peduncle extremely elongate. Stylocerite and scaphocerite reduced. Carpocerite elongate. First pereiopod asymmetrical; chela compressed, faces smooth; dactylus without cylinder and plunger. Second pereiopod with carpus 5-segmented. Third to fifth pereiopod with dactylus simple. Pleura of sixth abdominal segment without articulated plate (Banner & Banner, 1973).

Key to species of the genus of *Automate* from China seas

1. Antennal scale with lateral margin somewhat sinuous ........ *A. anacanthopusoides*
-. Antennal scale lateral margin convex, concave, or nearly straight, not sinuous ........ 2
2. Rostrum absent ........ *A. spinosa*
-. Rostrum present ........ *A. dolichognatha*

### 2.6.1 *Automate anacanthopusoides* Wang & Sha, 2017 (Figs. 2.106 and 2.107)

*Automate anacanthopus*–. Ledoyer, 1970: 127, 149, pl. 17, 24A.
*Automate anacanthopusoides* Wang & Sha, 2017a: 31, figs. 1, 2.

**Material Examined** MBM 270274, 1ovigerous♀, CL 4.9 mm, TL 13.7 mm, Beibu Gulf, depth 42 m, Zhang Bao-lin Coll., 11 November 1960; MBM 210082, 1ovigerous♀, CL 4.2 mm, TL 13.5 mm, Beibu Gulf, depth 29.9 m, Ma Xiutong Coll., 4 December 1959; MBM 210101, 1ovigerous♀ ,CL 4.1 mm, TL 13.4 mm, Beibu Gulf, depth 55 m, Sun Fu-zeng Coll., 13 January 1962; MBM 270166, 1♂, CL 2.2 mm, TL 7.8 mm, Beibu Gulf, depth 51 m, Zhang Bao-lin Coll., 4 November 1960; MBM 270170, 1♂, CL 4.2 mm, TL 13.5 mm, Beibu Gulf, depth 21 m, 19 October 1962; MBM 210095, 1♂, CL 2.6 mm, TL 7.7 mm, 1ovigerous♀, CL 4.2 mm, TL 11.5 mm, Beibu Gulf, depth 30.7 m, Fan Zhengang Coll., 19 April 1960; MBM 210076, 1ovigerous♀, CL 3.8 mm, TL 10.9 mm, Beibu Gulf, depth 22 m, Sun Fu-zeng Coll., 16 February 1960; MBM 210067, 1ovigerous♀, CL 3.8 mm, TL 11.1 mm, Beibu Gulf, depth 58 m, Sun Fu-zeng Coll.; MBM 210099, 2♂♂, CL 3.6–4.1 mm, TL 11.1–12.9 mm, Beibu Gulf, depth 58 m, 12 October 1962; MBM 210076, 1ovigerous♀, CL 4.8 mm, TL 13.5 mm, Beibu Gulf, depth 26 m, Sun Fu-zeng Coll., 8 February 1960; MBM 210071, 1ovigerous♀, CL 4.1 mm, TL 12.9 mm, Beibu Gulf, depth 65.5 m, Fan Zhengang Coll., 21 April 1960; MBM 210061, 1ovigerous♀, CL 4.2 mm, TL 13.1 mm, Beibu Gulf, depth 57 m, Zhang Bao-lin Coll., 13 July 1960;

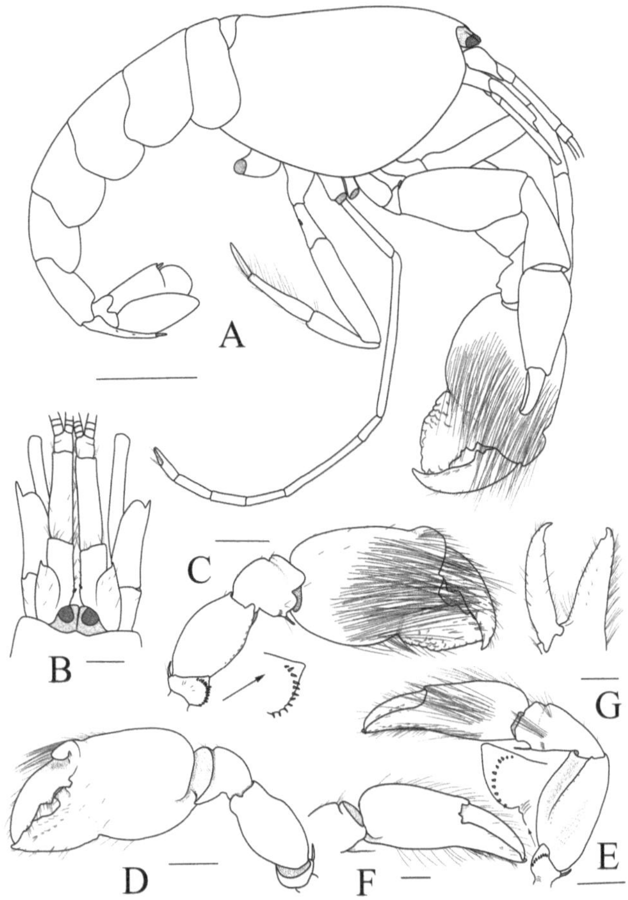


**Fig. 2.106** *Automate anacanthopusoides* Wang & Sha, 2017. MBM 270274, holotype, female, CL 4.9 mm. **a**, entire animal, lateral view; **b**, anterior part of carapace, dorsal view; **c**, left major cheliped; mesial view; **d**, same, lateral view; **e**, right minor cheliped, mesial view; **f**, minor chela, lateral view; **g**, same, fingers, lateral view (scale bars: **a** = 2 mm; **b**, **e**–**g** = 0.5 mm; **c**, **d** = 1 mm)

MBM 210085, 1ovigerous♀, CL 4.1 mm, TL 12.8 mm, Beibu Gulf, depth 57m, 18 August 1962; MBM 270243, 1ovigerous♀, CL 3.9 mm, TL 12.3 mm, Beibu Gulf, depth 24.5 m, Ma Xiu-tong Coll., 3 December 1959; MBM 210078, 1ovigerous♀, CL 2.9 mm, TL 8.7 mm, Beibu Gulf, depth 65.5 m, Fan Zhen-gang Coll., 21 April 1960; MBM 210066, 1ovigerous♀, CL 2.7 mm, TL 8.5 mm, Beibu Gulf, depth 49 m, Fan Zhen-gang Coll., 19 April 1960; MBM 270169, 1ovigerous♀, CL 4.1 mm, TL 12.7 mm, Beibu Gulf, depth 28 m, Hôńg Coll., 14 February 1960; MBM 210059, 1♂, CL 2.7 mm, TL 8.5 mm, Beibu Gulf, depth 52 m, Fan Zhen-gang Coll., 14 February 1960; MBM 210059, 1ovigerous♀, CL 2.2 mm, TL 7.5 mm, Beibu Gulf, depth 40.5 m, Sun Fu-zeng Coll., 19 April 1960; MBM 270275, 1♂, CL 4.1 mm, TL 12.5 mm, Beibu Gulf, depth 35 m, Shen Shou-peng Coll., 19 October 1960.

**Description** Carapace glabrous. Frontal region of carapace with moderately shallow orbital concavity on either side of rostrum. Rostrum triangular, tip far not reaching to level of antero-lateral margin of carapace, not covering eye-stalks; rostral carina and orbital hoods absent. Eye-stalls exposed dorsally and laterally; cornea developed. Pterygostomian angle rounded, continuous with branchiostegal; cardiac notch conspicuous.

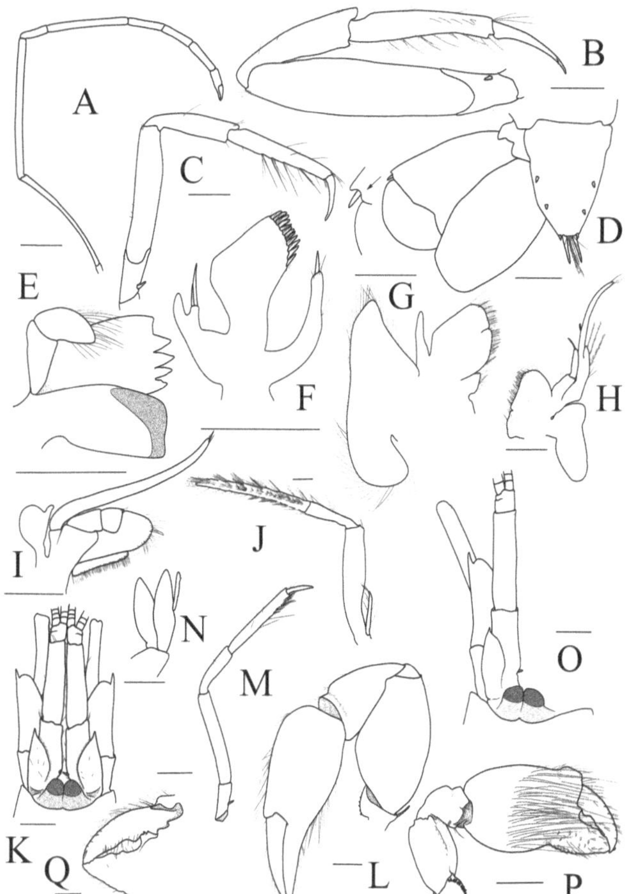


**Fig. 2.107** *Automate anacanthopusoides* Wang & Sha, 2017. MBM 270274, holotype, female, CL 4.9 mm. **a**, left second pereiopod; **b**, right third pereiopod; **c**, right fourth pereiopod; **d**, left uropod and telson; **e**, left mandible; **f**, left maxillule; **g**, left maxilla; **h**, left first maxilliped; **i**, left second maxilliped; **j**, maxilliped. MBM 210082, paratype, female, CL 4.2 mm, **k**, anterior part of carapace, dorsal view; l, minor cheliped, outer view; **m**, fifth pereiopod; **n**, second pleopod. MBM 210101, paratype, female, CL 4.1 mm, **o**, anterior part of carapace, dorsal view; **p**, major cheliped, mesial view; **q**, same, fingers, lateral view (scale bars: **a–o**, **q** = 0.5 mm; **p** = 1 mm)

Pleon with first to fifth pleura rounded postero-ventrally. Telson much shorter than uropods, 1.3 times as long as wide at base, dorso-lateral margin bearing two pair of moderately small spines, anterior and posterior pair situated about 1/2 and 3/4 length of telson, respectively; posterior margin slightly convex, bearing two pairs of distal spines, with mesial about seven times as long as lateral. Second pleopod without appendix masculine both sexes. Uropod with endopod slightly longer than exopod; exopod with postero-lateral corner sub-acute tooth, slightly produced, accompanied with one spine; diaeresis sinuous, without distinct projection and tooth.

Antennular peduncle slender; visible part of first segment about 3 times as long as wide, dorso-mesial margin with one small, somewhat anteriorly curved spine, ventro-mesial carina not terminating anteriorly in acute tooth; second segment much longer than visible part of first segment; stylocerite acute, far not reaching distal end of first segment of antennular peduncle. Antenna with basicerite bearing strong ventro-lateral tooth; carpocerite slender, distinctly over-reaching distal end of antennular peduncle; scaphocerite oval, reaching to about middle of second segment of antennular peduncle, lateral margin slightly concave, disto-lateral tooth distinctly beyond anterior margin of blade.

Mandible with two-segmented palp; incisor process bearing distally 4 blunt, strong, large teeth; molar process well developed. Maxillule with two endites, distal endite with numerous large spine-like setae on mesial margin, basal endite with one spine-like setae on distal margins; palp bilobed, upper lobe rounded, without seta, lower lobe with two robust seta. Maxilla with slender tapering palp; distal endite bilobed, densely setose medially; proximal endite with single lobe, lanceolate, distally with setae; scaphognathite well developed. First maxilliped with two-segmented palp; distal endite with medial margin slightly concave, with dense setae; proximal endite small; exopod with well-developed flagellum, distally with long setae; palp twosegmented. Second maxilliped with dactylar segment narrow, medial margin with moderately short spiniform setae; propodal segment slightly elongate, rounded distally, distomedial border with long setae; carpus short; coxa with large epipod; exopod with long flagellum, distally with long setae. Third maxilliped slender; arthrobranch normally developed; coxa with hook-like epipod, exopod short, tip hardly reaching to mid-length of antepenultimate segment; antepenultimate near as long as ultimate segment, superior margin with several moderately short setae; penultimate segment about half length of antepenultimate segment; ultimate segment about 9 times as long as greatest depth, with numerous transverse rows of stiff setae on ventral surface, spine unarmed on distal and dorsal surface.

First pereiopod robust and strongly asymmetrical in shape and size. Major cheliped with fingers distinctly shorter than palm, bearing 1–2 large teeth on cutting edge, inner margin half-length of pollex on mesial face tuberculate and lower margin slightly rugose. Palm rectangular, inner margin of about half length bearing densely long setae. Carpus vase-shaped, dorsal surface with one moderately deep transverse groove near distal margin, and ventro-distal surface with one robust spine. Merus about twice as long as wide at middle, bearing transverse rows of short setae on ventro-mesial margin. Ischium bearing one curving row of small spines on mesial face, and one robust spine on dorsal distal margin. Minor cheliped with fingers crossing, cutting edge of dactylus bearing two inconspicuous teeth near tip. Carpus about half-length of chela. Merus about twice as long as wide at middle. Merus and ischium similar to that of major cheliped.

Second pereiopod extremely slender; carpus five-segmented, second segment longer than other segments, ratio of segments from proximal to distal approximately equal to 1.0: 3.0: 1.4: 1.1: 1.4.

Third pereiopod moderately stout, with dactylus conical, longer than half-length of propodus, slightly curved, bearing several short setae near tip. Propodus about 3/5 length of merus, ventral margin bearing long robust setae, without spine. Carpus longer than half-length of merus, disto-ventral margin without spines. Merus about five times as long as wide, unarmed. Ischium bearing one spine on ventral margin. Fourth pereiopod similar to third pereiopod, but moderately slender. Fifth pereiopod similar to third and fourth pereiopod, propodus bearing rows of transverse setae on ventral margin.

**Remarks** Anker & Komai (2004) divided the genus Automate into three informal species groups: *A. dolichognatha* species group, *A. evermanni* species group and *A. hayashii* species group. *A. anacanthopusoides* is assigned to the *A. evermanni* species group for having the following features: the rostrum very short; the ventromesial carina on the first segment of the antennular peduncle unarmed; the stylocerite not reaching the distal margin of the basal segment of the antennular peduncle; the scaphocerite narrow; the second maxilliped with podobranch; the palm of the major chela sub-rectangular; the propodus of the third pereiopod without spine; the dactylus of the third pereiopod sub-spatulate; the uropodal endopod slightly exceeding the exopod, and the diaeresis on the uropodal exopod devoid of teeth. The *A. evermanni* species group contains another five species exclude *A. anacanthopusoides*: *A. evermanni* Rathbun, 1901, *A. rectifrons* Chace, 1972, *A. rugosa* Coutière, 1902, *A. branchialis* Holthuis & Gottlieb, 1958 and *A.*

*anacanthopus* De Man, 1910. The differences between *A. anacanthopusoides* and its similar species are discussed in Wang & Sha (2017).

**Distribution** Beibu Gulf, the South China Sea; probable also from Tulear, Madagascar.

### 2.6.2 *Automate dolichognatha* De Man, 1888 (Fig. 2.108)

*Automate dolichognatha* De Man, 1888b: 529, pl. 22, fig. 5; Lanchester, 1901: 564, pl. 34, fig.3; Coutière, 1903: 74, fig. 8; Banner DM & Banner AH, 1973: 299–303, fig. 1.

*Automate gardineri*–. Coutière, 1902: fig. 8; 1903: 72, figs. 1–7; 1905: 854, fig. 127, 128; Holthuis & Gottlieb, 1958: 17, fig. 6; Miyake & Miya, 1966: 137, fig. 2; Banner AH & Banner DM, 1966b: 37, fig. 8.

*Automate kingsleyi*–. Hay, 1917: 72.

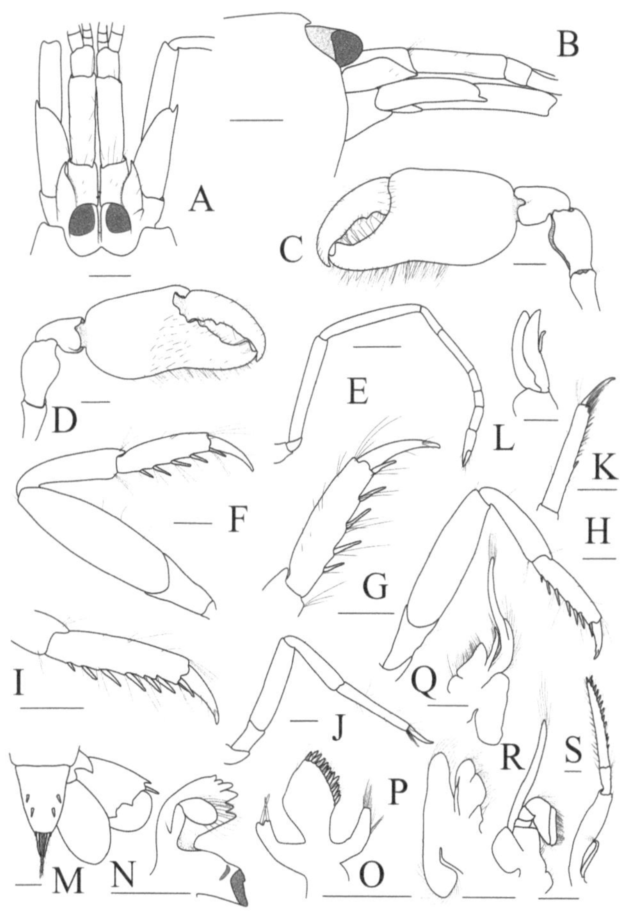


**Fig. 2.108** *Automate dolichognatha* De Man, 1888. MBM 270027, female, CL 4.1 mm, **a**, anterior part of carapace, dorsal view; **b**, same, lateral view; **c** & **d**, major cheliped; **e**, right second pereiopod; **f**, right third pereiopod; **g**, same, propodus and dactylus; **h**, right fourth pereiopod; **i**, same, propodus and dactylus; **j**, right fifth pereiopod; **k**, same, propodus and dactylus; **l**, right second pleopod; **m**, telson and right uropod; **n**, left mandible; **o**, left maxillule; **p**, left maxilla; **q**, left first maxilliped; **r**, left second maxilliped; **s**, left third maxilliped (scale bars: **a**, **b**, **f**–**r** = 0.5 mm; **c**–**e** = 1 mm)

*Automate haightæ*–. Boone, 1931: 184, fig. 22.
*Automate johnsoni*–.Chace, 1955: 13, fig. 7.
*Automate* sp.–. De Man, 1911: 140, fig. 2.

**Material Examined** MBM 270027, 2ovigerous♀♀, CL 4.1 mm, TL 13.8 mm, Weizhou Island, Beibu Gulf, 12 February 1956; MBM 157885, 1ovigerous♀, CL 5.2 mm, TL 14.1 mm, north of Xinying, Hainan, 10 May 1958.

**Description** Carapace glabrous. Frontal region of carapace with moderately shallow orbital concavity on either side rostrum. Rostrum broad, short triangular, tip not reaching to level of antero-lateral margin of carapace, not covering eye-stalls; rostral carina and orbital hoods absent. Eye-stalls exposed dorsally and laterally; cornea developed. Pterygostomian angle rounded, continuous with branchiostegal; cardiac notch conspicuous.

Telson much shorter than uropods, 1.2 times as long as wide at base, dorso-lateral margin bearing two pair of moderately large spines, anterior and posterior pair situated about 0.5 and 0.6 length of telson, respectively; posterior margin only slightly convex, bearing two pairs of spines, with mesial about six times as long as lateral.

Antennular peduncle slender; dorso-mesial margin of first segment with one small, somewhat anteriorly curved spine; second segment much longer than visible part of first segment; stylocerite acute, reaching near distal end of first segment of antennular peduncle. Antenna with basicerite bearing strong ventrolateral tooth; carpocerite slender, distinctly overreaching end of antennular peduncle; scaphocerite oval, reaching to about 2/3 length of second segment of antennular peduncle, lateral margin nearly straight, disto-lateral tooth distinctly beyond anterior margin of blade.

Mandible with two-segmented palp; incisor process bearing distally 6 blunt, strong, irregular teeth; molar process well developed. Maxillule with two endites, distal endite with numerous large spine-like setae on mesial margin, basal endite with several spine-like setae on distal margins; palp bilobed, upper and lower lobe both with one robust seta. Maxilla with slender tapering palp, distally with one robust seta; distal endite deeply bilobed, densely setose medially; proximal endite with single lobe, lanceolate, distally with setae; scaphognathite well developed. First maxilliped with two-segmented palp; distal endite with medial margin slightly concave, with densely setae; proximal endite small, with densely setae; exopod with developed flagellum, distally with long setae. Second maxilliped with dactylar segment narrow, medial margin with moderately short spiniform setae; propodal segment slightly elongate, rounded distally, distomedial border with long setae; carpus short; coxa with large epipod; exopod with long flagellum, distally with long setae. Third maxilliped slender; arthrobranch normally developed; coxa with hook-like epipod; exopod short, tip slightly reaching about half-length of antepenultimate segment; antepenultimate segment near as long as ultimate segment, superior margin with several moderately short setae; penultimate segment about 1/3 length of antepenultimate segment; ventral margin of ultimate segment setose; superior margin armed with transverse rows of moderately small, but strong spines, distally with one small, strong spine.

Major cheliped with ischium unarmed; merus more than twice times as long as wide at middle, margin not rugose, unarmed; carpus robust, cup-shaped, slightly elongate; low margin of chela bearing moderately long dense setae about distal 2/3 length, and one shallow contour near base of pollex; inner margin of palm bearing thin setae near base of pollex, palm rectangular, slightly compressed; cutting margin of dactylus armed with about two small teeth at base; tip of pollex strongly curved, cutting margin bearing one moderately large tooth near tip; tips of fingers crossing when closed. Minor cheliped lost.

Second pereiopod moderately slender; carpus five-segmented, second segment longer than other segments, ratio of segments from proximal to distal approximately equal to 0.7: 1.2: 0.5: 0.3: 0.6.

Third pereiopod robust, with dactylus about half-length of propodus, slightly curved, bearing several short setae near at tip. Propodus near as long as carpus, ventral margin bearing row of four large spins and distal one pair spines. Carpus distinctly shorter than merus, disto-ventral margin only bearing moderately long setae, without spines. Merus about 4 times as long as wide at base, unarmed. Ischium unarmed. Fourth pereiopod similar to third pereiopod, ventral margin of propodus bearing six spines. Fifth pereiopod much slender than third and fourth pereiopod, propodus bearing grooming brush.

Second pleopod without appendix masculine both sexes. Uropod with endopod distinctly longer than exopod; exopod with two posterior projections bordering a shallow notch on diaeresis, postero-lateral corner acute, strongly produced, accompanied with rather stout spine near as long as posterior margin of exopod.

**Remarks** *A. dolichognatha* De Man, 1888 is the type species of the genus. The present specimens are definitely identified as *A. dolichognatha* for the presence of the small, sub-triangular rostrum, not reaching distal end of the anterior margin of the carapace; the reduced scaphocerite not reaching distal end of the second segment of the antennular peduncle; the length-wide ratio of the merus of the first chelipeds; the presence of 4–6 spines on the ventral margin of the propodus of the third and fourth pereiopod.

**Distribution** pantropical, except for eastern Atlantic.

### 2.6.3 *Automate spinosa* Wang & Sha, 2017 (Figs. 2.109 and 2.110)

*Automate spinosa* Wang & Sha, 2017a: 37, figs. 4–5.

**Material Examined** MBM 159184, 1 ovigerous female (CL 4.9 mm, TL 14.1 mm), Sanya Gulf, 2 December 1997. MBM 270172, 1 male (CL 3.1 mm, TL 9.5 mm), Beibu Gulf, depth 42m, Hôñg Coll., 24 October 1962. MBM 107686, 2 ovigerous female (CL 4.7 mm, TL 13.6 mm), Sanya Gulf, 2 April 1958.

**Description** Carapace glabrous. Frontal region of carapace deeply recessed posterior to eyes, nearly straight mesially, without rostrum or median lobe; rostral carina and orbital hoods absent. Eye-stalls exposed dorsally and laterally, cornea developed. Pterygostomian angle rounded, continuous with branchiostegal; cardiac notch conspicuous.

Pleon with first to fifth pleura rounded postero-ventrally. Telson much shorter than uropods, 1.4 times as long as wide at base, dorso-lateral margin bearing two pair of moderately large spines, anterior and posterior pair situated about 1/2 and 3/4 length of telson, respectively; posterior margin only slightly convex, bearing two pairs of spines, with mesial about four times as long as lateral.

Antennular peduncle slender; dorso-mesial margin of first segment with 1–2 small, somewhat anteriorly curved spines; with second segment much longer than visible part of first; stylocerite acute, not reaching distal end of first segment of antennular peduncle. Antenna with basicerite bearing ventro-lateral tooth;

carpocerite slender, just reaching distal end of antennular peduncle; scaphocerite oval, overreaching middle of second segment of antennular peduncle, lateral margin slightly concave, distolateral tooth distinctly beyond anterior margin of blade.

Mandible with two-segmented palp; incisor process bearing distally 5 blunt, strong, large teeth and one small, sub-acute tooth; molar process well developed. Maxillule with two endites, distal endite with numerous large spine-like setae on inner margin; palp bilobed, upper lobe rounded, with one robust seta, lower lobe with two robust seta. Maxilla with slender tapering palp bearing one robust seta on tip; distal endite bilobed, densely setose medially; proximal endite

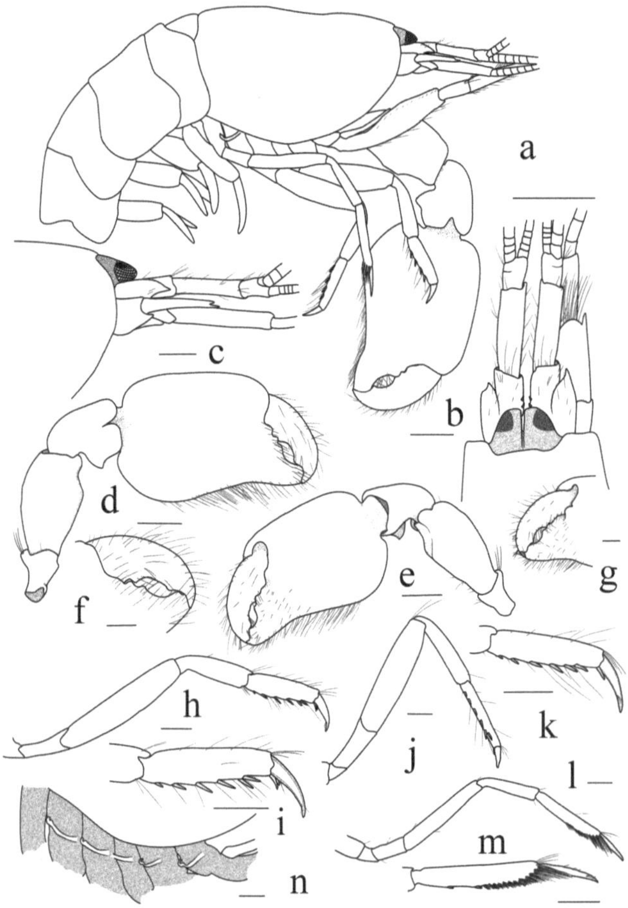


**Fig. 2.109** *Automate spinosa* Wang & Sha, 2017. MBM 159184, holotype, female, CL 4.9 mm, **a**, entire animal, lateral view; **b**, anterior part of carapace, dorsal view; **c**, same, lateral view; **d** & **e**, major cheliped, lateral view; **f** & **g**, same, fingers, lateral view; **h**, right third pereiopod; **i**, same, propodus and dactylus; **j**, right fourth pereiopod; **k**, same, propodus and dactylus; **l**, right fifth pereiopod; **m**, same, propodus and dactylus; **n**, hook-shaped epipod of third maxilliped to fourth pereiopod (scale bars: a = 2 mm; c, d = 1 mm; b, e–m = 0.5 mm)

with single lobe, lanceolate, distally with setae; scaphognathite well developed. First maxilliped with two-segmented palp; distal endite with medial margin slightly concave, with dense setae; proximal endite small; exopod with well-developed flagellum, distally with long setae. Second maxilliped with dactylar segment narrow, medial margin with moderately short spiniform setae; propodal segment slightly elongate rounded distally, distomedial border with long setae; carpus short; ischiomerus short, basis long, stout; coxa with large epipod; exopod well developed, with long flagellum, distally with long setae. Third maxilliped slender; arthrobranch normally developed; coxa with hook-like epipod, exopod slightly not reaching to mid-length of antepenultimate segment; antepenultimate segment near as long as ultimate segment, dorsal margin with several moderately short setae; penultimate segment about 1/3 length of antepenultimate segment; ventral margin of ultimate segment setose; superior margin armed with transverse rows of moderately small, but strong spines, distally with one small, strong spine; basipodite with epipod hook-shaped.

First pereiopod robust and strongly asymmetrical in shape and size. Major cheliped with fingers distinctly shorter than palm, bearing one or two large teeth on cutting edge, inner side of pollex bearing thin, short setae. Palm rectangular,

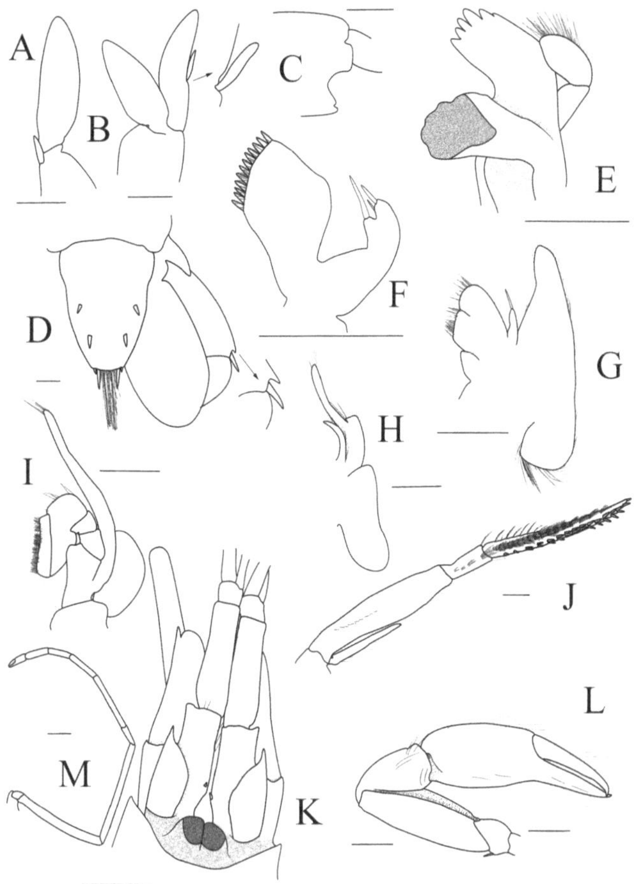


**Fig. 2.110** *Automate spinosa* Wang & Sha, 2017. MBM 159184, holotype, female, CL 4.9 mm, **a**, right first pleopod; **b**, right second pleopod; **c**, sixth abdominal segment, lateral view; **d**, telson and right uropod; e, right mandible; **f**, right maxillule; **g**, right maxilla; **h**, right first maxilliped; **i**, right second maxilliped; **j**, right third maxilliped. paratype, MBM 270172, female, CL 3.1 mm., **k**, anterior part of carapace; **l**, minor cheliped; **m**, left second pereiopod (scale bar = 0.5 mm).

lower margin bearing moderately dense, long setae. Carpus vase-shaped. Merus about three times as long as wide at base, slightly produced near distal. Ischium bearing only setae on superior distal margin. Minor cheliped with fingers slightly shorter than palm; tip of fingers crossing when closed. Carpus about 1/2 length of chela. Merus about three times as long as wide at middle. Merus and ischium similar to that of major cheliped.

Second pereiopod slender; carpus five-segmented, second segment longer other segments, ratio of segments from proximal to distal approximately equal to 0.4: 1.3: 0.7: 0.5: 0.5.

Third pereiopod moderately stout, with dactylus slightly longer than half-length of propodus, slightly curved, bearing short setae near tip. Propodus near as long as carpus, ventral margin bearing 6 spines and distal one pair of spines. Carpus about half-length of merus, distal ventral margin without spine. Merus about six times as long as wide at base. Ischium unarmed. Fourth pereiopod similar to third pereiopod. Fifth pereiopod slender than third and fourth pereiopod; propodus distinctly longer than carpus, with ventral margin bearing transverse rows of setae and two small spines; dactylus about half-length of propodus.

Second pleopod with endopod without appendix masculine both sexes. Uropod with endopod slightly distinctly longer than exopod; exopod with postero-lateral corner acute, strongly produced, accompanied with one spine slightly shorter than posterior margin of exopod.

**Distribution** The South China Sea.

**Remarks** *A. spinosa* is assigned to the *A. evermanni* species group on account of having the following characters: the absence of the rostrum; the stylocerite not reaching the distal end of basal segment of the antennular peduncle; the scaphocerite not reaching the distal end of the second segment of antennular peduncle (Anker & Komai 2004). The present new species is most similar to *A. rectifrons* for the nearly straight frontal region of the carapace and the lacking of rostrum. Morphologically, *A. spinosa* specifically differs from *A. rectifrons* in the broad disto-lateral tooth of the scaphocerite distinctly beyond the distal segment of palp of the mandible rather than narrow (Chace 1972, Fig. 24G); the ischium of the major cheliped unarmed with spines on superior distal margin; the propodus of the third and fourth pereiopod bearing 6 spines ventrally rather than only robust setae (Chace 1972, Fig. 24O, P); the dactylus of posterior three pereiopod long than half-length of the propodus rather than shorter than half-length of the propodus (Chace 1972, Fig. 24O, P), and in the dorsal spines of the telson not very small as shown by Chace (1792, Fig. 24C). Geographically, A. rectifrons occurs from the Atlantic Ocean (Chace 1972; Anker & Komai 2004), while the present new species is collected from the Western Pacific.

## 2.7 Genus *Batella* Holthuis, 1955

**Diagnose** Body not unusually compressed. Carapace without high carina throughout length of dorsal midline Rostrum distinct, acute in dorsal and lateral views. Eyes largely concealed in dorsal view, visible in anterior view. Abdomen without triangular plate articulated at posteroventral angle of sixth segment. Telson not terminating posteriorly in triangular tooth. Mandible with molar process, but without palp. Chelipeds similar, not necessarily equal, carried extended with movable finger dorsal or lateral, not ventral; major chela without molar-like tooth on cutting edge of dactylus. Second pereiopod with carpus 5-segmented. Pereiopods without strap-like epipods. Second pleopod with appendix masculina not overreaching exopod (Chace, 1988).

### 2.7.1 *Batella parvimanus* (Spence Bate, 1888) (Fig. 2.111)

*Cheirothrix parvimanus*.– Spence Bate, 1888: 533; Pl. 96, fig. 2–2b.
*Batella bifurcate*.– Miya & Miyake, 1968: 116; figs 2–4.
*Batella parvimanus*.– Banner DM & Banner AH, 1982: 16–16; fig. 3a–e; Chace, 1988: 67, fig. 8; Liu, 2008: 715.

**Material Examined** This species was reported by Chace (1988) and Liu (2008) from Hong Kong, but not found in the present specimens.

**Description** Rostrum acute, distinctly longer than wide at base, overreaching distal margin of first antennular segment. Orbital teeth distally acute, shorter than rostrum. Eyes visible in dorsal and lateral view. Pterygostomial angle triangular.

Antennular peduncle stout, with distally acute stylocerite hardly reaching distally margin of second segment; visible part of first segment slightly longer than second segment; third segment nearly as long as second segment. Antenna with basicerite bearing dorsolateral and ventrolateral tooth; antennal scale with broad blade far overreaching antennular peduncle, lateral teeth acute, small; carpocerite hardly reaching to distal margin of antennular peduncle.

Right cheliped lacking. Left cheliped moderately compressed. Dactylus about 2/3 length of

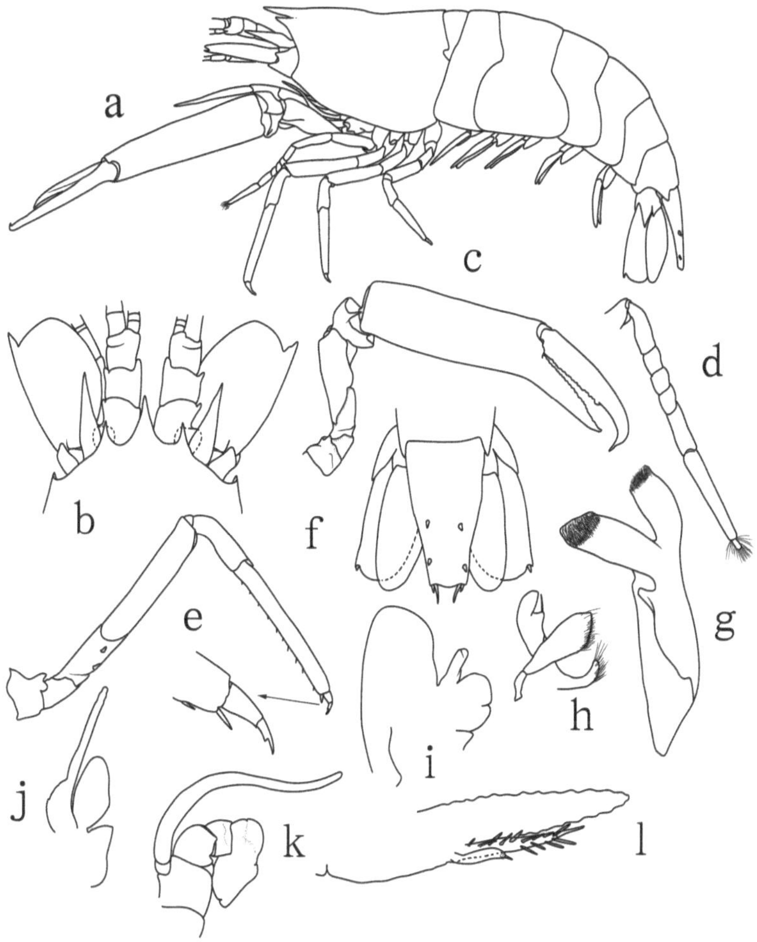


**Fig. 2.111** *Batella parvimanus* (Spence Bate, 1888). (Adapted from Chace, 1988). **a**, entire animal, lateral view; **b**, anterior part of carapace, dorsal view; **c**, left first pereiopod; **d**, chela and carpus of second pereiopod; **e**, third pereiopod; **f**, uropods and telson; **g**, right mandible; **h**, right maxillule; **i**, right maxilla; **j**, right first maxilliped; **k**, right second maxilliped; **l**, endopod of second pleopod

palm, far longer than fixed finger, with tip acute and strong curved distally; cutting edge dentate at proximal 2/3 length; fixed finger with tip acute, but not curved distally; cutting edge dentate on entire length. Palm rectangular, smooth. Carpus cup-shaped. Merus unarmed, obliquely articulated with ischium. Ischium unarmed.

Second pereiopod slender. Fingers about 1/10 length of palm, almost obscured by distal tuft of long setae. Carpus 5-segmented, with ratio subequal to 10: 6: 6: 6: 9. Merus slightly shorter than carpus.

Third pereiopod slender. Dactylus biunguiculate, inferior unguis much shorter and smaller than superior unguis. Propodus longer than merus, with ventral margin bearing about 9 small spinules and distal one pair slender spines. Carpus about 1/2 length of merus, with dorsal margin projecting as sub-acute tooth. Merus unarmed. Ischium armed with two spines on ventral margin.

Uropod slightly longer than telson. Telson about 1.7 times as long as anterior width; dorsal surface bearing two pairs of spines, situated approximately on 1/2 and 3/4 length of telson, respectively; posterior margin slightly convex, bearing two pairs of slender posterolateral spines.

**Remarks** *B. parvimanus* is reported by Chace (1988) from the East China Sea.

**Distribution** The East China Sea; Philippines; Torres Strait off Cape York; Australia.

## 2.8 Genus *Bermudacaris* Anker and Iliff, 2000

**Diagnose** Carapace slightly compressed laterally, smooth. Rostrum broadly rounded. Orbital teeth absent. Pterygostomial angle rounded. Eyestalks sub-conical, anterior portion visible in dorsal and lateral views; corneal pigmentation reduced or absent. Mandible with normally developed incisor and molar processes, and with 2-segmented palp. Ultimate segment of third maxilliped distally armed with slender spines. Chelipeds sub-symmetrical in shape and equal in size carried extended and elevated; dactylus situated ventrally; chelae laterally compressed, without significant armature; adhesive discs and linea impressa absent. Second pereiopods with carpus 5-segmented, and second segment longest. Third pereiopods with dactylus simple, propodus armed with spines on ventral margin. Sixth abdominal segment without articulated plate. Second pleopods of male with appendix interna and appendix masculina. Telson distally truncated, with 2 pairs of spines on postero-lateral margin (Anker & Iliff, 2000).

### 2.8.1 *Bermudacaris* sp. (Fig. 2.112)

**Material Examined** MBM 285974, 1, CL 5.1 mm, TL 14.3 mm, 36°0.683′N, 123°1.011′E, the Yellow Sea, Sta. 3600-6, Jiang Wei & Zhang Jun-long Coll., July 2008.

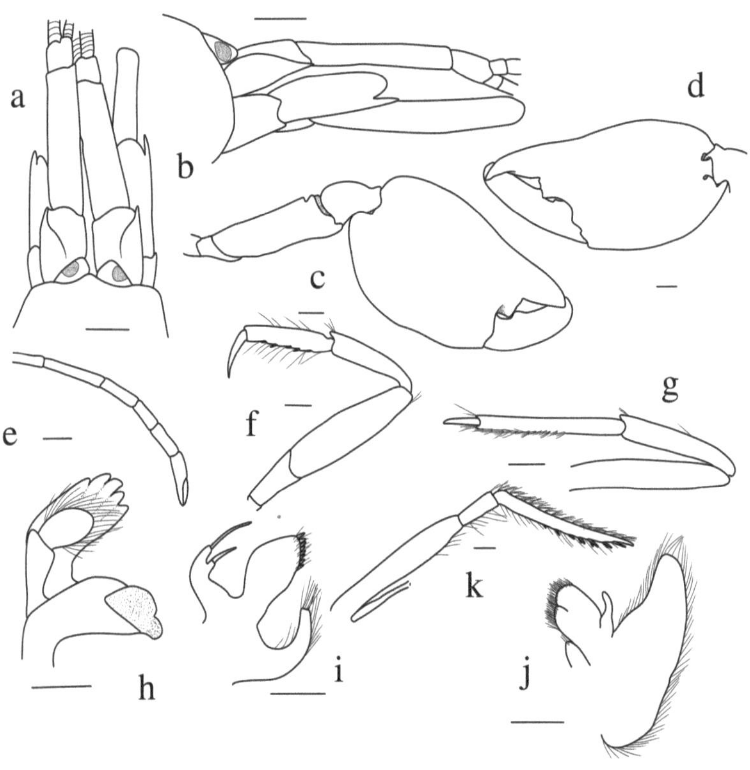


**Fig. 2.112** *Bermudacaris* sp. **a**, anterior part of carapace, dorsal view; **b**, same, lateral view; **c**, right major cheliped; **d**, major chela; **e**, chela and carpus of second pereiopod; **f**, third pereiopod; **g**, fifth pereiopod; **h**, left mandible; **i**, left maxillule; **j**, left maxilla; **k**, third maxilliped (scale bar = 0.5 mm)

**Description** Body smooth, glabrous. Rostrum triangular, shorter than width at base. Orbital teeth absent. Pterygostomial angle rounded.

Antennular peduncle slender, with distally acute stylocerite reaching to distal margin of first segment; visible part of first segment about 2/5 length of second segment; third segment shortest, about 1/3 length of first segment. Antenna with basicerite bearing slight dorsolateral and ventrolateral tooth; antennal scale with broad blade slightly shorter than lateral tooth, reaching to mid-length of second antennular segment; carpocerite slightly overreaching distal margin of antennular peduncle.

Left cheliped lacking. Right cheliped robust, with chela rectangular. Fingers longer than 1/2 length of palm; tips rounded; cutting edges of dactylus and fixed finger armed with one moderately large tooth respectively, leaving gaps when closed. Palm smooth, without depression or ridge. Carpus cup-shaped. Merus slender, about 3.2 times as long as wide; dorsal margin bearing one small tooth sub-distally.

Second pereiopod with carpus probably 4-segmented, ratios sub-equal to 1.8: 1.2: 0.8: 1.

Third pereiopod moderately stout. Dactylus slender, simple, longer than 1/2 length of propodus, slightly curved. Propodus nearly as long as carpus, with ventral margin bearing 5 small spines. Carpus shorter than merus, with dorsal margin projecting as sub-acute tooth. Merus about 4.5 times as long as wide at base, unarmed. Ischium unarmed with spine on ventral margin. Fifth pereiopod slender, with dactylus about 1/4 length of propodus, and propodus distinctly longer than carpus.

**Remarks** Species of *Bermudacaris* is firstly reported from China seas. The present specimen differs from other species of *Bermudacaris* in the following characters: the rostrum triangular, overreaching distal margin of the first antennular segment; base of the first antennular segment with small spine, and the second segment more than twice as long as first segment; the antennal scale not reaching to distal margin of the second antennular segment, and the blade shorter than the lateral tooth; the merus of the cheliped unarmed with spines on ventral margin, the palm rectangular, smooth; the fingers of chela with cutting edges armed with large tooth, leaving gaps when closed; the third pereiopod robust, unarmed with spine on ventral margin of the ischium, the propodus nearly as long as the carpus, and the dactylus longer than 1/2 length of the propodus. Unfortunately, the present specimen is broken, so it is identified as *Bermudacaris* sp. tentatively.

**Distribution** The Yellow Sea.

## 2.9 Genus *Betaeus* Dana, 1852

**Diagnose** Rostrum and orbital hoods absent; with anterior margin of carapace rounded or depressed. Eyes invisible in both lateral and dorsal view. Pterygostomial angle rounded. Stylocerite longer than first antennular segment. Chela cylindrical, showing slight asymmetry, carried in an inverted position with propodal finger uppermost. Chelipeds carried extended; cutting edges of fingers often armed with heavy molar processes; often with simple arrangement of meshing teeth. Dentition various. Second pereiopod with carpus 5-segmented. Third to fifth pereiopod with dactylus simple or biunguiculate; Pleura of sixth abdominal segment articulated. Males bearing appendix masculina (Banner & Banner, 1973).

Key to spices of genus of Betaeus from China seas

-. Fingers of major chela with gaps when closed ........ *B. pingi*

-. Fingers of major chela without gaps when closed ........ *B. granulimanus*

### 2.9.1 *Betaeus granulimanus* Yokoya, 1927 (Fig. 2.113)

*Betaeus granulimanus* Yokoya, 1927: 173, pl. 7, figs. 17–22; Nakazawa and Kubo, 1947: 778, fig. 2246; Utinomi, 1956: 59, pl. 29, fig. 1; Kubo, 1965: 613, fig. 966; Nishimura and Suzuki, 1971: 84, Pl. 27, fig. 8; Liu, 2008: 715.

*Betaeus yokoyaiv.*– Kubo, 1936b: 50; Plate 15, figs. A–G.

*Betaeus murayamai.*– Yokoya, 1936: 132; Fig. 2.

**Material Examined** MBM 285973, 1♂, CL 22.5 mm, TL 56.7 mm, Xiamen, Fujian, Ma Xiutong Coll., 1956.

**Description** Body moderately compressed, glabrous. Rostrum and Orbital teeth absent. Eyes covered by carapace. Pterygostomial angle rounded.

Antennular peduncle slender, with distally acute stylocerite reaching to 1/3 length of second segment; visible part of first segment slightly longer than second segment; first segment bearing

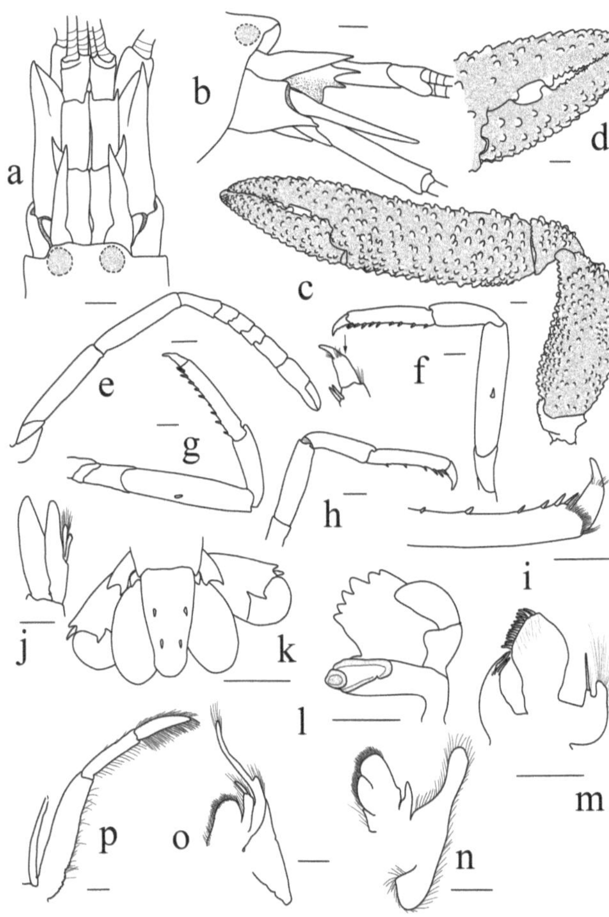


**Fig. 2.113** *Betaeus granulimanus* Yokoya, 1927. MBM 285973, male, CL 22.5 mm, **a**, anterior part of carapace, dorsal view; **b**, same, lateral view; **c**, left first pereiopod; **d**, same, fingers; **e**, right second pereiopod; **f**, left third pereiopod; **g**, left fourth pereiopod; **h**, left fifth pereiopod; **i**, same, dactylus and propodus; **j**, right second pleopod; **k**, uropod and telson; **l**, right mandible; **m**, right maxillule; **n**, right maxilla; **o**, first maxilliped; **p**, third maxilliped (scale bar = 2 mm)

strong lateral tooth distally; third segment distinctly shorter than second segment. Antenna with basicerite bearing strong ventrolateral tooth, not reaching to distal margin of first antennular segment; antennal scale with narrow blade slightly shorter than lateral tooth, not reaching to distal margin of antennular peduncle; carpocerite far overreaching antennular peduncle.

Major cheliped compressed, with all surfaces densely scattered with small well-developed tubercles. Fingers slightly shorter than palm; cutting edges bearing small teeth, leaving gap when closed. Carpus cup-shaped. Merus stout, nearly as long as palm. Minor cheliped lacking.

Second pereiopod moderately stout, with carpus 5-segmented, ratios sub-equal to 2.2: 1: 1: 1: 2. Merus nearly as long as ischium.

Third and fourth pereiopod moderately stout. Dactylus simple, bearing tufts of short setae on lateral surface. Propodus slightly shorter than merus, ventral margin bearing 8 spines and distally one pair. Carpus about 2/3 length of merus, with dorsal margin slightly projecting distally. Merus robust, armed with one spine on ventral margin. Ischium unarmed.

Second pleopod with appendix masculina distinctly longer than appendix interna. Uropod slightly longer than telson. Telson about 1.8 times as long as anterior margin width; Dorsal surface bearing two pairs of spines, situated on 1/3 and 2/3 length of telson, respectively; posterior margin rounded, without posterolateral spines.

**Remarks** Only one male specimen is collected from China seas, and the minor cheliped is lacking. Because of the major cheliped scattered with well-developed small tubercles, the male specimen is identified as *B. granulimanus*. The make specimen is generally agreed with the previous description, except for the telson not bearing posterolateral spines.

**Distribution** The South China Sea; Japan.

### 2.9.2 *Betaeus pingi* Yu, 1930 (Fig. 2.114)

*Betaeus Pingi* Yu, 1930: 454; Figs 1–3.

**Material Examined** This species was reported by Yu (1930) and De Grave & Fransen (2011) from Yantai, Shandong, but not found in the present specimens.

**Description** Body moderately compressed, glabrous. Rostrum and Orbital teeth absent. Eyes covered by carapace.

Antennular peduncle slender, with distally acute stylocerite hardly reaching to mid-length of second segment; visible part of first segment longer than second segment, bearing strong lateral tooth distally; third segment distinctly shorter than second segment. Antenna with basicerite bearing strong lateral tooth; antennal scale with narrow blade slightly shorter than lateral tooth, reaching to distal margin of antennular peduncle; carpocerite slightly longer than antennal scale.

Cheliped asymmetrical. Major cheliped compressed, with all surfaces densely scattered with small well-developed tubercles. Fingers nearly as long as palm; cutting edges of dactylus and fixed finger armed with two teeth respectively, not leaving gaps when closed. Minor cheliped similar to major cheliped in shaped, but cutting edges of finger unarmed with tooth.

Second pereiopod slender, with carpus 5-segmented, first segment longest, 2–4 segments subequal in length, fifth segment slightly shorter than first segment.

Third and fourth pereiopod with dactylus simple, about 1/4 length of propodus. Propodus nearly as long as merus, bearing 7 spines and distally one pair on ventral margin. Carpus distinctly longer than merus. Merus armed with one large spine on ventral margin.

**Remarks** *B. pingi* is originally described from Shandong.

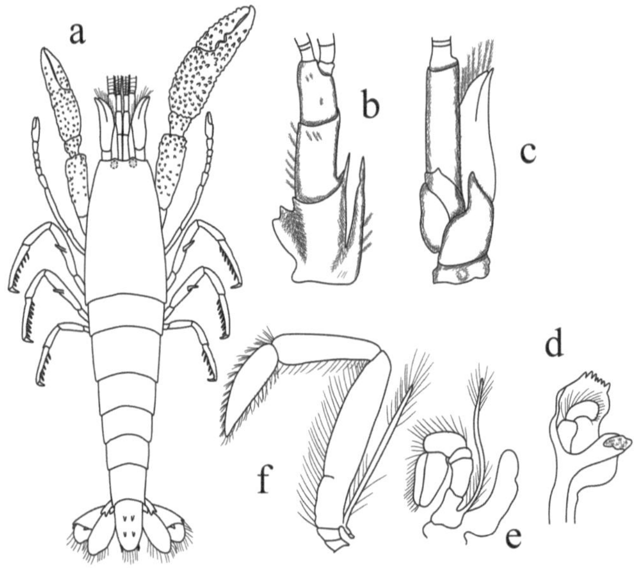


**Fig. 2.114** *Betaeus pingi* Yu, 1930. (Adapted from Yu, 1930). **a**, entire animal, dorsal view; **b**, antennular peduncle; **c**, antenna; **d**, mandible; **e**, second maxilliped; **f**, third maxilliped.

**Distribution** Yantai, Shandong.

## 2.10 Genus *Metalpheus* Coutière, 1908

**Diagnose** General body form similar to *Alpheus*. Eyes normally covered by orbital hoods. Antennular peduncle short and relatively heavy. Basicerite and carpocerite of antennal peduncles also massive; antennal scale may be reduced. Incisor process of mandible expanded, margin with numerous teeth. Chelipeds *like Alpheus,* with both major and minor chelipeds varying in form with sex and maturity; major chela with at most slight to moderate sculpturing. Second pereiopod with carpus 5-segmented. Third to fifth pereiopod with dactylus biunguiculate. Telson similar to *Alpheus* (Banner & Banner, 1982).

### 2.10.1 *Metalpheus paragracilis* (Coutière, 1897) (Figs. 2.115 and 2.116)

*Alpheus paragracilis*.– Coutière, 1897a: 304; 1905: 833, Pl. 76, fig. 22.
*Crangon paragracilis*.– Banner, 1953: 96, fig. 33.
? *Alpheus aglaopheniae*.– Borradaile, 1900: 417; Pl. 38, figs. 13a–f.
*Metalpheus paragracilis*.– DM. & A.H. Banner, 1982: 282, fig. 86; Chace, 1988: 69.

**Material Examined** MBM 107648, 1♂, CL 4.5 mm, TL 11.6 mm, Jinyin Island, Xisha Islands, Wang Yong-liang & Ren Xian-qiu Coll., 15 May 1980; MBM 107647, 8ovigerous♀♀, CL 3.6–5.4 mm, TL 9.7–13.8 mm, Jinyin Island, Xisha Islands, Wang Yong-liang & Ren Xian-qiu Coll., 5 May 1980; MBM 107649, 1♂, CL 2.9 mm, TL 8.1 mm, Yongxing Island, Xisha Islands, 11–13 June 1980; MBM 107646, 1ovigerous♀, CL 3.4 mm, TL 9.8 mm, East Island, Xisha Islands, 28–31 May 1980; MBM 158065,

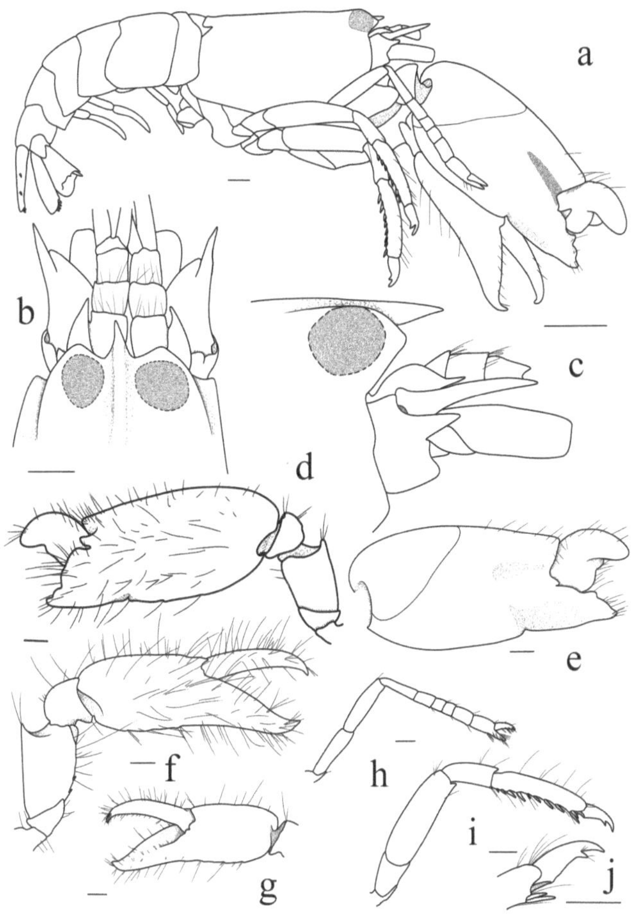


**Fig. 2.115** *Metalpheus paragracilis* (Coutière, 1897). MBM 107648, male, CL 4.5 mm, **a**, entire animal, lateral view; **b**, anterior part of carapace, dorsal view; **c**, same, lateral view; **d**, right major cheliped; **e**, major chela; **f**, left minor cheliped; **g**, minor chela; **h**, right second pereiopod; **i**, right third pereiopod; **j**, same, dactylus and anterior part propodus (scale bar = 0.5 mm)

1♂, CL 4.9 mm, TL 12.9 mm, Bei Reef, Fan Zhen-gang & Xu Jie-shan Coll., 23 March 1958; MBM 157991, 1ovigerous♀, CL 5.8 mm, TL 13.2 mm, Yongxing Island, Xisha Islands, 16 May 1958.

**Description** Body moderately compressed, glabrous. Rostrum about twice as long as wide at base, reaching to first antennular segment; rostral carina feebly developed, extending to posterior margin of eyes. Orbital hoods with rounded teeth, far shorter than rostrum. Eyes covered by carapace. Pterygostomial angle rounded.

Antennular peduncle stout, with distally acute stylocerite reaching to mid-length of second segment; visible part of first segment longer than second segment; third segment sub-equal to second segment. Antenna with basicerite bearing strong dorsolateral tooth almost reaching to distal margin of second antennular segment; antennal scale with narrow blade distinctly shorter than lateral tooth, slightly overreaching distal margin of antennular peduncle; carpocerite longer than lateral tooth of antennal scale.

Major cheliped compressed. Fingers about 1/3 length of palm; tip of dactylus rounded in females and in immature males, acute in larger males. Medial face of palm bearing three rounded pro-

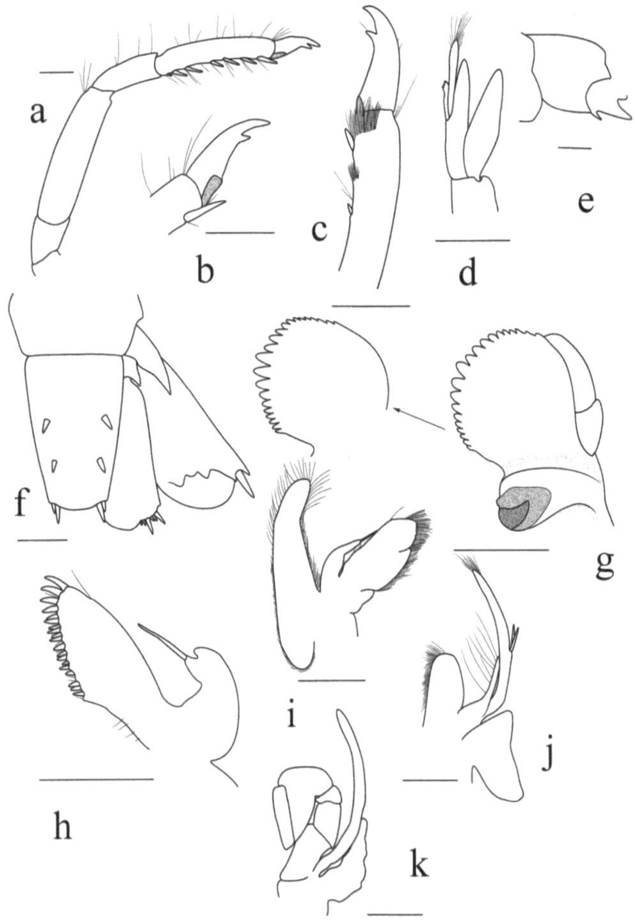


**Fig. 2.116** *Metalpheus paragracilis* (Coutière, 1897). MBM 107648, male, CL 4.5 mm, **a**, right fourth pereiopod; **b**, dactylus and distal part of propodus of fourth pereiopod; **c**, dactylus and distal part of propodus of fifth pereiopod; **d**, second pleopod; **e**, sixth abdominal segment; **f**, right uropod and telson; **g**, right mandible; **h**, right maxillule; **i**, right maxilla; **j**, right first maxilliped; **k**, right second maxilliped (scale bar = 0.5 mm)

trusions below dorsal ridge that continuing obliquely from plaque crest to near *linea impressa*. Merus about 1.5 times as long as broad, with dorsal margin terminating in acute projecting; ventromesial margin unarmed, rounded distally. Minor chela not showing sexual dimorphism, with fingers nearly as long as palm. Ventral margin of palm bearing 5 slight notches fringed with tufts of short stiff setae. Merus similar to that of major cheliped, but ventromesial margin bearing two small spines proximally.

Second pereiopod slender, with fingers slightly shorter than palm. Carpus 5-segmented, with ratios sub-equal to 3.6: 2: 1: 1.1: 2.1.

Third pereiopod moderately stout, with dactylus biunguiculate, about 2/5 length of propodus. Propodus nearly as long as merus, ventral margin bearing 7–9 slender spines and distally one pair. Carpus 1/2 length of merus, with ventral margin terminating in acute tooth. Merus about 3 times as long as wide, ventral margin terminating in large acute tooth. Ischium unarmed.

Second pleopod in male with elongate appendix masculina far overreaching distal margin of endopod. Uropods longer than telson, with posterior margin of endopod bearing about 6 irregular spines. Telson 1.5 times as long as anterior margin width; dorsal surface bearing two pairs of spines, situated at 1/3 and 2/3 length of telson, respectively; posterior margin rounded, bearing two slender posterolateral spines.

**Remarks** The present specimens are generally agreed with the previous descriptions of *M. paragracilis*, except for the merus of major cheliped without spines on ventral margin, and the merus of minor cheliped with ventral margin only bearing two small spines proximally rather than several spines on entire length.

**Distribution** The South China Sea, Taiwan; from the Red Sea and Madagascar across the Indo-Pacific to Hawaii and the Societies Islands.

## 2.11 Genus *Prionalpheus* Banner & Banner, 1960

**Diagnose** Rostrum small. Supra-, extra-, and infra-corneal teeth absent. Anterior carapace projecting to cover corneas from dorsal and lateral view, flat, not inflated to form orbital hoods. Antennules and antennae normal for family. Mandibles without molar process; incisor process greatly expanded and consisting of five strong teeth, the most anterior tooth of left mandible developed as an awl-shaped process that slides within almost closed tube formed by corresponding portion of right mandible. Cheliped similar in form to those of *Alpheopsis equalis* Coutière. Second pereiopod with carpus 3- or 4-segmented. Last three pereiopod slender, elongate with biunguiculate dactylus. Pleura of sixth abdominal somite articulated. Telson and uropods normal (Banner & Banner, 1960).

### 2.11.1 *Prionalpheus mortoni* Bruce, 1990 (Fig. 2.117)

*Prionalpheus mortoni* Bruce, 1990: 647, fig. 18; Liu, 2008: 715; De Grave & Fransen, 2011: 400.

*Prionalpheus* sp. aff. *triarticulatus*.– Bruce, 1990: 628, figs. 10–12.

**Material Examined** This species was reported by Bruce (1990) and Liu (2008) from Hong Kong, but not found in the present specimens.

**Description** Body moderately compressed, glabrous. Rostrum triangular, slightly broader than long, not reaching to distal margin of first antennular segment; lateral margins bearing setae. Pterygostomial angle rectangular.

Antennular peduncle stout, with distally acute stylocerite reaching to distal margin of second antennular segment; first segment with strong distoventral tooth medially; second segment slightly longer than wide, shorter than first segment; third segment nearly as long as wide, about 1/2 length of first segment. Antenna with stout basicerite bearing large acute ventrolateral tooth; antennal scale with broad blade much shorter than distolateral tooth; carpocerite robust, distinctly overreaching distal margin of blade.

Major and minor cheliped symmetrical in shape and size. Fingers about 1/2 length of palm; tips curved; cutting edges of fixed finger feebly dentate, with a distinct groove to accommodate cutting edge of dactylus, bearing short setae laterally. Palm moderately compressed, with ventral margin irregular fringed with long setae. Carpus short and stout. Merus about 2.2 times longer than distal width, with dorsal margin bearing 3 strong spines; ventral margin feebly excavate. Ischium about 3/5 length of merus, with dorsal margin bearing one strong spine distally.

Second pereiopod with fingers nearly as long as palm. Carpus 3-segmented, with ratios subequal to 6: 1: 2

Fifth pereiopod stout, with dactylus biunguiculate, about 1/4 length of propodus. Propodus longer than merus, ventral margin bearing 5 spinules and distally one pair. Carpus robust, 0.82 times as long as propodus. Merus slightly longer than ischium, unarmed. Ischium about 1/2 length of merus, unarmed.

Telson about 0.8 times as long as anterior margin wide; lateral margins straight; dorsal surface bearing two pairs of small spines, situated at 0.68 and 0.85 length of telson, respectively; posterior

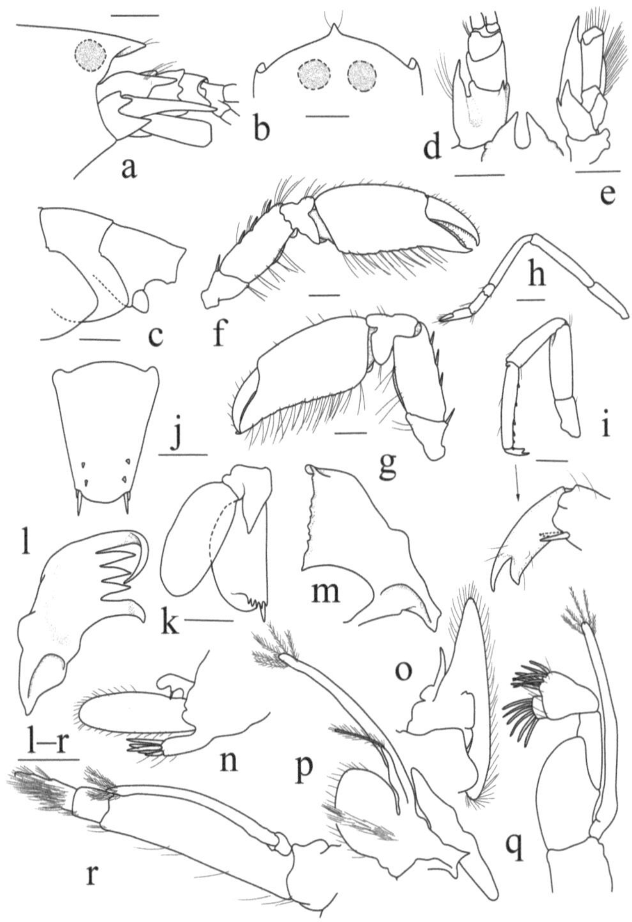


**Fig. 2.117** *Prionalpheus mortoni* Bruce, 1990. (Adapted from Bruce, 1990) **a**, anterior part of carapace, lateral view; **b**, same, dorsal view; **c**, fifth and sixth abdominal segments; **d**, antennular peduncle, ventral view; **e**, same, dorsal view; **f**, right first pereiopod; **g**, left first pereiopod; **h**, second pereiopod; **i**, fifth pereiopod; **j**, telson; **k**, uropod; **l**, left mandible; **m**, right mandible; **n**, maxillule; **o**, maxilla; **p**, first maxilliped; **q**, second maxilliped; **r**, third maxilliped (scale bar = 0.5 mm)

margin convex, bearing two pairs of slender posterolateral spines.

**Remarks** *P. mortoni* is originally described from Hong Kong.

**Distribution** The South China Sea (Hong Kong).

## 2.12 Genus *Rugathanas* Anker & Jeng, 2007

**Diagnose** Body stout. Carapace smooth, not setose. Rostrum with tip acute, with or without sub-distal tooth on ventral margin. Extra-corneal tooth distally acute, strong, beyond anterior margin of eyes; supra-corneal teeth well-developed; infra-corneal teeth reduced or absent. Orbital hoods absent; eyes visible in dorsal and lateral views. Pterygostomial angle protruding, not acute. Cardiac notch well developed. Antennules with minute teeth on distal margin of first segment; ventromesial carina of first segment without tooth. Antenna with stout basicerite, with or without distolateral tooth; scaphocerite short, broadly oval-shaped. Mandible with palp, incisor process greatly expanded, distally with at least 15 teeth. Chelipeds sexually dimorphic, chelae without linea impressa on palm or cavity-tooth system on fingers. Male chelipeds subequal or unequal, sub-symmetrical or asymmetrical, with major cheliped situated either on left or right side, carried flexed; fingers unarmed or armed with simple teeth; chela enlarged, oval, compressed, slightly convex mesially, flattened laterally, dorsal and ventral margins rugose; linea impressa absent. Female chelipeds equal, symmetrical; fingers unarmed; chela varying from oval-shaped to slender, with smooth or rugose margins. Second pereiopod with carpus 4- or 5-segmented. Third pereiopod with dactylus simple. Second pleopod with appendix interna and appendix masculina in males, with appendix interna only in females. Sixth abdominal segment with rounded articulated plate at posteroventral angle. Telson with 2 pairs of dorsal spines and 2 pairs of posterolateral spines (Anker & Jeng, 2007).

## 2.12.1 *Rugathanas borradailei* (Coutière, 1903) (Fig. 2.118)

*Arete borradailei* Coutière, 1903: 80, figs. 19–24; Coutière, 1905: 861, fig. 133; Jacquotte, 1964: 180; Banner AH & Banner DM, 1966a: 152; Chace, 1988: 61; Sha & Liu, 2008: 7, figs. 1–6.

*Arete ghardaqensis* Ramadan, 1936: 15, Pl. 1, figs. 2–8.

*Athanas ghardaquensis* Banner AH & Banner DM, 1981: 42; Banner DM & Banner AH, 1981: 73.

*Athanas polynesia* Banner AH & Banner DM, 1966a: 152, fig. 7; Banner DM & Banner AH, 1982: 302, fig. 93.

*Rugathanas borradailei*.– Anker & Jeng, 2007: 458, figs. 1–3, 6a.

**Material Examined** 1♀, CL 2.14 mm, TL 8.87 mm, 80C-200, Yongxing Island, Xisha Islands, Wang Yong-liang & Ren Xian-qiu Coll., 11–13 June 1980; 1♀, CL 1.33 mm, TL 5.8 mm, 58C-661, Dengqing Island, Xisha Islands, Fan Zhen-gang & Xu Jie-shan Coll., 3 May 1958.

**Description** Body small. Rostrum with apex curved upwards, almost reaching to mid-length of second antennular segment; ventral margin armed with an obtuse tooth sub-distally. Supra-corneal tooth developed, extending to mid-length of cornea; extra-corneal tooth well-developed, beyond distal margin of cornea; infra-corneal tooth absent.

Antennular peduncle stout, with distally acute stylocerite reaching to mid-length of second segment; second segment the shortest, wider than long. Scaphocerite beyond distal margin of third antennular article.

In female chelipeds almost symmetrical, carried forward. Fingers 0.37 times as long as palm; cutting edges bearing short setae. Carpus rounded, with distal end encompassing base of palm. Merus 2.4 times as long as wide, unarmed. Ischium with dorsal margin bearing one medial and one distal spine, respectively.

Second pereiopod slender, with fingers as long as palm. Carpus 4-segmented, with ratios subequal to 2.0: 1.5: 1.0: 2.2. Merus 5 times as long as wide. Ischium unarmed.

Third pereiopod slender, with dactylus simple, about 1/2 length of propodus. Propodus armed with 4 small spines on ventral margin. Carpus as long as ischium. Ischium with dorsal margin bearing one spine distally. Fourth pereiopod similar to third pereiopod, but propodus armed with 6 small spines on ventral margin. Fifth pereiopod similar to the third. Pleurobranchs on pereiopods 1–5; strap-like epipods on pereiopods 1–2.

**Remarks** Anker & Jeng (2007) establish the genus *Rugathanas* for this species and *R. verrucosus*. Additionally, they have given a detailed description of *R. borradailei* and discussed the sexual dimorphism of this species. The present specimens are generally agreed with that description.

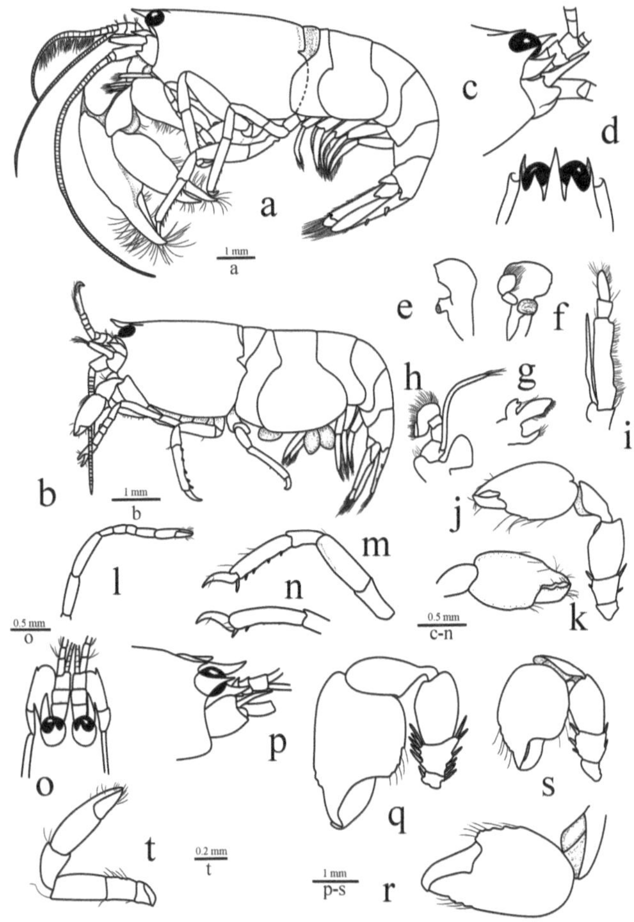


**Fig. 2.118** *Rugathanas borradailei* (Coutière, 1903) (adapted from Anker & Jeng, 2007) **a**, male, entire animal, lateral view; **b**, ovigerous female, entire animal, lateral view. Ovigerous female, **c**, anterior part of carapace, lateral view; **d**, anterior margin of carapace, dorsal view; **e** & **f**, mandible; **g**, maxillule; **h**, second maxilliped; **i**, third maxilliped. Male, **j**, left cheliped, lateral view; **k**, same, chela, mesial view; **l**, second pereiopod; **m**, third pereiopod; **n**, dactylus and propodus of fifth pereiopod. Male, **o**, anterior part of carapace, dorsal view; **p**, same, lateral view; **q**, major cheliped, lateral view; **r**, same, chela and carpus, mesial view; **s**, minor cheliped, mesial view. Female, **t**, first pereiopod

**Distribution** Xisha Islands, the South China Sea; Red Sea; Eastern Africa; Madagascar; Maldives; Philippines; Australia; Somoa Islands.

## 2.13 Genus *Salmoneus* Holthuis, 1955

**Diagnose** Body not unusually compressed. Rostrum triangular. Abdomen of usual form, posterolateral angle of sixth abdominal segment without articulated plate. Telson not terminating posteriorly in triangular tooth. Eyes at least partially concealed from dorsal view. Mandible with palp and molar process. First pereiopod asymmetric in size and shape; major cheliped carried in flexed position, without molar-like tooth on dactylus. Second pereiopod with carpus 5-segmented. Anterior four pereiopods bearing strap-like epipods. Second pleopod with masculina appendix not overreaching exopod (Chace, 1988).

Key to species of the genus Salmoneus from China seas

1. Minor chela with fingers bearing large teeth on proximally 1/3–1/2 length of cutting edges .............. *S. rostratus*
-. Minor chela with fingers only bearing small teeth........................................................................2
2. Carapace with carina ...........................................................................................................3
-. Carapace without carina.........................................................................................................5
3. Carapace with lateral carina arising from lateral margin of rostrum...............................................*S. cristatus*
-. Carapace with lateral carina arising from tip of orbital hoods, or without lateral carina..................................4
4. Carapace without lateral carina..........................................................................................*S. auroalatus*
-. Carapace with lateral carina............................................................................................. *S. tricristatus*
5. Rostrum distinctly longer than wide at base ..........................................................................*S. pusillus*
-. Rostrum shorter than wide at base ..........................................................................................6
6. Third pereiopod with dactylus slender, 6–10 times as long as wide, about 1/2 length of propodus ...................7
-. Third pereiopod with dactylus normal, 3–5 times as long as wide, about 1/3 length of propodus ......................8
7. Ventral margin of rostrum with one sub-distal tooth; second pereiopod with ischium bearing spine on ventral margin ...........................................................................................................*S. alpheophilus*
-. Ventral margin of rostrum without tooth; second pereiopod with ischium unarmed on ventral margin ...........................................................................................................*S. falcidactylus*
8. Lateral margins of rostrum convex ....................................................................... *S. nhatrangensis*
-. Lateral margins of rostrum concave ..........................................................................................9
9. Third pereiopod ischium armed with 2–3 spines on ventral margin and propodus bearing 3–6 spines on ventral margin ...........................................................................................................*S. serratidigitus*
-. Third pereiopod ischium armed with 1 spine on ventral margin and propodus bearing 2 spines on ventral margin ........................................................................................................... *S. sibogae*

## 2.13.1 *Salmoneus alpheophilus* Anker & Marin, 2006 (Fig. 2.119)

*Salmoneus alpheophilus* Anker & Marin, 2006: 310–313, figs. 12, 13; Liu, 2008: 715.

**Material Examined** This species was reported by Liu (2008) and Anker & Marin (2006) from the South China Sea, but not found in the present specimens.

**Description** Carapace setose. Rostrum slender, much longer than wide at base, reaching to 3/4 length of second antennular peduncle; ventral margin bearing one tooth sub-distally; lateral margins slightly concave. Rostral carina feebly developed. Extra-corneal teeth relatively small, acute, directed anteriorly or antero-mesially; margin between rostrum base and extra-corneal teeth broadly U-shaped. Pterygostomial angle rounded.

Antennular peduncle stout, second segment slightly shorter than visible part of first segment, sub-equal to third segment; ventromesial carina of first segment with acute tooth; stylocerite distally acute, about reaching to 2/3 length of second segment. Antenna with basicerite bearing ventrolateral tooth; antennal scale with anterior margin of blade concave, slightly longer than disto-lateral tooth. Carpocerite short, slightly overreaching middle length of scaphocerite.

Major cheliped elongate, with chela subcylindrical, somewhat swollen. Fingers shorter than palm; cutting edge of fixed finger bearing 14 sub-triangular, regularly spaced teeth, distally unarmed, with tip strongly curved; cutting edge of dactylus bearing 13 teeth, most distal part unarmed, with tip strongly curved. Palm with proximoventrally with deep complex groove, continuing as notch on mesial face. Carpus cup-shaped. Merus slender, distally slightly widening, ventrally depressed. Ischium armed with strong spine on ventral margin. Minor cheliped slender, with chela simple, not enlarged. Fingers

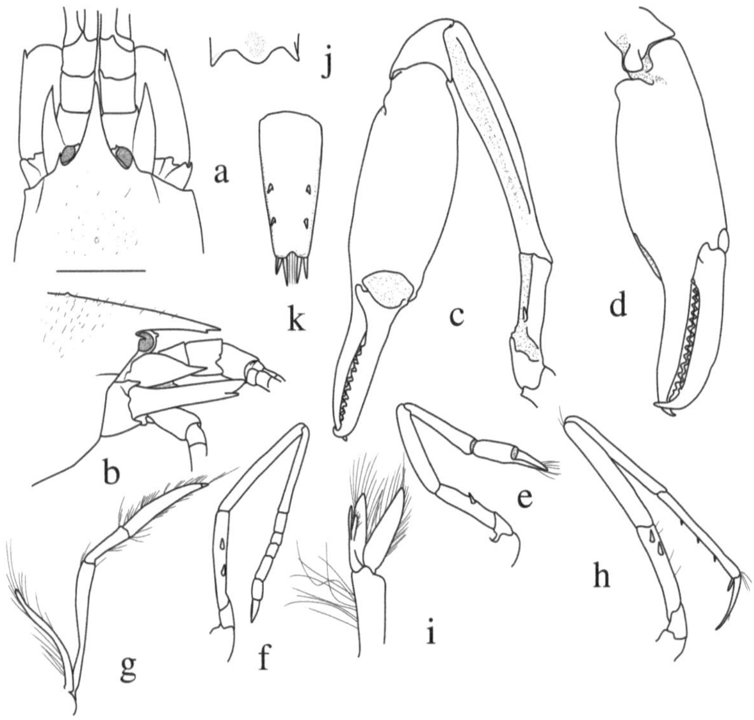


**Fig. 2.119** *Salmoneus alpheopilus* Anker & Marine, 2006. (Adapted from Anker & Marine, 2006). **a**, anterior part of carapace, dorsal view; **b**, same, lateral view; **c**, major cheliped; **d**, major chela; **e**, minor cheliped; **f**, second pereiopod; **g**, third maxilliped; **h**, third pereiopod; **i**, second pleopod; **j**, posterior part of sixth abdominal segment, ventral view; **k**, telson (scale bar = 1 mm)

nearly as long as palm, cutting edges unarmed. Carpus sub-equal to merus. Ischium elongate, shorter than merus, ventral margin armed with spine.

Second pereiopod slender, carpus 5-segmented, with ratios about 7: 2: 1: 1.5: 2.5. Ischium armed with two spines on ventral margin.

Third and fourth pereiopod similar in shape, with dactylus simple, about 1/2 length of propodus. Propodus slightly longer than carpus, ventral margin bearing 3 slender spinules and distally one spine. Carpus more slender and shorter than merus, ventral margin bearing one spine distally. Merus slender, 1.3 times as long as ischium, about 6 times as long as wide at base. Ischium armed with two spines on ventral margin.

Uropod slightly longer than telson. Second pleopod with appendix masculina appendix masculina shorter than appendix interna, bearing slender spines on tip and lateral margins. Telson more than twice as long as wide proximally; basal width less than twice width of posterior margin; dorsal surface bearing two pairs of spines, inserted approximately at 1/2 and 3/4 length of telson, respectively; posterior margin with shallow U-shaped median cleft, furnished with four long plumose setae; two pairs of posterolateral spines slender.

**Remarks** *S. alpheophilus* is originally described from the South China Sea.

**Distribution** The South China Sea; Vietnam.

### 2.13.2 *Salmoneus auroculatus* Anker & Marin, 2006 (Fig. 2.120)

*Salmoneus auroculatus* Anker & Marin, 2006: 300–304, figs. 4–6, 7a; Liu, 2008: 715

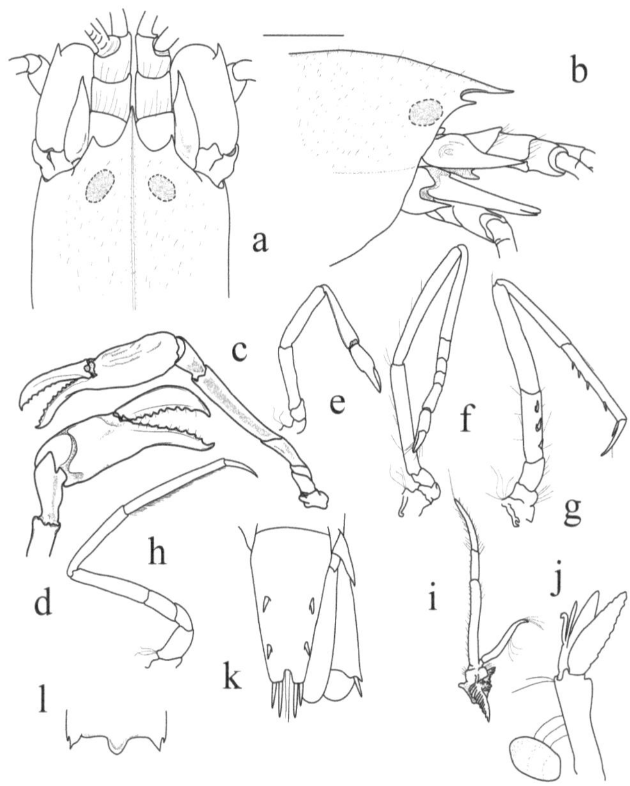


**Fig. 2.120** *Salmoneus auroculatus* Anker & Marine, 2006 (Anker & Marine, 2006) **a**, anterior part of carapace, dorsal view; **b**, same, lateral view; **c**, left major cheliped; **d**, major chela; **e**, right minor cheliped; **f**, second pereiopod; **g**, third pereiopod; **h**, fifth pereiopod; **i**, third maxilliped; **j**, second pleopod; **k**, left uropod and telson; **l**, posterior part of sixth abdominal segment, ventral view (scale bar = 1 mm)

**Material Examined** This species was reported by Liu (2008) and Anker & Marin (2006) from the South China Sea, but not found in the present specimens.

**Description** Carapace slightly setose, with inconspicuous pits, appearing finely granular, laterally with shallow oblique depression and slight suture starting from lateral margin proximal to base of antenna. Rostrum slender, overreaching distal margin of first antennular segment, longer than wide at base; lateral margins slightly concave; rostral carina well developed, extending to 3/4 length of carapace. Extra-corneal teeth acute; margins between rostrum base and extra-corneal teeth broadly concave. Pterygostomial angle rounded.

Antennular peduncle stout, three segments sub-equal to each other; ventromesial carina of first segment with acute tooth; distally acute stylocerite reaching to distal margin of second antennular segment. Antenna with basicerite bearing strong ventrolateral tooth; antennal scale with broad blade slightly beyond distolateral tooth, anterior margin convex; carpocerite short, reaching to about 3/4 length of scaphocerite.

Major pereiopod robust, with chela sub-cylindrical. Tip of fingers curved; cutting edges of dactylus and fixed finger bearing 9 and 11 rounded-triangular, regularly spaced teeth, respectively. Palm slightly longer than fingers, proximoventrally with deep groove, continuing by oblique groove across lateral to dorsal surface. Carpus somewhat elongate, vase-shaped, ventrally depressed; distally with two pronounced lobs. Merus elongate, not widening distally, with feebly marked rounded lobes; flattened ventrally. Ischium unarmed. Minor cheliped slender, with chela simple, not particularly enlarged. Fingers slightly shorter than palm, cutting edges unarmed. Carpus nearly as long as merus, sub-cylindrical; flattened ventrally, distally widening. Ischium elongate, nearly as long as merus.

Second pereiopod slender, with fingers equal to palm. Carpus 5-segmented, , with ratios about 4.5: 1: 0.6: 0.6: 1.2.

Third and fourth pereiopod similar, slender. Dactylus simple, less than half length of propodus. Propodus slightly longer than merus, ventral margin bearing 4 small spines and distally one. Carpus shorter than merus, ventral margin bearing one spinule distally. Merus about 1.4 times as long as ischium, unarmed. Ischium armed with three spines on ventral margin.

Second pleopod with appendix masculina longer than appendix interna, with slender spines along mesial margin. Uropods slightly longer than telson. Telson about twice as long as wide proximally; dorsal surface bearing two pairs of spines, inserted approximately at 1/2 and 3/4 length of telson, respectively; posterior margin with deep U-shaped median notch, furnished two long setae; two pairs of posterolateral spines slender.

**Remarks** *S. auroculatus* is originally described from the South China Sea.

**Distribution** The South China Sea; Vietnam.

### 2.13.3 *Salmoneus cristatus* (Coutière, 1897) (Figs. 2.121, 2.122, 2.123, and 2.124)

*Jousseaumea cristata* Coutière, 1897: 234.
*Salmoneus cristatus*–. Holthuis, 1958: 18, fig. 7; Wang & Sha, 2016c: 1586, figs. 1–4.

**Material Examined** MBM 107734, 2ovigerous♀♀, CL 5.5 mm, TL 12.5–14.5 mm, Yongxing Island, Xisha Islands, Fang Xiu-bin Coll., 11–13 June 1980; MBM 107735, 1ovigerous♀, CL 4.6 mm, TL 11.5 mm, Yongxing Island, Xisha Islands, Fang Xiu-bin Coll., 11–13 June 1980; MBM 107729, 1ovigerous♀, CL 5.2 mm, TL 12.5 mm, Yongxing Island, Xisha Islands, Fang Xiu-bin Coll., 9 May 1980.

**Description** Carapace slightly setose. Rostrum triangular, at base wider than long, variously reaching to end of second segment to up to 3/4 length of third segment of antennular peduncle. Lateral margins of rostrum continued on carapace to posterior of level of orbital hoods as strong lateral ridges, giving rostral base a plateau-like appearance. Rostral carina extending posteriorly about 4/5 length of carapace. Triangular orbital hoods small, acute, directed straight forward, lateral margins concave, without carina. Pterygostomial angle rounded. Cardiac notch well developed. Eyes well developed, cornea partly visible dorsally and laterally.

Antennular peduncle moderately stout; first segment longest; second segment slightly wider than long; third segment nearly as long as second segment; stylocerite reaching to about 1/2 length of third segment of antennular peduncle, acute distally; lateral flagellum biramous, shorter ramus composed of at least three segments bearing tufts of aesthetascs; proximal fused portion composed of at least four segments. Basicerite bearing strong distodorsal tooth, slightly shorter than distolateral tooth; scaphocerite oval, anterior margin of blade rounded; distolateral tooth strong, acute, distinctly shorter than anterior margin of blade, reaching about 1/2 length of third segment of antennular peduncle; carpocerite moderately stout, reaching about end of second segment of antennular peduncle. Mandible with incisor process bearing distally 8 rounded irregular teeth, with third and fourth larger than others, molar process typical for genus. Maxillule with two endites, distal endite with numerous large, spine-like setae on inner margin, basal endite with long setae on distal margins; palp bilobed, lower lobe with one robust plumose seta, upper lobe rounded. Maxilla with slender tapering palp;

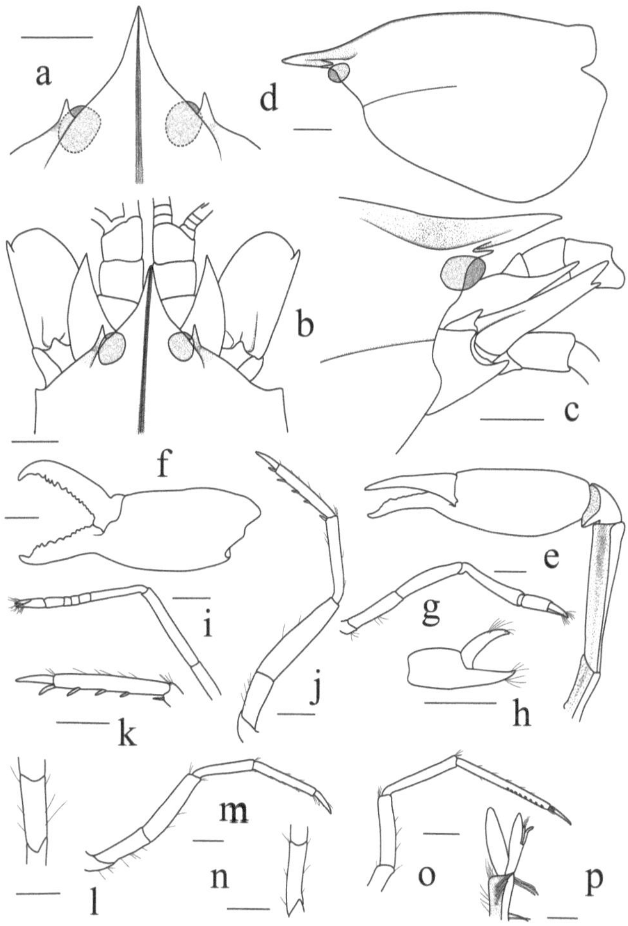


**Fig. 2.121** *Salmoneus cristatus* (Coutière, 1897). MBM 107734, ovigerous female, CL 4.5 mm, **a**, rostrum and orbital hoods, dorsal view; **b**, anterior part of carapace, dorsal view; **c**, same, lateral view; **d**, carapace, lateral view; **e**, left major cheliped; **f**, major chela; **g**, right minor cheliped; **h**, minor chela; **i**, right second pereiopod; **j**, right third pereiopod; **k**, same, propodus and dactylus; **l**, same, ischium, ventral view; **m**, right fourth pereiopod; **n**, same, ischium, ventral view; **o**, left fifth pereiopod; **p**, right second pleopod (scale bar = 0.5 mm, e & g share same scale)

distal endite bilobed, but not deeply, densely setose medially; proximal endite with single lobe, lanceolate, distally with setae; scaphognathite well developed. First and second maxilliped typical for genus. Third maxilliped moderately slender; ultimate segment very setose, distally tapering, bearing 2 subapical spines; arthrobranch normally developed.

First chelipeds very asymmetrical. Major chela subcylindrical, without distinct grooves on surface; cutting edge of dactylus and pollex bearing teeth along full length of cutting edge, finger tips corneous, crossing when closed; palm longer than fingers; carpus cup-shaped; merus slender, elongate, ventrally flattened; ischium slightly elongate, without spines, ventrally flattened. Minor cheliped slender; chela simple, not particularly enlarged, fingers subequal to palm, cutting edge unarmed; carpus subcylindrical, slightly longer than merus; merus and ischium unarmed, merus slightly longer than ischium.

Second pereiopod slender; carpus 5-segmented, length ratio subequal to 1.7: 0.4: 0.3: 0.3: 1.1; chela simple, fingers sub-equal to palm.

Third pereiopod slender, with dactylus simple, slightly and gradually curved, about 0.4 times as

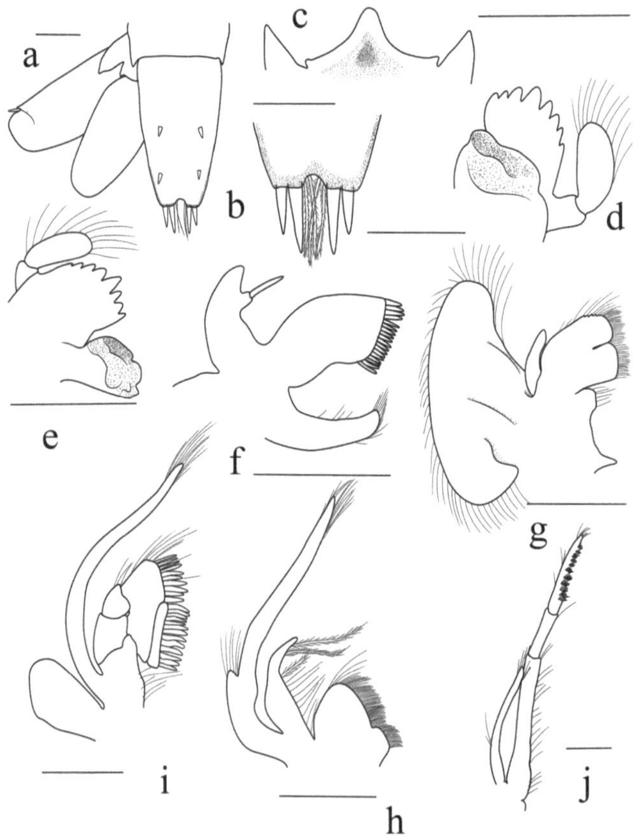


**Fig. 2.122** *Salmoneus cristatus* (Coutière, 1897). MBM 107734, ovigerous female, CL 4.5 mm, **a**, telson and left uropod; **b**, telson, distal portion; c, preanal plate of sixth abdominal somite, ventral view; **d** & **e**, left mandible; **f**, left maxillule; **g**, left maxilla; **h**, left first maxilliped; **i**, left second maxilliped; **j**, third maxilliped (scale bar = 0.5 mm)

long as propodus. Propodus slightly shorter than merus, armed with 3 spines along ventral margin, and distally one pair of spines. Carpus sub-equal to merus, with 1 distal spine on ventral margin. Merus about twice as long as ischium, unarmed. Ischium unarmed. Fourth pereiopod similar to third. Fifth pereiopod more slender, ischium unarmed; merus, carpus, and propodus similar to those of third pereiopod, slightly different, carpus without distal spine on ventral margin; propodus with at least 9 rows of short setae along ventral margin; dactylus more slender than that of third and fourth pereiopods. Sixth abdominal somite without articulate pleura; preanal plate rounded posteriorly.

Endopod of second pleopod with appendix masculina distinctly shorter than appendix interna.

Telson about 1.7 times as long as wide at base, dorsal surface with two pairs of spines situated at some distance from lateral margin, anterior and posterior pair situated at about 1/2 and 4/5 length of telson, respectively; posterior margin with shallow and moderately large median notch bearing two single or two pairs of long plumose setae, and two pairs of slender posterolateral spines, lateral one about 3/4 length of mesial. Uropods slightly shorter than telson; diaeresis with one strong, acute spine; distolateral spine shorter.

**Remarks** *S. cristatus* is originally described from Djibouti. Holthuis (1958) and Bruce (1976) record the species from Israel and Kenya, respectively. Moreover, Banner & Banner (1966b, 1981) report this species from Djibouti and Thailand. The *S. cristatus* species-group is

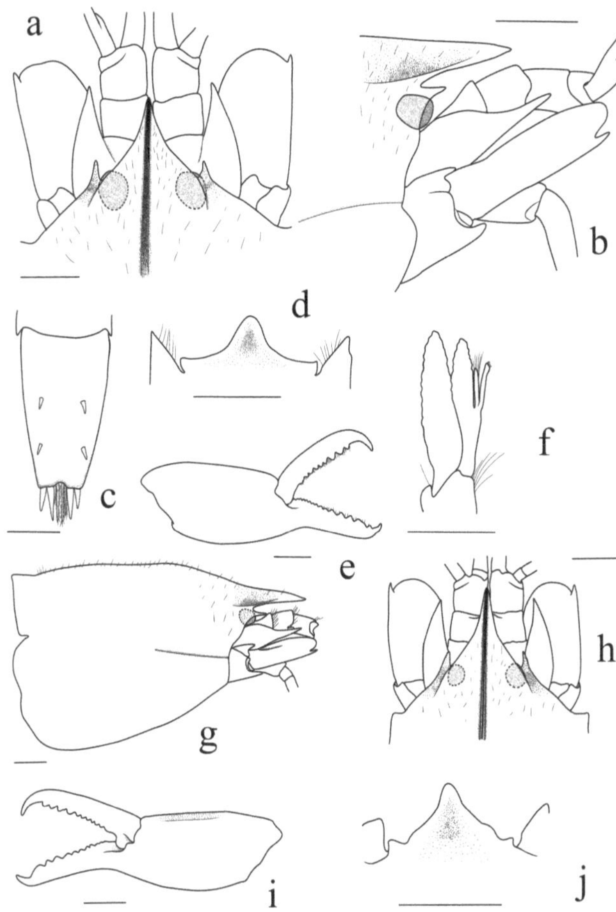


**Fig. 2.123** *Salmoneus cristatus* (Coutière, 1897). MBM 107734, ovigerous female, CL 5.5 mm, **a**, anterior part of carapace, dorsal view; **b**, same, lateral view; **c**, telson; **d**, preanal plate of sixth abdominal segment, ventral view; **e**, major chela. MBM 107735, male, CL 4.6 mm, **f**, second pleopod; **g**, carapace, lateral view; **h**, anterior part of carapace, dorsal view; **i**, left major chela; **j**, preanal plate of sixth abdominal segment, ventral view (scale bar = 0.5 mm)

erected by Anker & Marin (2006) and characterized by the presence of the developed mediodorsal carina extending from the rostrum to or beyond the midlength of the carapace. The *S. cristatus* species-group contains five species now. *S. cristatus* can be easily separated from the other four species by the presence of the "shoulder" on the carapace. The present specimens are in general agreed with the previous descriptions, except for the following slight differences characters: the posterior margin of the telson with broad U-shaped notch bearing one or two pair of long plumose setae in the present material rather than the V-shaped notch as described in Banner & Banner (1966b, fig. 10j); both sides of the posterior margin of the telson armed with two spines, and the one side armed with three spines as shown in Banner & Banner (1966b, fig. 10j) not found in the present material. Anker (2003b) also reports this species from Australia, and the description is slightly different from the present material in the following three characters: the eyes partly visible both in dorsal and lateral view, versus the eyes invisible both in dorsal and lateral view according to Anker (2003b, fig. 6A, B); the ischium of the third pereiopod unarmed in the present material, while that armed with one moveable spine on the ventral margin (Anker, 2003b, fig. 6H); the posterior margin of the telson bearing a broad U-shaped notch, whereas it very

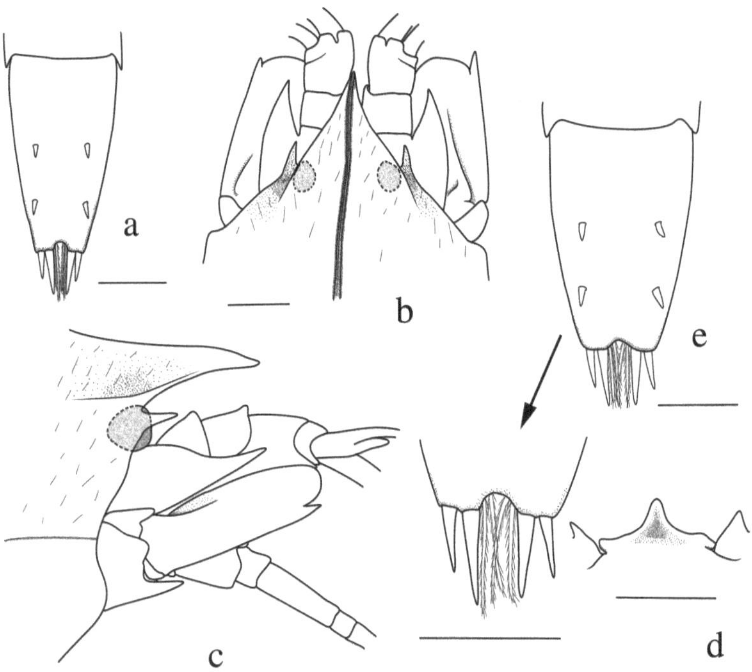


**Fig. 2.124** *Salmoneus cristatus* (Coutière, 1897). MBM 107735, ovigerous female, CL 4.6 mm, **a**, telson. MBM 107729, ovigerous female, CL 5.2 mm, **b**, anterior part of carapace, dorsal view; **c**, same, lateral view; **d**, preanal plate of sixth abdominal segment, ventral view; **e**, telson (scale bar = 0.5 mm)

narrow and deep in Anker (2003b, fig. 6J). Those differences should be taken as intra-specific difference of *S. cristatus*. Interestingly, the present ovigerous female specimens with endopod of second pleopod bear both the appendix masculina and the appendix interna. The same character is also found in *S. hispaniolensis* Anker, 2010 (cf. Anker, 2010, fig. 1h), *S. camaroncito* Anker, 2010 (cf. Anker, 2010, fig. 5h), *S. degravei* Anker, 2010 (cf. Anker, 2010, fig. 7g) and *S. auroculatus* Anker & Marin, 2006 (cf. Anker & Marin, 2006, fig. 5i). Additionally, *S. carvachoi* Anker, 2007 is reported by Oliveira et al. (2015) as a possible simultaneous hermaphrodite, also bearing the appendix masculina in ovigerous female specimens. Therefore, *S. cristatus* could also be a simultaneous hermaphrodite.

**Distribution** Xisha Islands, South China Sea; Djibouti; Kenya; Israel; Thailand; Ashmore Reef, Coral Sea, Queensland, Australia.

### 2.13.4 *Salmoneus falcidactylus* Anker & Marin, 2006 (Fig. 2.125)

*Salmoneus falcidactylus* Anker & Marin, 2006: 304–307, figs. 8, 9; Liu, 2008: 715

**Material Examined** This species was reported by Liu (2008) and Anker & Marin (2006) from the South China Sea, but not found in the present specimens.

**Description** Carapace glabrous, laterally with shallow oblique depressions. Rostrum slender, reaching to distal 3/4 length of second antennular segment, longer than wide at base; lateral margins shallowly concave; rostral carina absent. Extra-corneal teeth acute; concavity between rostrum base and extra-corneal teeth deep, U-shaped. Pterygostomial angle slightly protruding, rounded.

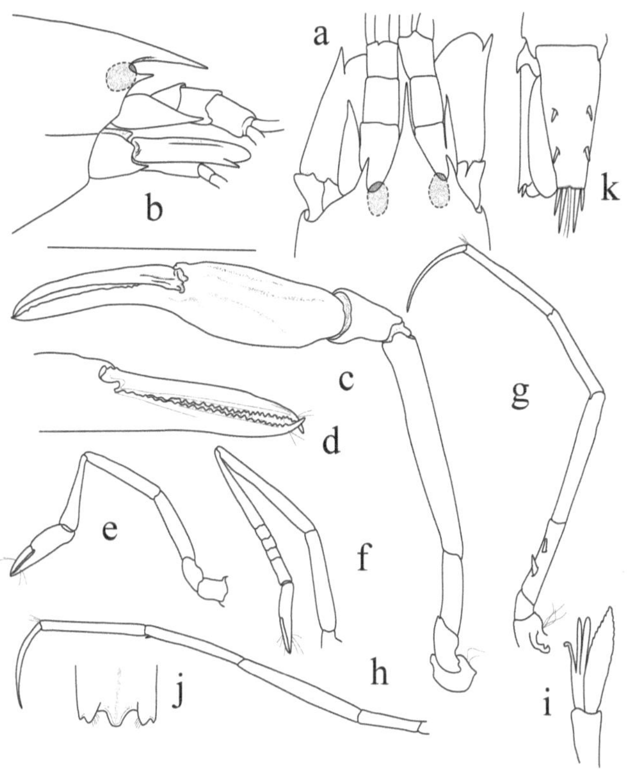


**Fig. 2.125** *Salmoneus falcidactylus* Anker & Marine, 2006. (Adapted from Anker & Marine, 2006). **a**, anterior part of carapace, dorsal view; **b**, same, lateral view; **c**, left major cheliped; **d**, same, fingers; **e**, right minor cheliped; **f**, second pereiopod; **g**, third pereiopod; **h**, fifth pereiopod; **i**, second pleopod; **j**, posterior part of sixth abdominal segment, ventral view; **k**, telson (scale bar = 1 mm)

Antennular peduncle not particularly stout; second segment short than visible part of first segment, longer than third segment; ventromesial carina of first segment with acute tooth; stylocerite slightly overreaching mid-length of second antennular segment. Antenna with basicerite bearing strong ventrolateral tooth; antennal scale moderately broad, with anterior margin of blade convex, not beyond distolateral tooth.

Major cheliped slender, with chela sub-cylindrical. Fingers longer than palm, tips of fingers acute and strongly curved; cutting edges of dactylus and fixed finger each bearing at least 20 very small, subtriangular, regularly spaced teeth. Palm proximoventrally with deep groove; dorsal and lateral surface with longitudinal ridges and depressions; ventral surface slightly flattened. Minor cheliped slender, with chela simple, not particularly enlarged. Fingers slightly shorter than palm, cutting edges unarmed. Carpus slightly longer than merus, sub-cylindrical, distally widening. Ischium elongate, nearly as long as merus.

Second pereiopod slender, with chela simple, fingers sub-equal to palm. Carpus 5-segmented, length ratio subequal to 5.2: 1: 1: 1: 1.5.

Third and fourth pereiopod similar. Dactylus simple, extremely slender, about 4/5 length of propodus, strongly curved. Propodus longer than carpus; ventral margin only armed with one pair of spines distally. Carpus slender, shorter than merus, unarmed. Merus about 1.5 times as long as ischium. Ischium armed with 2 spines on ventral margin.

Second pleopod with appendix masculina longer than appendix interna, hardly reaching distal margin of endopod, tip bearing two slender spines. Uropod slightly longer than telson. Telson about 2.3 times as long as proximal width; dorsal surface bearing two pairs of spines, situated at about 1/2 and 3/4 length of telson, respectively; posterior margin with very shallow median notch, bearing two long setae; posterolateral spines slender.

**Remarks** *S. falcidactylus* is originally described from the South China Sea.

**Distribution** The South China Sea; Vietnam.

### 2.13.5 *Salmoneus nhatrangensis* Anker & Marin, 2006 (Fig. 2.126)

*Salmoneus nhatrangensis* Anker & Marin, 2006: 296–300, figs. 1–3; Liu, 2008: 715.

**Material Examined** This species was reported by Liu (2008) and Anker & Marin (2006) from the South China Sea, but not found in the present specimens.

**Description** Carapace slightly setose, with small pits appearing somewhat granular; laterally

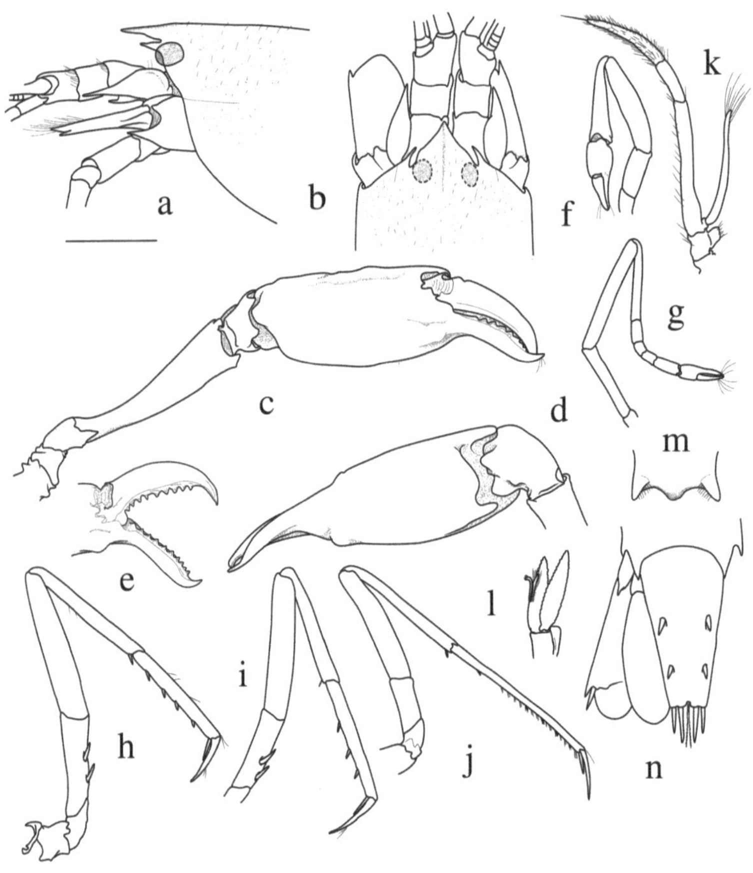


**Fig. 2.126** *Salmoneus nhatrangensis* Anker & Marine, 2006. (Adapted from Anker & Marine, 2006). **a**, anterior part of carapace, lateral view; **b**, same, dorsal view; **c**, right major cheliped; **d**, major chela; **e**, same, fingers; **f**, left minor cheliped; **g**, second pereiopod; **h**, third pereiopod; **i**, fourth pereiopod; **j**, fifth pereiopod; **k**, third maxilliped; **l**, second pleopod; **m**, posterior part of sixth abdominal segment; **n**, left uropod and telson (scale bar = 1 mm)

with shallow oblique depression and slight suture starting from lateral margin proximal to base of antenna. Rostrum wider than long at base, not reaching to distal margin of first antennular segment; lateral margins distinctly convex proximally; rostral carina moderately developed, extending to posterior margin of eyes. Extra-corneal teeth acute; notch between rostrum base and extra-corneal teeth deep.

Antennular peduncle stout, second segment shorter than visible part of first segment; ventromesial carina of first segment with acute tooth; stylocerite distally subacute, reaching to distal margin of second antennular segment. Antenna with basicerite bearing strong ventrolateral tooth; antennal scale relatively broad, with anterior margin of blade strongly convex, beyond disto-lateral tooth; carpocerite short, reaching to about 3/4 length of scaphocerite.

Major cheliped robust, with chela sub-cylindrical. Fingers about 3/5 length of palm, tip strongly curved, crossing when fingers closed; cutting edges of dactylus and fixed finger bearing 9 and 10 rounded regularly spaces teeth, respectively. Palm proximoventrally with complex deep groove; dorsal surface with elongate depression. Carpus cup-shaped, distally with several pronounced lobes. Merus elongate, distally widening; ventrally depressed, with rounded distal lobes. Ischium short, unarmed. Minor cheliped slender, with chela simple, not particularly enlarged. Fingers as long as palm, cutting edges unarmed. Carpus slightly longer than merus, cylindrical. Ischium slightly elongate, about 1/2 length of merus.

Second pereiopod slender, with chela simple, fingers slightly longer than palm. Carpus 5-segmented, length ratio subequal to 4: 1: 0.7: 0.7: 1.4.

Third and fourth pereiopod slender. Dactylus simple, less than half length of propodus. Propodus slightly shorter than carpus, ventral margin bearing 4 spines and distally one slender spine. Carpus slender, shorter than merus, ventral margin bearing 1 spine distally. Merus about 1.7 times as long as ischium, unarmed. Ischium armed with 2 spines on ventral margin.

Second pleopod with appendix masculina slightly shorter than appendix interna, with row of slender spines along mesial margin. Uropod slightly longer than telson. Telson about twice as long as wide proximally; dorsal surface bearing two pairs of spines, situated at about 1/2 and 3/4 length of telson, respectively; posterior margin with shallow median notch, bearing two long setae; posterolateral spines slender.

**Remarks** *S. nhatrangensis* is originally described from the South China Sea.

**Distribution** The South China Sea; Vietnam.

### 2.13.6 *Salmoneus pusillus* Anker & Marin, 2006 (Fig. 2.127)

*Salmoneus pusillus* Anker & Marin, 2006: 307–310, figs. 10, 11; Liu, 2008: 715.

**Material Examined** This species was reported by Liu (2008) and Anker & Marin (2006) from the South China Sea, but not found in the present specimens.

**Description** Carapace sparsely setose, without pits, laterally with slight suture starting from lateral margin proximal to base of antenna, dorsally with minute tubercle. Rostrum slender, longer than wide at base, reaching to distal margin of second antennular segment; ventral margin bearing conspicuous tooth sub-distally; lateral margins slightly concave; rostral carina inconspicuous. Extra-corneal teeth acute; margins between base of rostrum and extra-corneal teeth deeply incised, U-shaped. Pterygostomial angle rounded.

Antennular peduncle moderately stout, second segment shorter than visible part of first and third segment; distally acute stylocerite reaching to distal margin of second segment. Antenna with basicerite bearing strong ventrolateral tooth;

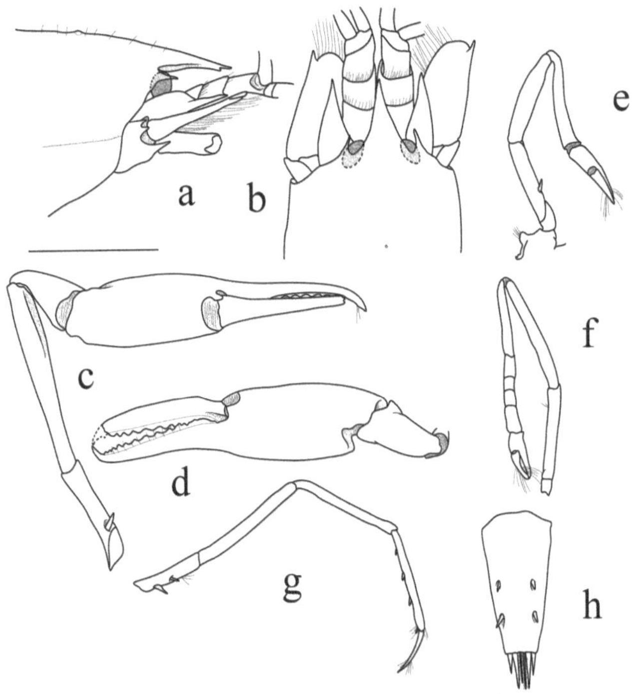


**Fig. 2.127** *Salmoneus pusillus* Anker & Marine, 2006. (Adapted from Anker & Marine, 2006) **a**, anterior part of carapace, lateral view; **b**, same, dorsal view; **c**, left major cheliped; **d**, major chela; **e**, right minor cheliped; **f**, second pereiopod; **g**, third pereiopod; **h**, telson (scale bar = 1 mm)

antennal scale with anterior margin of blade convex, shorter than distolateral tooth; carpocerite reaching to about 3/4 length of scaphocerite.

Major cheliped slender, with chela sub-cylindrical. Fingers slightly longer than or sub-equal to palm; tips curved; cutting edges of both dactylus and fixed finger armed with 11–12 teeth. Palm with proximoventrally with deep groove. Carpus vase-shaped, distally with broad lobes. Merus elongate, distally not widening; flattened ventrally. Ischium elongate, ventral margin armed with one strong spine. Minor cheliped slender, with chela simple, not particularly enlarged. Fingers slightly longer than palm, cutting edges unarmed. Carpus sub-cylindrical, slightly longer than merus; distally slightly widening. Ischium elongate, slightly shorter than merus; ventral margin armed with one spine.

Second pereiopod slender, with chela simple, fingers longer than palm. Carpus 5-segmented, length ratio subequal to 5: 2: 1: 1.2: 2.2.

Third and fourth pereiopod slender. Dactylus simple, slender, about half length of propodus. Propodus nearly as long as carpus, ventral margin armed with 3 spinules and 1 distally slender spine. Carpus sub-equal to merus, ventral margin bearing one small spinule distally. Merus slender, unarmed. Ischium armed with spine on ventral margin.

Second pleopod with slender appendix masculina, almost twice as long as appendix interna, with small spines distally. Uropods slightly longer than telson. Telson about twice as long as wide proximally; dorsal surface bearing two pairs of spines, situated at about 1/2 and 3/4 length of telson, respectively; posterior margin straight, with two long plumose and two posterolateral slender spines.

**Remarks** *S. pusillus* is originally described from the South China Sea.

**Distribution** The South China Sea; Vietnam.

### 2.13.7 *Salmoneus rostratus* Barnard, 1962 (Fig. 2.128)

*Salmoneus rostratus* Barnard, 1962: 240, fig. 1; Anker & Marin, 2006: 313, figs. 7b, c, 14; Liu, 2008: 715.

**Material Examined** This species was reported by Liu (2008) and Anker & Marin (2006) from the South China Sea, but not found in the present specimens.

**Description** Carapace smooth. Rostrum hardly reaching to middle length of second antennular segment; ventral margin bearing one acute tooth sub-distally; short supraorbital tooth present; rostral carina extending to posterior margin of orbital hoods; one blunt tooth present on middle length of carapace medially. Pterygostomial angle rounded.

Antennular peduncle moderately slender, second segment sub-equal to visible part of first segment, distinctly longer than third segment; distally acute stylocerite slightly overreaching distal margin of first segment. Antennal with basicerite bearing strong ventrolateral tooth; antennal scale with distal margin of blade convex, beyond distolateral tooth; carpocerite reaching to middle length of third antennular segment.

Major cheliped elongate, with chela subcylindrical. Fingers slightly longer than palm, with tip of dactylus strongly curved; cutting edges of both dactylus and fixed finger armed with two large teeth and one small proximally. Palm with mesial face bearing a ridge distally extending proximally half-length of palm,

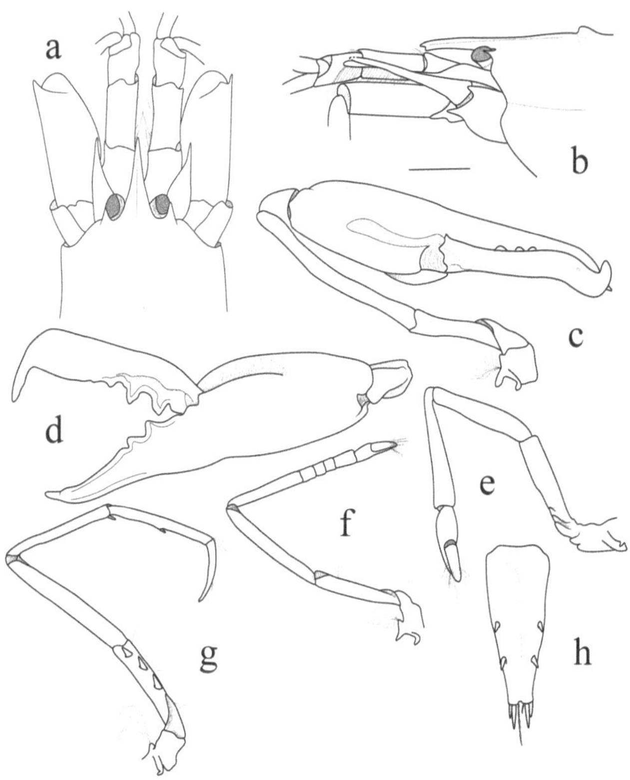


**Fig. 2.128** *Salmoneus rostratus* Barnard, 1962 (adapted from Anker & Marine, 2006). **a**, anterior part of carapace, lateral view; **b**, same, dorsal view; **c**, right major cheliped; **d**, major chela; **e**, left minor cheliped; **f**, second pereiopod; **g**, third pereiopod; **h**, telson (scale bar = 1 mm)

between ridge shallow grooves present ventrally. Carpus cup-shaped. Merus extremely slender, gently curved, unarmed. Ischium about 1/2 length of merus, unarmed. Minor cheliped with chela simple, not enlarged. Carpus elongate, distinctly longer than merus. Merus slightly longer than ischium, unarmed.

Second pereiopod slender, with chela simple, fingers slightly longer than palm. Carpus 5-segmented, length ratio subequal to 5: 21: 0.7: 0.7: 1.3.

Third pereiopod and fourth pereiopod slender. Dactylus simple, curved, slightly longer than 1/2 length of propodus. Propodus nearly as long as carpus, ventral margin bearing one spinule. Carpus about 3/4 length of merus, ventral margin bearing one slender spinule distally. Merus unarmed. Ischium armed with 3 spines on ventral margin.

Telson more than twice as long as wide proximally; dorsal surface bearing two pairs of spines, situated at about 1/2 and 3/4 length of telson, respectively; posterior margin straight, with U-shaped notch medially and two long plumose; two posterolateral spines slender.

**Remarks** *S. rostratus* is reported by Anker & Marin (2006) from the South China Sea.

**Distribution** The South China Sea; Madagascar; New Guinea; Vietnam.

### 2.13.8 *Salmoneus* cf. *serratidigitus* (Coutière, 1896) (Figs. 2.129 and 2.130)

*Jousseaumea serratidigitus* Coutière, 1896: 382.
*Salmoneus serratidigitus*–. Banner & Banner, 1981: 58 (in part.), figs. 7a–g, 8; Wang & Sha, 2016c: 1592, figs. 5–6.
(?) *Jousseaumea sibogae*–. De Man, 1910: 303.
(?) *Salmoneus sibogae*–. Holthuis, 1958: 20, fig. 8.

**Material Examined** MBM 107731, 1ovigerous♀, CL 4.1 mm, TL 11.5 mm), Yongxing Island, Xisha Islands, Fang Xiu-bin Coll., 11–13 June 1980; MBM 107732, 1ovigerous♀, CL 4.5 mm, TL 11.8 mm, Dong Island, Xisha Islands, Fang Xiu-bin Coll., 28–31 May 1980; MBM 107733, 1ovigerous♀, CL 4.9 mm, TL 14.9 mm, Yongxing Island, Xisha Islands, Fang Xiu-bin Coll., 11–13 June 1980.

**Description** Carapace slightly smooth. Rostrum triangular, at base slightly wider than long, reaching end of second segment of antennular peduncle, without carina. Triangular orbital hoods small, acute, directed straight forward. Pterygostomial angle rounded. Cardiac notch well developed. Eyes well developed, cornea partly visible dorsally and laterally.

Antennular peduncle moderately stout; first segment longest; second segment slightly wider than long; third segment nearly as long as second segment; stylocerite reaching to about 1/3 length of third segment of antennular peduncle, acute distally; lateral flagellum biramous, shorter ramus composed of at least three segments bearing tufts of aesthetascs; proximal fused portion composed of at least four segments. Basicerite bearing strong distodorsal tooth, slightly shorter than distolateral tooth; scaphocerite oval, anterior margin of blade rounded; distolateral tooth strong, acute, distinctly shorter than anterior margin of blade, reaching about end of third segment of antennular peduncle; carpocerite moderately stout, reaching about end of second segment of antennular peduncle.

Major cheliped lost in all present specimens. Minor cheliped slender; chela simple, not particularly enlarged, fingers sub-equal to palm, cutting edge unarmed; carpus subcylindrical, nearly as long as merus; merus and ischium unarmed, merus slightly longer than ischium.

Second pereiopod slender; carpus 5-segmented, length ratio subequal to 2.6: 0.6: 0.4: 0.4: 0.8; chela simple, fingers slightly longer than palm.

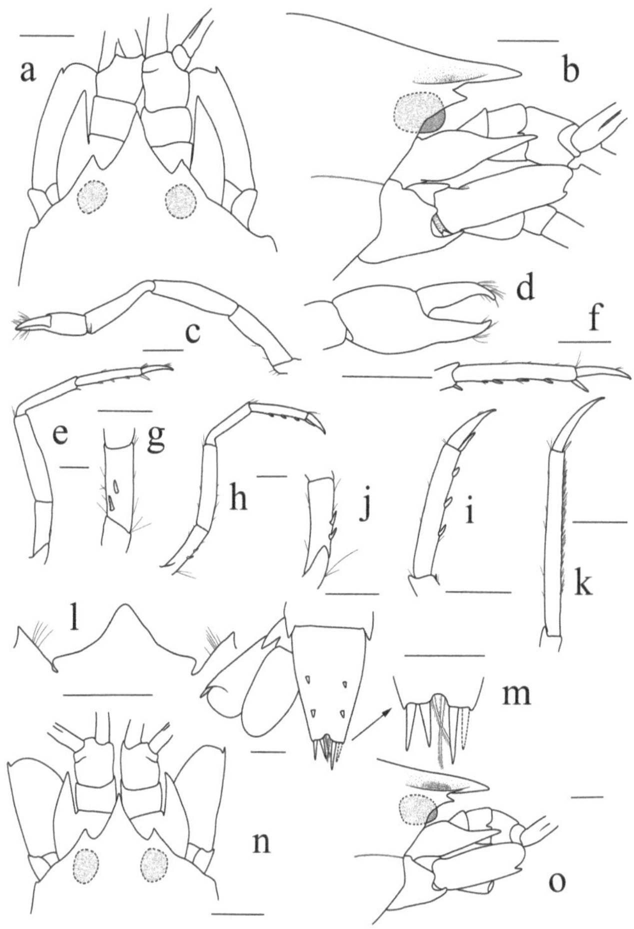


**Fig. 2.129** *Salmoneus* cf. *serratidigitus* (Coutière, 1896). MBM 107733, ovigerous female, CL 4.9 mm, **a**, anterior part of carapace; dorsal view; **b**, same, lateral view; **c**, left minor cheliped; **d**, same, chela; **e**, left third pereiopod; **f**, same, propodus and dactylus; **g**, same, ischium; **h**, left fourth pereiopod; **i**, same, propodus and dactylus; **j**, same, ischium; **k**, left fifth pereiopod, propodus and dactylus; **l**, preanal plate of sixth abdominal somite; **m**, telson. MBM 107731, ovigerous female, CL 4.1 mm, **n**, anterior part of carapace, dorsal view; **o**, same, lateral view (scale bar = 0.5 mm)

Third pereiopod slender, with dactylus simple, slightly and gradually curved, about 0.46 times as long as propodus. Propodus shorter than merus, armed with 3–4 spines along ventral margin, and distally one pair of larger spines. Carpus slightly shorter than merus, with 1 distal spine on ventral margin. Merus about twice as long as ischium, unarmed, about 5 times as long as wide at base. Ischium armed with 2 moderately large spines on ventral margin. Fourth pereiopod similar to third; propodus armed with 2–3 spines and one pair of moderately large spines distally on ventral margin. Fifth pereiopod more slender; propodus with rows of short setae, and 8 small spines along ventral margin, distally armed with one pair of moderately large spines; dactylus more slender than that of third and fourth pereiopods.

Sixth abdominal somite without articulate pleura; preanal plate rounded posteriorly. Endopod of second pleopod with appendix masculina distinctly longer than appendix interna. Telson about 1.5 times as long as wide at base, dorsal surface with two pairs of moderately small spines situated at some distance from lateral margin, anterior and posterior pair situated at about 1/2 and 4/5 length of telson, respectively; poste-

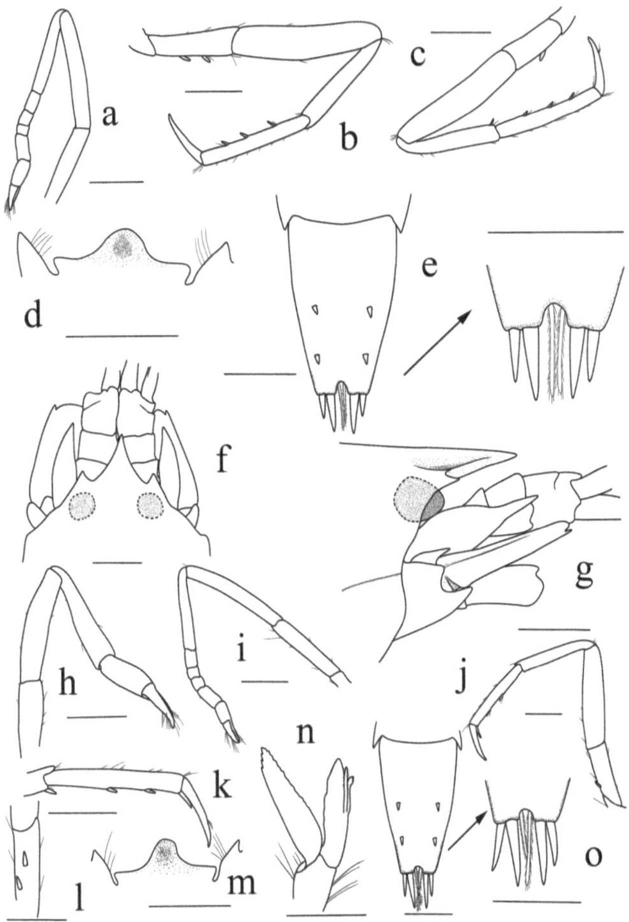


**Fig. 2.130** *Salmoneus* cf. *serratidigitus* (Coutière, 1896). MBM 107731, ovigerous female, CL 4.1 mm, **a**, left second pereiopod; **b**, right third pereiopod; **c**, right fourth pereiopod; **d**, preanal plate of sixth abdominal segment; **e**, telson. MBM 107732, ovigerous female, CL 4.5 mm, **f**, anterior part of carapace, dorsal view; **g**, same, lateral view; **h**, right minor cheliped; **i**, left second pereiopod; **j**, left third pereiopod; **k**, same, propodus and dactylus; **l**, same, ischium; **m**, preanal plate of sixth abdominal segment; **n**, second pleopod; **o**, telson (scale bar = 0.5 mm)

rior margin with moderately deep and large median notch bearing two long plumose setae, and two pairs of slender posterolateral spines, lateral at about 3/4 length of mesial. Uropods slightly shorter than telson; diaeresis with one strong acute spine; distolateral spine shorter.

**Remarks** *S. serratidigitus* is originally described from Djibouti. Liu (2008) list this species from the South China Sea. The species is morphologically very similar to *S. sibogae* De Man, 1910, and *S. sibogae* has even been synonymized with S. serratidigitus by Banner & Banner (1981). However, these two species are reported to differ in their colorful in life by Anker (2003b) and Anker et al. (2015). The differences between this species and *S. sibogae* are: in *S. sibogae* the rostrum overreaches the end of the second segment of the antennular peduncle; the third segment of the antennular peduncle is much longer than the second segment; the notch on the posterior margin of the telson is deeper (Banner & Banner, 1973). The present specimens are identified as *S. serratidigitus* for the rostrum being nearly as long as wide at the base; eyes completely invisible in dorsal view and partly visible in lateral view; the ischium of the third pereiopod armed with two spines on the ventral margin

(Anker, 2003). For *S. serratidigitus* is a variable species, the present material is taken as *S.* cf. *serratidigitus* herein. As Anker et al. (2015) state: "*S. serratidigitus* needs a thorough taxonomic revision", and a molecular phylogeny research may be helpful to resolve this problem.

**Distribution** Xisha Islands, Hong Kong, South China Sea; Indo-West Pacific: from the Red Sea and Madagascar to Japan, Indonesia, Australia and French Polynesia.

### 2.13.9 *Salmoneus* cf. *sibogae* (De Man, 1910) (Figs. 2.131, 2.132, and 2.133)

*Jousseaumea sibogae* De Man, 1910: 303.
*Salmoneus sibogae*–. Banner & Banner, 1982: 305, fig. 94; Wang & Sha, 2016c: 1596, figs. 7–9.

**Material Examined** MBM 158179, 1♂, CL 5.1 mm, TL 16.8 mm, Yongxing (Wude) Island, Xisha Islands, Fan Zhen-gang, Xu Jie-shan Coll., 2 April 1958; MBM 107728, 2♀♀, CL 5.5 mm, TL 17.3 mm, Shanhu Island, Xisha

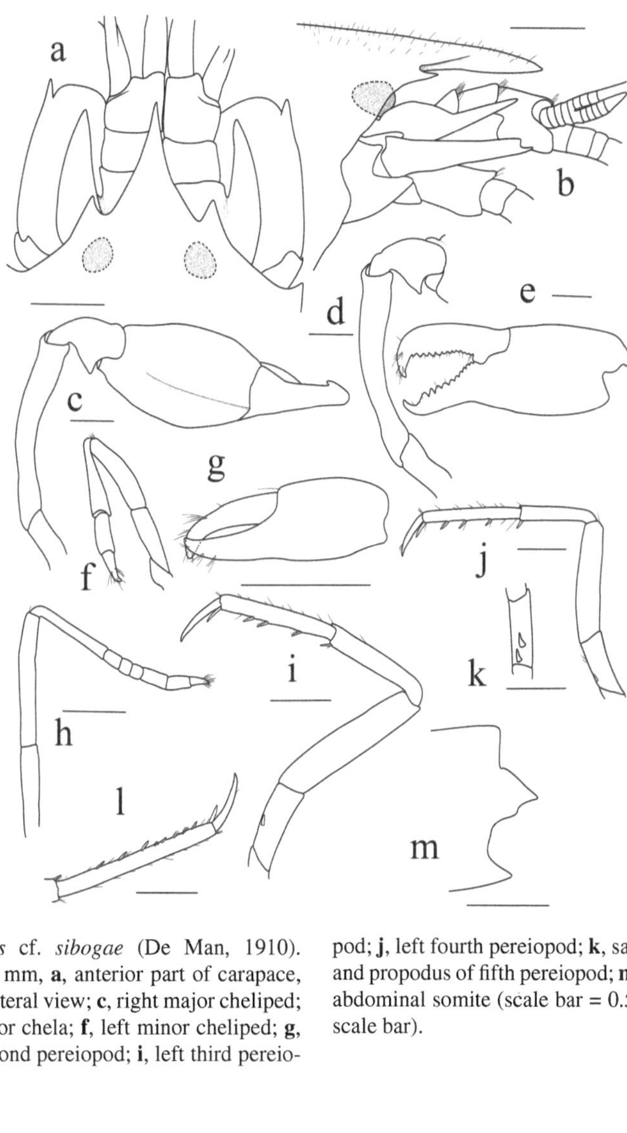


**Fig. 2.131** *Salmoneus* cf. *sibogae* (De Man, 1910). MBM 284966, CL 4.2 mm, **a**, anterior part of carapace, dorsal view; **b**, same, lateral view; **c**, right major cheliped; **d**, same, merus; e, major chela; **f**, left minor cheliped; **g**, minor chela; **h**, left second pereiopod; **i**, left third pereiopod; **j**, left fourth pereiopod; **k**, same, ischium; **l**, dactylus and propodus of fifth pereiopod; **m**, lateral margin of sixth abdominal somite (scale bar = 0.5 mm, **c** & **f** share same scale bar).

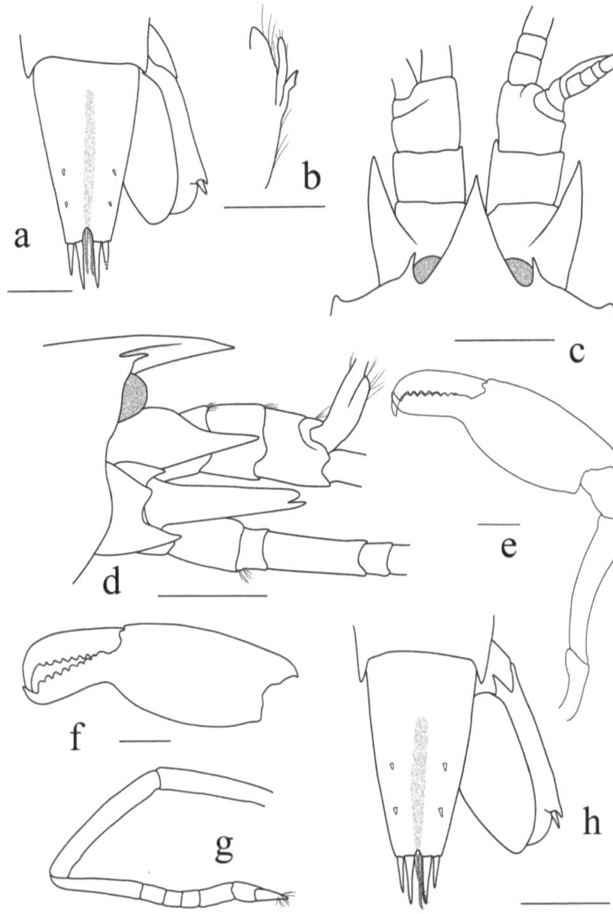


**Fig. 2.132** *Salmoneus* cf. *sibogae* (De Man, 1910), MBM 284966, male, CL 4.2 mm, **a**, telson and right uropod; **b**, endopod of second pleopod. MBM 284964, male, CL 3.1 mm, **c**, anterior part of carapace, dorsal view; **d**, same, lateral view; **e**, major cheliped; **f**, major chela; **g**, second pereiopod; **h**, telson and right uropod (scale bar = 0.5 mm)

Islands, Fang Xiu-bin Coll., 19–21 May 1980; MBM 284966, 1♂, CL 4.2 mm, TL 10.5 mm, Celiang Tan, Fan Zhen-gang, Xu Jie-shan Coll., 28 April 1958; MBM 284964, 1♂, CL 3.1 mm, TL 8.3 mm, Fan Zhen-gang Coll., 5 August 1958.

**Description** Carapace slightly setose. Rostrum triangular, longer than wider at base, reaching variously to near middle of second segment to up to about 3/4 length of third segment of antennular peduncle. Triangular orbital hoods small, acute, directed straight forward, without carina. Pterygostomial angle rounded. Cardiac notch well developed. Eyes well developed, cornea partly visible dorsally and laterally.

Antennular peduncle moderately stout; first segment longest; second segment slightly wider than long; third segment nearly as long as second segment; stylocerite reaching to about 3/4 length of third segment of antennular peduncle, acute distally; lateral flagellum biramous, shorter ramus composed of at least three segments bearing tufts of aesthetascs; proximal fused portion composed of at least five segments. Basicerite bearing strong distodorsal tooth, distinctly shorter than distolateral tooth; scaphocerite oval, anterior margin of blade rounded; distolateral

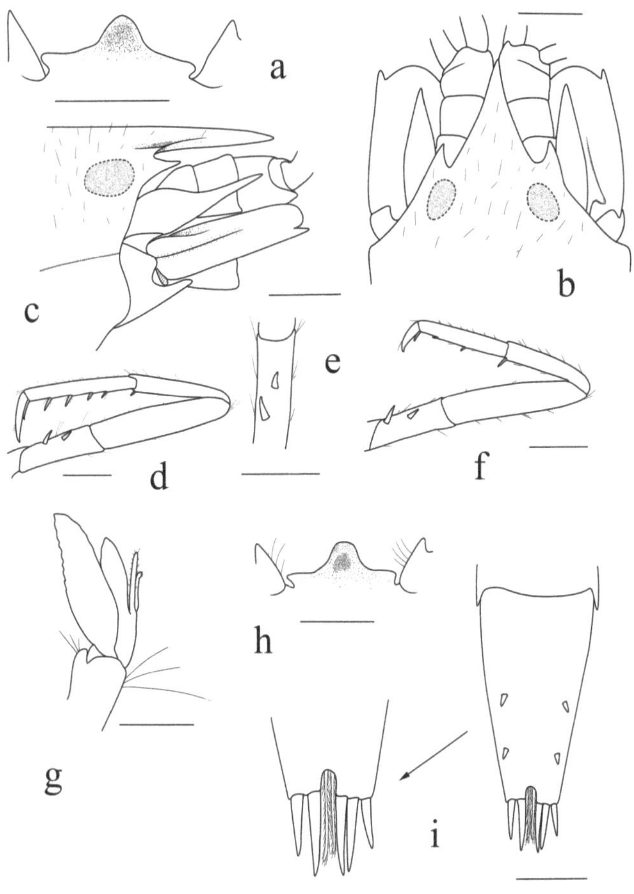


**Fig. 2.133** *Salmoneus* cf. *sibogae* (De Man, 1910), MBM 284966, male, CL 4.2 mm, **a**, preanal plate of sixth abdominal segment, ventral view. MBM 158179, male, CL 5.1 mm, **b**, anterior part of carapace, dorsal view; **c**, same, lateral view; **d**, left third pereiopod; **e**, same, ischium, ventral view; **f**, left fourth pereiopod; **g**, second pleopod; **h**, preanal plate of sixth abdominal somite, ventral view; **i**, telson (scale bar = 0.5 mm)

tooth strong, acute, slightly shorter than anterior margin of blade, reaching nearly end of third segment of antennular peduncle; carpocerite moderately stout, reaching to about end of second segment of antennular peduncle.

First chelipeds very asymmetrical. Major chela subcylindrical, without distinct grooves on surface; cutting edge of dactylus and pollex bearing teeth along full length of cutting edge, finger tips corneous, crossing when closed; palm slightly longer than fingers; carpus cup-shaped, elongate; merus slender, elongate, about 5.6 times as long as wide at base, ventrally flattened; ischium slightly elongate, without spines, ventrally flattened. Minor cheliped slender; chela simple, not particularly enlarged, fingers slightly longer than palm, cutting edge unarmed; carpus subcylindrical, slightly shorter than merus; merus and ischium unarmed, merus longer than ischium.

Second pereiopod slender; carpus 5-segmented, length ratio subequal to 2.1: 0.6: 0.4: 0.4: 0.9; chela simple, fingers sub-equal to palm.

Third pereiopod slender, with ischium armed with 1–2 moderately large movable spines; merus about twice as long as ischium, about 4.8 times as long as wide at base, unarmed; carpus shorter than merus, with 1 distal spine on ventral margin; propodus slightly shorter than merus, armed with 3 spines along ventral margin, and distally one pair of spines; dactylus simple, slightly and gradually curved, about 0.47 times as long as propodus. Fourth pereiopod similar to third, ischium armed with 2 spines on ventral margin. Fifth pereiopod more slender; propodus with rows of short setae and at least 5 small spines along ventral margin, and distally one pair of larger spines; dactylus more slender than that of third and fourth pereiopods.

Endopod of second pleopod with appendix masculina distinctly longer than appendix interna.

Sixth abdominal somite without articulate pleura; preanal plate truncate posteriorly. Telson about 1.8 times as long as wide at base, dorsal surface with two pairs of moderately small spines situated at some distance from lateral margin, anterior and posterior pair situated at about 1/2 and 4/5 length of telson, respectively; posterior margin with moderately deep and large median notch bearing two long plumose setae, and two pairs of slender posterolateral spines, lateral slightly shorter than mesial. Uropods slightly shorter than telson; diaeresis with one strong acute spine; distolateral spine shorter.

**Remarks** *S. sibogae* is firstly described by De Man from Indonesia. Banner & Banner (1978) report this species from Hong Kong. Banner & Banner (1982) re-descript the species based on one ovigerous female and one male specimen collected from Heron Island. The present specimens are generally similar to the re-description made by Banner & Banner (1982), except for the first segment of the carpus of the second pereiopod always longer than the sum of the posterior four segments, and the posterior margin of the telson bearing a moderately shallow notch rather than a deep V-shaped notch. Therefore, the present material should be taken as S. cf. sibogae. As for *S. serratidigitus*, a thorough taxonomic revision of *S. sibogae* is required.

**Distribution** Xisha Islands, Hong Kong, South China Sea; Red Sea; Indonesia; Micronesia; Kiribati; Fiji; French Polynesia; Australia; Philippines.

### 2.13.10 *Salmoneus tricristatus* Banner, 1959 (Figs. 2.134 and 2.135)

*Salmoneus tricristata* Banner, 1959: 131, fig. 1.
*Salmoneus tricristatus*–. Banner AH & Banner DM, 1968b: 270; Wang & Sha, 2016c: 1601, figs. 10–11.

**Material Examined** MBM 284951, 1♂, CL 5.1 mm, TL 14.3 mm, Shi Island, Xisha Islands, Fan Zhen-gang & Xu Jie-shan Coll., 6 April 1958; MBM 107730, 1♂, CL 3.3 mm, TL 8.7 mm, Yongxing Island, Xisha Islands, Fang Xiu-bin Coll., 11–13 June 1980.

**Description** Carapace slightly setose. Rostrum triangular, longer than wide at base, reaching to nearly middle of third segment of antennular peduncle. Triangular orbital hoods moderately large, acute, directed straight forward, about 1/3 length of rostrum, slightly overreaching end of first segment of antennular peduncle. Rostrum carina slightly uneven near middle, but without an obtuse protuberance, extending backward to about 11/14 length of carapace; another two carinas from base of orbital hood extending backward, parallel to rostrum carina, then joining it at about near 9/14 length of carapace, areas between rostrum carina and other two carinas distinctly depressed. Pterygostomial angle rounded. Cardiac notch well developed. Eyes well developed, cornea invisible dorsally and laterally.

Antennular peduncle moderately stout; first segment longest; second segment slightly wider than long; third segment nearly as long as second segment. Stylocerite slightly overreaching end of second segment of antennular peduncle, acute distally; lateral flagellum biramous, shorter ramus composed of at least three segments bearing tufts of aesthetascs; proximal fused portion composed of at least five segments. Basicerite bearing strong, rounded distodorsal tooth, distinctly shorter than distolateral tooth; scaphocerite oval, anterior margin of blade rounded;

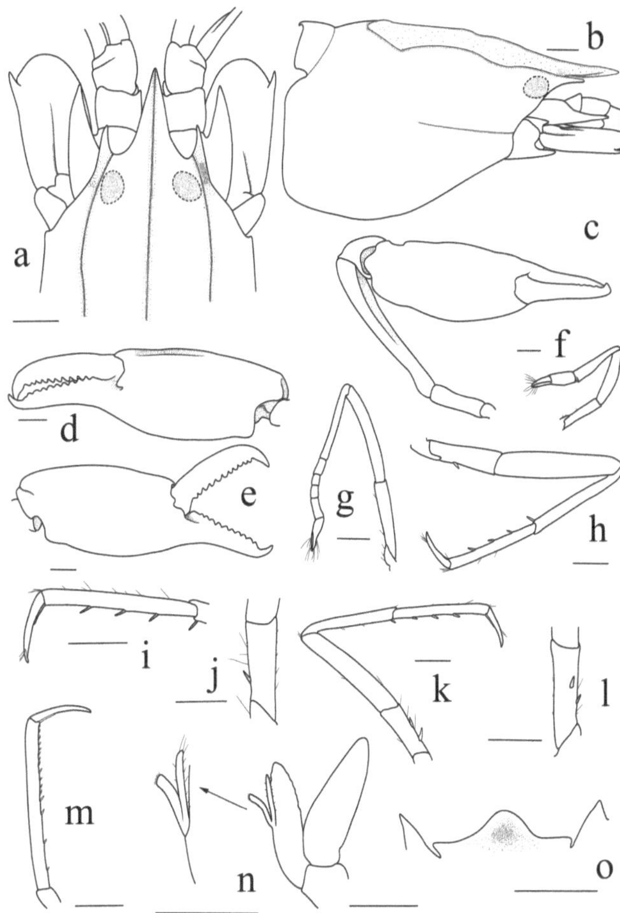


**Fig. 2.134** *Salmoneus tricristatus* Banner, 1959. MBM 284965, male, CL 5.1 mm, **a**, anterior part of carapace, dorsal view; **b**, carapace, lateral view; **c**, major cheliped; **d**, major chela, mesial view; **e**, same, lateral view; **f**, minor cheliped; **g**, right second pereiopod; **h**, right third pereiopod; **i**, same, propodus and dactylus; **j**, same, ischium; **k**, right fourth pereiopod; **l**, same, ischium, ventral view; **m**, fifth pereiopod, propodus and dactylus; **n**, second pleopod; **o**, preanal plate of sixth abdominal segment, ventral view (scale bar = 0.5 mm, **c** & **f** share same scale bar)

distolateral tooth strong, acute, distinctly shorter than anterior margin of blade, reaching near end of third segment of antennular peduncle; carpocerite moderately stout, reaching about end of second segment of antennular peduncle.

Mandible with incisor process bearing distally 7 rounded irregular teeth, with third and fourth larger than others, molar process typical for genus. Maxillule with two endites, distal endite with numerous large, spine-like setae on inner margin, basal endite with long setae on distal margins; palp bilobed, lower lobe with one robust plumose seta, upper lobe rounded. Maxilla and first and second maxilliped typical for genus (fig. 11e, f). Third maxilliped moderately slender; ultimate segment very setose, distally tapering, bearing 2 subapical spines; arthrobranch normally developed.

First chelipeds very asymmetrical. Major chela subcylindrical, slightly depressed on upper surface, but without any grooves or depressions on inner and outer face; cutting edge of dactylus and pollex bearing corneous teeth along full length of cutting edge, finger tips corneous, crossing when closed; palm longer than fingers; carpus cup-shaped, elongate; merus slender, elongate, about 8.7 times as long as wide at base, ventrally flattened; ischium slightly elongate,

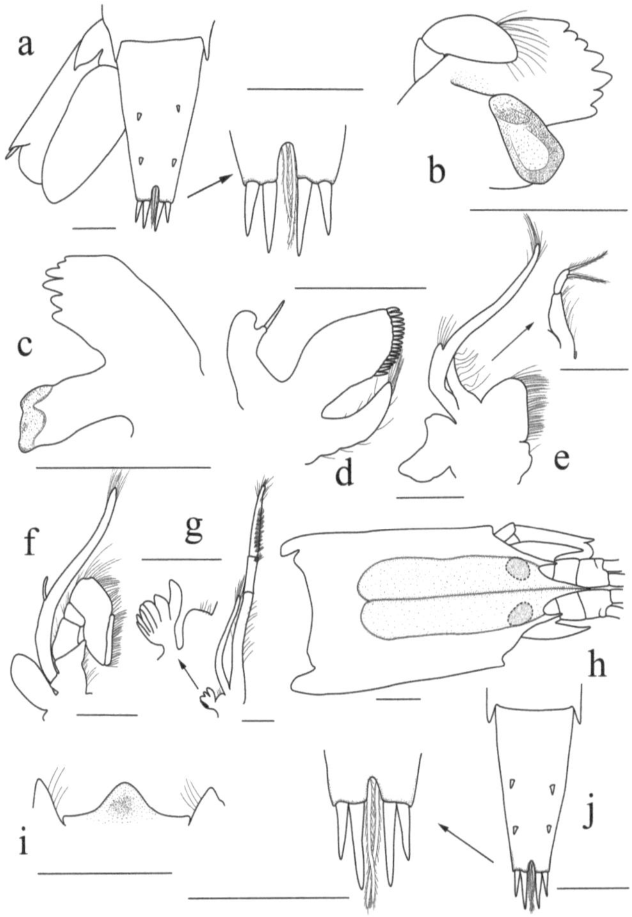


**Fig. 2.135** *Salmoneus tricristatus* Banner, 1959. MBM 284965, male, CL 5.1 mm, **a**, telson; **b** & **c**, left mandible; **d**, left first maxilla; **e**, left first maxilliped; **f**, left second maxilliped; **g**, left third maxilliped. MBM 107730, male, CL 3.3 mm, **h**, carapace, dorsal view; **i**, preanal plate of sixth abdominal segment, ventral view; **j**, telson (scale bar = 0.5 mm)

ventrally without spines. Minor cheliped slender; chela simple, not particularly enlarged, fingers nearly as long as palm, cutting edge unarmed; carpus subcylindrical, slightly longer than merus; merus and ischium unarmed, merus longer than ischium.

Second pereiopod slender; carpus 5-segmented, length ratio subequal to 2.4: 0.5: 0.3: 0.4: 0.8; chela simple, fingers sub-equal to palm.

Third pereiopod slender, with dactylus simple, slightly ad gradually curved, about 0.4 times as long as propodus. Propodus slightly shorter than merus, armed with 3 spines along ventral margin, and distally one pair of spines. Carpus slightly shorter than merus, with 1 distal spine on ventral margin. Merus about twice as long as ischium, about 5.7 times as long as wide at base, unarmed. Ischium armed with 1 moderately large movable spine. Fourth pereiopod similar to third, ischium armed with 2 spines on ventral margin. Fifth pereiopod more slender, propodus with rows of short setae and at least 5 small spines along ventral margin; dactylus more slender than that of third and fourth pereiopods.

Sixth abdominal somite without articulate pleura; preanal plate rounded posteriorly. Endopod of second pleopod with appendix masculina distinctly longer than appendix interna. Telson about 1.7 times as long as wide at base, dorsal surface with two pairs of moderately small spines situated at some distance from lateral margin, anterior and posterior pair situated at about 1/2 and 3/4 length of telson, respectively; posterior margin with moderately narrow and deep median notch bearing two long plumose setae, and two pairs of slender posterolateral spines, lateral slightly shorter than mesial. Uropods slightly shorter than telson; diaeresis with one strong acute spine; distolateral spine shorter.

**Remarks** *S. tricristatus* is one newly recorded species from the China seas, and it is first described by Banner (1959) from the Caroline Archipelago. This species is re-described by Miya (1972) from Japan and by Banner & Banner (1973) from Australia in detail, respectively. Anker et al. (2015) reported it from Indonesia, and presented photos of living specimens. *S. tricristatus* is easily recognized by the dorsal margin of the carapace bearing three strong, longitudinal or oblique, carinas (Anker et al., 2015). The differences between the present material and previous descriptions (Banner, 1959; Miya, 1972; Banner & Banner, 1973) are listed below: the rostrum of the present specimens reaching the middle of the third segment of the antennular peduncle, versus only slightly overreaching the end of the second segment in previous descriptions; the rostral carina extending from the tip to the posterior 3/4 length of the carapace, and joining together with the orbital hood carinas, versus varying in extension to the posterior 2/3 or 1/2 of the length of the carapace, and not joining together with the orbital hood carinas (see Banner, 1959, fig. 1a; Miya, 1972, figs. a, b; Banner & Banner, 1973, fig. 13a); the middle part of the rostral carina only slightly uneven in the present material similar to that of Anker et al. (2015, fig. 22b), not bearing the rounded projection described by Miya (1972, fig. a) and Banner & Banner (1973, fig. 13b); the orbital hood carina arising from about the middle of the notch between the rostrum and the orbital hoods in the present specimens, versus arising from the tip of the orbital hoods (see Banner, 1959, fig. 1a; Banner & Banner, 1973, fig. 13a) or from the base of the orbital hoods (see Miya, 1972, fig. b); the major chela only slightly flattened on the dorsal surface, and bearing the inconspicuous, longitudinal depression in the present material as in Miya (1972), versus bearing ridges and depressions on the surfaces in Banner & Banner (1973, fig. 13c, d); the ischium of the third and fourth pereiopods bearing 1 and 2 moveable small spines, respectively, versus not bearing any spines in Banner (1959, fig. 1d).

**Distribution** Xisha Islands, South China Sea; Red Sea; France; Lucipara Islands (Banda Sea); Madagascar; Indonesia; Philippines; Australia; Japan; Caroline Islands; Marshall Islands; Society Islands.

## 2.14 Genus *Stenalpheops* Miya, 1997

**Diagnose** Body compressed, elongate. Rostrum short, tip acute. Orbital hoods absent. Pterygostomial angle rounded. Cardiac notch present. Cornea of eyes largely exposed dorsally and laterally, with dull-pointed and naked anteromedial protuberance of eyestalk. Antennular peduncle slender, with stylocerite strongly developed. Antennal scale with narrow blade, shorter than strong lateral tooth. Mandible with 2-segmented palp. First pereiopod slender, symmetrical, Carpus of second pereiopod 5-segmented. Dactylus of last three pereiopods simple. Abdomen of usual form, posterolateral angle of sixth abdominal segment with articulated plate (Miya, 1997).

Key to species of the genus *Stenalpheops* from China seas

1. Major chela sub-chelar shape in adult male; Uropod with exopod bearing caudal appendix on posterior margin ........................................ *S. anacanthus*

-. Major chela normal in adult male; Uropod with exopod not bearing caudal appendix on posterior margin ........................................ *S. koreanus*

## 2.14.1 *Stenalpheops anacanthus* Miya, 1997 (Figs. 2.136 and 2.137)

*Stenalpheops anacanthus* Miya, 1997: 145–161, (in part) figs. 2D, 3A, E–G; Wang & Sha, 2017b: 1617, figs. 1–10.

*Cavipelta yamashitai* Hayashi, 1998b: 229–238, (in part) figs. 1–3, 4C, D, F, H, K, L.

*Chelomalpheus crangonus* Anker, Jeng & Chan, 2001: 1053–1060, figs. 4–8.

**Material Examined** MBM 107708, 1♀, CL 5.4 mm, TL 13.5 mm, Cangkou, Qingdao, Zhang Feng-Xuan Coll., 15 October 1955; MBM 107120, 1♀, CL 5.3 mm, TL 14.5 mm, 1♂, CL 4.9 mm, TL13.5 mm, 2♀♀1♂, CL 5.1–5.6 mm, TL 15.2–16.3 mm, Cangkou, Qingdao, Fan Zhen-gang Coll., 19 October 1956; MBM 283908, 1♀, CL 5.2 mm, TL 14.9 mm, Cangkou, Qingdao, Liu Rui-yu Coll., 4 October 1952; MBM 107716, 1 ♂, CL 5.3 mm, TL 15.1 mm, 1♀, CL 4.2 mm, TL 11.1 mm, Cangkou, Qingdao, Zhang Feng-Xuan Coll., 7 January 1958; MBM 283914, 1♀, CL 7.9 mm, TL 20.9 mm, Zhanggezhuang, Qingdao, 10 September 2003. MBM 270049, 3♀♀, CL 6.1–7.2 mm, 1♂, CL 8.7 mm, TL 22.9 mm; MBM 107708, 2♀♀, CL 8.1 mm, TL 22.3 mm, 5 ♂♂,

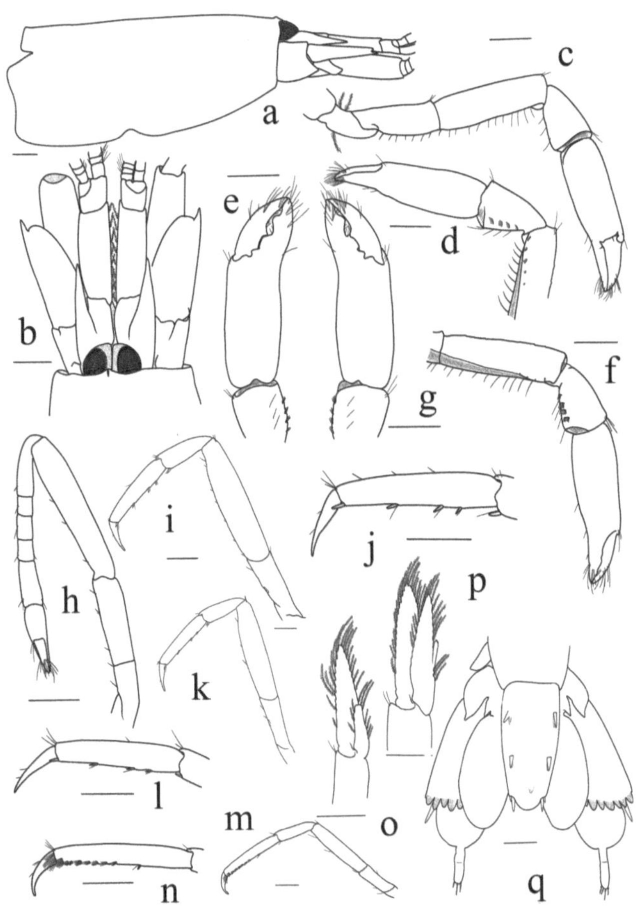


**Fig. 2.136** *Stenalpheops anacanthus* Miya, 1997. MBM 107720, female, CL 5.3 mm, **a**, anterior part of carapace, lateral view; **b**, same, dorsal view; **c**, right first pereiopod, lateral view; **d**, same, mesial view; **e**, same, chela; **f**, left first pereiopod, mesial view; **g**, same, chela; **h**, left second pereiopod; **i**, left third pereiopod; **j**, same, dactylus and propodus; **k**, left fourth pereiopod; **l**, same, dactylus and propodus; **m**, left fifth pereiopod; **n**, same, dactylus and propodus; **o**, right first pleopod; **p**, right second pleopod; **q**, uropods and telson (scale bar = 0.5 mm)

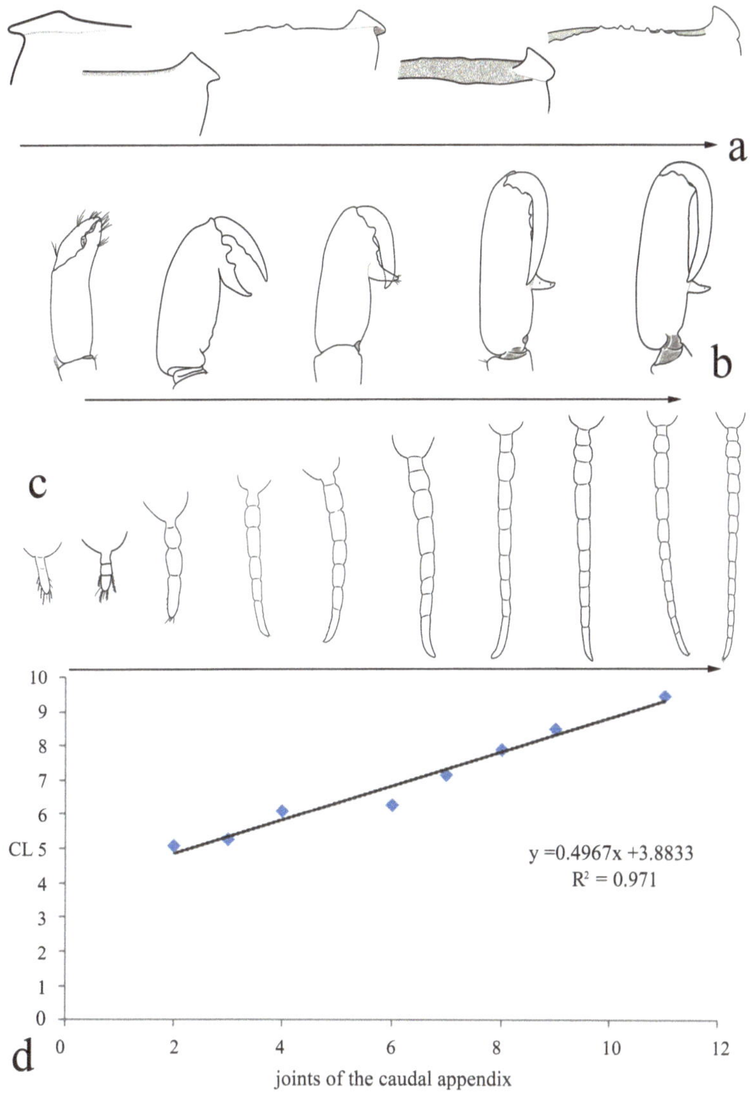


**Fig. 2.137** *Stenalpheops anacanthus* Miya, 1997: **a**, rostrum and carina, change correlated with growth, represented by the length of the carapace; **b**, shape of the chela changes with the length of the carapace; **c**, the number of segments of the caudal appendix changes with the length of the carapace; **d**, relationship between length of carapace and segment number of the caudal appendix

CL 5.2–7.1 mm, TL 14.9–18.9 mm, Cangkou, Qingdao, Zhang Feng-Xuan Coll., 15 October 1955; MBM 107720, 1♂, CL 7.5 mm, TL 21.8 mm, Cangkou, Qingdao, Fan Zheng-gang Coll., 19 October 1956; MBM 107702, 1♀, CL 8.4 mm, TL 21.1 mm, Cangkou, Qingdao, Zhang Feng-Xuan Coll., 23 January 1958; MBM 107707, 1♂, CL 8.1 mm, TL 21.5 mm, Nvgukou, Qingdao, 8 April 1951; MBM 107705, 1♀, CL 8.5 mm, TL 21.3 mm, Nvgukou, Qingdao, 30 November 1989; MBM 107712, 1♀, Huiquan Bay, Qingdao, Liu Rui-Yu Coll., 10 December 1950; MBM 107716, 1♀, CL 7.5 mm, TL 20.5 mm, 2 ♂♂, CL 7.1–10.0 mm, TL 18.5–24.5 mm, Cangkou, Qingdao, Zhang Feng-Xuan Coll., 7 January 1958; MBM 270049, 1 ♂, CL 9.5 mm, TL 25.3 mm; MBM 107708, 1♀, CL 5.6 mm, TL 17.2 mm, Cangkou, Qingdao, Zhang Feng-Xuan Coll., 15 October 1955; MBM 107720, 6♂♂, CL 6.0–6.1 mm, TL

15.5–15.9 mm, Cangkou, Qingdao, 1 Fan Zhengang Coll., 9 October 1956; MBM 107702, 1♂, CL 5.1 mm, TL 14.1 mm, Cangkou, Qingdao, Zhang Feng-Xuan Coll., 23 January 1958; MBM 107706, 1♂, CL 6.9 mm, TL 19.5 mm, Cangkou, Qingdao, Zhang Feng-Xuan Coll., 2 May 1957; MBM 283914, 1♀, CL 5.5 mm, TL 17.3 mm, Zhanggezhuang, Qingdao, 10 November 2003; MBM 107710, 1♂, CL 7.2 mm, TL 19.5 mm, Cangkou, Qingdao, 11 April 1956.

**Description** Body slightly compressed, elongate. Rostrum small, acute, carinate for short distance posteriorly, triangular, varying from small to larger, with tip curved backward; rostral process and orbital hoods lacking. Pterygostomial margin rounded and naked; cardiac notch well developed. Carapace with anterior margin shallowly concave near base of rostrum and then almost straight; two ridges varying from slightly developed for short distance posteriorly to stronger, and equipped with small teeth, about 1/3 length of carapace; anteroventral margin bluntly rounded, not produced anteriorly; inferior margin of carapace with a broad notch at middle part near basis of second pereiopod. Eyes well developed, cornea largely exposed dorsally and laterally, anterior margin bearing small, rounded median projection; cornea well pigmented.

Antennular peduncle elongate, first segment longer than wide, 2/3 length of second segment; second segment longest, about 3.5 times as long as wide at base; third segment 1/3 length of second. Stylocerite slender, not reaching middle of second segment of antennular peduncle. Outer flagellum with about 6 articles proximal to bifurcation. Antennal basicerite with sharp ventral tooth. Carpocerite slightly overreaching end of antennular peduncle, about 3 times as long as wide. Lateral margin of scaphocerite almost straight, sharp distal spine reaching middle of third segment; lamella moderately narrow, about 2.5 times as long as wide at base, slightly overreaching end of second segment of antennular peduncle.

Mandible with incisor process rather stout, bearing 5 teeth on distal margin, not similar in size; molar process bearing some ridges provided with patches of soft setae; palp 2-segmented, distal margin oblong, fringed with plumose setae. First maxilla with palp bilobed, distal lobe with 2 plumose setae, proximal lobe with apical plumose seta; distal endite subrectangular, with dense fringe of serrate setae along medial margin. Second maxilla with two endites; distal endite not bilobed, subrectangular, densely fringed with simple and plumose setae along medial margin; palp slender, not bearing setae; scaphognathite broadly auriculate. First maxilliped proximal endite bluntly pointed at distomedial corner, with apical plumose setae; distal endite elliptical; palp two-articulated, slender, distal segment with two long plumose apical setae; lobe a little broad and with distinctly separated exopod distally. Second maxilliped endopod 5-segmented, ischium markedly produced at distomedial corner, merus 1/2 length of ischium, palm greatly elongate; epipod well developed. Third maxilliped slender and elongate, with densely plumose setae on all articles; antepenultimate segment longest; penultimate segment about 1/3 length of antepenultimate segment; tip of ultimate segment bearing 2–3 spine-like setae, twice as long as penultimate segment; exopod not reaching end of antepenultimate segment, distally fringed with long plumose setae; coxa with 1 arthrobranch.

First pereiopods similar in shape and size, sexual dimorphism. Males in larval stage with normal chela, similar to female; fixed finger gradually bent backwards and with the length of the carapace increasing, until partially fused with palm, only tip free; dactylus varying from normal to strongly curved and bent backwards, cutting edge of fixed finger and dactylus varying from being equipped with large teeth to nearly smooth. Carpus cup-shaped, about 1.3 times as long as wide at distal end, about 4/7 length of palm, with six short tufts of setae along inferior margin. Merus about 3.6 times as long as wide at base, with about 6 immovable small spinules along superior margin, about 6 short tufts of setae along inferior margin. Superior margin of ischium bearing about 4 immovable small spinules, both merus and ischium flattened on ventral margins and extremely setose. Female chela normal, fingers gaping when closed; dactylus slightly curved; cutting edge of fixed finger and dactylus armed with strong teeth.

Second pereiopod slender, similar in shape and size. Fingers almost as long as palm. Carpus 5-segmented, with ratios about 2: 0.7: 0.8: 0.7: 1.4.

Third pereiopod moderately stout, laterally flattened. Dactylus slender, conical, distally sharpened, about 3/8 length of propodus, strongly curved to propodus, with inconspicuous angle approximately at middle of inferior margin. Propodus about 5.2 times as long as wide at base, with 3 small spines along ventral margin and one pair of spines at distal end. Carpus about 5/8 length of propodus, with 1 small spine at distal end of inferior margin. Merus about 5.3 times as long as wide at base, about 2.5 times as long as carpus, unarmed. Ischium with one rather strong movable spine approximately at middle of ventral margin. Fourth pereiopod similar to third pereiopod. Propodus with 3 movable spines on ventral margin and a pair at distal end. Fifth pereiopod similar to third and fourth, but much more slender, dactylus about 3/9 length of propodus, strongly hooked, curved to propodus. Ventral margin of propodus with brush of setae on distal portion. Ischium without movable spine.

Abdomen normal; sixth somite with large subtriangular movable plate. Second pleopod of male with appendix masculina far overreaching appendix interna. Uropodal exopod and endopod bearing spinules along distal margins; exopod with small, acute distolateral tooth and strong subterminal movable spines; lamella of 9–11 large, rounded teeth, variable in size, along full length of diaeresis. A long caudal appendix extends from outer margin of exopod of uropod in both male and female, caudal appendix symmetrical or not, composed of 2–13 segments, distal margins of caudal appendix with long plumose setae or short setae.

Telson narrow, tapering; about twice as long as wide at anterior end, lateral margin slightly convex; posterior margin strongly convex; dorsal surface with 2 pairs of moderately large spines, inserted rather laterally, approximately at 1/2 and 1/6 length of telson; posterolateral angle each with two pairs of spines, outer spines about 1/5 length of inner.

**Remarks** The material presents a growth series of *S. anacanthus*. With the carapace length increasing, three identification characters show obvious changes: the carina becomes stronger, and the tip of it curves slightly backwards in both sexes; the fingers of the first chela curve strongly backwards; and the fixed finger gradually reduces to the distal portion of the chela, until the chela becomes a sub-chelar shape; both the length and the number of segments of the caudal appendix increase, and the number of these segments varies from 2 to 13 in both sexes. In addition, the carapace length correlates with the number of segments of the caudal appendix in a way that is positively linear. The caudal appendix was only found in large males by both Miya (1997) and Hayashi (1998b), in those authors' respective collections. As the caudal appendix was not found in the females and in small males, this character was taken as both a case of sexual dimorphism and an element of allometric growth by Hayashi (1998b). Then Anker et al. (2001) found a female bearing the caudal appendix as well. In the present specimens, however, the same lengths and the same numbers of segments in the caudal appendix can be found both in the females and the males of the present specimens. Thus, we presume that neither the length, nor the segment number of the caudal appendix is elements of sexual dimorphism. In addition, ovigerous females are found in the type specimens of *S. anacanthus* Miya, 1997, in those of *Cavipelta yamashitai* Hayashi, 1998, as well as in those of *C. koreanus* Kim, 1998 – and in the present material as well. Therefore, the presence or absence of the caudal appendix should be taken as a character to identify different species in the genus Stenalpheops rather than as a phenomenon of sexual dimorphism. *S. anacanthus*, as the type species of *Stenalpheops*, is still a valid species. However, this species should be re-defined. All the large male type specimens of *S. anacanthus* Miya, 1997 and *Cavipelta yamashitai* Hayashi, 1998, which bear the caudal appendix, are now to be identified as *S. anacanthus*. Two characters, (1) the carina of the carapace; and (2) the length and segment number of the caudal appendix, vary with growth. The shape of the chela shows sexual dimorphism, and the shape of the first chela only varies with growth in the male (fig. 10b): the chela shape of the female is always normal; the first chela of the male finally becomes a sub-chela. As the characters of *S. crangonus* (Anker, Jeng & Chan, 2001) are identical with the middle stage in the growth series of *S. anacanthus* as herein established, *S. crangonus* is consequently synonymized with *S. anacanthus* in the present paper.

**Distribution** The Yellow Sea, Taiwan; Japan.

### 2.14.2 *Stenalpheops koreanus* (Kim, 1998) (Figs. 2.138, 2.139, 2.140, and 2.141)

*Chelomalpheus koreanus* Kim, 1998: 140–145, figs. 1–3.

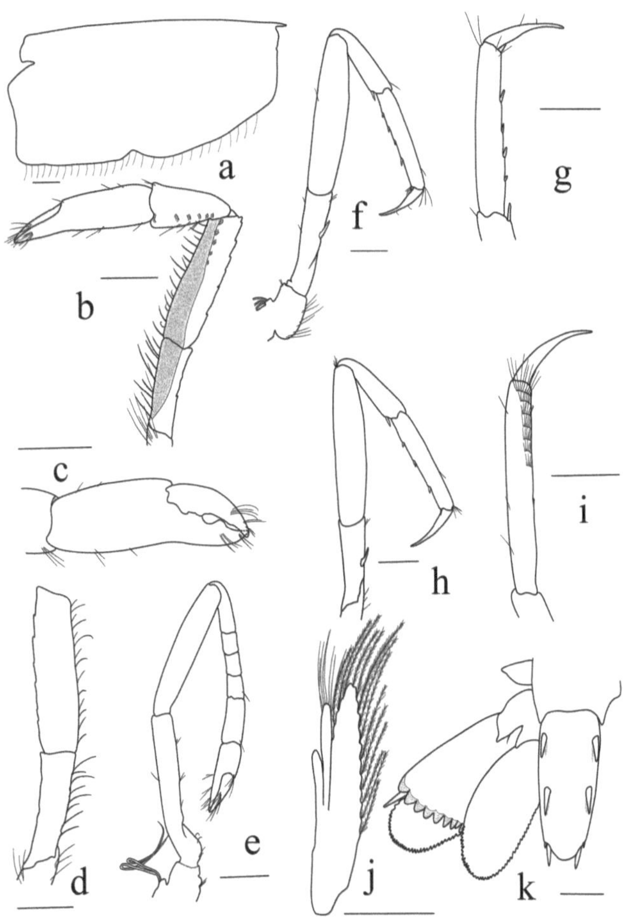


**Fig. 2.138** *Stenalpheops koreanus* (Kim, 1998). MBM 283912, male, CL 4.8 mm, **a**, carapace; **b**, left first pereiopod, mesial view; **c**, same, chela, lateral view; **d**, same, merus and ischium, lateral view; **e**, left second pereiopod; **f**, left third pereiopod; **g**, same, propodus and dactylus; **h**, left fourth pereiopod; **i**, left fifth pereiopod, propodus and dactylus; **j**, appendix interna and appendix masculina; **k**, telson and left uropod (scale bars = 0.5 mm)

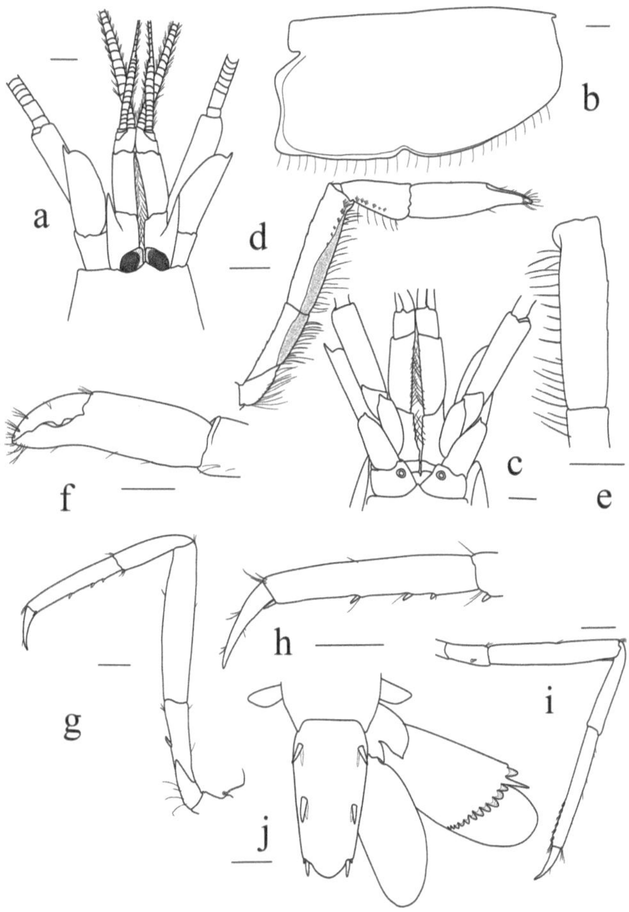


**Fig. 2.139** *Stenalpheops koreanus* (Kim, 1998). MBM 283913, ovigerous female, CL 6.1 mm, **a**, anterior part of carapace, dorsal view; **b**, carapace; **c**, anterior part of carapace, ventral view; **d**, right first pereiopod; **e**, same, merus; **f**, same, chela, mesial view; **g**, right third pereiopod; **h**, same, propodus and dactylus; **i**, right fifth pereiopod; **j**, right uropod and telson (scale bars = 0.5 mm)

*Stenalpheops anacanthus* Miya, 1997: 145–161, (in part.) figs. 2A–C, 3B–D, 4, 5 (Not *S. anacanthus* Miya, 1997).

*Cavipelta yamashitai* Hayashi, 1998b: 229–238, (in part.) fig. 4A, B, E, G.

*Stenalpheops koreanus* –. Anker, Jeng, Chan, 2001: 1061 ; Wang & Sha, 2017b: 1629, figs. 11–14.

**Material Examined** MBM 107708, 1♂5♀♀, CL 3.8–6.1 mm, TL 10.5–16.1 mm, Cangkou, Qingdao, Zhang Feng-Xuan Coll., 15 October 1955; MBM 107720, 1♂8♀♀, CL 4.2–5.3 mm, TL 11.2–17.1 mm, Cangkou, Qingdao, Fan Zhen-gang Coll., 19 October 1956; MBM 283912, 2♂♂4♀♀, CL 3.2–5.5 mm, TL 9.7–16.2 mm, Cangkou, Qingdao, 28 October 1954; MBM 283913, 1ovigerous♀, CL 6.1 mm, TL 16.5 mm, Cangkou, Qingdao, 30 August 1954; MBM 283902, 1♀, CL 3.6 mm, TL 10.1 mm; MBM 283903, 1♂, CL 5.3 mm, TL 15.5 mm, Cangkou, Qingdao, 30 July 1954; MBM 283904, 1♀, CL 3.1 mm, TL 8.2 mm, Cangkou, Qingdao, Liu Rui-Yu Coll., 13 September 1957; MBM 283905, 1♂1♀, CL 3.5–5.0 mm, TL 10.5–14.9 mm, Cangkou, Qingdao, 14 July 1954; MBM 283906, 3♀♀, CL 4.9–5.5 mm, TL 13.5–16.3 mm, Cangkou, Qingdao, Fan Zhen-gang Coll., 25 November 1954; MBM 283907, 2♀♀,

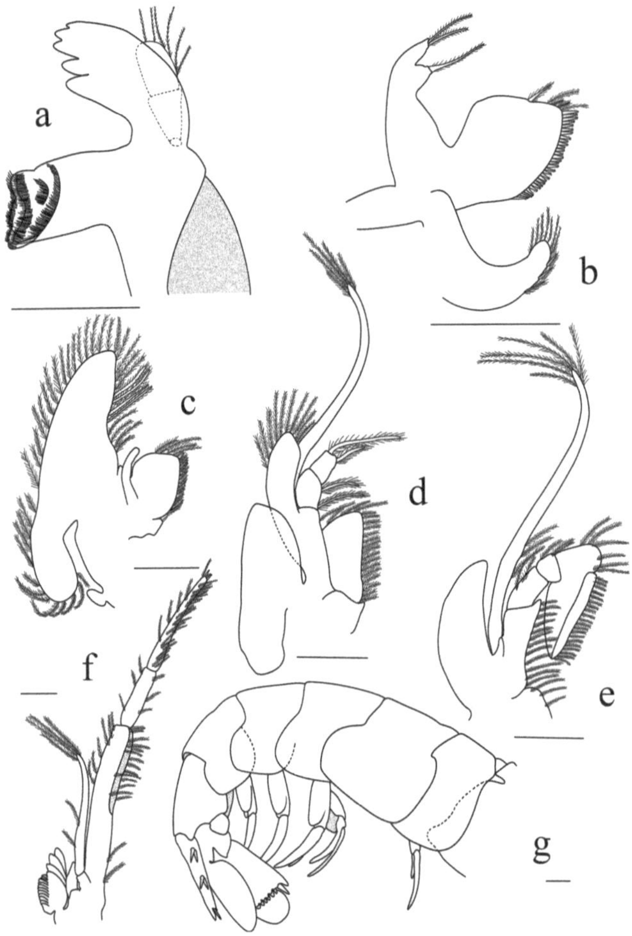


**Fig. 2.140** *Stenalpheops koreanus* (Kim, 1998). MBM 283903, male, CL 5.3 mm, **a**, left mandible; **b**, left maxillule; **c**, left maxilla; **d**, left first maxilliped; **e**, left second maxilliped; **f**, left third maxilliped; **g**, abdominal segments and tail fan in lateral view (scale bars = 0.5 mm)

CL 6.0–6.9 mm, TL 16.0–19.5 mm, Cangkou, Qingdao, Zhang Feng-Xuan Coll., 21 November 1953; MBM 283908, 1♂8♀♀, CL 4.7–6.2 mm, TL 14.1–18.5 mm, Cangkou, Qingdao, Liu Rui-Yu Coll., 4 October 1952; MBM 283909, 6 juveniles, CL 2.1–3.1 mm, TL 7.2–8.6 mm, Cangkou, Qingdao, 2 Liu Rui-Yu Coll., 2 August 1952; MBM 283910, 23juveniles, CL 3.7–4.3 mm, TL 10.1–13.2 mm, Cangkou, Qingdao, Liu Rui-Yu Coll., 4 October 1952; MBM 283911, 1♂, CL 4.2 mm, TL 10.2 mm, 2♀♀, CL 5.2–5.3 mm, TL 12.7–13.1 mm, Cangkou, Qingdao, Zhang Feng-Xuan Coll., 1956; MBM 107716, 1♂, CL 4.2 mm, TL 11.9 mm, Cangkou, Qingdao, Zhang Feng-Xuan Coll., 7 January 1958; MBM 283914, 1♀, CL 4.1 mm, TL 9.2 mm, Zhanggezhuang, Qingdao, 10 November 2003; MBM 283915, 2 ovigerous♀♀, CL 6.2 mm, TL 19.1 mm, Cangkou, Qingdao, 5 August 1954; MBM 283916, 2♂♂2♀♀, Cangkou, Qingdao, 11 April 1956.

**Description** Body slightly compressed, elongate. Rostrum small, acute, triangular in dorsal view, slightly carinate for short distance posteriorly; rostral process and orbital hoods lacking. Pterygostomial margin rounded and naked; car-

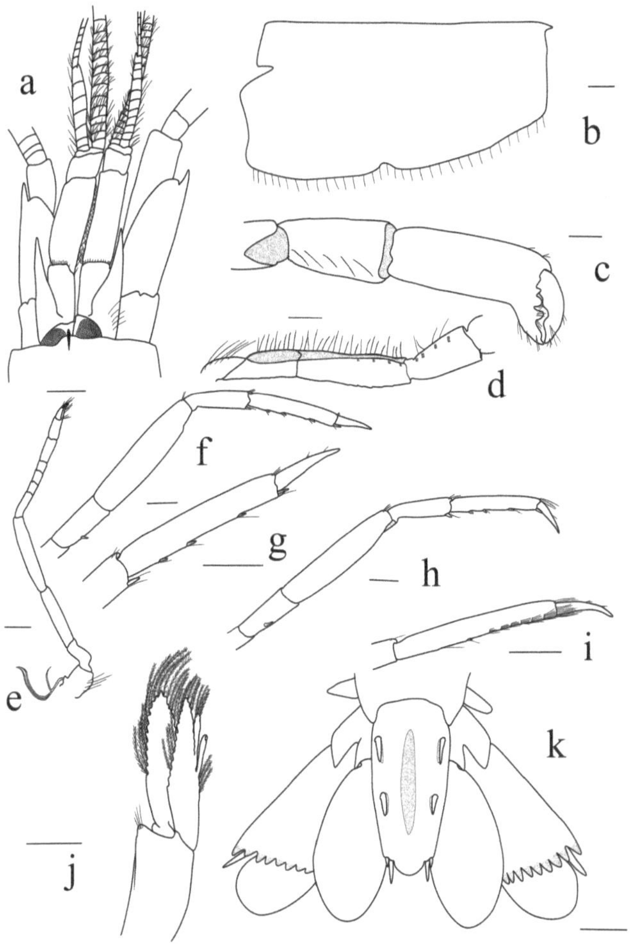


**Fig. 2.141** *Stenalpheops koreanus* (Kim, 1998). MBM 283903, male, CL 5.3 mm, **a**, anterior part of carapace, dorsal view; **b**, carapace; **c**, right first pereiopod, chela; **d**, same, carpus to ischium; **e**, right second pereiopod; **f**, right third pereiopod; **g**, same, propodus and dactylus; **h**, right fourth pereiopod; **i**, right fifth pereiopod, propodus and dactylus; **j**, right second pleopod; **k**, uropods and telson (scale bars = 0.5 mm)

diac notch well developed. Carapace smooth; antero-ventral margin bluntly rounded, not produced anteriorly; inferior margin of carapace with a broad notch at about middle part near basis of second pereiopod. Eyes well developed, cornea largely exposed dorsally and laterally; cornea well pigmented.

Antennular peduncle elongate, first segment longer than wide, 2/3 length of second segment; second segment longest, about 2.8 times as long as wide at base; third segment 1/3 length of second. Stylocerite slender, reaching 1/3 length of second segment of antennular peduncle. Outer flagellum with about 10 articles proximal to bifurcation. Antennal basicerite with sharp ventral tooth. Carpocerite overreaching end of third segment of antennular peduncle, about 3.3 times as long as wide. Lateral margin of scaphocerite slightly convex, sharp distal spine not overreaching end of antennular peduncle; lamella moderately narrow, about 3.0 times as long as wide at base, slightly overreaching end of second segment of antennular peduncle.

First pereiopods symmetrical in shape and size. Chela elongate, about 3.9 times as long as wide at base, bent below near base of fixed finger, both dactylus and fixed finger bent internally, more than 1/2 length of palm, palm subcylindrical, about 2.8 times as long as wide at base; dactylus shallowly arched along superior margin, bearing 2 blunt teeth on inferior margin; fixed finger with 1 blunt tooth at middle of superior margin, fitting space between 2 immovable teeth of dactylus, and another small tooth proximally. Carpus about twice as long as wide at distal end, slightly shorter than palm, with 6 short tufts of setae along inferior margin. Merus about 4 times as long as wide at base, with about 5 immovable small spinules along superior margin, about 6 short tufts of setae along inferior margin. Superior margin of ischium bearing about 4 immovable small spines.

Second pereiopods slender, similar in shape and size. Fingers slightly longer than palm. Carpus 5-segmented, with ratios of 1.7: 0.6: 0.7: 0.7: 1.4.

Third pereiopod moderately slender. Dactylus slender, distally sharpened, slightly less than 1/2 length of propodus, curved at right angle to propodus. Propodus about 5.7 times as long as wide at base, with 4 small spines along ventral margin and one pair of spines at distal end. Carpus slightly more than 1/2 length of propodus, with one small spine at distal end of inferior margin. Merus about 6.2 times as long as wide at base, twice as long as carpus, unarmed. Ischium with one rather strong movable spine on middle ventral margin. Fourth pereiopod similar to third pereiopod. Propodus with 3 movable spines on ventral margin and a pair at distal end. Fifth pereiopod similar to fourth, much more slender, dactylus about 3/7 length of propodus. Ischium without movable spine.

Abdomen normal; sixth somite with large, subtriangular movable plate. Uropodal exopod and endopod bearing spinules along distal margins; exopod with small, acute distolateral tooth and strong subterminal movable spines; lamella of 9–11 convex teeth, variable in size, along full length of diaeresis.

Telson narrow, tapering; about 2.5 times as long as wide at anterior end, lateral margin slightly convex; posterior margin strongly convex; dorsal surface with 2 pairs of moderately large spines, inserted rather laterally, approximately at 1/6 and 1/2 length of telson; posterolateral angles each with two pairs of spines, outer spines about 1/5 length of inner.

**Remarks** *Chelomalpheus koreanus* Kim, 1998 was synonymized with *Stenalpheops anacanthus* Miya, 1997 by Anker et al. (2001). However, as the males and ovigerous females in the type species of *C. koreanus*, *S. anacanthus*, and *C. yamashitai*, as well as in the present material, were found to be lacking the caudal appendix. *Chelomalpheus koreanus* Kim, 1998 is now considered a valid species, and transferred to *Stenalpheops*. *S. koreanus* is not showing considerable variation according to sex or growth. The shape of the first chela in the male is normal, as it is in the female. Therefore, the female and small male type specimens of *S. anacanthus* and *C. yamashitai* are all identified as *S. koreanus* herein. The genus *Stenalpheops* Miya, 1997 thus contains two species now: *S. anacanthus* and *S. koreanus*. *S. anacanthus* differs from *S. koreanus* (1) by the posterior margin of the uropodal exopods bearing a caudal appendix; (2) by the sexual dimorphism of the first chela, which gradually becomes sub-chela-shaped in male, whereas it retains its normal chela shape in the female; and (3) by the rostral carina becoming gradually stronger, with the tip of it slightly curving backwards.

**Distribution** The Yellow Sea; Korea; Japan.

## 2.15 Genus *Synalpheus* Spence Bate, 1888

**Diagnosis** Carapace usually with well-developed rostrum and orbital teeth. Eyes covered by orbital hoods. Pterygostomial angle produced to rounded or sub-acute tooth. Antennular peduncle with stylocerite well-developed. Antennal scale with blade usually shorter than lateral tooth; basicerite at least bearing ventrolateral teeth. Major cheliped with chela cylindrical, smooth, without sculpture. Minor cheliped much shorter than major cheliped, with fingers usually unarmed and fitting exactly when closed; carpus frequently elongate. Second pereiopod with carpus usually 5-segmented. Third to fifth pereiopod with dactylus biunguiculate or triunguiculate. Second endopod lacking appendix masculina in male. Sixth abdominal segment without articulated plate.

Key of the genus *Synalpheus* from China Sea

1. Without orbitorostral process........................................................2
-. With orbitorostral process........................................................6
2. Rostrum overreaching end of first antennular segment and much longer than orbital hoods; dactylus of minor chela tapering........................................................3
-. Rostrum much shorter than end of first antennular segment and sub-equal to orbital hoods; dactylus of minor chela spatulate........................................................5
3. Distodorsal of palm of major chela without any projecting; distal of merus of third pereiopods normal, without acute tooth........................................................*S. carinatus*
-. Distodorsal of palm of major chela with one subacute projecting; distal of merus of third pereiopods bearing an acute tooth........................................................4
4. Stylocerite reaching to about midlength of first antennular segment........................................................*S. odontophorus*
-. Stylocerite reaching or overreaching end of first antennular segment........................................................*S. stimpsonii*
5. Merus of major and minor chela with distodorsal margin rounded, without any projecting.........*S. pescadorensis*
-. Merus of major and minor chela with distodorsal margin with acute triangular tooth........................................................*S. liui*
6. Dactylus of third pereiopods triunguiculate, but with inferior unguis at times reduced to a subacute angle.........7
-. Dactylus of third pereiopods biunguiculate, never more than slight swelling as indication of third unguis........10
7. Posterior margin of sixth abdominal segment projecting as 7–10 small teeth........................................................
........................................................*S. mortenseni*
-. Posterior margin of sixth abdominal segment not projecting as teeth........................................................8
8. Spines on propodus of third pereiopods confined to distal quarter or less; dorsal spines of telson either greatly reduces or absent........................................................*S. demani*
-. Spines on propodus of third pereiopod with spines along entire length; dorsal spines of telson always present and conspicuous........................................................9
9. Merus of third pereiopod with spines on ventral margin........................................................*S. nilandensis*
-. Merus of third pereiopod without spines on ventral margin........................................................*S. fossor*
10. Sixth abdominal somite with acute posterior projection either side of base of telson........................................................11
-. Sixth abdominal somite without posterior projection either side of base of telson........................................................12
11. Rostrum much longer than orbital teeth, distolateral tooth of basicerite much shorter than stylocerite
........................................................*S. trispinosus*
-. Rostrum sub-equal to orbital teeth, distolateral tooth of basicerite sub-equal to stylocerite..............*S. bispinosus*
12. Distal of dactylus and fixed finger of minor chela with 3 or 2 teeth........................................................*S. jinqingensis*
-. Distal of dactylus and fixed finger of minor chela tapering, single tooth........................................................13
13. Basicerite usually without distodorsal tooth........................................................14
-. Basicerite with distodorsal tooth........................................................23
14. Dactylus of minor chela with definite longitudinal row of setae on crest or inner face........................................................15
-. Dactylus of minor chela bearing setae, but not in rows or patches........................................................16
15. Stylocerite shorter than first segment antennular peduncle, and short than distolateral tooth of basicerite; raw of setae on dactylus of minor chela medial to crest of finger; outer angles of telson small.................*S. lophodactylus*
-. Stylocerite overreaching end of first segment antennular peduncle, and sub-equal or longer than distolateral tooth of basicerite; row of setae on dactylus of minor chela on crest of finger; outer angles of telson usually heavy and strongly projecting........................................................
........................................................*S. hastilicrassus*

16. Telson with posterior angles rectangular ........................................ 17
-. Telson with posterior angles usually acutely produced ........................................ 21
17. Distolateral tooth of basicerite much longer than stylocerite ........................................ 18
-. Distolateral tooth of basicerite shorter than stylocerite ........................................ 20
18. Posterior margin of telson strongly protruding ........................................ *S. beibuensis*
-. Posterior margin of telson almost straight ........................................ 19
19. Tip of third maxilliped bearing only fine setae ........................................ *S. theano*
-. Tip of third maxilliped bearing circlet of short heavy spines ........................................ *S. neptunus*
20. Third pereiopods slender, with extensor tooth of dactylus longer and wider at base than flexor tooth ........................................ *S. biunguiculatus*
-. Third pereiopods stout, with extensor tooth of dactylus sub-equal to and much narrower at base than flexor tooth ........................................ *S. charon*
21. Rostrum long and slender, overreaching middle of second antennular segment, 5 or more times as long as wide at base ........................................ *S. gracilirostris*
-. Rostrum shorter and more triangular, hardly overreaching end of first antennular segment, not more than 3 times as long as wide at base ........................................ 22
22. Dactylus of third pereiopods with uniform taper, and with extensor tooth longer and often wider at base than flexor tooth ........................................ *S. tumidomanus*
-. Dactylus of third pereiopods with slight swelling on inferior surface proximal to flexor tooth, flexor tooth sub-equal to or slightly shorter than and wider at base than extensor tooth ........................................ *S. paraneomeris*
23. Merus of third pereiopods unarmed with spines ........................................ 24
-. Merus of third pereiopods armed with spines ........................................ 25
24. Rostrum distinctly shorter than orbital tooth; scaphocerite blade and lateral spine much reduced; posterior pair of spines inserted posterior on dorsal margin of telson ........................................ *S. brevrostrus*
-. Rostrum sub-equal to orbital teeth; scaphocerite blade and lateral spine normal developed; posterior pair of spines inserted middle on dorsal margin of telson ........................................ *S. coutierei*
25. Superior distal margin of palm of major chela with two conical, obtuse tubercles, directed forward and upturned ........................................ *S. bituberculatus*
-. Superior distal margin of palm of major chela with an acute tooth, straight or proximal to dactylus ........................................ 26
26. Extensor tooth of dactylus of third pereiopods less than half as long and less than half as wide at base as flexor tooth ........................................ 27
-. Extensor tooth of dactylus of third pereiopods sub-equal to or slightly shorter than and about half as wide at base as flexor tooth ........................................ 28
27. Fingers of minor chela compressed and sub-equal to or slightly shorter than palm ........................................ *S. neomeris*
-. Fingers of minor chela not compressed and about half length of palm ........................................ *S. iphinoe*
28. Rostrum slender, about 5 times as long as wide at base, reaching middle of second antennular segment ........................................ *S. modestus*
-. Rostrum shorter, less than 3 times as long as wide at base, hardly reaching end of first antennular segment ...... 29
29. Third pereiopod with terminal teeth of dactylus longer, 1/4-1/3 as long as segment ............... *S. streptodactylus*
-. Third pereiopod with terminal teeth of dactylus much reduced, 1/10-1/6 as long as segment ........................................ 30
30. Propodus third pereiopod with distinct superior curvature, ventral margin with only two spines, plus one or two terminal spines ........................................ *S. pococki*
-. Propodus of third pereiopod only slightly curved, ventral margin with numerous spines ..................... *S. iocasta*

### 2.15.1 *Synalpheus bispinosus* De Man, 1910 (Fig. 2.142)

*Synalpheus bispinosus* De Man, 1910: 302; 1911: 280, fig. 54; Banner DM & Banner AH, 1975: 346, fig. 18j–m; Wang & Sha, 2015: 359, fig. 1.

**Material Examined** MBM 270524, 1♀1♂, CL 10.1 mm, TL 30.1 mm, CL 8.5 mm, TL 24.0 mm, Beibu Gulf, Sta. 7502, depth: 32 m, 21 August 1962; MBM 159286, 1♂, CL 2.5 mm, TL 6.3 mm, Yezhu Island N-side, depth: 9 m, 21 March 1992; MBM 270607, 2♂, CL 4.5 mm, TL 11.5 mm, Beibu Gulf, Sta. 6218, depth: 62 m, Fan Zhen-gang Coll., 18 April 1960; MBM 270606, 1♀, 1♂, CL 6.8 mm, TL 17.5 mm; CL 6.1 mm, TL 16.5 mm, Beibu Gulf, Sta. 6276, depth: 26 m, Shen Shou-peng Coll., 7 July 1960; MBM 270475, 1♂, CL 4.5 mm, TL 11.5 mm, Beibu Gulf, Sta. 7902, depth: 59 m, 10 October 1962; MBM 270601,1♀, CL 5.1 mm, TL 12.8 mm, Beibu Gulf, Sta. 7503, depth: 47 m, Sun Fu-zeng Coll., 15 January 1962; MBM 107770, 1♂, CL 4.5 mm, TL 11.1 mm, Yalong Bay, Yezhu Island N-coast, depth: 8 m, from dead coral, Neumann Coll., 27 March 1992.

**Description** Carapace smooth, glabrous. Rostrum twice as long as wide at base, with lateral margins slightly convex, rounded tip bearing two short setae, reaching 1/2 length of visible part of first antennular segment. Orbital hoods three times as wide at base as rostrum, subequal to rostrum, with rounded tip bearing two short setae; notches between orbital hoods and rostrum

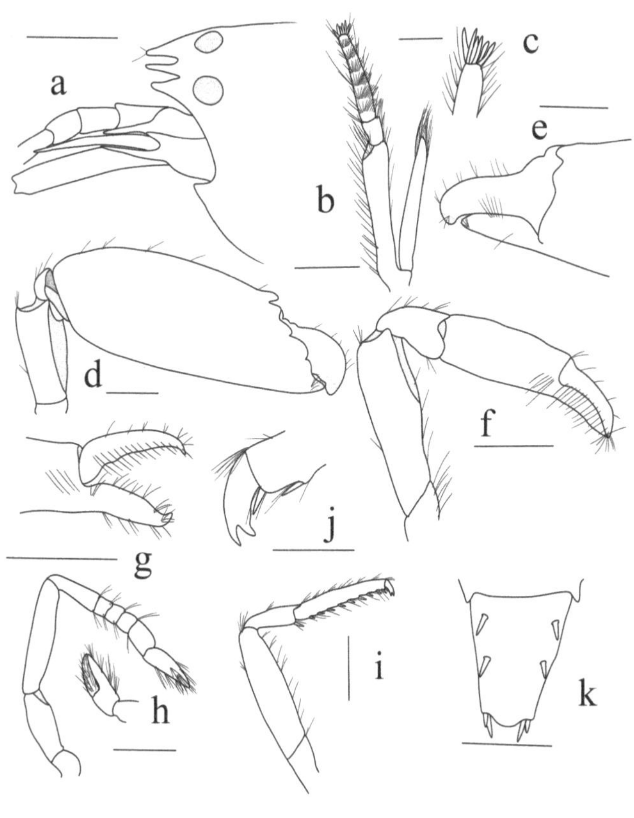


**Fig. 2.142** *Synalpheus bispinosus* De Man, 1910. MBM 157976, male, **a**, anterior part of carapace, lateral view; **b**, third maxilliped; **c**, same, tip; **d**, major cheliped; **e**, same, chela, medial view; **f**, minor cheliped; **g**, same, chela; **h**, second pereiopod and its chela; **i**, third pereiopod; **j**, same, distal propodus and dactylus; **k**, telson (scale bar: **a–b**, **d–i**, **k** = 1 mm; **c**, j = 0.25 mm)

V-shaped. Pterygostomial corner produced into blunt acute angle.

Abdominal segments smooth, glabrous, posterolateral margin of sixth pleura each with one acute tooth; first to third pleura with posterior corner rounded; fourth and fifth pleura with lower margin produced into acute angle. Telson about 1.2 times as long as wide at base; lateral margin slightly concave posteriorly; dorsal surface with median groove, with two pairs of large spines, inserted approximately at 1/6 and 1/2 length of telson; posterior margin strongly convex, fringed with long setae, posterolateral angle each with two pairs of spines, lateral 1/2 length of mesial, posterolateral corner right angle.

Antennular peduncle with distally acute stylocerite, latter reaching 1/3 of second antennular segment; second segment twice as long as wide; visible part of first segment longer than second and about twice as long as third. Antenna with basicerite with acute distodorsal tooth, 1/4-1/3 length of distolateral tooth, latter subequal to stylocerite; carpocerite beyond end of antennular peduncle by length of third segment; scaphocerite blade narrow, 4/5 length of third antennular segment, lateral spine with lateral margin slightly concave posteriorly, overreaching end of third segment, shorter than carpocerite.

Mouthpart not dissected. Third maxilliped far beyond antennular peduncle and carpocerite when extended; antepenultimate segment longest, about 5 times as long as wide; penultimate segment about 1.7 times as long as wide at base; tip of ultimate segment bearing crown of five-six spin-like setae, with several brush consisting of six-eight transverse rows of setae along ventral margin.

Major cheliped with chela cylindrical, not a bit compressed or twisted; dactylus distinctly longer than fixed finger; palm about 3.3 times as long as dactylus, distodorsal tooth palm subacute, slightly upturned. Carpus cup-shaped, with long setae distodorsally. Merus three times as long as wide at base, distodorsal margin ending in subacute tooth. Ischium stout, short. Minor chela with dactylus tapering, with flexor margin concave, distal single tooth, lateral margin bearing row of long setae; fixed finger tapering, with flexor surface obliquely convex, distal single tooth, lateral margin bearing row of long setae. Palm about 1.7 times as long as fingers, lateral face near dactylus bearing patch of stiff setae. Carpus cup-shaped proportionally longer that of major cheliped. Merus about 3.4 times as long as wide, longer than that of major cheliped, distodorsal margin rounded, without any projecting, ventral margin bearing row of short setae. Ischium with ventral margin bearing long setae.

Second pereiopod with fingers 1.5 times as long as palm. Carpus five-segmented, segment ratio subequal to 4: 1: 1: 1: 2, with first segment shorter than sum of last four segment, segments 2–4 slightly wider than long.

Third pereiopod with dactylus around 1/6 propodus length, biunguiculate, moderately slender, extensor tooth subequal to flexor tooth; extensor tooth as wide as flexor at base; notch between extensor and flexor tooth U-shaped. Propodus with usual seven spines along ventral margin plus one distal pair of spines near dactylus. Carpus about 4/11 merus length, with dorsal margin extended into an obtuse tooth, ventral margin with one distal spine. Merus longer than propodus, about 4.8 times as long as wide at base, without spines on ventral margin. Fourth pereiopod generally similar to third; but ventral margin of carpus without distal spine. Propodus of fifth pereiopod with six transversal rows of setae along ventral margin plus one pair of spines near dactylus.

**Variation** The smaller specimens maybe was an immature male having those following two differences: propodus of third pereiopod with ten spines along ventral margin rather than usual seven spines; notch between extensor and flexor tooth V-shaped.

**Remarks** This species was originally described by De Man (1910). The present specimens are agreed with the original description and re-descriptions. This species was closely related to *S. coutierei*, and the single morphological difference was the posterior margin of the sixth abdominal. Banner & Banner (1975) has questioned the validity of this species. Nevertheless, they has accepted this species, because of the stability of this character and this two species were not collected together. The Present specimens of this species are also not collected with *S. coutierei* together. Therefore, this character is also valid to separate this two species.

This species maybe an social snapping shrimp, similar to the other six *Synalpheus* shrimp that includes S. *regalis*; *S. filidigitus*; *S. fossor*; *S. hastilicrassus*; *S. n. neptunus*; *S.* aff. *neomeris* (Duffy & Macdonald, 1999; Didderen et al., 2006), because of there was a pair of larger specimens. The female having one "queen" character: first-fifth pleura with posterior corner rounded, and second pleura proportionally longer and wider, carried huge number of eggs, apparently more than the smaller ovigerous female.

**Distribution** South China Sea; Indonesia; Australia.

## 2.15.2 *Synalpheus bituberculatus* De Man, 1910 (Fig. 2.143)

*Synalpheus bituherculatus* De Man, 1910: 294; 1911: 276 , fig. 53; Banner AH & Banner DM, 1966b: 66, fig. 22; Miya, 1972: 57, Pl. 10; Banner DM & Banner AH, 1975: 307, fig. 8; Wang & Sha, 2015: 361.

**Material Examined** This species was reported by Banner & Banner (1978) and Liu (2008) from South China Seas, but not found in present specimens.

**Description** Rostrum distally acute, with tip curved upwards, reaching to distal margin of first antennular segment. Orbital teeth nearly as long as rostrum, but wider than rostrum at base.

Antennular peduncle with distally acute stylocerite slightly overreaching distal margin of first segment; second segment about 1.6 times as long as wide, longer than third segment. Antenna with basicerite bearing ventrolateral teeth, longer than

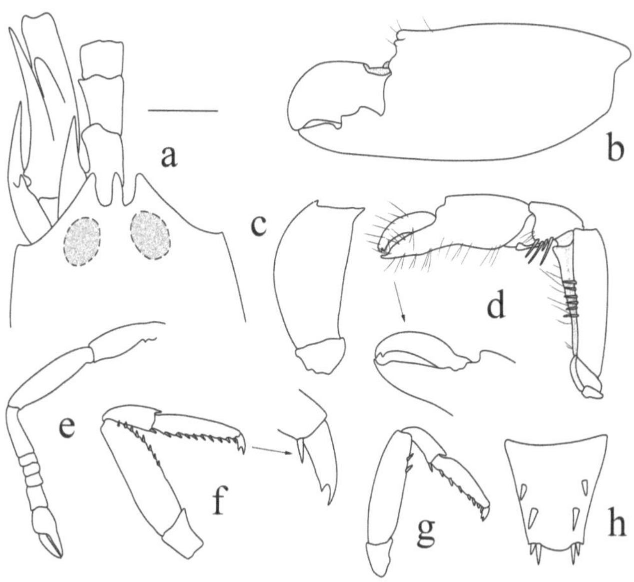


**Fig. 2.143** *Synalpheus bituberculatus* De Man, 1910. (Adapted from Banner & Banner, 1975) **a**, anterior part of carapace, dorsal view; **b**, major chela; **c**, merus and ischium of major cheliped; **d**, minor cheliped; **e**, second pereiopod; **f** & **g**, third pereiopod; **h**, telson

stylocerite; antennal scale with narrow blade shorter than lateral teeth, slightly overreaching distal margin of antennular peduncle; carpocerite 6 times as long as wide, overreaching distal margin of antennular peduncle.

Major cheliped with chela cylindrical, about 2.5–3.0 times as long as wide. Fingers 1/3 length of palm. Palm with dorsal margin terminating in two sub-acute tubercles, overhanging dactylus. Carpus cup-shaped. Merus twice as long as wide; with margins all terminating in triangular acute tooth. Ischium unarmed. Minor cheliped with chela 2.8–3.5 times as long as wide. Dactylus spatulate, terminating in two acute teeth. Carpus cup-shaped, bearing several slender spines on ventral margin. Merus 4 times as long as wide, with ventromesial margin bearing 6–9 slender spines. Ischium unarmed.

Second pereiopod with carpus 5-segmented, ratios sub-equal to 10: 2–3: 2–3: 2–3: 5–6.

Third pereiopod with dactylus biunguiculate, superior unguis about twice as long as inferior unguis, and nearly as wide at base as inferior unguis. Propodus shorter than merus, with ventral margin bearing 6–10 spines and distally one pair. Carpus with ventral margin bearing one spine distally; dorsal margin terminating in sun-acute tooth. Merus about 3 times as long as wide, with ventral margin bearing 6–9 spines; distally rounded. Ischium unarmed.

Telson about 4 times as long as posterior width; dorsal surface bearing two pairs spines, situated on 1/2 and 3/4 length of telson, respectively.

**Distribution** The South China Sea (Taiwan, Hong Kong); Indian Ocean; Singapore; Thailand; Indonesia; Japan; Philippines; Australian.

### 2.15.3 *Synalpheus biunguiculatus* (Stimpson, 1860) (Figs. 2.144 and 2.145)

*Alpheus biunguiculatus* Stimpson, 1860: 31; *Synalpheus biunguiculatus* –. Banner, 1953: 32, fig. 10a–j; Wang & Sha, 2015: 361, figs. 2–3.

**Material Examined** 1♂, CL 3.1 mm, TL 8.2 mm, Dengqing island, San Zhen-gang & Xu Jie-shan Coll., 3 May 1958; 2♂, CL 4.2 mm, TL 10.1 mm, Zhubi Reef, Nansha Islands, Sta. 8, depth: 2 m, 12 May 2004; MBM 157909, 1♀( non-ovigerous), CL 4.1 mm, TL 10.5 mm.

**Description** Carapace smooth, glabrous. Rostrum acute, twice as long as wide at base; with lateral margins slightly concave; acute tip bearing two short setae, slightly overreaching end of first antennular segment. Orbital hoods acute, much wider at base than rostrum, about 2/3 length of rostrum, notches between orbital hoods and rostrum V-shaped. Pterygostomial corner produced into blunt acute angle. Cardiac notch well developed.

Abdominal segments smooth, glabrous. Telson about 1.4 times as long as wide at base; lateral margin slightly concave posteriorly; dorsal surface with two pairs of spines, inserted approximately at 1/2 and 3/4 length of telson; posterior margin strongly convex, fringed with long setae, posterolateral angle each with two pairs of spines, lateral about 1/4 length of mesial, posterolateral corner right angle.

Antennular peduncle stout, with distally acute stylocerite, latter hardly reaching end of second antennular segment; second segment wider than long; visible part of first segment longer than second, later slightly longer than third. Antenna with basicerite without distodorsal tooth, distolateral tooth acute, distinctly shorter than stylocerite; carpocerite beyond end of antennular peduncle by 1/2 length of third segment; scaphocerite blade much wider, shorter than antennular peduncle, lateral spine with lateral margin slightly concave posteriorly, slightly overreaching end of third segment, shorter than carpocerite.

Third maxilliped far beyond antennular peduncle and carpocerite when extend; antepenultimate segment longest; penultimate segment about 1.8 times as long as wide at base; tip of ultimate segment bearing crown of five spine-like setae, with several brush transverse rows of setae along ventral margin.

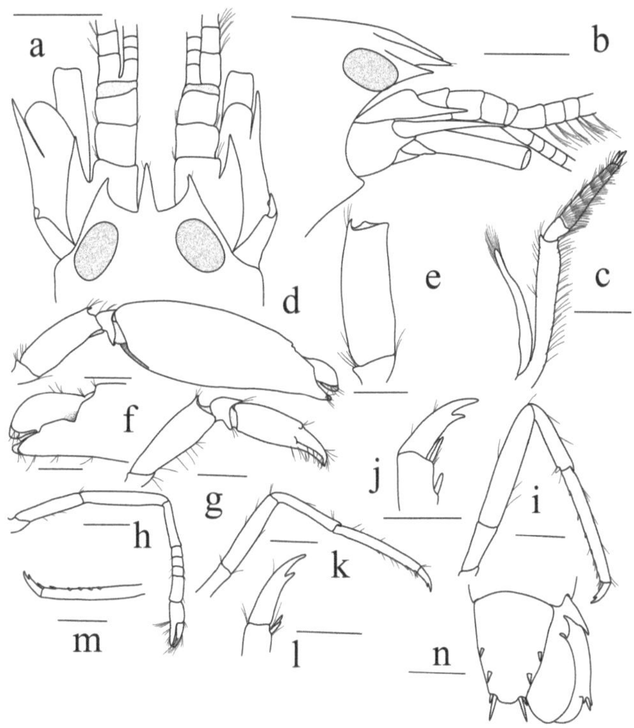


**Fig. 2.144** *Synalpheus biunguiculatus* (Stimpson, 1860), male, CL 2.7 mm. **a**, anterior part of carapace, dorsal view; **b**, same, lateral view; **c**, third maxilliped; **d**, left major cheliped; **e**, same, merus and ischium; **f**, same, fingers; **g**, right minor cheliped; **h**, second pereiopod; **i**, third pereiopod; **j**, same, distal part of propodus and dactylus; **k**, fourth pereiopod; **l**, same, distal part of propodus and dactylus; **m**, propodus and dactylus of fifth pereiopod; **n**, right uropod and telson (scale bar: **a**–**i**, **k**, **m**–**n** = 0.5 mm; **j**, **l** = 0.25 mm)

Major cheliped with dactylus distinctly longer than fixed finger. Palm about 2.8 times as long as dactylus, superior distal margin of palm at most with one obtuse tooth. Carpus cup-shaped, with long setae distodorsally. Merus3.1 times as long as wide at base, distodorsal margin ending in acute triangular tooth. Minor chela with dactylus as long as fixed finger, distally slightly curved. Palm about 1.8 times as long as fingers. Carpus cup-shaped proportionally longer that of major cheliped. Merus about 3.5 times as long as wide, longer than that of major cheliped, distodorsal margin ending in acute triangular tooth.

Second pereiopod with fingers slightly longer than palm. Carpus five-segmented, segment ratio subequal to 10: 2: 2: 1.7: 4.2, with first segment sub-equal to sum of last four segment.

Third pereiopod with dactylus slender, about1/5 propodus length, biunguiculate, flexor tooth distinctly smaller than extensor, 1/3 length of extensor tooth and 2/3 width of extensor tooth; notch between extensor and flexor tooth V-shaped. Propodus with four-six spines along ventral margin plus one distal pair of spines near dactylus. Carpus about 1/2 merus length, with dorsal margin extended into an obtuse tooth, ventral margin with one distal spine. Merus slender, slightly longer than propodus, about 5.4 times as long as wide at base, without spines on ventral margin. Fourth pereiopod generally similar to third; but ventral margin of carpus without distal spine. Propodus of fifth generally similar to that of fourth and third pereiopod, but with 4–5 transversal rows of spine-like setae and two spines along ventral margin plus one pair of spines near dactylus.

**Remarks** This species is originally reported by Stimpson, and that original description much simple and without figures to illustrate. Therefore,

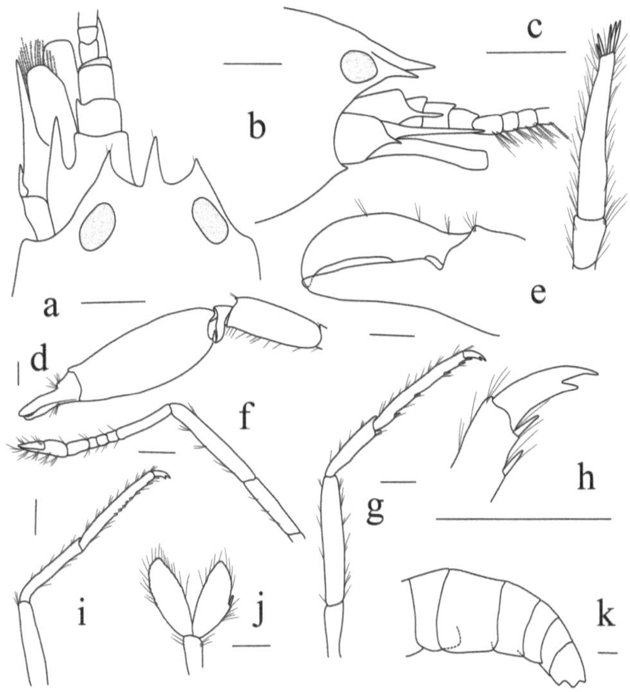


**Fig. 2.145** *Synalpheus biunguiculatus* (Stimpson, 1860). MBM 157909, female, CL 4.1 mm, **a**, anterior part of carapace, dorsal view; **b**, same, lateral view; **c**, third maxilliped, tip; **d**, major cheliped; **e**, same, fingers; **f**, second pereiopod; **g**, third pereiopod; **h**, same, distal part of propodus and dactylus; **i**, fifth pereiopod; **j**, second pleopod; **k**, abdominal segments, lateral (scale bar = 0.5 mm)

this species is controversial (Banner, 1953). The present specimens are generally agree with the re-description made by Banner (1953), except for the following characters: the stylocerite not overreaching the end of the second segment of the antennular peduncle; the ventral margin of fifth pereiopod as usual with 4–5 transversal rows of spine-like setae rather than with no trace of brush; the anterior pair dorsal spines of the telson inserted about 1/2 length of the telson rather 0.3 length of that. Additionally, one specimen was much different with others in following characters: the merus of the major chela was more slender, about 4.6 times as long as wide; the fingers of the major chela much longer, slender and a little compressed, with palm about 2.2 time as long as fingers. As species in this genus always variable in kinds of characters, those differences should be taken variation of the *S. biunguiculatus*.

**Distribution** China Seas; Hanauma Bay, Oahu; Oahu off Waikiki, off Nanakui, Maui off Kalama Park; off Hawaii at Halape.

### 2.15.4 *Synalpheus brevrostrus* Wang & Sha, 2015 (Figs. 2.146, 2.147, and 2.148)

*Synalpheus brevrostrus* Wang & Sha, 2015: 364, figs. 4–6.

**Material Examined** MBM 283384, 1♂, CL 6.5 mm, TL 12.5 mm, Jinqing Island, Xisha Islands, 16°27′57″N; 111°44′17″E, Fan Zhen-gang & Xu Jie-shan Coll., 25 April 1958; MBM 283383, 1♀, CL 5.5 mm, TL 15.0 mm, Xisha Islands, Fan Zhen-gang & Xu Jie-shan Coll., 28–29 April 1958; MBM 283384, 1♀, CL 5.6 mm, TL 14.8 mm, Jinqing Island, Xisha

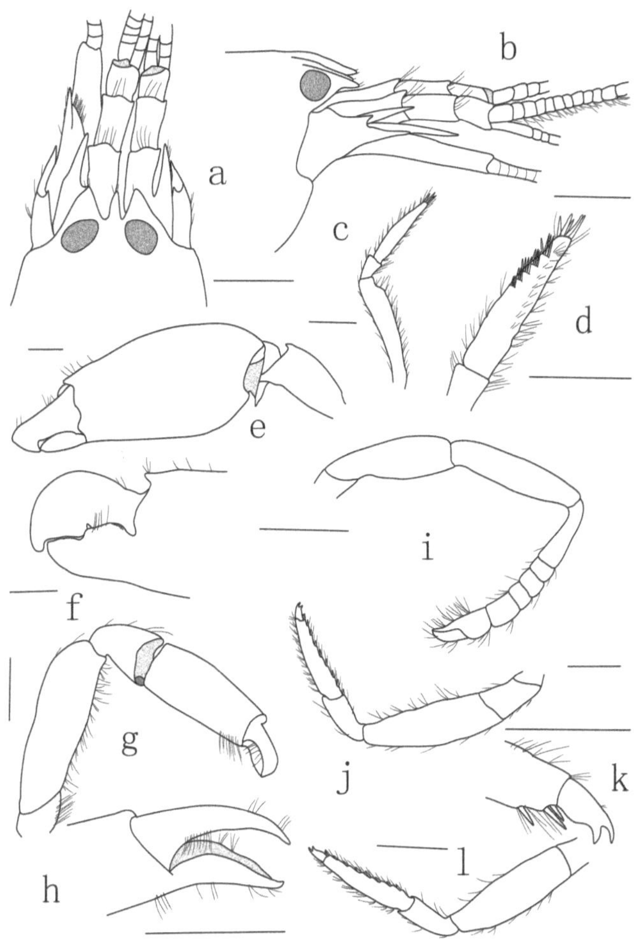


**Fig. 2.146** *Synalpheus brevrostrus* Wang & Sha, 2015. MBM 284484, male, TL 12.5 mm, **a**, anterior part of carapace, dorsal view; **b**, same, lateral view; **c**, third maxilliped; **d**, same, ultimate segment; **e**, major cheliped; **f**, same, fingers; **g**, minor cheliped; **h**, same, fingers; **i**, second pereiopod; **j**, third pereiopod; **k**, same, distal part of propodus and dactylus; **l**, fourth pereiopod (scale bar: **a–c**, **e–j**, **l** = 1 mm; **d**, **k** = 0.5 mm)

Islands, (16°27′57″N; 111°44′17″E), Fan Zhengang & Xu Jie-shan Coll., 25 April 1958.

**Description** Carapace smooth, glabrous. Rostrum twice as long as wide at base, with lateral margins slightly convex; rounded tip bearing two short setae, hardly reaching 1/2 length of visible part of first antennular segment. Orbital hoods about 2.7 times as wide at base as rostrum, distinctly longer than rostrum, with rounded tip bearing two short setae; notches between orbital hoods and rostrum deep, narrow, V-shaped. Pterygostomial corner produced into blunt acute angle.

Abdominal segments smooth, glabrous, posterolateral margin of sixth pleura with 3–4 small teeth, sex bimorphology, first pleura of male with posterior corner distinctly produced ventrally into anteriorly directed hook; second pleura with posterior corner rounded; third to fifth with lower margin produced into acute angle; first pleura of female with posterior corner rounded. Telson about 1.3 times as long as wide at base; lateral margin slightly concave posteriorly; dorsal surface without median groove, with two pairs of large spines, inserted approximately at 1/3 and

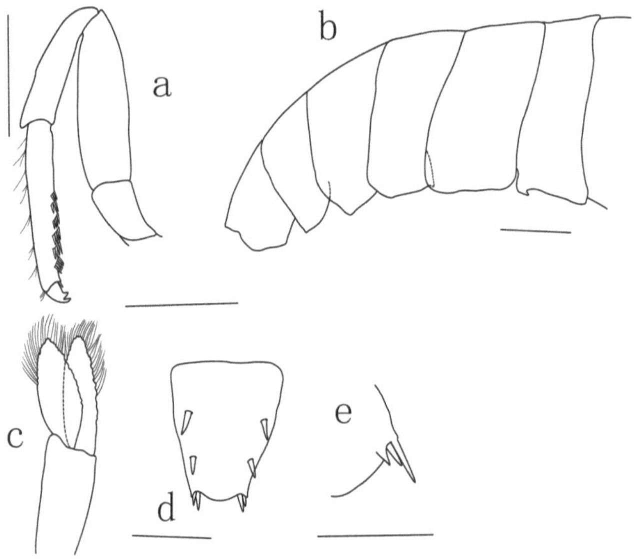


**Fig. 2.147** *Synalpheus brevrostrus* Wang & Sha, 2015. MBM 284484, male, TL 12.5 mm, **a**, fifth pereiopod; **b**, abdominal segments, lateral view; **c**, telson; **d**, right uropod, distolateral margin of exopod (scale bar = 1 mm)

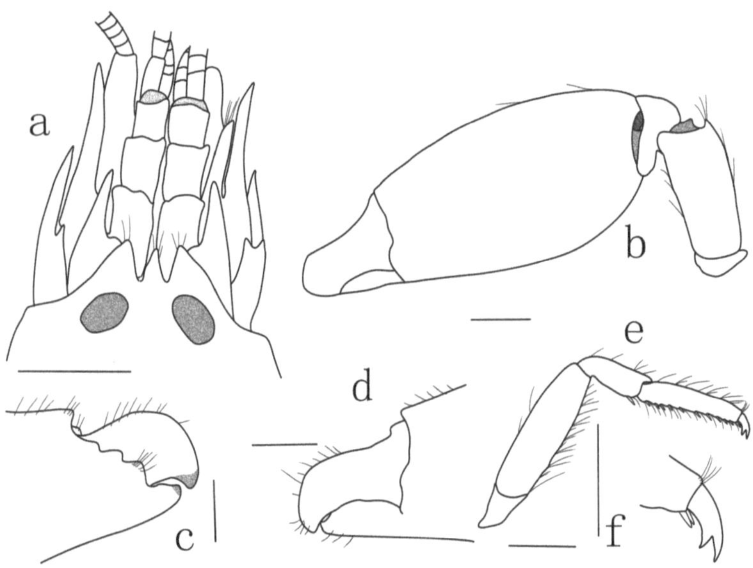


**Fig. 2.148** Synalpheus *brevrostrus* Wang & Sha, 2015. MBM 283384, female, TL 14.8 mm, **a**, anterior part of carapace, dorsal view; **b**, major cheliped; **c** & **d**, same, fingers; **e**, third pereiopod; **f**, same, distal part of propodus and dactylus (scale bar: **a**–**e** = 1 mm; **f** = 0.5 mm)

2/3 length of telson; posterior margin moderately convex, fringed with long setae, posterolateral angle each with two pairs of spines, lateral 2/3 length of mesial, posterolateral margins each produced into acute tooth, but distinctly shorter than lateral spine.

Antennular peduncle slender, with distally acute stylocerite, latter overreaching end of first antennular segment, about 1/6-1/5 length of second; second segment 1.8 times as long as wide; visible part of first segment slightly longer than second and about twice as long as third. Antenna with basicerite with acute distodorsal tooth, 1/8 length of distolateral tooth, latter hardly reaching end of first segment, distinctly shorter than stylocerite; carpocerite beyond end of antennular peduncle by 1/3 length of third segment; scaphocerite blade narrow, 4/5 length of second antennular segment, lateral spine moderately developed, with lateral margin slightly concave posteriorly, almost overreaching end of second segment.

Mouthpart not dissected. Third maxilliped far beyond antennular peduncle and carpocerite when extended; antepenultimate segment longest, about 5.2 times as long as wide; penultimate segment about 1.7 times as long as wide at base; tip of ultimate segment bearing crown of five-six spine-like long setae, with several brush consisting of six-eight transverse rows of setae along ventral margin.

Major cheliped with dactylus longer than fixed finger. Palm about 2.8 times as long as dactylus, distodorsal tooth of palm subacute, straightforward. Carpus cup-shaped, with long setae distodorsally; chela cylindrical, not a bit compressed or twisted. Merus four times as long as wide at base, distodorsal margin ending in subacute tooth. Ischium stout. Minor chela with lateral face near dactylus bearing patch of stiff setae; dactylus tapering, with flexor margin concave, distal single tooth, lateral margin near palm bearing row of long setae; fixed finger tapering, with flexor surface obliquely convex, distal single tooth. Palm about 2.5 times as long as fingers. Carpus cup-shaped proportionally longer that of major cheliped. Merus about 3.3 times as long as wide, longer than that of major cheliped, distodorsal margin rounded, without any projecting, ventral margin bearing row of short setae. Ischium with ventral margin bearing long setae.

Second pereiopod with fingers subequal to palm. Carpus five-segmented, segment ratio subequal to 4.3: 1: 1: 1: 2, with first segment shorter than sum of last four segment, segments 2–4 wider than long.

Third pereiopod stout; dactylus around 3/14 propodus length, biunguiculate, moderately slender, ventral margin near to flexor tooth concave, extensor tooth slightly thicker than flexor tooth at base, subequal; notch between extensor and flexor tooth V-shaped. Propodus with usual seven spines along ventral margin plus one distal pair of spines near dactylus. Merus longer than propodus, about 3.8 times as long as wide at base, without spines on ventral margin; carpus about 1/3 merus length, with dorsal margin extended into an obtuse tooth, ventral margin with one distal spine.

Fourth pereiopod generally similar to third; but ventral margin of carpus without distal spine. Fifth pereiopod with merus about three times as long as wide; carpus about 5/7 merus length, ventral margin without distal spine; propodus as long as merus, with six transversal rows of setae along ventral margin plus one pair of spines near dactylus.

**Remarks** *S. brevrostrus* can be unequivocally distinguished from other species by the deep, narrow notches between rostrum and orbital hoods; the rostrum distinctly shorter than the orbital hoods, and distinctly lower than the orbital hoods in lateral view; the reduced scaphocerite blade and lateral spine. The differences between this species and its similar species are discussed in Wang & Sha (2015).

**Distribution** Xisha Islands, the South China Sea.

### 2.15.5 *Synalpheus carinatus* (De Man, 1888) (Figs. 2.149 and 2.150)

*Alpheus carinatus* De Man, 1888b: 508, Pl. 22, fig. 2.
*Synalpheus carinatus* –. De Man, 1911: 210, Pl. 6, fig. 23. Banner AH & Banner DM, 1968b: 272; Banner DM & Banner AH, 1975: 283, figs. 1, 2n; Wang & Sha, 2015: 368, figs. 7–8.
*Svnalpheus carinatus var. binongcensis* De Man, 1909a: 111; 1911: 211, Pl. 6, fig. 23a.
*Svnalpheus carinatus* var. *ubianensis* De Man, 1909a: 111; 1911: 212.
Not *Synalpheus carinatus* Pearson, 1905: 83, Pl. 2, figs. 9 [= *S. trispinosus* De Man].

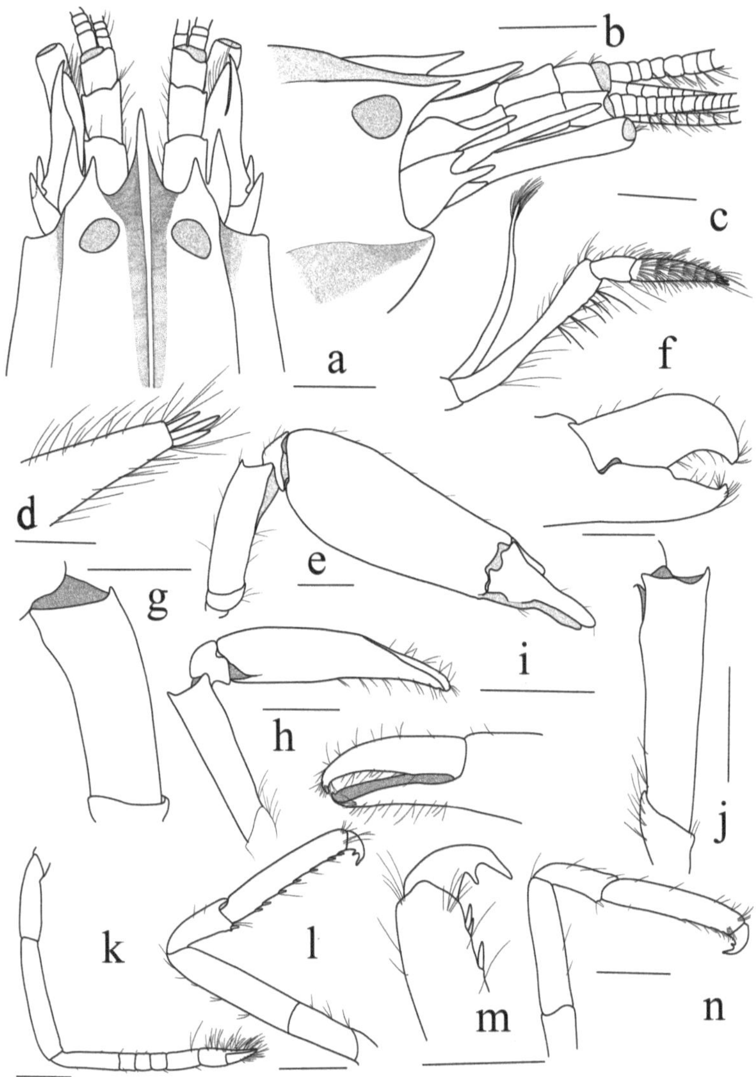


**Fig. 2.149** *Synalpheus carinatus* (De Man, 1888). MBM 158008, female, CL 7.1 mm, **a**, anterior part of carapace, dorsal view; **b**, same, lateral view; **c**, third maxilliped; **d**, same, tip; **e**, right major cheliped; **f**, same, fingers; **g**, merus of major cheliped; **h**, left minor cheliped; **i**, same, fingers; **j**, merus of minor cheliped; **k**, second pereiopod; **l**, third pereiopod; **m**, same, distal part of propodus and dactylus; **n**, fourth pereiopod (scale bar: **a–c**, **e–l**, **n** = 1 mm; **d**, **m** = 0.5 mm)

**Material Examined** MBM 158008, 1♀1♂, CL 7.1 mm, TL 12.5 mm, Yongxing Island, Xisha Islands, 16.8°N; 112.3°E, 15 April 1957.

**Diagnosis** Carapace smooth, glabrous. Rostrum depressed, slightly concave in lateral view; acute tip reaching middle of second antennular segment. Rostrum and carapace bearing strong median carina, extending almost entire length of carapace. Orbital hoods half-length of rostrum, with rounded tip bearing two short setae. Pterygostomial corner produced into blunt acute angle. Cardiac notch well developed.

Telson about 1.5 times as long as wide at base; dorsal surface with two pairs of moderately large

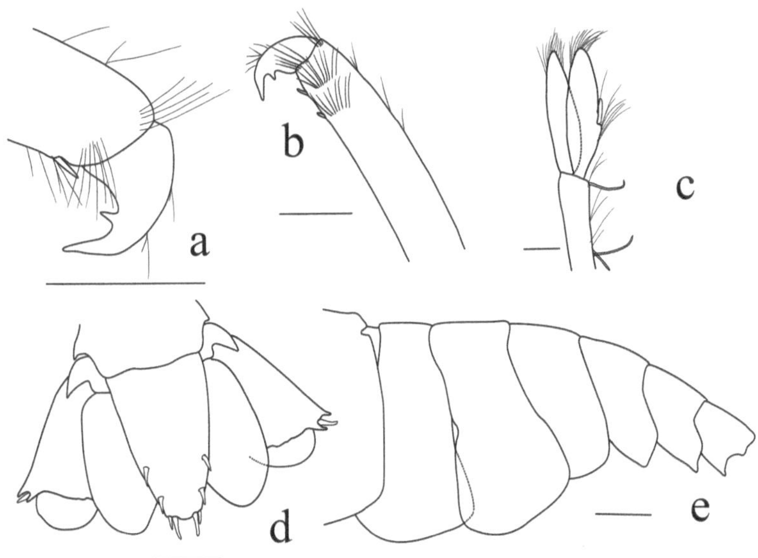


**Fig. 2.150** *Synalpheus carinatus* (De Man, 1888). MBM 158008, female, CL 7.1 mm, **a**, distal part of propodus and dactylus of fourth pereiopod; **b**, distal part of propodus and dactylus of fifth pereiopod; **c**, second pleopod; **d**, uropods and telson; **e**, abdominal segments, lateral view (scale bar: **a**–**c** = 0.5 mm; **d**, **e** = 1 mm)

spines, inserted approximately at 3/5 and 4/5 length of telson; posterior margin moderately produced, fringed with long setae, posterolateral angle each with two pairs of spines, lateral about 1/2 length of mesial, posterolateral margins not produced.

Antennular peduncle moderately stout, with distally acute stylocerite, latter reaching 1/3 length of second antennular segment; second segment about 1.5 times as long as wide; visible part of first segment longer than second and about twice as long as third. Basicerite with acute distodorsal tooth, distolateral tooth reaching about 2/3 length of visible part of first segment of antennular, distinctly shorter than stylocerite; carpocerite beyond end of antennular peduncle by 1/3 length of third segment; scaphocerite blade narrow, reaching end of third antennular segment, lateral spine developed, distally curved inward, overreaching end of antennular peduncle, shorter than carpocerite.

Third maxilliped far beyond antennular peduncle and carpocerite when extended; antepenultimate segment longest, about twice as long as ultimate segment; penultimate segment about 2.5 times as long as wide at base; tip of ultimate segment bearing crown of four spine-like setae, with transverse rows of setae along ventral margin.

Major cheliped with chela slightly compressed. Dactylus as long as than fixed finger, distally slightly curved. Palm about 1.8 times as long as dactylus, distodorsal of palm without any projecting. Carpus cup-shaped, with setae distodorsally. Merus about 3.5 times as long as wide at base, distodorsal margin with acute triangular tooth. Minor chela with dactylus and fixed finger tapering, distal single tooth. Palm about 1.5 times as long as fingers. Merus about 4.8 times as long as wide at base, distodorsal margin with acute triangular tooth, ventral margin bearing short setae; carpus cup-shaped, proportionally longer that of major cheliped.

Second pereiopod with fingers about 1.25 times as long as palm. Carpus five-segmented, segment ratio subequal to 4: 1: 1: 1: 2, with first segment slightly shorter than sum of last four segment.

Third pereiopod with dactylus strongly curved, at right angles to propodus, biunguiculate, extensor tooth about twice as long as flexor tooth, notch between extensor and flexor tooth V-shaped. Propodus with six spines along ventral margin plus one distal pair of spines near dactylus. Carpus about 1/2 length of merus, with dorsal margin extended into an obtuse tooth, ventral margin with one distal spine. Merus shorter than propodus, about 4 times as long as wide at base. Fourth pereiopod generally similar to third; ventral margin of propodus with four spines. Fifth pereiopod similar to third and fourth pereiopods, propodus with distally two transversal rows of setae and two spines on ventral margin.

**Remarks** *S. carinatus* is original described by De Man (1888), and re-described by him (1911) with another two varieties (*Synalpheus carinatus* var. *binongcensis* De Man, 1909; *Synalpheus carinatus* var. *ubianensis* De Man, 1909). Banner (1958, 1975) and Banner & Banner (1968, 1975) have given excellent discussions about the variations in this species. The present two specimens collected from Xisha Islands, South China Sea, plainly are *S. carinatus*. However, the collections slightly differ from the re-description made by Banner & Banner (1975) in the following characters: the stylocerite reaching just 1/3 length rather than middle of second segment of the antennular peduncle; the distolateral tooth of the basicerite distinctly shorter than the stylocerite rather than subequal to it; the carpus of the major and minor chela normal in form, not bearing acute teeth either in dorsal or ventral margin; the merus of major and minor margin slender than normal species in this genus, but the ventromesial margin not bearing acute teeth; the ventral margin of the propodus in third pereiopods bearing only 6 spines, less than 12–14 spines.

**Distribution** China Seas; Indonesia; Malaysian Archipelago; Caroline; Marshall; Gilbert Islands.

### 2.15.6 *Synalpheus charon* (Heller, 1861) (Fig. 2.151)

*Alpheus charon* Heller, 1861: 272, Pl. 3, figs. 21, 22; 1865: 107; Paulson, 1874: 104, Pl. 8, fig. 4.

*Synalpheus charon* –. De Man, 1911: 245, figs. 37; Banner, 1953: 37, fig. 11; Banner & Banner, 1967: 262; Banner DM & Banner AH, 1975: 369, fig. 25; Wang & Sha, 2015: 370, figs. 9–10.

*Synalpheus charon charon* Banner, 1956: 331.

*Synalpheus charon obscurus* Banner, 1956: 329, fig. 5.

*Synalpheus prolificus* Bate, 1888: 556, Pl. 99, fig. 4.

*Synalpheus helleri* De Man, 1911: 245, fig. 37.

**Material Examined** CL 4.1–5.3 mm, TL 10.2–14.1 mm. MBM 233243, 1♂, Dong Island, Xisha Islands, Sta. 3, 10 June 1975; MBM 107741, 1ovigerous♀1♂, Yongxing Island, Xisha Islands, Fang Xiu-bin Coll., 11–13 June 1980; MBM 107650), 1ovigerous♀1♂, Jinqing Island, Xisha Islands, Fan Zhen-gang & Xu Jie-shan Coll., 25 April 1958; MBM 107755, 1♂, Yongxing Island, Xisha Islands, Fang Xiu-bin Coll., 11-13 June 1980; MBM 107736, 1♂, Jinyin Island, Xisha Islands, Fang Xiu-bin Coll., 14 May 1980; MBM 107743, 2ovigerous♀2♂, Zhaoshu Island, Xisha Islands, Fang Xiu-bin Coll., 11 May 1980; MBM 107724, 1 immature specimen1♂, Shi Island, Xisha Islands, Fang Xiu-bin Coll., 10 May 1980; MBM 107223, 1ovigerous♀1♂, Sanya, Dadonghai, depth: 5–9 m, Li Xin-zheng Coll., 3 December 1997; MBM 158865, 1♂, Nansha Islands, 22 April 1987; MBM 158746, 1♂; MBM 158434, 1♂, Yezhu Island N-side, 22 March 1992; MBM 158033, 1ovigerous♀, Fan Zhen-gang & Xu Jie-shan Coll., 28–29 April 1958; MBM 157745, 4ovigerous♀9♂; MBM 158071, 1ovigerous♀, Dengqing island, Fan Zhen-gang & Xu Jie-shan Coll., 3 May 1958; MBM 157986, 1♂, Dengqing island, Fan Zhen-gang & Xu Jie-shan Coll., 3 May 1958.

**Diagnosis** Carapace smooth, glabrous. Rostrum about twice as long as wide at base, triangular tip not reaching end of first antennular segment. Orbital hoods distinctly shorter than rostrum, about twice as wide at base as rostrum. Pterygostomial corner produced into acute angle. Cardiac notch well developed.

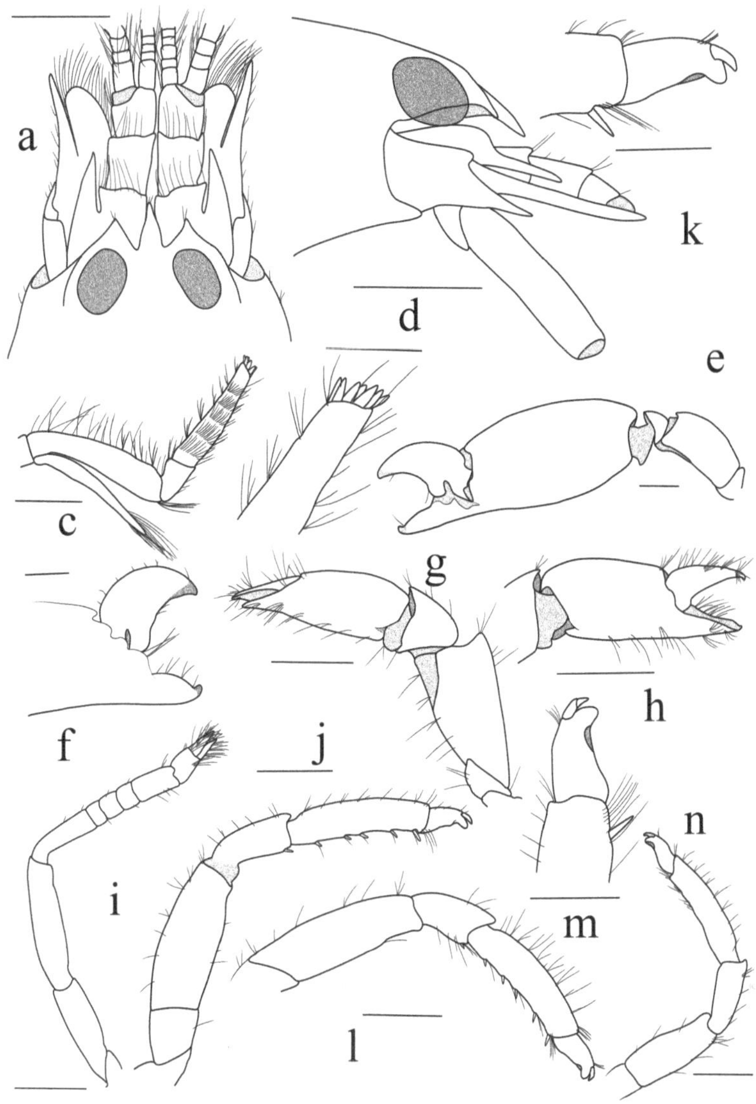


**Fig. 2.151** *Synalpheus charon* (Heller, 1861). MBM 107743, female, CL 4.9 mm, **a**, anterior part of carapace, dorsal view; **b**, same, lateral view; **c**, third maxilliped; **d**, same, tip; **e**, left major cheliped; **f**, same, fingers; **g**, right minor cheliped; **h**, same, chela; **i**, second pereiopod; **j**, third pereiopod; **k**, same, distal part of propodus and dactylus; **l**, fourth pereiopod; **m**, same, distal part of propodus and dactylus; **n**, fifth pereiopod; **o**, uropods and telson (scale bar: **a**–**c**, **e**–**j**, **l**, **n**, **o** = 1 mm; **d**, **k**, **m** = 0.5 mm)

Telson about 1.25 times as long as wide at base; dorsal surface with two pairs of moderately small spines, inserted approximately at 1/2 and 3/4 length of telson; posterior margin strongly produced, fringed with long setae, posterolateral angle each with two pairs of spines, lateral about 1/3 length of mesial, posterolateral margins each not produced.

Antennular peduncle slender, with distally acute stylocerite, latter reaching 2/3 length of second antennular segment; second segment about 1.25 times as long as wide at base; visible part of first segment slightly longer than second and less twice as long as third. Basicerite without distodorsal tooth, distolateral tooth distinctly shorter than stylocerite, overreaching end of first antennular segment; carpocerite far overreaching end of third antennular segment; scaphocerite blade moderately broad, overreaching end of third antennular segment, lateral spine developed, overreaching end of antennular peduncle, distinctly longer than carpocerite.

Third maxilliped far beyond antennular peduncle and carpocerite when extended; antepenultimate segment longest; penultimate segment about less twice as long as wide at base; tip of ultimate segment bearing crown of short six spine-like setae, with transverse rows of long setae along ventral margin.

Major cheliped with dactylus slightly longer than fixed finger. Palm about 1.7 times as long as dactylus, distodorsal of palm with one small obtuse projecting. Carpus cup-shaped, with setae distodorsally. Merus about 2.4 times as long as wide at base, distodorsal margin with one subacute triangular tooth. Ischium stout. Minor chela with dactylus and fixed finger tapering, distal single tooth. Palm about 1.3 times as long as fingers. Carpus cup-shaped, proportionally longer that of major cheliped. Merus about 3 times as long as wide at base, distodorsal margin with one subacute triangular tooth

Second pereiopod with fingers slightly longer than palm. Carpus five-segmented, segment ratio subequal to 5: 1: 1.2: 1.3: 2.3, with first segment slightly shorter than sum of last four segments. Ischium shorter than merus.

Third pereiopod moderately stout. Dactylus about 3/10 length of propodus, slightly curved to propodus, biunguiculate, extensor tooth with acute tip, near as long as flexor tooth, flexor tooth much wider at base than extensor tooth, inferior margin of flexor tooth with excavate "pocket". Propodus with five spines along ventral margin plus one. Merus near as long as propodus, about 3 times as long as wide at base; carpus about 1/2 length of merus, with dorsal margin extended into an obtuse tooth, ventral margin with one distal spine. Fourth pereiopod generally similar to third; merus near as long as propodus, ventral margin of propodus with five spines. Fifth pereiopod similar to third and fourth pereiopods, ventral margin of carpus without one distal spine; ventral margin of propodus with distal three spines.

**Remarks** This species is originally described by Heller. It is readily recognized by the dactylus of the posterior three pereiopods (Banner & Banner 1975). The present specimens are generally agreed with the re-descriptions by Miya (1972) and Banner & Banner (1975), and also variable in the size and shape of the rostrum (Banner & Banner 1967).

**Distribution** China Seas; Suez Canal and eastern Africa to Philippines, Indonesia, Australia, and most of Pacific islands, but not including Japan, Hawaii, or Society Islands.

### 2.15.7 *Synalpheus coutierei* Banner, 1953 (Figs. 2.152, 2.153, and 2.154)

*Synalpheus coutierei* Banner, 1953: 36; Banner AH & Banner DM, 1966b: 62, fig. 20; Banner DM & Banner AH, 1975: 343, fig. 18a–i; Wang & Sha, 2015: 373, figs. 11–13. *Alpheus biunguiculatus* De Man, 1888a: 502, Pl. 21, fig. 6; Bate, 1888: 562, Pl. 101, fig. 4.

*Synalpheus biunguiculatus* Coutière, 1898: 232, figs. 1–4; 1905: 873, Pl. 71, fig. 8.

*Synalpheus biunguiculatus* var. *exilipes* Coutière, 1905: 874, fig. 10; Balss, 1921: 9.

*Synalpheus exilipes* Johnson, 1962: 51.

**Material Examined** MBM 158469, 1♂, CL 8.2 mm, TL 21.3 mm; MBM 157752, 1♂, CL 5.1 mm, TL 12.1 mm, 13 May 1958; MBM 107754, 2♀1♂, CL 5.8 mm, TL 15.9 mm, CL 4,6 mm, TL 12.1 mm, Xisha Islands, Fang Xiubin Coll., 19–21 May 1980; MBM 107767, 1♀, CL 5.7 mm, TL 15.1 mm, Xisha Islands, Li Xinzheng Coll., 14 April 1975; MBM 158838, 1♀1♂, CL 6.0 mm, TL 14.9 mm, CL 5.3 mm, TL 12.8 mm, Xinying, 28 July 1957; MBM 107763, 1♂, CL 4.7 mm, TL 10.5 mm, Lelong, 25

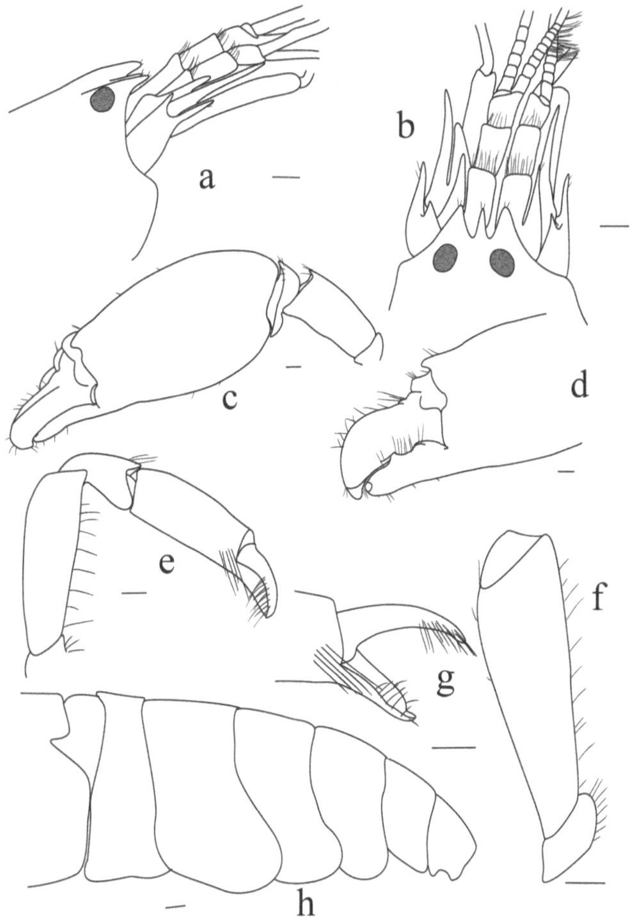


**Fig. 2.152** *Synalpheus coutierei* Banner, 1953. MBM 158469, female, CL 8.1 mm, **a**, anterior part of carapace, lateral view; **b**, same, dorsal view; **c**, major cheliped; **d**, same, fingers; **e**, minor cheliped; **f**, same, merus and ischium; **g**, same, fingers; **h**, abdominal segments, lateral view (scale bar = 0.5 mm)

November 1990; MBM 107761, 1♂, CL 4.8 mm, TL 10.8 mm, Zhaoshu Island, Xisha Islands, 11 May 1980; MBM 107771, 1♀1♂, CL 8.3 mm, TL 21.1 mm, CL 7.2 mm, TL 16.5 mm, Yalong Bay, Yezhu Island N-coast, depth: 8 m, from dead coral, Neumann Coll., 27 March 1992; MBM 158006, 2♀1♂, CL 5.7 mm, TL 13.2 mm, CL 4.5 mm, TL 10.3 mm; MBM 158184, 1♂, CL 5.1 mm, TL 11.3 mm, Yongxing Island, Xisha Islands, 12–13 June 1958.

**Description** Carapace smooth, glabrous. Rostrum twice as long as wide at base, with lateral margins slightly convex, rounded tip bearing two short setae, reaching 1/2 length of visible part of first antennular segment. Orbital hoods three times as wide at base as rostrum, subequal to rostrum, with rounded tip bearing two short setae, notches between orbital hoods and rostrum V-shaped. Pterygostomial corner produced into blunt acute angle. Cardiac notch well developed.

Abdominal segments smooth, glabrous. Telson about 1.3 times as long as wide at base; lateral margin slightly concave posteriorly; dorsal surface with median groove, with two pairs of large spines, inserted approximately at 1/5 and 1/2 length of telson; posterior margin strongly convex, fringed with long setae, posterolateral angle each with two pairs of spines, lateral 1/2

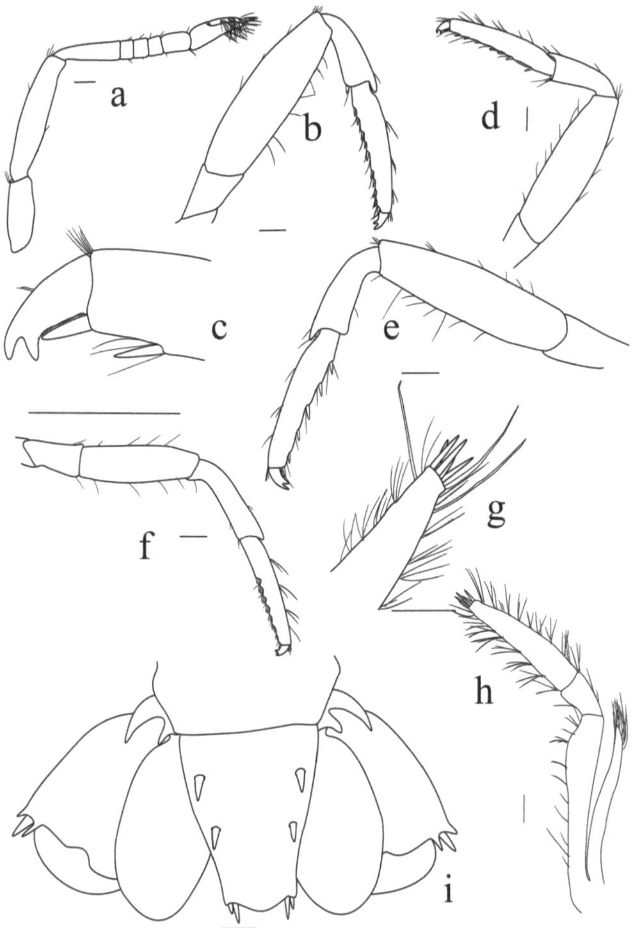


**Fig. 2.153** *Synalpheus coutierei* Banner, 1953. MBM 158469, female, CL 8.1 mm, **a**, second pereiopod; **b**, third pereiopod; **c**, same, distal part of propodus and dactylus; **d** & **e**, fourth pereiopod; **f**, fifth pereiopod; **g**, tip of third maxilliped; **h**, third maxilliped; **i**, uropods and telson (scale bar = 0.5 mm)

length of mesial, posterolateral corner right angle.

Antennular peduncle with distally acute stylocerite, latter reaching 1/3–1/2 of second antennular segment; second segment 1.5 as long as wide; visible part of first segment longer than second and about twice as long as third. Antenna with basicerite with acute distodorsal tooth, 1/4–1/3 length of distolateral tooth, latter subequal to stylocerite; carpocerite beyond end of antennular peduncle by length of third segment; scaphocerite blade narrow, overreaching second antennular segment, lateral spine with lateral margin slightly concave posteriorly, overreaching end of third segment, shorter than carpocerite.

Mouthpart not dissected. Third maxilliped far beyond antennular peduncle and carpocerite when extended; antepenultimate segment longest, about six times as long as wide; penultimate segment about 1.8 times as long as wide at base; tip of ultimate segment bearing crown of five-six spin-like setae, with several brush consisting of 6–8 transverse rows of setae along ventral margin.

Major cheliped with chela cylindrical, not a bit compressed or twisted. Dactylus distinctly longer than fixed finger. Palm about 2.5 times as long as dactylus, distodorsal tooth palm subacute, slightly upturned. Carpus cup-shaped, with long setae distodorsally. Merus three times as long as

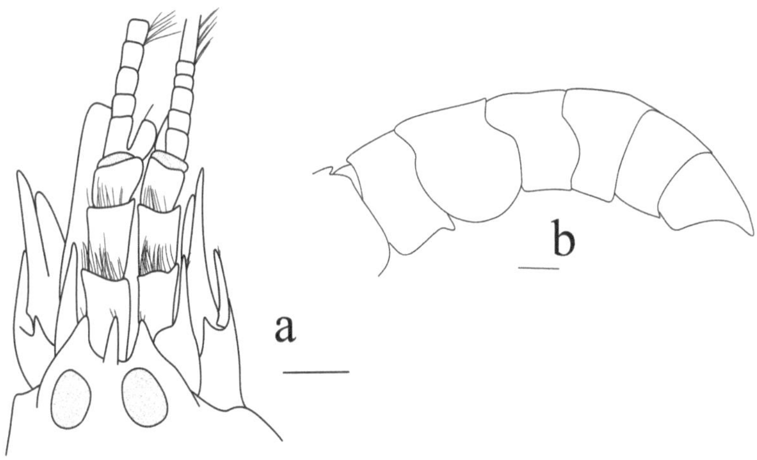


**Fig. 2.154** *Synalpheus coutierei* Banner, 1953. MBM 107754, male, CL 5.8 mm, **a**, anterior part of carapace, dorsal view; **b**, abdominal segments (scale bar = 0.5 mm)

wide at base, distodorsal margin ending in subacute triangular tooth. Ischium stout. Minor chela with dactylus tapering, with flexor margin concave, distal single tooth, lateral margin bearing row of long setae; fixed finger tapering, with flexor surface obliquely convex, distal single tooth, lateral margin bearing row of long setae. Palm about 1.9 times as long as fingers, lateral face near dactylus bearing patch of stiff setae. Carpus cup-shaped proportionally longer that of major cheliped. Merus about 3.5 times as long as wide, longer than that of major cheliped, distodorsal margin rounded, without any projecting, ventral margin bearing row of short setae. Ischium with ventral margin bearing long setae.

Second pereiopod with fingers 1.5 times as long as palm. Carpus five-segmented, segment ratio subequal to 4:1:1:1: 2, with first segment shorter than sum of last four segment, segments 2–4 slightly wider than long.

Third pereiopod with dactylus around 1/6 propodus length, biunguiculate, moderately slender, extensor tooth subequal to flexor tooth; extensor tooth as wide as flexor at base; notch between extensor and flexor tooth U-shaped. Propodus with seven spines along ventral margin plus one distal pair of spines near dactylus. Carpus about 1/3 merus length, with dorsal margin extended into an obtuse tooth, ventral margin with one distal spine. Merus longer than propodus, about four times as long as wide at base, without spines on ventral margin. Fourth pereiopod generally similar to third; but ventral margin of carpus without distal spine. Propodus of fifth generally similar to that of fourth and third pereiopod, but with six transversal rows of setae along ventral margin plus one pair of spines near dactylus.

**Remarks** Only one of specimens differs from others in: the relative length of rostrum and orbital hoods, which the rostrum distinctly shorter than the orbital hoods and slightly lower than the orbital hoods when viewed laterally; the right third pereiopod with nine spines along the ventral margin of propodus. This species most similar to *S. bispinosus*, but can be distinguished from *S. bispinosus* by lacking the acute tooth on the posterolateral margin of sixth pleura.

**Distribution** The South China Sea; From Suez Canal and Persian Gulf south to Mozambique and eastward across the Indo-Pacific.

### 2.15.8 *Synalpheus demani* Borradaile, 1900 (Figs. 2.155 and 2.156)

*Synalpheus demani* Borradaile, 1900: 416; De Man, 1911: 257, fig. 42; Banner AH & Banner DM, 1968b: 274; Miya, 1972: 60, Pl. 11;

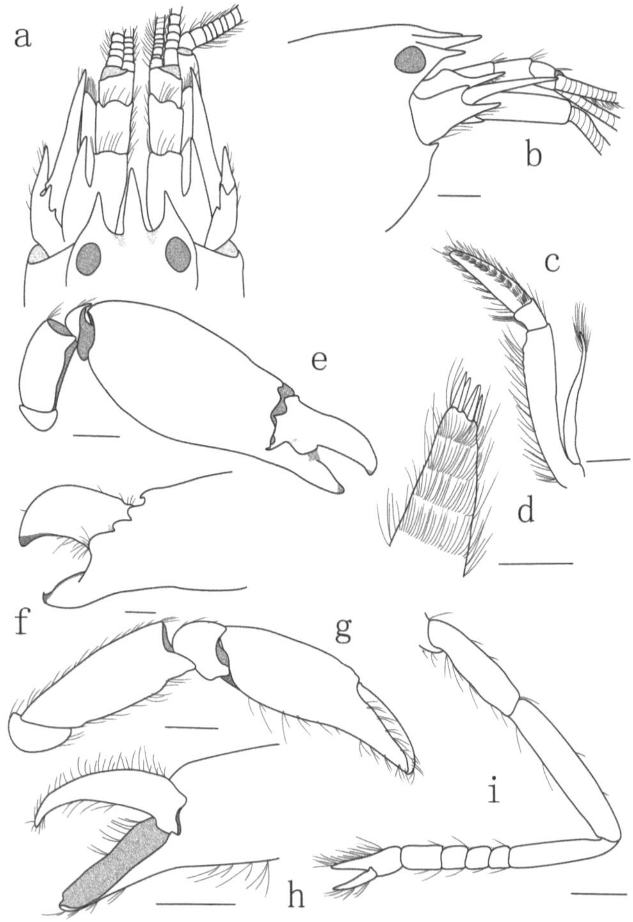


**Fig. 2.155** *Synalpheus demani* Borradaile, 1900. MBM 107803, male, **a**, anterior part of carapace, dorsal view; **b**, same, lateral view; **c**, third maxilliped; **d**, same, tip; **e**, right major cheliped; **f**, same, fingers; **g**, left minor cheliped; **h**, same, fingers; **i**, second pereiopod (scale bar: **a**–**c**, **e**–**i** = 1 mm; **d** = 5 mm)

Banner DM & Banner AH, 1975: 324, fig. 13; Wang & Sha, 2015: 376, figs. 14–15.

*Alpheus triunguiculatus* De Man, 1888b: 504, Pl. 22, fig. 1.

*Synalpheus spiniger* Bate, 1888: 560, Pl. 100, fig. 3.

*Synalpheus broekii* Nobili, 1901: 2.

**Material Examined** MBM 157933, 2♂, CL 4.1–5.9 mm, TL 10.1–13.5 mm, Luhuitou, Sanya, 16.8°N; 112.3°E, Liao Yu-lin Coll., 30 December 1959; MBM 107803, 2♀3♂, Yalong Bay, Yezhu Island N-coast, depth: 8 m, from *Comanthus sp.*, Neumann Coll.; MBM 107800, 1♂, CL 8.2 mm, TL 21.1 mm, Yalong Bay, Yezhu Island, 19 November 1990; MBM 158952, 5♀2♂, CL 6.1–10.1 mm, TL 14.3–26.2 mm, Xincun, 19 April 1958; MBM 107801, 2♂, CL 8.1–9.1 mm, TL 20.2–24.5 mm, Yezhu Island, Wang Yong-qiang Coll., 19 November 1990; MBM 158952, 1♂, 5 immature specimens, CL 7.2 mm, TL 15.2 mm, Xincun, 19 April 1958.

**Description** Carapace smooth, glabrous. Rostrum about 3.4 times as long as wide at base, acute tip not reaching end of first antennular segment. Orbital hoods 2/3 length of rostrum. Pterygostomial corner produced into acute angle. Cardiac notch well developed.

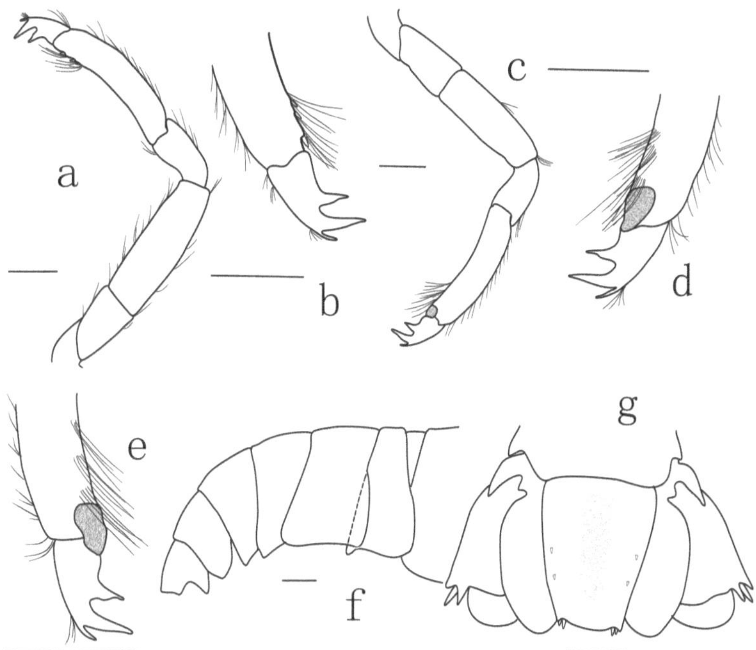


**Fig. 2.156** *Synalpheus demani* Borradaile, 1900. MBM 107803, male, **a**, third pereiopod, lateral; **b**, same, distal part of propodus and dactylus; **c**, fourth pereiopod; **d**, same, distal part of propodus and dactylus; **e**, distal part of propodus and dactylus of fifth pereiopod; **f**, abdominal, lateral view; **g**, uropods and telson (scale bar = 1 mm)

Telson about 1.3 times as long as wide at base; dorsal surface with two pairs of small spines, inserted approximately at 1/2 and 2/3 length of telson; posterior margin slightly produced, fringed with long setae, posterolateral angle each with two pairs of spines, lateral about 2/3 length of mesial, posterolateral margins each produced into acute small tooth, distinctly shorter than lateral spines.

Antennular peduncle moderately stout, with distally acute stylocerite, latter reaching about 1/3 length of second antennular segment; second segment about 1.5 times as long as wide at base; visible part of first segment longer than second and about twice as long as third. Basicerite with acute distodorsal tooth, distolateral tooth slightly shorter than stylocerite, distolateral margin bearing setae, superior margin of basicerite bearing third short subacute tooth medially; carpocerite slightly overreaching end of third antennular segment; scaphocerite blade moderately broad, reaching middle of third antennular segment, lateral spine developed, reaching end of antennular peduncle, distinctly shorter than carpocerite.

Third maxilliped far beyond antennular peduncle and carpocerite when extended; antepenultimate segment longest; penultimate segment about 1.5 times as long as wide at base; tip of ultimate segment bearing crown of five spine-like setae, with transverse rows of setae along ventral margin.

Major cheliped with dactylus slightly longer than fixed finger, distally slightly curved. Palm about 1.8 times as long as dactylus, distodorsal of palm with one subacute projecting, slightly curved to dactylus. Carpus cup-shaped, with setae distodorsally. Merus about 2.7 times as long as wide at base, distodorsal margin rounded, without triangular tooth. Ischium stout. Minor chela with dactylus and fixed finger tapering, distal single tooth. Palm about 1.3 times as long as fingers. Carpus cup-shaped, proportionally longer that of major cheliped. Merus about three times as long as wide at base, distodorsal margin rounded, ventral and dorsal margin bearing fine short setae.

Second pereiopod with fingers about 1.5 times as long as palm. Carpus five-segmented, segment ratio subequal to 5: 1: 1: 1: 1.8, with first segment slightly longer than sum of last four segments.

Third pereiopod moderately stout, dactylus slightly curved to propodus, triunguiculate, middle tooth about three times as long as extensor tooth, flexor tooth about 1.5 times as long as extensor tooth, notch between extensor and middle tooth V-shaped, middle tooth distinctly much wider than extensor tooth at base, notch between middle tooth and flexor tooth U-shaped, middle tooth slight wider at base than flexor tooth. P propodus slightly curved, distal with 3 spines and dense patches of setae. Merus near as long as propodus, about 3.7 times as long as wide at base; carpus about 1/2 length of merus, with dorsal margin extended into an obtuse tooth, ventral margin without distal spine. Fourth pereiopod generally similar to third; merus shorter than propodus, ventral margin of propodus distally bearing patches of setae. Fifth pereiopod similar to third and fourth pereiopods.

**Remarks** *S. demani* is original described by De Man (1888), and put in the genus *Alpheus* with the name *Alpheus triunguiculatus*. Borradaile (1900) lists this species under the genus *Synalpheus*, and give it a new name *S. demani*. This species is original described based on two male specimens collected from Loyalty Islands. It is re-described or illustrated by Miya (1972), Banner & Banner (1975), Chace (1988) and Bruce (1989). The present materials are similar from each other, and agreed well with the description made by Banner & Banner (1975), except the length of the rostrum. The major chela in the specimens is agreed with the description by Miya (1972), but the second tooth on the superior margin of basicerite is not mentioned in that description. In addition, this species is found only have four segments in the carpus of second pereiopod in Japan Sea (Miya, 1972), which is not found in present collections. The length of the rostrum is taken variable in Chace (1988), which reaching to overreaching the end of the first segment of the antennular peduncle, but in all present specimens it not reaching that end.

**Distribution** China Seas; Red Sea, Japan, Philippines, Indonesia, Australia, and Marshall and Loyalty islands.

### 2.15.9 *Synalpheus fossor* (Paul'son, 1875) (Fig. 2.157)

*Alpheus fossor* Paul'son, 1875: 103, Pl. 13, fig. 5.
*Synalpheus fossor* –. Coutière, 1905: 872, Pl. 70, fig. 6; Tattersall, 1921: 374; Banner DM & Banner AH, 1975: 335, fig. 16; Wang & Sha, 2015: 379.
*Synalpheus fossor* var. *propinqua* De Man, 1911: 250, fig. 39.
*Synalpheus bakeri* Coutière, 1908: 199.
*Synalpheus bakeri* var. *stormi* De Man, 1911: 253, fig. 40; Banner AH & Banner DM, 1966b: 53, fig. 15.
*Alpheus sp. varietas* De Man, 1897: 741, Pl. 35.

**Material Examined** This species was reported by Banner & Banner (1978) and Liu (2008) from South China Seas, but not found in present specimens.

**Description** Rostrum distally acute, slender, reaching from mid-length of first antennular segment to 1/4 length of second segment; tip curved upwards, bearing short setae. Orbital teeth variable from reaching 3/5 length of rostrum to nearly as long as rostrum; tips curved upwards, bearing short setae.

Antennular peduncle with distally acute stylocerite reaching mid-length of second segment; second segment slightly short than visible part of first segment; third segment about half-length of second segment. Antenna with basicerite bearing strong dorsolateral teeth nearly as long as stylocerite; carpocerite about 6 times as long as wide, overreaching distal margin of antennular peduncle.

Major cheliped with chela cylindrical, about 2.5 times as long as wide. Fingers 1/3 length of palm. Palm with dorsal margin terminating in

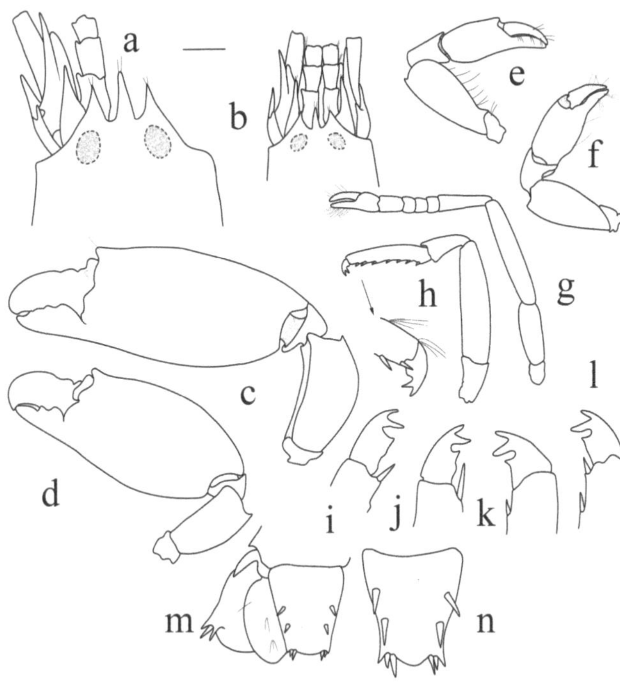


**Fig. 2.157** *Synalpheus fossor* (Paul'son, 1875) (adapted from Banner & Banner, 1975). **a** & **b**, anterior part of carapace, dorsal view; **c** & **d**, major cheliped; **e** & **f**, minor cheliped; **g**, second pereiopod; **h**, third pereiopod; **i–l**, same, dactylus; **m**, left uropod and telson; **n**, telson (scale bar = 1 mm, **i–l** without scale bar, other share same scale bar)

1–2 rounded (sometimes acute) tubercles. Carpus cup-shaped. Merus about twice as long as wide; dorsal margin terminating in acute teeth. Minor cheliped with chela 2.4–2.6 times as long as wide. Merus similar to that of major cheliped.

Second pereiopod with carpus 5-segment, ratios sub-equal to 10: 2: 2: 2: 5.

Third pereiopod with dactylus triunguiculate, medial unguis wider at base than superior unguis, but nearly as long as it, forming V-shaped notch at base; inferior unguis smaller than medial unguis, usually forming U-shaped notch at base. Propodus about 4.7–7.0 times as long as wide, with ventral margin bearing 6–10 spines and distally one pair. Carpus with ventral margin bearing one spine distally; dorsal margin terminating in sub-acute tooth. Merus 3.2–4.1 times as long as wide, unarmed with spines on ventral margin.

**Distribution** Hong Kong, South China Sea; Red Sea; Maldive Archipelago; Indonesia; Thailand; Philippines; Australian.

### 2.15.10 *Synalpheus gracilirostris* De Man, 1910 (Fig. 2.158)

*Synalpheus gracilirostris* De Man, 1910: 291; 1911: 269, fig. 49; Banner DM & Banner AH, 1975: 372, fig. 26; Wang & Sha, 2015: 379, fig. 16.

**Material Examined** MBM 107784, 1♂, CL 3.6 mm, TL 7.9 mm, Slalao. 4 April 1992; MBM 159244, 1♂, CL 3.4 mm, TL 7.5 mm; MBM 158033, 1ovigerous♀, CL 4.2 mm, TL 8.2 mm, Nansha Islands, 2 August 1944.

**Description** Carapace smooth, glabrous. Rostrum about 4.5 times as long as wide at base, 7 times as long as wide at middle, acute tip reaching about 2/3 length of second antennular segment. Orbital hoods shorter than rostrum, not reaching end of first antennular segment.

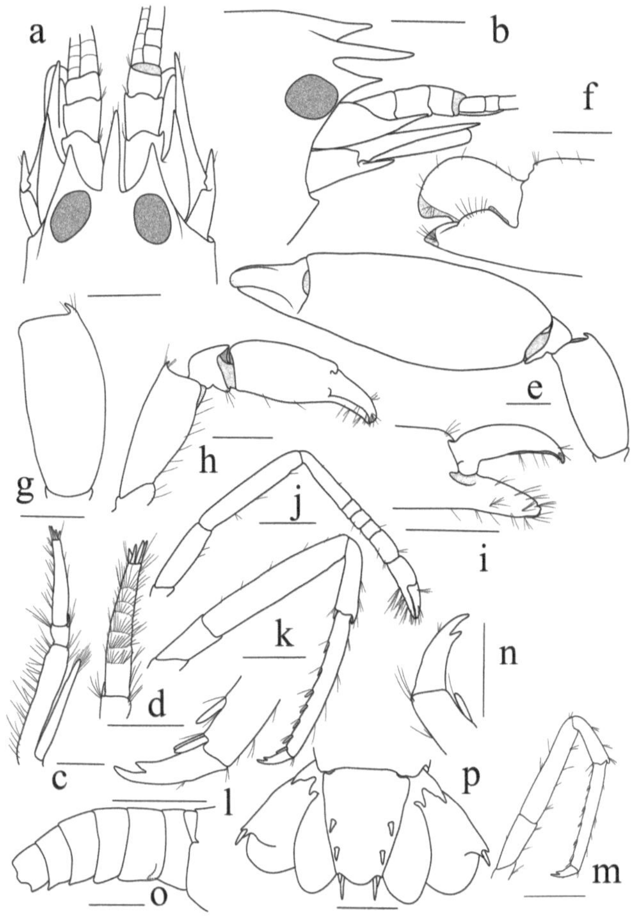


**Fig. 2.158** *Synalpheus gracilirostris* De Man, 1910. MBM 157909, male, CL 2.5 mm, **a**, anterior part of carapace, dorsal view; **b**, same, lateral view; **c**, third maxilliped; **d**, same, tip; **e**, left major cheliped; **f**, same, fingers; **g**, merus of major cheliped; **h**, right minor cheliped; **i**, same, fingers; **j**, second pereiopod; **k**, third pereiopod; **l**, same, distal part of propodus and dactylus; **m**, fourth pereiopod; **n**, same, distal part propodus and dactylus; **o**, abdominal segments, lateral view; **p**, uropods and telson (scale bar: **a–k**, **m**, **o**, **p** = 0.5 mm; **l**, **n** = 0.25 mm)

Pterygostomial corner produced into acute angle. Cardiac notch well developed.

Telson about 1.3 times as long as wide at base; dorsal surface with two pairs of moderately large spines, inserted approximately at 3/7 and 5/7 length of telson; posterior margin strongly produced, fringed with long setae, posterolateral angle each with two pairs of spines, lateral about 1/3 length of mesial, posterolateral margins each produced into small acute tooth, distinctly shorter than contiguous short spines.

Antennular peduncle moderately slender, with distally acute stylocerite, latter reaching middle of second antennular segment; second segment near as long as wide at base; visible part of first segment longer than second and about twice as long as third. Basicerite without distodorsal tooth, distolateral tooth distinctly shorter than stylocerite, not reaching end of first antennular segment; carpocerite overreaching end of third antennular segment; scaphocerite blade moderately narrow, hardly reaching end of third antennular segment, lateral spine developed, overreaching end of antennular peduncle, longer than carpocerite.

Third maxilliped far beyond antennular peduncle and carpocerite when extended; antepenultimate segment longest; penultimate segment about 1.4 times as long as wide at base; tip of ultimate segment bearing crown of five long spine-like setae, with transverse rows of setae along ventral margin.

Major cheliped with dactylus as long as fixed finger, distally slightly curved. Palm about 2.8 times as long as dactylus, distodorsal of palm only slightly projecting. Carpus cup-shaped, with setae distodorsally. Merus about 3.5 times as long as wide at base, distodorsal margin with one acute triangular tooth. Ischium stout. Minor chela with dactylus and fixed finger tapering, distal single tooth. Palm about 1.9 times as long as fingers. Carpus cup-shaped, proportionally longer that of major cheliped. Merus about 3.8 times as long as wide at base, distodorsal margin with 1 acute triangular tooth.

Second pereiopod with fingers slightly longer than palm. Carpus five-segmented, segment ratio subequal to 5.6: 1: 1: 1: 2.6, with first segment near as long as sum of last four segments.

Third pereiopod moderately slender, dactylus about 1/4 length of propodus, slightly curved to propodus, biunguiculate, extensor tooth about twice as long as flexor tooth, notch between extensor and flexor tooth V-shaped, flexor tooth near as wide at base as extensor tooth. Propodus with 6 spines along ventral margin plus one distal pair of spines near dactylus. Carpus less 1/2 length of merus, with dorsal margin extended into an obtuse tooth, ventral margin with 1 distal spine. Merus longer than propodus, about 5.7 times as long as wide at base. Fourth pereiopod generally similar to third; merus near as long as propodus; ventral margin of propodus with 5 spines plus one distal pair of spines.

**Remarks** *S. gracilirostris* is originally described by De Man (1910). The present specimens are generally agreed with the original description. Banner & Banner (1975) have re-described this species based on Australian specimens, but it was variable in the following characters: basicerite with distolateral tooth sub-equal to stylocerite; basicerite bearing the distodorsal tooth; the distal superior margin of the palm in major chela bearing one acute projecting; posterolateral margins of the telson each produced in one acute tooth.

**Distribution** The South China Sea; Timor in Indonesia; southern Philippines; Red Sea; eastern Africa; Réunion; Mauritius; Australia.

### 2.15.11 *Synalpheus hastilicrassus* Coutière, 1905 (Fig. 2.159)

*Synalpheus hastilicrassus* Coutière, 1905: 875,Pl. 72, fig. 12; De Man, 1911: 263, fig. 45; Banner DM & Banner AH, 1975: 353, fig. 21; Wang & Sha, 2015: 381, fig. 17.

*Synalpheus hastilicrassus acanthitelsoniformis* De Man, 1920: 108; 1922: 29, fig. 14.

*Synalpheus acanthitelsonis* Coutière, 1905: 875, fig. 13; De Man, 1911: 256, fig. 46. Gravely, 1930: 77, Pl. I, fig. 1; Banner AH & Banner DM, 1966a: 58, fig. 18.

**Material Examined** CL 3.2–6.1 mm, TL 6.8–12.1 mm. MBM 270350, 1ovigerous♀, Beibu Gulf, Sta. 6167, depth: 57 m, Zhang Bao-lin Coll., 8 October 1960; MBM 270351, 1♂, Beibu Gulf, Sta. 6167, depth: 57 m, Zhang Bao-lin Coll., 8 October 1960; MBM 270076, 1ovigerous♀, Jinqing Island, Xisha Islands, 23 May 1974; MBM 270709, 1ovigerous♀1♂, Linchang, 3 December 1990; MBM 270599, 1ovigerous♀1♂, Beibu Gulf, Sta. 7502, depth: 32 m, 15 April 1962; MBM 270608, 2♂, Beibu Gulf, Sta. 7305, depth: 64 m, 18 October 1965; MBM 107764, 2ovigerous♀3♂, Yalong Bay, Yezhu Island, depth: 6–11 m, 20 November 1990; MBM 107766, 1♂, Yalong Bay, Yezhu Island, Neumann Coll., 17 November 1990; MBM 159244, 1ovigerous♀1♂, Sanya, Dadonghai, depth: 7–10 m, Li Xin-zheng Coll., 21 November 1997; MBM 107785, 1♂, Shalao, 4 April 1992; MBM 158489, 1ovigerous♀; MBM 158728, 1♂, Jinqing Island, Xisha Islands, Fan Zhen-gang & Xu Jie-shan Coll.,

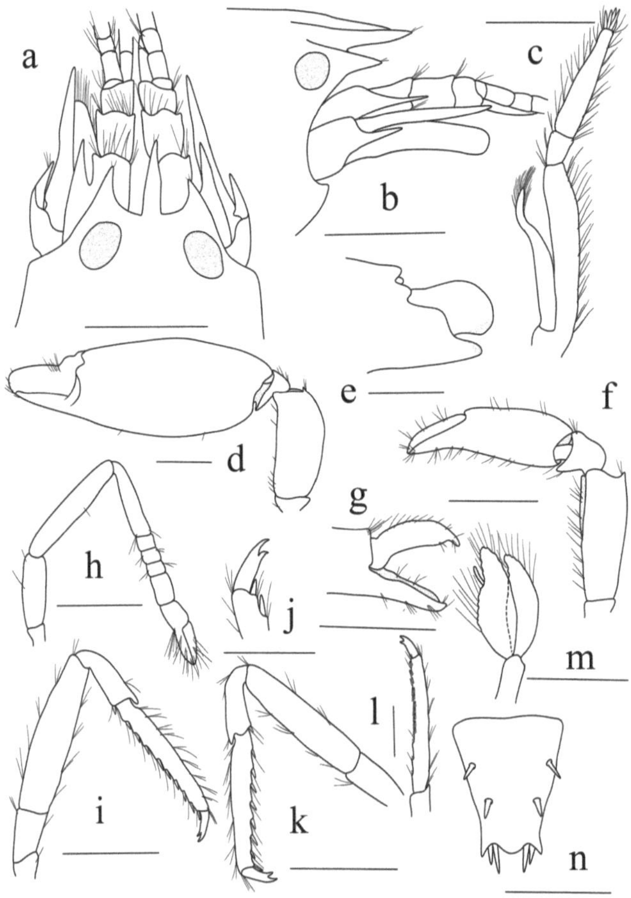


**Fig. 2.159** *Synalpheus hastilicrassus* Coutière, 1905. MBM 159279, female, CL 4.1 mm, **a**, anterior part of carapace, dorsal view; **b**, same, lateral view; **c**, third maxilliped; **d**, major cheliped; **e**, same, fingers; **f**, minor cheliped; **g**, same, fingers; **h**, second pereiopod; **i**, third pereiopod; **j**, same, distal part of propodus and dactylus; **k**, fourth pereiopod; **l**, propodus and dactylus of fifth pereiopod; **m**, second pleopod; **n**, telson (scale bar: **a**–**i**, **k**–**n** = 1 mm; **j** = 0.5 mm)

25 April 1958; MBM 159299, 3ovigerous♀4♂; MBM 270600, 1ovigerous♀2♂, Beibu Gulf, Sta. 7403, depth: 38 m, Sun Fu-zeng Coll., 21 August 1962; MBM 159310, 3ovigerous♀2♂ and 1 immature specimen, Nanhai, Xincun, depth: 1–3 m, 26 March 1992; MBM 159279, 4ovigerous♀, Yezhu Island, 27 March 1992; MBM 158223, 2♂, Liu Rui-Yu Coll., 26 December 1954; MBM 159206, 1ovigerous♀, Sanya, Dadonghai, depth: 5–9 m, Li Xin-zheng Coll., 3 December 1997; MBM 158541, 2ovigerous♀, Yalong Bay, Yezhu Island, Li Xin-zheng Coll., 19 March 1992; MBM 158499, 1ovigerous♀.

**Diagnosis** Carapace smooth, glabrous. Rostrum about 3.2 times as long as wide at base, acute tip overreaching end of first antennular segment. Orbital hoods shorter than rostrum, longer than wider at base. Pterygostomial corner produced into acute angle. Cardiac notch well developed.

Telson about 1.2 times as long as wide at base; dorsal surface with two pairs of spines, inserted approximately at 1/3 and 2/3 length of telson; posterior margin strongly produced, fringed with long setae, posterolateral angle each with two pairs of large spines, lateral slightly shorter than mesial, posterolateral margins each produced into large acute tooth, slightly shorter than or sub-equal to contiguous spines.

Antennular peduncle moderately slender, with distally acute stylocerite, latter hardly reaching middle of second antennular segment; second segment about 1.5 times as long as wide at base; visible part of first segment longer than second and less twice as long as third. Basicerite with or without distodorsal tooth, distolateral tooth variously from distinctly shorter than to sub-equal to stylocerite; carpocerite far overreaching end of third antennular segment; scaphocerite blade moderately narrow, reaching middle of third antennular segment, lateral spine developed, overreaching end of antennular peduncle, much shorter than carpocerite.

Third maxilliped far beyond antennular peduncle and carpocerite when extended; antepenultimate segment longest; penultimate segment about 1.5 times as long as wide at base; tip of ultimate segment bearing crown of five long spine-like setae, with transverse rows of setae along ventral margin.

Major cheliped with dactylus as long as fixed finger, distally slightly curved. Palm about 2.3 times as long as dactylus, distodorsal of palm with at most an obtuse projecting. Carpus cup-shaped, with setae distodorsally. Merus about 3 times as long as wide at base, distodorsal margin with acute triangular tooth. Ischium stout. Minor chela with dactylus and fixed finger tapering, superior surface of dactylus bearing row of moderately short setae, extending almost full length of dactylus distal single tooth. Palm about 1.4 times as long as fingers. Carpus cup-shaped, proportionally longer that of major cheliped. Merus about 4 times as long as wide at base, distodorsal margin with one acute triangular tooth.

Second pereiopod with fingers about 1.5 times as long as palm. Carpus five-segmented, segment ratio subequal to 5: 1: 1: 1: 1.6, with first segment slightly longer than sum of last four segments.

Third pereiopod moderately slender, dactylus about 1/4 length of propodus, slightly curved to propodus, biunguiculate, extensor tooth slightly longer and slightly wider at base than flexor tooth. Propodus with 8–9 spines along ventral margin plus one distal pair of spines near dactylus. Carpus about 1/3 length of merus, with dorsal margin extended into an obtuse tooth, ventral margin with one distal spine. Merus slightly longer than propodus, about 5.3 times as long as wide at base. Fourth pereiopod generally similar to third; ventral margin of propodus with 7 spines plus one distal pair of spines. Fifth pereiopod similar to third and fourth pereiopods, ventral margin of carpus without distal one spine; ventral margin of propodus with 9 transversal rows of spine-like setae and 3 spines.

**Remarks** This species is rather complex (Banner & Banner, 1975). It is readily recognized by the posterior lateral margins of the telson producing into large tooth, slightly shorter than the contiguous spines. The present specimens are generally agreed with the re-descriptions (Banner & Banner, 1975; Chace, 1988), and the marked variation is found in the distodorsal tooth of the basicerite (present or absent). This variation lies within that recorded by Banner & Banner.

**Distribution** China Seas; Maldive Archipelago; Gulf of Manaar; Indonesia; Philippiines, Caroline and Marshall Islands; Fiji; Australia.

### 2.15.12 *Synalpheus iocasta* De Man, 1909 (Figs. 2.160 and 2.161)

*Synalpheus iocasta* De Man, 1909a: 119; 1911: 235, fig. 33; Banner DM & Banner AH, 1975: 368, fig. 24i–n; Wang & Sha, 2015: 383, figs. 18–19.

**Material Examined** MBM 270422, 1♂, CL 4.1 mm, TL 11.5 mm, Beibu Gulf, Sta. 7802, depth: 67 m, Sun Fu-zeng Coll., 10 April 1962; MBM 270587, 1♂, CL 4.2 mm, TL 11.3 mm, Beibu Gulf, Sta. 6266, depth: 50.5 m, Shen Shou-peng Coll., 13 July 1960; MBM 270423, 1♂, CL 4.1 mm, TL 11.1 mm, Beibu Gulf, Sta. 7303, depth: 34 m, 20 April 1960.

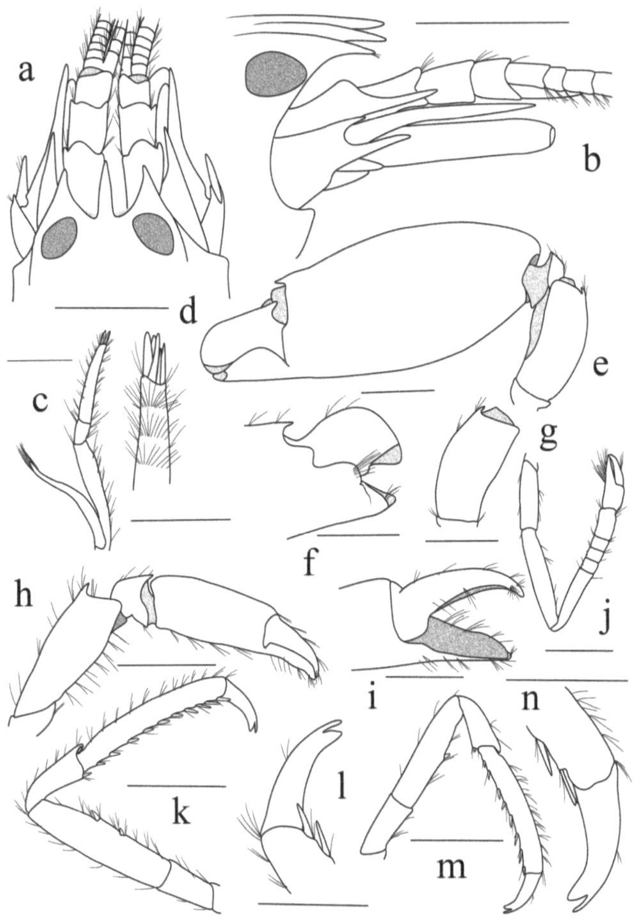


**Fig. 2.160** *Synalpheus iocasta* De Man, 1909. MBM 270587, female, CL 4.1 mm, **a**, anterior part of carapace, dorsal view; **b**, same, lateral view; **c**, third maxilliped; **d**, same, tip; **e**, left major cheliped; **f**, same, fingers; **g**, merus of major cheliped; **h**, right minor cheliped; **i**, same, fingers; **j**, second pereiopod; **k**, third pereiopod; **l**, same, distal part of propodus and dactylus; **m**, fourth pereiopod; **n**, same, distal part of propodus and dactylus (scale bar: **a**–**c**, **e**–**k**, **m** = 1 mm; **d**, **l**, **n** = 0.5 mm)

**Description** Carapace smooth, glabrous. Rostrum about 2.8 times as long as wide at base, acute tip not reaching end of first antennular segment. Orbital hoods slightly shorter than rostrum. Pterygostomial corner produced into acute angle. Cardiac notch well developed.

Telson about 1.2 times as long as wide at base; dorsal surface with two pairs of moderately large spines, inserted approximately at 1/2 and 5/7 length of telson; posterior margin slightly produced, fringed with long setae, posterolateral angle each with two pairs of spines, lateral about 1/4 length of mesial, posterolateral margins not produced into acute tooth.

Antennular peduncle moderately slender, with distally acute stylocerite, latter hardly reaching middle of second antennular segment; second segment about 1.25 times as long as wide at base; visible part of first segment longer than second and about twice as long as third. Basicerite with acute distodorsal tooth, about 1/3 length of distolateral tooth, distolateral tooth distinctly shorter than stylocerite, hardly reaching end of first antennular segment; carpocerite far overreaching end of third antennular segment; scaphocerite blade moderately narrow, reaching 3/4 length of

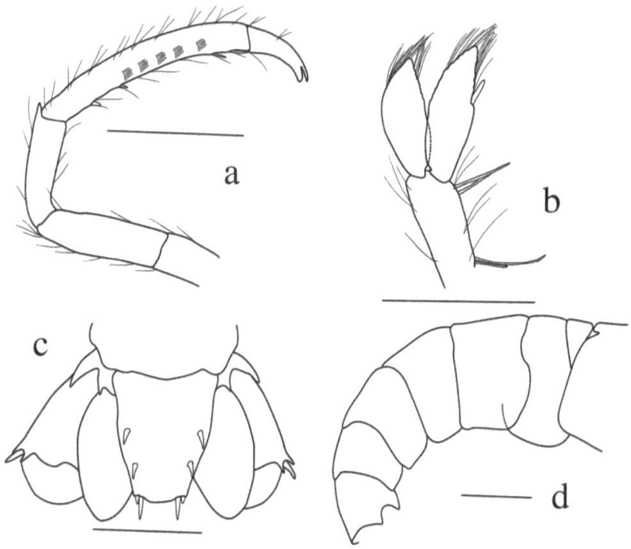


**Fig. 2.161** *Synalpheus iocasta* De Man, 1909. MBM 270587, female, CL 4.1 mm, **a**, fifth pereiopod; **b**, second pleopod; **c**, uropods and telson **d**, abdominal segments, lateral view (scale bar = 1 mm)

third antennular segment, lateral spine developed, overreaching end of antennular peduncle, much shorter than carpocerite.

Third maxilliped far beyond antennular peduncle and carpocerite when extended; antepenultimate segment longest; penultimate segment about 1.3 times as long as wide at base; tip of ultimate segment bearing crown of five long spine-like setae, with transverse rows of setae along ventral margin.

Major cheliped with dactylus near as long as fixed finger, distally slightly curved. Palm about 2.5 times as long as dactylus, distodorsal of palm with one acute projecting, straightly, not curved to dactylus. Carpus cup-shaped, with setae distodorsally. Merus about 2.5 times as long as wide at base, distodorsal margin with one acute triangular tooth. Ischium stout. Minor chela with dactylus and fixed finger tapering, distal single tooth. Palm about 1.7 times as long as fingers. Carpus cup-shaped, proportionally longer that of major cheliped. Merus about 3.6 times as long as wide at base, distodorsal margin with one acute triangular tooth.

Second pereiopod with fingers about 1.5 times as long as palm. Carpus five-segmented, segment ratio subequal to 6: 1: 1: 1: 2.5, with first segment longer than sum of last four segments.

Third pereiopod moderately slender, dactylus about 2/7 length of propodus, slightly curved to propodus, biunguiculate, extensor tooth near as long as flexor tooth, notch between extensor and flexor tooth V-shaped, flexor tooth much wider at base than extensor tooth. Propodus with nine spines along ventral margin plus one distal pair of spines near dactylus. Carpus less 1/2 length of merus, with dorsal margin extended into an obtuse tooth, ventral margin with one distal spine. Merus shorter than propodus, about 5 times as long as wide at base, ventral margin with 2 spines. Fourth pereiopod generally similar to third; merus shorter than propodus, ventral margin with one moderately large spine; ventral margin of propodus with seven spines plus one distal pair of spines. Fifth pereiopod similar to third and fourth pereiopods, ventral margin of carpus without distal one spine; ventral margin of propodus with five transversal rows of spine-like setae and four spines.

**Remarks** This species is generally similar to *Synalpheus pococki* Coutière, 1898, but can be distinguished by the propodus of the third pereiopods (De Man, 1911; Banner & Banner, 1975). De Man (1909) originally described this species. The present specimens are generally agreed with the original description, except the following two characters: the posterolateral margins of the telson not produced into acute teeth rather than "the outer angles of the posterior margin are spiniform, half as long as the contiguous short spinules"; the posterior margin of the telson only slightly produced rather than "the posterior margin rather prominent in the middle, almost semicircular". Those two characters are variable in this genus, and not wise to take it as a new species.

**Distribution** The South China Sea; Indonesia; Australia; Philippines.

### 2.15.13 *Synalpheus iphinoe* De Man, 1909 (Figs. 2.162, 2.163, and 2.164)

*Synalpheus iphinoë* De Man, 1909a: 116; Banner DM & Banner AH, 1985: 43; Wang & Sha, 2015: 386, figs. 20–22.
*Synalpheus jedanensis* De Man, 1909a: 117.

**Material Examined** CL 3.1–5.9 mm, TL 6.7–11.9 mm. MBM 270449, 52ovigerous♀, 10 immature specimens, Beibu Gulf, Sta. 6277, depth: 35 m, Shen Shou-peng Coll., 19 October 1960; MBM 270447, 30ovigerous♀17♂, Beibu Gulf, Sta. 6277, depth: 13.5 m, Fan Zhen-gang

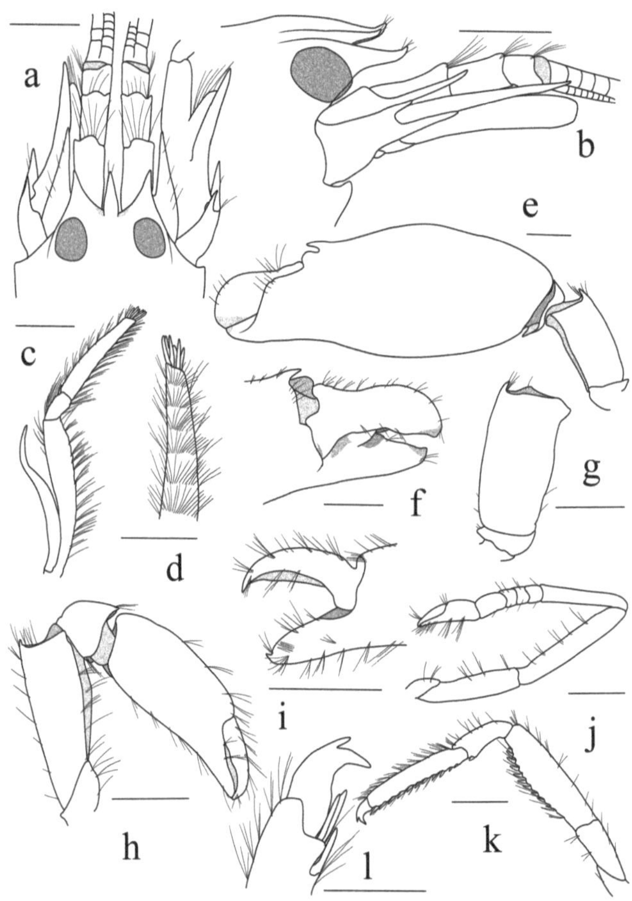


**Fig. 2.162** *Synalpheus iphinoe* De Man, 1909. MBM 158695, female, CL 5.7 mm, **a**, anterior part of carapace, dorsal view; **b**, same, lateral view; **c**, third maxilliped; **d**, same, tip; **e**, right major cheliped; **f**, same, fingers; **g**, merus and ischium of major cheliped; **h**, left minor cheliped; **i**, same, fingers; **j**, second pereiopod; **k**, third pereiopod; **l**, same, distal part of propodus and dactylus (scale bar: **a**–**c**, **e**–**k** = 1 mm; **d**, **l** = 0.5 mm)

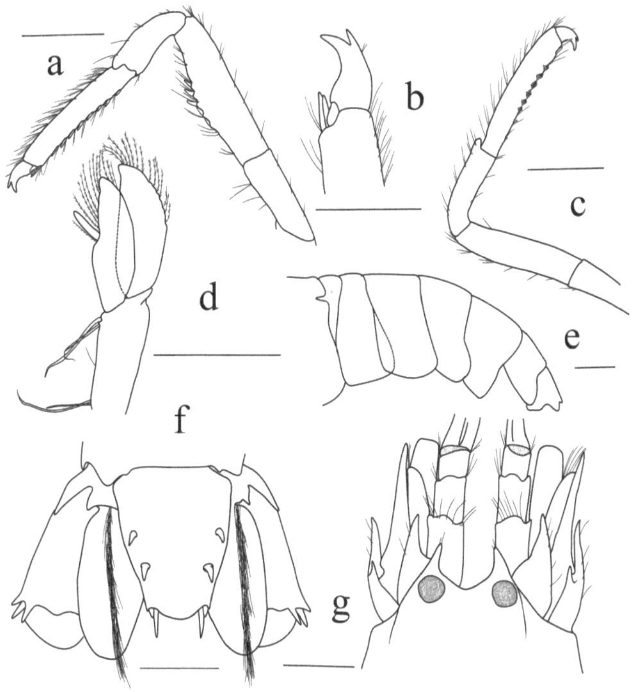


**Fig. 2.163** *Synalpheus iphinoe* De Man, 1909. MBM 158695, female, CL 5.7 mm, **a**, fourth pereiopod; **b**, same, distal part of propodus and dactylus; **c**, fifth pereiopod; **d**, second pleopod; **e**, abdominal segment; **f**, uropods and telson. MBM 270539, female, CL, 5.9 mm, **g**, anterior part of carapace, dorsal view (scale bar: **b** = 0.5 mm; **a**, **c–g** = 1 mm)

Coll., 9 July 1960; MBM 270595, 3ovigerous♀6♂, 14 immature specimens, Beibu Gulf, Sta. 6277, depth: 31.3 m, Fan Zhen-gang Coll., 11 May 1960; MBM 270584, 3♂, Beibu Gulf, Sta. 6283, depth: 25 m, Fan Zhen-gang Coll., 11 February 1960; MBM 270199, 1♂, Beibu Gulf, Sta. 6282, depth: 28 m, Fan Zhen-gang Coll., 11 February 1960; MBM 270598, 1♂, Beibu Gulf, Sta. 6276, depth: 30 m, Fan Zhen-gang Coll., 5 December 1959; MBM 270239, 4ovigerous♀3♂, Beibu Gulf, Sta. 6284, depth: 18.5 m, Fan Zhen-gang Coll., 18 May 1960; MBM 270581, 3ovigerous♀4♂, Beibu Gulf, Sta. 6237, depth: 50.5 m, Fang Xiu-bin Coll., 14 April 1960; MBM 270194, 1♂, Beibu Gulf, Sta. 7502, depth: 32 m, Sun Fu-zeng Coll., 15 January 1962; MBM 270585, 2♂, Beibu Gulf, Sta. 6267, depth: 32 m, Zhang Bao-lin Coll., 5 November 1960; MBM 270439, 4ovigerous♀1♂, Beibu Gulf, Sta. 6280, depth: 27 m, Ma Xiu-tong Coll., 5 December 1959; MBM 270451, 1ovigerous♀1♂, Beibu Gulf, Sta. 6261, depth: 42 m, Fan Zhen-gang Coll., 4 December 1959; MBM 270527, 17 specimens, Beibu Gulf, Sta. 7603, depth: 44 m, Sun Fu-zeng Coll., 18 August 1962; MBM 270539, 3♀(two ovigerous)2♂, Beibu Gulf, Sta. 7602, depth: 26 m, 18 August 1962; MBM 270583, 1ovigerous♀3♂, Beibu Gulf, Sta. 6269, depth: 30 m, Zhang Bao-lin Coll., 6 November 1960; MBM 270597, 2ovigerous♀7♂, Beibu Gulf, Sta. 6282, depth: 28 m, Fan Zhen-gang Coll., 9 July 1960; MBM 270583, 4♀(three ovigerous)27♂, Beibu Gulf, Sta. 6277, depth: 35 m, Fan Zhen-gang Coll., 11 February 1960; MBM 270579, 11♀(10ovigerous)9♂, Beibu Gulf, Sta. 6276, depth: 29 m, 6 November 1960; MBM 270126, 49♂, Beibu Gulf, Sta. 6277, depth: 35 m, Fan Zhen-gang & Xu Feng-shan Coll., 9 March 1960; MBM 158690, 1ovigerous♀2♂, Nansha Islands, Sta. 55, depth: 46 m, 8 June 1960; MBM 158364, 1♂, Nansha Islands, Sta. 64, depth: 44 m, 9 June 1960; MBM 270445, 23ovigerous♀27♂, Beibu

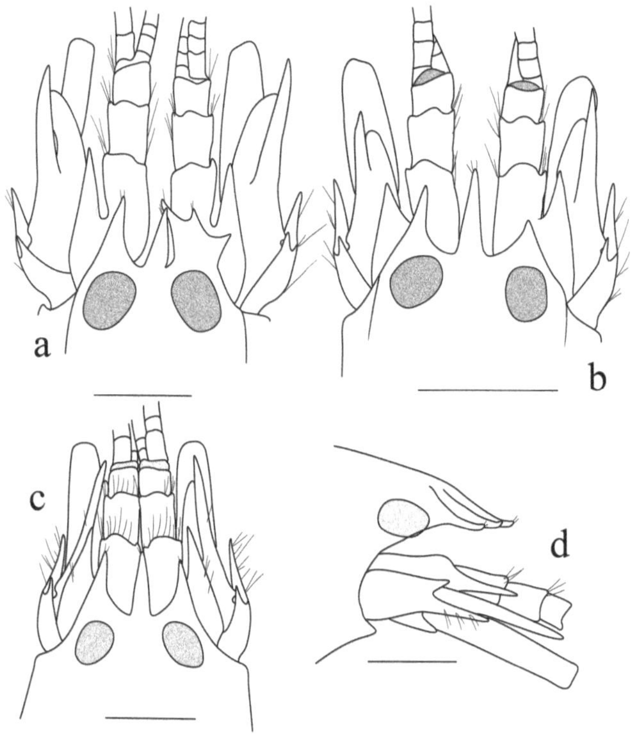


**Fig. 2.164** *Synalpheus iphinoe* De Man, 1909. MBM 158690, male, CL 4.9 mm, **a**, anterior part of carapace, dorsal view. MBM 270126, male, CL 4.5 mm, **b**, anterior part of carapace, dorsal view. MBM 270126, male, CL 5.1 mm, **c**, anterior part of carapace, dorsal view; **d**, same, lateral view (scale bar = 1 mm)

Gulf, Sta. 6277, depth: 35 m, Shen Shou-peng Coll., 19 October 1960; MBM 270582, 2ovigerous♀4♂, Beibu Gulf, Sta. 6249, depth: 31 m, Zhang Bao-lin Coll., 7 November 1960; MBM 158846, 1♂, Nansha Islands, Sta. Z 13, depth: 50 m, 28 April 1986; MBM 270440, 1♂, Beibu Gulf, Sta. 7604, depth: 58 m, Sun Fu-zeng Coll., 12 January 1962; MBM 270377, 1♂, Beibu Gulf, Sta. 6265, depth: 66.3 m, Fan Zhen-gang Coll., 14 July 1960.

**Description** Carapace smooth, glabrous. Conical rostrum about 3 times as long as wide at base, acute tip overreaching middle of first antennular segment. Orbital hoods slightly longer than rostrum, diverge slightly outward, both rostrum and orbital hoods distinctly curved upturned at tips. Pterygostomial corner produced into acute angle; cardiac notch well developed.

Telson about 1.3 times as long as wide at base; dorsal surface with two pairs of spines, inserted approximately at 2/5 and 3/5 length of telson; posterior margin strongly produced, fringed with long setae, posterolateral angle each with two pairs of spines, lateral about 1/3 length of mesial, posterolateral margins each produced into acute small tooth, but distinctly shorter than lateral spines. Media-dorsal of endopod bearing dense setae, coves all length of it.

Antennular peduncle slender, with distally acute stylocerite, latter reaching about 1/4 length of second antennular segment; second segment about 1.7 times as long as wide at base; visible part of first segment longer than second and more than twice as long as third. Basicerite with acute distodorsal tooth, less 1/2 length of distolateral tooth, latter distinctly shorter than stylocerite, not reaching end of first antennular segment; carpocerite overreaching end of third antennular segment; scaphocerite blade moderately narrow, reaching about 1/2 length of third antennular segment, lateral spine developed, overreaching end of antennular peduncle, hardly shorter than carpocerite.

Third maxilliped far beyond antennular peduncle and carpocerite when extended; antepenultimate segment longest; penultimate segment about 1.7 times as long as wide at base; tip of ultimate segment bearing crown of moderately long six spine-like setae, with transverse rows of long setae along ventral margin.

Major cheliped with dactylus slightly longer than fixed finger. Palm about 2.3 times as long as dactylus, distodorsal of palm with one strong acute projecting, slightly upturned at base and curved to dactylus at tip. Carpus cup-shaped, with setae distodorsally. Merus about 2.6 times as long as wide at base, distodorsal margin with one acute triangular tooth. Minor chela with dactylus and fixed finger tapering, distal single tooth. Palm about twice as long as fingers. Carpus cup-shaped, proportionally longer that of major cheliped. Merus about 3 times as long as wide at base, distodorsal margin with one acute triangular tooth.

Second pereiopod with fingers about 1.3 times as long as palm. Carpus five-segmented, segment ratio subequal to 7: 1: 1: 1: 2.7, with first segment distinctly longer than sum of last four segments.

Third pereiopod moderately slender, dactylus about 1/5 length of propodus, slightly curved to propodus, biunguiculate, extensor tooth about 1/3 length of flexor tooth, notch between extensor and flexor tooth V-shaped, flexor tooth much wider at base than extensor tooth. Propodus with 11 spines along ventral margin plus one distal pair of spines near dactylus. Carpus less 1/2 length of merus, with dorsal margin extended into an obtuse tooth, ventral margin with one distal spine. Merus longer than propodus, about 4 times as long as wide at base, ventral margin with 10 moderately large spines. Fourth pereiopod generally similar to third; merus slightly longer than propodus, ventral margin with five moderately large spines; ventral margin of propodus with 8 spines plus distal one pair spines. Fifth pereiopod similar to third and fourth pereiopods, but ventral margin of carpus without one distal spine; merus slightly shorter than propodus, ventral margin without spines; ventral margin of propodus with 6 transversal rows of spine-like setae and three spines.

**Remarks** *S. iphinoe* and *S. jedanensis* are described by De Man (1909) based on 7 and 6 species, respectively. They are morphologically very similar to *S. neomeris*, and separated from each other mainly by the minor chela (De Man, 1909). The morphological characters used by De Man to separate the three species are: setae on the frontal teeth; the proportions of three segments of the antennular peduncle; the proportions of the small chela; the number of spines on the merus of the third pereiopods; the posterolateral margins each produced to spines or not. However some of these characters are variable in most species of alpheids, so they put *S. iphinoe* and *S. jedanensis* under the name *S. iphinoe* on the basis of page priority. Banner & Banner (1985) present "*S. iphinoe* be separated from *S. neomeris* on basis of small size, the probable heavier small chela, and the upturned tips to the frontal teeth". The present materials have following differences from *S. neomeris*: the total length of the specimens was varied from 6.5 to 13.2 mm, contrasted to the longest specimen of *S. neomeris* was about 30 mm; the fingers of the minor chela in *S. neomeris* subequal to palm or slightly shorter than it, in *S. iphinoe* the ratio of fingers/palm length of minor chela was about 0.5; the frontal teeth of part of our specimens were upturned at tips. The present specimens slightly variable in the shape of the rostrum, but this variation was not enough to separate them into two species. Additionally, three of our specimens were aberrant at the frontal teeth: one of them lost the rostrum; the right orbital hood bearing three teeth at tip; the right orbital hoods of the other specimen was reduced.

**Distribution** Beibu Gulf, South China Sea; Sapeh-Strait; Southern Molo-Strait; Banda-anchorage; Roma-Island; East cost of Aru-islands.

### 2.15.14 *Synalpheus jinqingensis* Wang & Sha, 2015 (Figs. 2.165, 2.166, and 2.167)

*Synalpheus jinqingensis* Wang & Sha, 2015: 390, figs. 23–25.

**Material Examined** MBM 283385, 1♀, CL 4.6 mm, TL 12.5 mm, Jinqing Island, Xisha Islands, 16°27′57″N, 111°44′17″E, Fan Zhen-gang & Xu Jie-shan Coll., 25 April 1958; MBM 283385, 1♂, CL 4.5 mm, TL 10.5 mm, Jinqing Island, Xisha Islands, 16°27′57″N; 111°44′17″E, Fan Zhen-gang & Xu Jie-shan Coll., 25 April 1958; MBM 283386, 1♀, CL 4.1 mm, TL 10.5 mm, Jinqing Island, Xisha Islands, 16°27′57″N; 111°44′17″E, Fan Zhen-gang & Xu Jie-shan Coll., 25 April 1958.

**Description** Carapace smooth, glabrous. Rostrum 1.7 as long as wide at base, with lateral margins slightly convex, rounded tip bearing two short setae, reaching about 2/3 length of visible part of first antennular segment. Orbital hoods about 1.5 times as wide at base as rostrum, distinctly shorter than rostrum, with rounded tip bearing two short setae; notches between orbital hoods moderately deep, V-shaped. Pterygostomial corner produced into blunt acute angle.

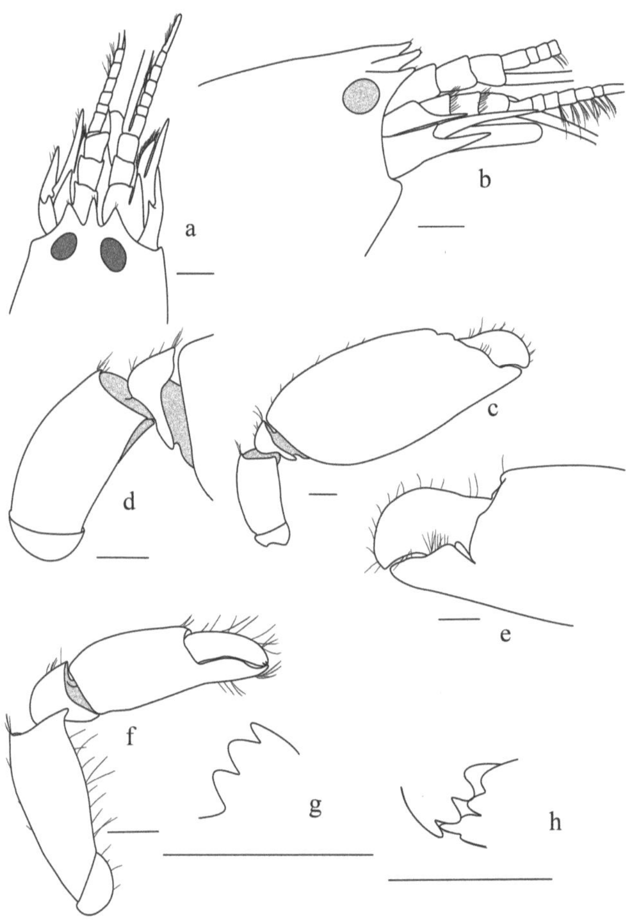


**Fig. 2.165** *Synalpheus jinqingensis* Wang & Sha, 2015. MBM 283385, female, CL 4.6 mm, **a**, anterior part of carapace, dorsal view. **b**, same, lateral view; **c**, major cheliped; **d**, same, merus and ischium; **e**, same, fingers; **f**, minor cheliped; **g**, same, tip of fixed finger; **h**, same, tips of fingers (scale bar = 0.5 mm)

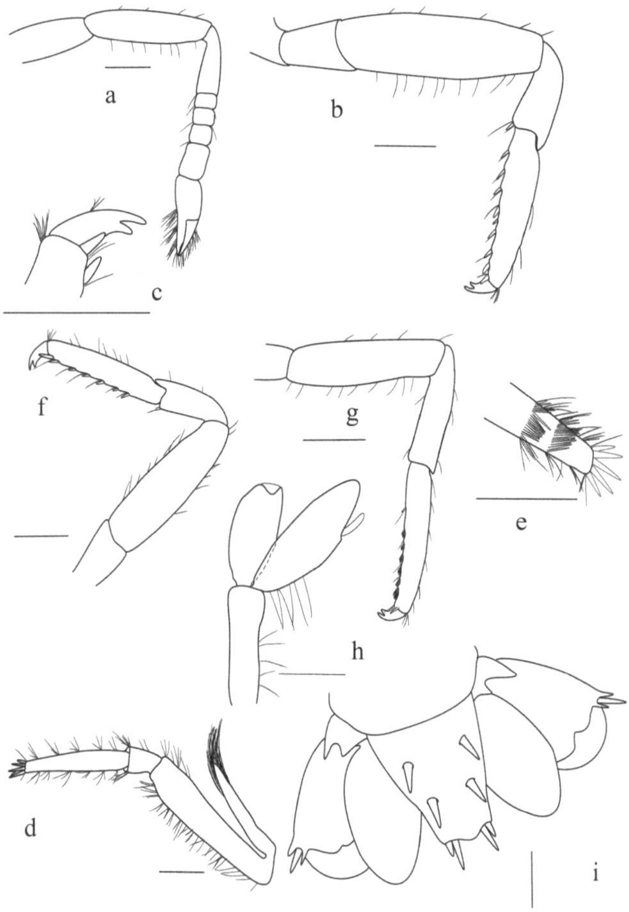


**Fig. 2.166** *Synalpheus jinqingensis* Wang & Sha, 2015. MBM 283385, female, CL 4.6 mm, **a**, second pereiopod; **b**, third pereiopod; **c**, same, distal part of propodus and dactylus; **d**, third maxilliped; **e**, same, tip; **f**, fourth pereiopod; **g**, fifth pereiopod; **h**, second pleopod; **i**, uropods and telson (scale bar = 0.5 mm)

Abdominal segments smooth, glabrous. Telson about 1.1 times as long as wide at base; lateral margin slightly concave posteriorly; dorsal surface with median groove, with two pairs of large spines, inserted approximately at 1/4 and 1/2 length of telson; posterior margin strongly convex, fringed with long setae, posterolateral angle each with two pairs of spines, lateral 1/2 length of mesial, posterolateral margins each produced into acute tooth, but distinctly shorter than lateral spine.

Antennular peduncle slender, with distally acute stylocerite, latter overreaching middle of second antennular segment, about 1/6-1/5 length of second; second segment 1.3 times as long as wide; visible part of first segment slightly longer than second, latter sub-equal to third. Antenna with basicerite with acute distodorsal tooth, 1/7 length of distolateral tooth, latter hardly overreaching middle of third segment, distinctly longer than stylocerite; carpocerite beyond end of antennular peduncle by 1/3 length of third segment; scaphocerite blade narrow, slightly longer than antennular peduncle, lateral spine with lateral margin slightly concave posteriorly, slightly shorter than carpocerite.

Mouthpart not dissected. Third maxilliped far beyond antennular peduncle and carpocerite when extended; antepenultimate segment longest, about 4.6 times as long as wide; penultimate segment about 1.3 times as long as wide at base; tip of ultimate segment bearing crown of five

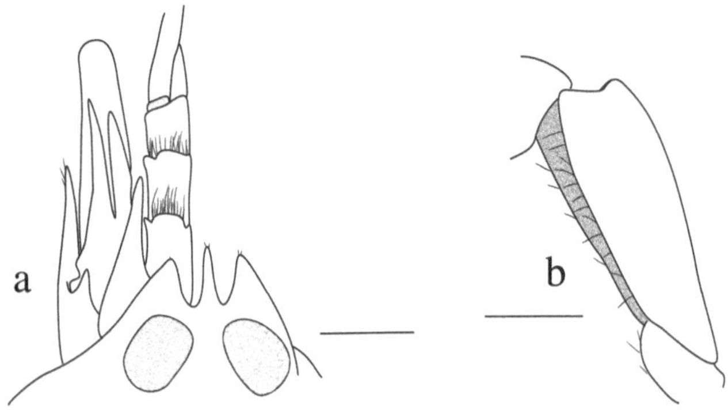


**Fig. 2.167** *Synalpheus jinqingensis* Wang & Sha, 2015. MBM 283385, female, CL 4.5 mm, **a**, anterior part of carapace, dorsal view; **b**, merus of minor cheliped (scale bar = 0.5 mm)

spin-like long setae, with several brush consisting of six-eight transverse rows of setae along ventral margin.

Major cheliped with chela cylindrical, not a bit compressed or twisted. Dactylus longer than fixed finger. Palm about 3.2 times as long as dactylus, distodorsal margin of palm almost without projecting. Carpus cup-shaped, with long setae distodorsally. Merus 2.2 times as long as wide at base, distodorsal margin ending in acute tooth. Minor chela with dactylus spatulate, with flexor margin slightly convex, distal three similarly teeth; fixed finger spatulate, with flexor surface obliquely convex, distal four teeth, with one longest, two middle and one shortest. Palm about 1.6 times as long as fingers. Carpus cup-shaped proportionally longer that of major cheliped. Merus about 3.2 times as long as wide, longer than that of major cheliped, distodorsal margin ending in acute tooth, ventral margin bearing short setae.

Second pereiopod with fingers slightly shorter than palm. Carpus five-segmented, segment ratio subequal to 4 1 :1 :1 : 2, with first segment shorter than sum of last four segment, segments 2–4 wider than long; dorsal margin of palm convex.

Third pereiopod stout, dactylus around 1/6 propodus length, biunguiculate, moderately slender, extensor tooth slightly longer than flexor tooth, sub-wide flexor tooth at base; notch between extensor and flexor tooth V-shaped. Propodus with usual eight spines along ventral margin plus one distal pair of spines near dactylus. Carpus about 1/3 merus length, with dorsal margin extended into an obtuse tooth, ventral margin with one distal spine. Merus longer than propodus, about four times as long as wide at base, without spines on ventral margin. Fourth pereiopod generally similar to third; but ventral margin of carpus without distal spine. Fifth pereiopod with merus about 3.5 times as long as wide; carpus about 5/7 merus length, ventral margin without distal spine; propodus slightly shorter than merus, with one moveable spine and seven transversal rows of setae along ventral margin plus one pair of spines near dactylus.

**Distribution** Xisha Islands, the South China Sea.

**Remarks** *S. jinqingensis* can be unequivocally distinguished from other species by the tip of spatulate dactylus and fixed finger of minor chela armed with three and two teeth, respectively. There are four species in Indo-Pacific waters bearing teeth on tip of fingers of minor chela, including *S. sciro, S. pescadorensis, S. antenor, S. liui*. The differences between S. jinqingensis and its similar species are discussed in Wang & Sha (2015).

### 2.15.15 *Synalpheus liui* Wang & Sha, 2015 (Figs. 2.168, 2.169, and 2.170)

*Synalpheus liui* Wang & Sha, 2015: 394, figs. 26–28.

**Material Examined** MBM 283382, non-ovigerous individual, CL 3.2 mm, TL 8.1 mm, Zhubi Jiao, Nansha Islands, Sta. 9, 10°54.997′N, 114°05.813′E, depth: 3 m, from coral reef, Wang Yong-liang Coll., 5 May 2004. MBM 283382, non-ovigerous individual, CL 4.3 mm, TL 9.6 mm, Zhubi Jiao, Nansha Islands, Sta. 9, 10°54.997′N, 114°05.813′E, depth: 3 m, from coral reef, Wang Yong-liang Coll., 5 May 2004.

**Description** Carapace smooth, glabrous. Rostrum with lateral margins slightly convex, rounded tip reaching 1/2 length of visible part of first antennular segment, bearing two short setae, slightly upturned. Orbital hoods about 1.5 as wide at base as rostrum, length almost equal to rostrum, rounded tip bearing two short setae, slightly upturned. Pterygostomial corner produced into blunt acute angle.

Abdominal segments smooth, glabrous, posterolateral margin of sixth pleura each with one blunt tooth, posterior corner of first pleura bluntly produced ventrally; second to fifth pleura with lower margin produced into acute angle. Telson about 1.1 as long as wide at base; lateral margins slightly concave posteriorly; dorsal surface slightly convex, with median groove, with two pairs of large spines, inserted approximately at 2/5 and 1/2 length of telson; posterior margin not produced, slightly concave, fringed with long setae, posterolateral angle each with two pairs of spines, lateral 1/2 length of mesial, posterolateral margins forming right angles. Eyes concealed in lateral and dorsal view.

Antennular peduncle with distally acute stylocerite, latter hardly reaching distal margin of first antennular segment; second segment distinctly longer than wide; visible part of first segment longer than second and about twice as long as third segment. Antenna with basicerite with acute

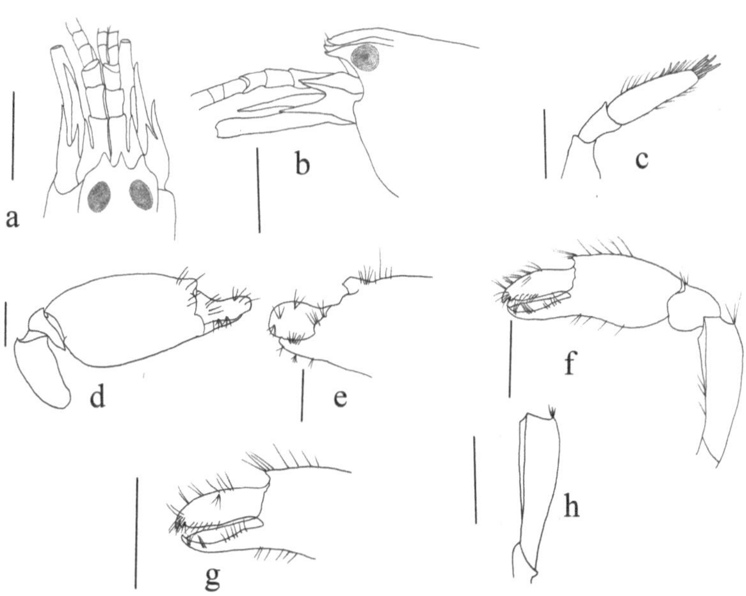


**Fig. 2.168** *Synalpheus liui* Wang & Sha, 2015. MBM 283382, male, CL 3.2 mm, **a**, anterior part of carapace, dorsal view; **b**, same, lateral view; **c**, left third maxilliped, lateral view; **d**, major cheliped, dorsal view; **e**, same, distal part of palm and fingers, lateral view; **f**, minor cheliped, lateral view; **g**, same, distal part of palm and fingers, lateral view; **h**, merus of minor cheliiped, lateral view (Scale bar: **a**, **b**, **d**–**h** = 1 mm; **c** = 0.5 mm)

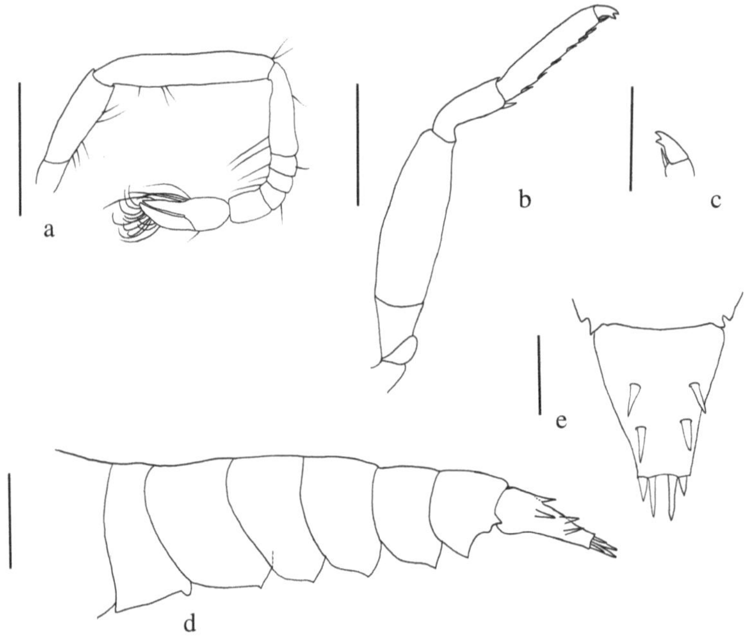


**Fig. 2.169** *Synalpheus liui* Wang & Sha, 2015. MBM 283382, male, CL 3.2 mm, **a**, left second pereiopod; **b**, left third pereiopod; **c**, same, dactylus **d**, abdominal segments and telson, lateral view; **e**, telson (Scale bar: **a**, **b**, **d** = 1 mm; **c**, **e** = 0.5 mm)

distodorsal tooth, 1/2 length of distolateral tooth, subequal to stylocerite, distolateral tooth distinctly longer than stylocerite, overreaching middle of second antennular segment; carpocerite beyond end of antennular peduncle by the length of third segment; scaphocerite blade narrow, reaching end of second antennular segment, lateral spine robust, with lateral margin slightly concave posteriorly, overreaching antennular peduncle, shorter than carpocerite.

Mouthparts not dissected. Third maxilliped far beyond antennular peduncle and carpocerite when extended; antepenultimate segment longest, about five times as long as wide; penultimate segment about 2.5 times as long as wide at base; tip of ultimate segment bearing crown of five-six spin-like long setae, with several brush consisting of six-eight transverse rows of setae along ventral margin. Origin of marginal setae on pleopod two exopod close to base.

Major cheliped with chela cylindrical, not a bit compressed or twisted. Dactylus distinctly longer than fixed finger. Palm about 2.6 times as long as dactylus, distodorsal margin produced into prominent tubercle with little subacute ventrally directed tooth. Carpus cup-shaped, with setae distodorsally. Merus 3.8 times as long as wide at base, distodorsal margin with acute triangular tooth, inferointernal margin with acute tooth. Minor chela with dactylus excavate and denticulate, with flexor margin slightly convex, distal with one large tooth medially, smaller tooth adjacently, lateral margin bearing row of long setae; fixed finger tapering distally, with flexor surface obliquely convex, distal single tooth, lateral margin bearing row of long setae. Palm 1.5 times as long as fingers. Carpus cup-shaped, proportionally longer than that of major cheliped. Merus about 4.4 times as long as wide, longer than that of major cheliped, distodorsal margin with acute triangular tooth.

Second pereiopod with fingers 1.2 times as long as palm. Carpus five-segmented, segment ratio subequal to 4.7:1:1:1:2.1, with first segment little shorter than sum of last four segments.

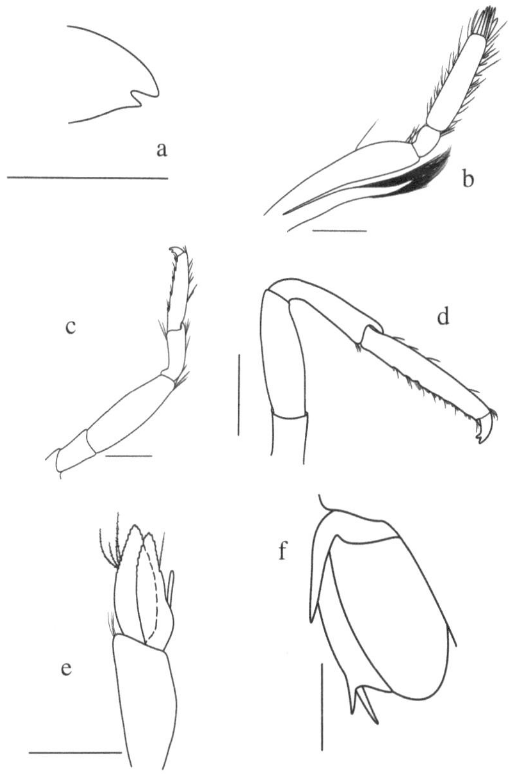


**Fig. 2.170** *Synalpheus liui* Wang & Sha, 2015. MBM 283382, male, CL 3.2 mm, **a**, tip of dactylus of minor chela; **b**, left third maxilliped; **c**, left fourth pereiopod; **d**, left fifth pereiopod; **e**, left second pleopod; **f**, left uropod. (Scale bar = 0.5 mm)

Third pereiopod stout, dactylus around 1/6 propodus length, biunguiculate, moderately slender, extensor tooth 1.5 times length of flexor tooth, thicker at base than flexor tooth, notch between extensor tooth and flexor tooth V-shaped. Propodus with usual seven spines along ventral margin plus one distal pair of spines near dactylus. Carpus about 1/2 merus length, with dorsal margin extended into an obtuse tooth, ventral margin with one long distal spine. Merus longer than propodus, about three times as long as wide, without spines on ventral margin. Fourth pereiopod generally similar to third; but ventral margin of carpus without distal spine. Fifth pereiopod stout, with merus about three times as long as wide, without spines on ventral margin; carpus about 7/9 length of merus, ventral margin without distal spine; propodus slightly longer than merus, with one moveable spine and seven-nine transverse rows of short setae along on ventral margin plus one distal pair of spines near dactylus, dactylus similar that of three and four pereiopod.

**Distribution** Zhubi Reef, Nansha Island.

**Remarks** *S. liui* can be unequivocally distinguished by the following characters: posterolateral margin of sixth pleura each with one blunt tooth; the palm of major cheliped with distodorsal margin produced into prominent tubercle with little subacute ventrally directed tooth; the longer distodorsal tooth of basicerite, which subequal to the stylocerite, and the longer tooth of uropods of protopodite, which was 1/2 length of outer uropod. There are another nine species in the Indo-Pacific waters with the sixth pleura was armed with two, three, four or seven teeth. The differences between *S. liui* and its similar species are discussed in Wang & Sha (2015).

## 2.15.16 *Synalpheus lophodactylus* Coutière, 1908 (Fig. 2.171)

*Synalpheus lophodactylus* Coutière, 1908: 197; 1921: 417, Pl. 61, fig. 11; Banner, 1957: 195; Banner DM & Banner AH, 1975: 350, fig. 20; Wang & Sha, 2015: 383, fig. 29.

**Material Examined** MBM 158724, 2♀1♂, CL 3.9 mm, TL 8.1 mm, Shalao, depth; 2 m, 4 April 1992.

**Diagnosis** Carapace smooth, glabrous. Rostrum about 2.5 as long as wide at base, rounded tip bearing two short setae, reaching end of visible part of first antennular segment. Orbital hoods about 2.4 times as wide at base as rostrum, shorter than rostrum, with rounded tip bearing two short setae. Pterygostomial corner produced into blunt acute angle. Cardiac notch well developed.

Telson about 1.2 times as long as wide at base; dorsal surface with two pairs of moderately large spines, inserted approximately at 1/3 and 3/5 length of telson; posterior margin moderately produced, fringed with long setae, posterolateral

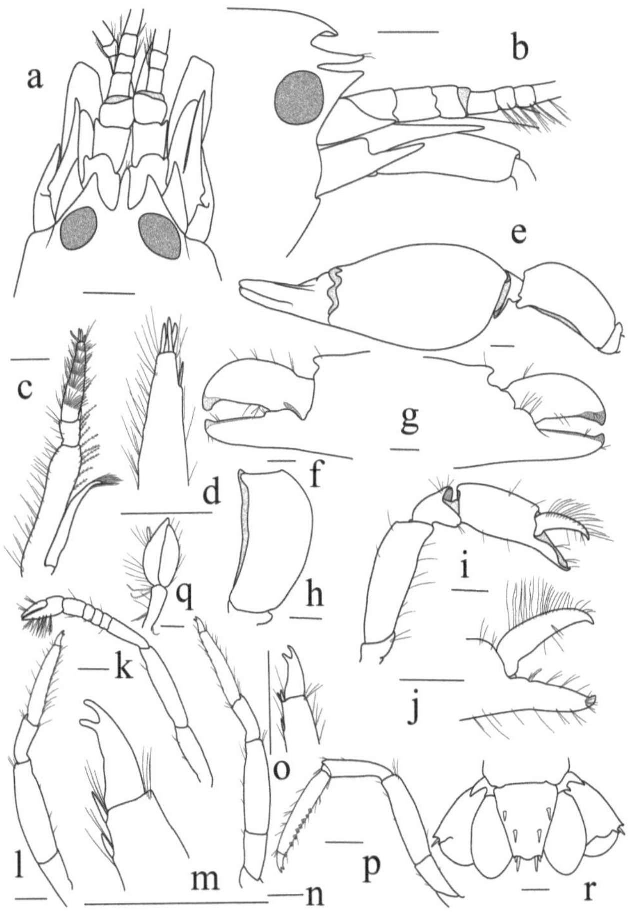


**Fig. 2.171** *Synalpheus lophodactylus* Coutière, 1908. MBM 158724, female, CL 3.9 mm, **a**, anterior part of carapace, dorsal; **b**, same, lateral view; **c**, third maxilliped; **d**, same, tip; **e**, left major cheliped; **f** & **g**, same, distal part of palm and fingers; **h**, merus of major cheliped; **i**, right minor cheliped; **j**, same, fingers; **k**, second pereiopod; **l**, third pereiopod; **m**, same, distal part of propodus and dactylus; **n**, fourth pereiopod; **o**, same, distal part propodus and dactylus; **p**, fifth pereiopod; **q**, second pleopod; **r**, uropods and telson (scale bar = 0.5 mm)

angle each with two pairs of spines, lateral 1/3 length of mesial, posterolateral margins each produced into acute small tooth, distinctly shorter than lateral spines.

Antennular peduncle stout, with distally acute stylocerite, latter hardly reaching end of first antennular segment; second segment almost as long as wide; visible part of first segment longer than second and about twice as long as third. Basicerite without distodorsal tooth, distolateral tooth reaching middle of second segment of antennular, distinctly longer than stylocerite; carpocerite beyond end of antennular peduncle by total length of third segment; scaphocerite blade narrow, not reaching end of third antennular segment, lateral spine developed, reaching end of antennular peduncle, shorter than carpocerite.

Third maxilliped far beyond antennular peduncle and carpocerite when extended; antepenultimate segment longest; penultimate segment about 1.5 times as long as wide at base; tip of ultimate segment bearing three spine-like setae, with transverse rows of setae along ventral margin.

Major cheliped with chela cylindrical. Dactylus as long as fixed finger. Palm about 1.7 times as long as dactylus, distodorsal of palm with one triangular tubercle, slightly upturned, bearing fine setae. Carpus cup-shaped, with setae distodorsally. Merus about three times as long as wide at base, with dorsal margin convex medially, distodorsal margin without triangular tooth. Minor chela with dactylus tapering, superior surface bearing row of long setae, extending almost full length of dactylus; fixed finger tapering, distal single tooth. Palm about 1.2 times as long as fingers. Carpus cup-shaped, proportionally longer that of major cheliped, about 1/2 length of palm. Merus about 2.8 as long as wide, longer than that of major cheliped, distodorsal margin without any projecting, ventral margin bearing short setae

Second pereiopod with fingers about twice as long as palm. Carpus five-segmented, segment ratio subequal to 5: 1: 1: 1: 3, with first segment shorter than sum of last four segment; dorsal margin of palm convex.

Third pereiopod stout, dactylus around 1/5 propodus length, biunguiculate, extensor tooth as long as flexor tooth, and as wide at base as flexor tooth, notch between extensor and flexor tooth U-shaped. Propodus with eight spines along ventral margin plus one distal pair of spines near dactylus, dorsal margin with fine setae. Carpus about 1/2 length of merus, with dorsal margin extended into an obtuse tooth, ventral margin with one distal spine. Merus longer than propodus, about 4 times as long as wide at base. Fourth pereiopod generally similar to third; ventral margin of carpus without distal spine. Fifth pereiopod similar to third and fourth pereiopods, carpus about 4/5 merus length, ventral margin without distal spine; propodus slightly longer than merus, with seven transversal rows of setae and distally pair of spines on ventral margin.

**Remarks** *S. lophodactylus* is originally described based on one male specimen, and the description is rather simple and without figures to illustrate (Coutière, 1908). Coutière (1921) has re-examined the same specimen, and adds figures (1921: pl. 61, fig. 11). Unfortunately, the major chela has lost. Banner & Banner (1975) recheck the type specimen at the Muséum d'Histoire Naturelle in Paris, and re-descript this species. The re-descriptions are well agreed with the description made by Coutière (1921), excepting for the following characters: the length of the distolateral tooth of basicerite; the length of the brush of setae on the superior margin of dactylus (minor chela). The present specimens are all in small size, collected from South China Sea in depth 2 m, agree with the re-description made by Banner & Banner (1975).

**Distribution** China Seas; Chagos Archipelago; Indian Ocean; Marshall Islands; Australian.

### 2.15.17 *Synalpheus modestus* De Man, 1909 (Fig. 2.172)

*Synalpheus modestus* De Man, 1909a: 115; Banner DM & Banner AH, 1985: 46; Wang & Sha, 2015: 399, fig. 30.

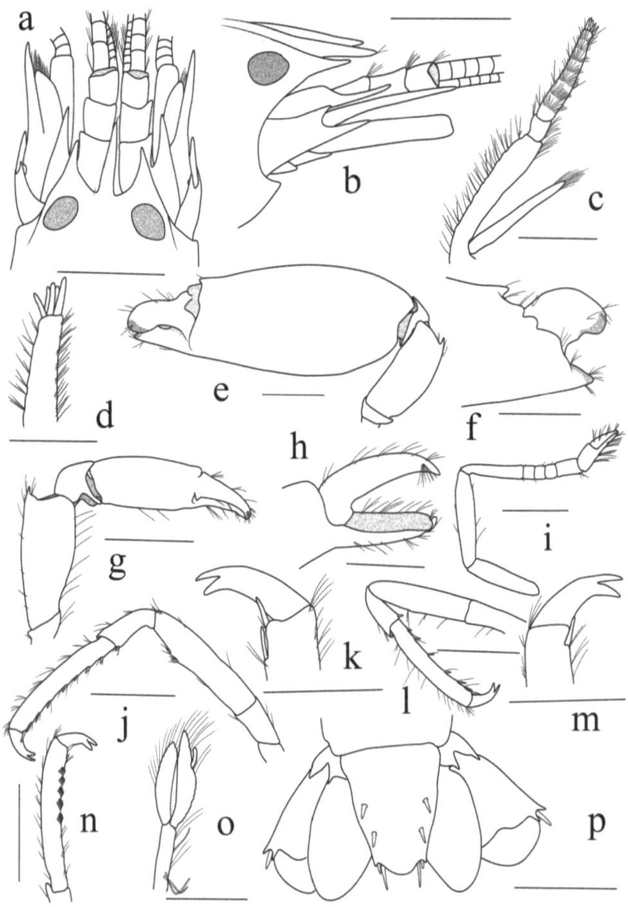


**Fig. 2.172** *Synalpheus modestus* De Man, 1909. MBM 158475, female, CL 4.1 mm, **a**, anterior part of carapace, dorsal view; **b**, same, lateral view; **c**, third maxilliped; **d**, same, tip; **e**, left major cheliped; **f**, same, distal part of palm and fingers; **g**, right minor cheliped; **h**, same, fingers; **i**, second pereiopod; **j**, third pereiopod; **k**, same, distal part of propodus and dactylus; **l**, fourth pereiopod; **m**, same, distal part of propodus and dactylus; **n**, propodus and dactylus of fifth pereiopod; **o**, second pleopod; **p**, uropods and telson (scale bar: **a–c**, **e–g**, **i–j**, **l**, **o–p** = 1 mm; **d**, **h**, **k**, **m**, **n** = 0.5 mm)

**Material Examined** MBM 158475, 1ovigerous♀, CL 4.1 mm, TL 10.2 mm, Xincun, depth: 1–3 m, 26 March 1992; MBM 159310, 1♂, CL 3.5 mm, TL 8.9 mm; MBM 159278, 2ovigerous♀3♂, CL 3.1–3.9 mm, TL 8.1–10.2 mm; MBM 159244, 1♂; MBM 107778, 2♂, CL 3.4 mm, TL 8.9 mm, Shalao, depth: 6 m, Li Xinzheng Coll.; MBM 158542, 1♂, Xincun, depth: 1–3 m, 26 March 1992.

**Description** Carapace smooth, glabrous. Rostrum about 4.2 times as long as wide at base, acute tip reaching middle of second antennular segment. Orbital hoods distinctly shorter than rostrum, not reaching end of first antennular segment. Pterygostomial corner produced into acute angle. Cardiac notch well developed.

Telson about 1.3 times as long as wide at base; dorsal surface with two pairs of moderately small spines, inserted approximately at 2/5 and 3/5 length of telson; posterior margin strongly produced, fringed with long setae, posterolateral angle each with two pairs of spines, lateral about 1/4 length of mesial, posterolateral margins each produced into acute small tooth, but distinctly shorter than lateral spines.

Antennular peduncle slender, with distally acute stylocerite, latter overreaching middle of second antennular segment; second segment about 1.25 times as long as wide at base; visible part of first segment longer than second and about twice as long as third. Basicerite with acute distodorsal tooth, distolateral tooth distinctly shorter than stylocerite, reaching end of first antennular segment; carpocerite overreaching end of third antennular segment; scaphocerite blade moderately narrow, reaching about 3/4 length of third antennular segment, lateral spine developed, far overreaching end of antennular peduncle, distinctly longer than carpocerite.

Third maxilliped far beyond antennular peduncle and carpocerite when extended; antepenultimate segment longest; penultimate segment about 1.4 times as long as wide at base; tip of ultimate segment bearing crown of moderately long five spine-like setae, with transverse rows of long setae along ventral margin.

Major cheliped with dactylus slightly longer than fixed finger. Palm about 2.9 times as long as dactylus, distodorsal of palm with one strong subacute projecting. Carpus cup-shaped, with setae distodorsally. Merus about three times as long as wide at base, distodorsal margin with one acute triangular tooth. Minor chela with dactylus and fixed finger tapering, distal single tooth. Palm about 1.7 times as long as fingers. Carpus cup-shaped, proportionally longer that of major cheliped. Merus about 3.5 times as long as wide at base, distodorsal margin with one acute triangular tooth.

Second pereiopod with fingers about 1.2 times as long as palm. Carpus five-segmented, segment ratio subequal to 5: 1: 1: 1: 2.2, with first segment slightly shorter than sum of last four segment; dorsal margin of palm convex.

Third pereiopod moderately slender, dactylus about 3/11 length of propodus, slightly curved to propodus, biunguiculate, extensor tooth near as long as flexor tooth, notch between extensor and flexor tooth V-shaped, flexor tooth wider at base than extensor tooth. Propodus with seven spines along ventral margin plus one distal pair of spines near dactylus. Carpus about 2/5 length of merus, with dorsal margin extended into an obtuse tooth, ventral margin with one distal spine. Merus near as long as propodus, about 4 times as long as wide at base, ventral margin with 2 spines. Fourth pereiopod generally similar to third; merus shorter than propodus, ventral margin of propodus with six spines. Fifth pereiopod similar to third and fourth pereiopods, ventral margin of carpus without one distal spine; ventral margin of propodus with five transversal rows of spine-like setae and two spines.

**Remarks** *S. modestus* is originally described based on single specimen, and it most similar to *S. streptodactylus*. Banner & Banner (1985) have ever discussed the validity of the species, and considered the slender rostrum of this species to be unique in this genus. The slender rostrum most similar to another the Indo-Pacific species (*S. gracilirostris*), which also described by De Man (1910), but they could be easily distinguished by the spines on the merus of the third pereiopod. The present specimens are generally agreed with the original description, excepting for the projecting situated on the distal-superior margin of the palm in the major chela, which is sub-acute rather than acute in the original description.

**Distribution** Xincun, South China Sea; Banda Reef.

### 2.15.18 *Synalpheus mortenseni* Banner & Banner, 1985 (Figs. 2.173 and 2.174)

*Synalpheus mortenseni* Banner DM & Banner AH, 1985: 47; fig. 5; Wang & Sha, 2015: 401, figs. 31–42.

**Material Examined** MBM 270604, 1♀, CL 5.4 mm, TL 14.3 mm, Beibu Gulf, Sta. 6229, depth: 41 m, 8 July 1960.

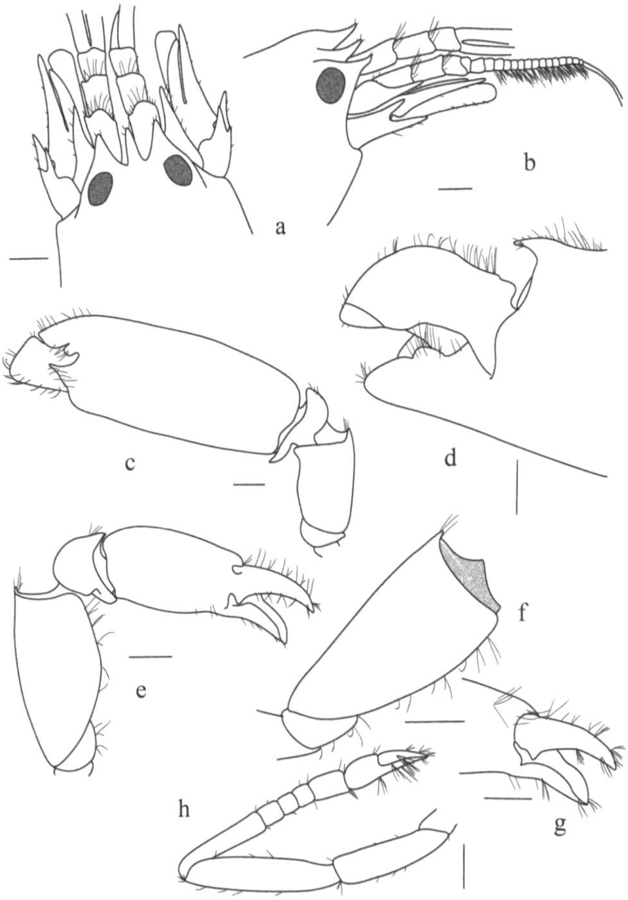


**Fig. 2.173** *Synalpheus mortenseni* Banner & Banner, 1985. MBM 270604, male, CL 5.4 mm, **a**, anterior part of carapace, dorsal view; **b**, same, lateral view; **c**, major cheliped; **d**, same, distal part of palm and fingers; **e**, minor cheliped; **f**, same, merus; **g**, fingers of minor chela; **h**, second pereiopod (scale bar = 0.5 mm)

**Description** Carapace smooth, glabrous. Rostrum twice as long as wide at base, with lateral margins slightly convex, rounded tip bearing two short setae, overreaching 1/2 length of visible part of first antennular segment. Orbital hoods about 2.6 times as wide at base as rostrum, equal to rostrum, with rounded tip bearing two short setae; both rostrum and orbital hoods tips upturned; notches between orbital hoods and rostrum deep, narrow, V-shaped. Pterygostomial corner produced into blunt acute angle. Cardiac notch well developed.

Abdominal segments smooth, glabrous, posterolateral margin of sixth pleura each with one triangular tooth, margin between two teeth bearing eight subacute teeth; first-fifth pleura of male with posterior corner rounded. Telson as long as wide at base; lateral margin slightly concave posteriorly; dorsal surface without median groove, with two pairs of large spines, inserted approximately at 1/3 and 3/5 length of telson; posterior margin strongly produced, fringed with long setae, posterolateral angle each with two pairs of spines, lateral 2/3 length of mesial, posterolateral margins each produced into acute tooth, but shorter than lateral spines.

Antennular peduncle slender, with distally acute stylocerite, latter hardly reaching middle of second antennular segment; second segment 1.25 times as long as wide; visible part of first segment slightly longer than second and about twice

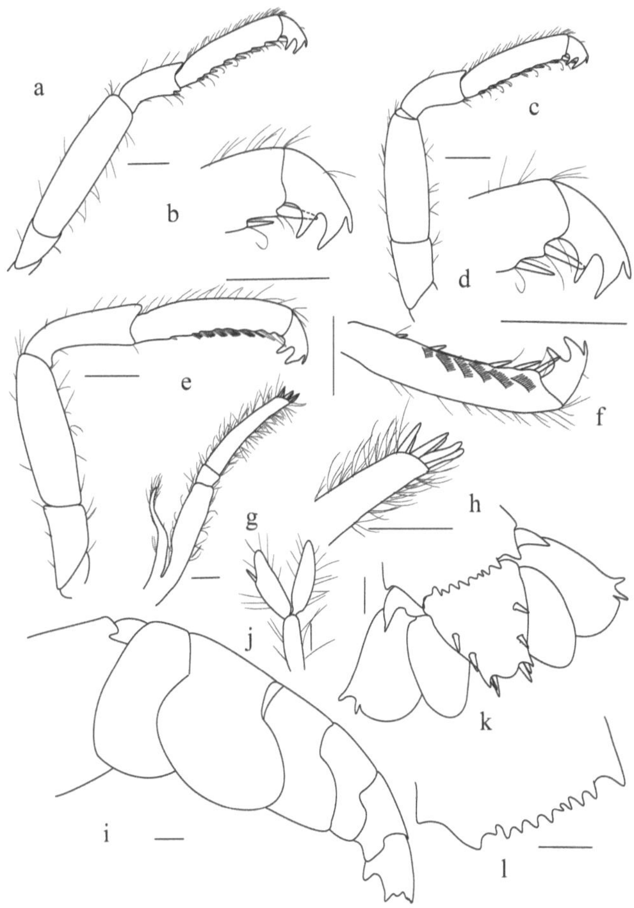


**Fig. 2.174** *Synalpheus mortenseni* Banner & Banner, 1985. MBM 270604, male, CL 5.4 mm, **a**, third pereiopod; **b**, same, distal part of propodus and dactylus; **c**, fourth pereiopod; **d**, same, distal part of propodus and dactylus; **e**, fifth pereiopod; **f**, same, distal part of propodus and dactylus; **g**, third maxilliped; **h**, same, tip; **i**, abdominal segments, lateral view; **j**, second pleopod; **k**, uropods and telson; **l**, posterior margin of sixth abdominal segment (scale bar = 0.5 mm)

as long as third. Antenna with basicerite with acute distodorsal tooth, about 1/2 length of distolateral tooth, latter overreaching end of first segment, slightly shorter than stylocerite; carpocerite beyond end of antennular peduncle by 1/4 length of third segment; scaphocerite blade narrow, hardly reaching end of third antennular segment, lateral spine developed, with lateral margin slightly concave posteriorly, overreaching end of antennular peduncle, slightly shorter than carpocerite.

Third maxilliped far beyond antennular peduncle and carpocerite when extended; antepenultimate segment longest, about five times as long as wide; penultimate segment as long as wide at base; tip of ultimate segment bearing crown of five spin-like long setae, with several brush consisting of six-eight transverse rows of setae along ventral margin.

Major cheliped with chela cylindrical, not a bit compressed or twisted. Dactylus slightly longer than fixed finger. Palm about 2.8 times as long as dactylus, distodorsal of palm with swollen tubercle, bearing in turn a smaller distinct and acute tooth. Carpus cup-shaped, with long setae distodorsally. Merus 1.6 times as long as wide at base, with ventral margin convex medially, distodorsal margin ending in subacute triangular tooth. Minor chela with dactylus tapering, with flexor margin concave, distal single tooth; fixed finger tapering, distal single tooth. Palm about 1.5 times as long as fingers. Carpus cup-shaped, proportionally longer that of major cheliped. Merus about twice as long as wide, longer than that of major cheliped, distodorsal margin with subacute triangular tooth, ventral margin bearing short setae.

Second pereiopod with fingers 1.5 times as long as palm. Carpus five-segmented, segment ratio subequal to 5: 1: 1: 1: 2, with first segment equal to sum of last four segments.

Third pereiopod stout, dactylus around 1/5 propodus length, triunguiculate, extensor tooth smallest, about 1/3 length of medial tooth, medial tooth five times as wide at base as extensor tooth, notch between extensor and medial tooth V-shaped; flexor tooth slightly shorter than medial tooth, 6/7 width at base of medial tooth, notch between medial and flexor tooth U-shaped. Propodus with usual seven spines along ventral margin plus one distal pair of spines near dactylus. Carpus about 1/3 merus length, with dorsal margin extended into an obtuse tooth, ventral margin with one distal spine. Merus longer than propodus, about 3.8 times as long as wide at base, without spines on ventral margin. Fourth pereiopod generally similar to third; ventral margin of carpus with one distal spine. Fifth pereiopod similar to third and fourth pereiopods, but ventral margin of carpus without distal spine, with merus about 2.7 times as long as wide; carpus about 3/4 merus length, ventral margin without distal spine; propodus as long as merus, with six transversal rows of setae and five spines along ventral margin.

**Remarks** *S. mortenseni* is originally described based on 13 specimens, and the variations between the type specimens and the differences between its similar species are also well discussed (Banner & Banner, 1985). Another species *Synalpheus septemspinosus* De Man, 1910 is also bearing about 7 spines on the posterior margin of the sixth abdominal somite, but it can well distinguished from *S. mortenseni* by the merus of third pereiopod bearing several spines along the ventral margin of the third pereiopods. The present material is agreed well with the description of the female holotype in Banner & Banner (1985), excepting for the number of the spines of the posterior margin of sixth abdominal segment.

**Distribution** Beibu Gulf, the South China Sea; Ambon, Indonesia.

### 2.15.19 *Synalpheus neomeris* (De Man, 1897) (Figs. 2.175 and 2.176)

*Alpheus minor neptunus* De Man, 1888: 272.
*Alpheus neomeris* De Man, 1897: 734, fig. 61a, d, e.
*Synalpheus neomeris*–. Coutière, 1905: 869, fig. 1; De Man, 1911: 212, fig. 24; Banner DM & Banner AH, 1975: 357, fig. 22; Wang & Sha, 2015: 404, figs. 33–34.
*Synalpheus gravieri* Coutière, 1905: 870, fig. 2; Pearson, 1905: 82; De Man, 1911: 216, fig. 25; Gurney, 1927: 261, fig. 64; Banner AH & Banner DM, 1966b: 48, fig. 13; Miya, 1972: 66, Pl. 13.
*Alpheus prolificus* Ortmann, 1890: 484.
*Alpheus neptunus*–. Bate, 1888: 563, Pl. 101, fig. 2 [Not *A. neptunus* Dana, 1852].

**Material Examined** CL 5.9–13.1 mm, TL 15.1–30.1 mm. MBM 270310, 2ovigerous♀2♂, depth: 60 m; MBM 270380, 1ovigerous♀1♂, Beibu Gulf, Sta. 6179, depth: 16 m, Sun Fu-zeng Coll., 13 February 1960; MBM 270379, 1ovigerous♀2♂, Beibu Gulf, Sta. 6251, depth: 51 m, Shen Shou-peng Coll., 9 October 1960; MBM 270378, 2♂, Beibu Gulf, Sta. 6267, depth: 35 m, Sun Fu-zeng Coll., 10 February 1960; MBM 270376, 1♂, Beibu Gulf, Sta. 6273, depth: 46 m, Shen Shou-peng Coll., 25 October 1960; MBM 270375, 1ovigerous♀1♂, Beibu Gulf, Sta. 6273, depth: 46 m, Shen Shou-peng Coll., 25 October 1960; MBM 270374, 2♂, Beibu Gulf, Sta. 6263, depth: 59 m, 19 October 1960; MBM 270373, 2♂, Beibu Gulf, Sta. 6249, depth: 35 m, Sun

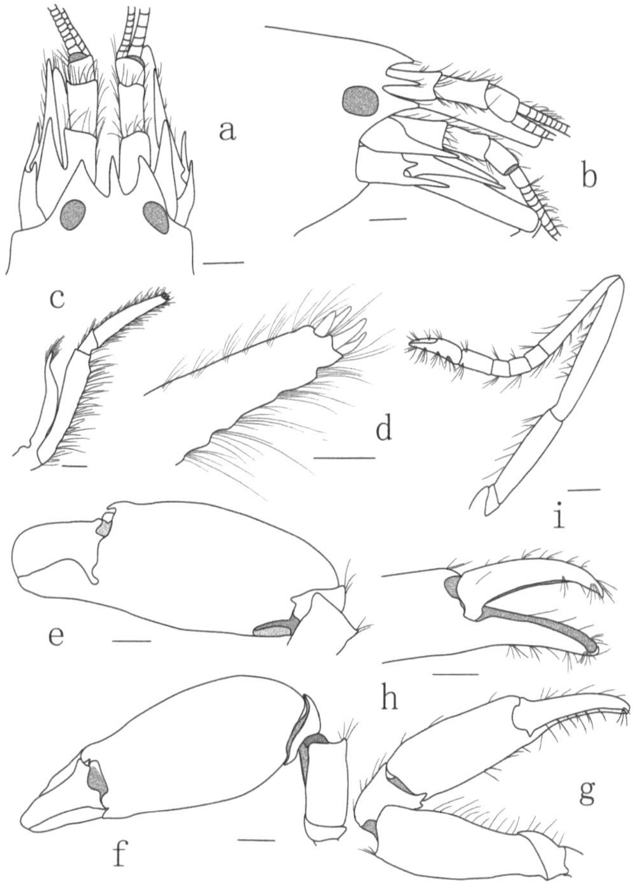


**Fig. 2.175** *Synalpheus neomeris* (De Man, 1897). MBM 279528, female CL 7.5 mm, **a**, anterior part of carapace, dorsal view; **b**, same, lateral view; **c**, third maxilliped; **d**, same, tip; **e**, major chela; **f**, major cheliped; **g**, minor cheliped, lateral; **h**, same, fingers; **i**, second pereiopod (scale bar = 1 mm, **g** and **h** share same scale bar)

Fu-zeng Coll., 14 April 1960; MBM 270561, 1ovigerous♀, Beibu Gulf, Sta. 6237, depth: 45.5 m, Sun Fu-zeng Coll., 13 December 1959; MBM 270439, 2ovigerous♀7♂, Beibu Gulf, Sta. 6280, depth: 27 m, Ma Xiu-tong Coll., 5 December 1959; MBM 270568, 4 immature specimens, Beibu Gulf, Sta. 6277, depth: 31.3 m, Fan Zhen-gang Coll., 17 May 1960; MBM 270403, 1♂, Beibu Gulf, Sta. 6270, depth: 32.8 m, Fan Zhen-gang Coll., 7 July 1960; MBM 270417, 1♂, Beibu Gulf, Sta. 6249, depth: 35 m, Sun Fu-zeng Coll., 14 April 1960; MBM 270418, 1♂, Beibu Gulf, Sta. 6223, depth: 30 m, Sun Fu-zeng Coll., 20 May 1960; MBM 270421, 3♂, Beibu Gulf, Sta. 7404, depth: 46 m, Sun Fu-zeng Coll., 15 April 1962; MBM 270420, 1ovigerous♀1♂, Beibu Gulf, Sta. 7404, depth: 48 m, 21 August 1962; MBM 270238, 1ovigerous♀7♂, Beibu Gulf, Sta. 6239, depth: 63.3 m, 4 December 1959; MBM 270420, 1♂, Beibu Gulf, Sta. 7404, depth: 48 m, 21 August 1962; MBM 107834, 1ovigerous♀1♂, the East China Sea, Sta. V-3, depth: 130 m, 29 August 1976,Tang Zhi-chan Coll.; MBM 107831, 1♂, the East China Sea, Sta. IV-6, depth: 112 m, Liu Rui-Yu Coll., 29 August 1976; MBM 107835, 3♂, the East China Sea, Sta. V-7, depth: 126 m, Tang Zhi-chan & Xu Feng-shan Coll., 21 September 1976; MBM 270202, 1ovigerous♀2♂, Beibu Gulf, Sta. 6262, depth: 48 m, Shen Shou-peng Coll., 15 October 1960; MBM 270382, 1♂, Beibu Gulf, Sta. 6251, depth: 50 m, Fan Zhen-

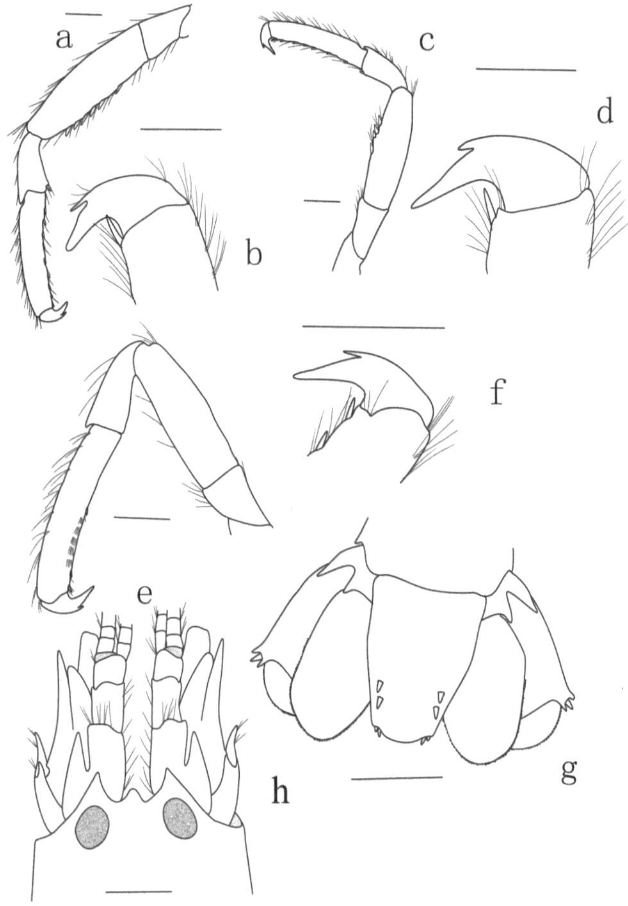


**Fig. 2.176** *Synalpheus neomeris* (De Man, 1897). MBM 279528, female CL 7.5 mm, **a**, third pereiopod; **b**, same, distal part of propodus and dactylus; **c**, fourth pereiopod; **d**, same, distal part of propodus and dactylus; **e**, fifth pereiopod; **f**, same, distal part of propodus and dactylus; **g**, uropods and telson. MBM 158504, female, CL 6.9 mm, **h**, anterior part of carapace, dorsal view (scale bar = 1 mm)

gang Coll., 6 July 1960; MBM 270370, 1♂; MBM 270369, 2♂, Beibu Gulf, Sta. 6179, depth: 15 m, Shen Shou-peng Coll., 14 July 1960; MBM 270368, 1ovigerous♀2♂, Beibu Gulf, Sta. 6249, Zhang Bao-lin Coll., 11 November 1960; MBM 270367, 1ovigerous♀, 1♂, Beibu Gulf, Sta. 6249, Zhang Bao-lin Coll., 11 November 1960; MBM 270366, 2ovigerous♀4♂, Beibu Gulf, Sta. 6249, Zhang Bao-lin Coll., 11 November 1960; MBM 270365, 2♂, Beibu Gulf, Sta. 6249, depth: 33.5 m, Sun Fu-zeng Coll., 13 April 1960; MBM 270364, 2ovigerous♀2♂, Beibu Gulf, Sta. 6249, depth: 33.5 m, Sun Fu-zeng Coll., 13 April 1960; MBM 270448, 2♂, Beibu Gulf, Sta. 6271, depth: 44 m, Fan Zhen-gang Coll. 17 May 1960; MBM 270381, 5♂, Beibu Gulf, Sta. 6251, depth: 50 m, Fan Zhen-gang Coll., 6 July 1960; MBM 270419, 4ovigerous♀5♂, Beibu Gulf, Sta. 7304, Sun Fu-zeng Coll., 21 January 1962; MBM 270419, 1ovigerous♀5♂, Beibu Gulf, Sta. 7304, depth: 50 m, Sun Fu-zeng Coll., 20 August 1962; MBM 270401, 1♂, 2 immature specimens, Beibu Gulf, Sta. 6239, depth: 57 m, Fan Zhen-gang Coll., 5 February 1960; MBM 270563, 1♂, Beibu Gulf, Sta. 6230, depth: 64 m, Fan Zhen-gang Coll., 18 April 1960; MBM 270397, 1♂, Beibu Gulf, Sta. 6270, depth: 32.8 m, Fan Zhen-gang Coll., 7 July 1960; MBM 270400, 2♂, Beibu Gulf, Sta. 6217, depth:

60 m, Sun Fu-zeng Coll., 9 February 1960; MBM 270530, 1ovigerous♀2♂, Beibu Gulf, Sta. 7305, depth: 64 m, 18 October 1962; MBM 270402, 2♂, Beibu Gulf, Sta. 6240, depth: 60 m, Fan Zhen-gang Coll., 6 December 1959; MBM 270398, 1ovigerous♀3♂, Beibu Gulf, Sta. 6222, depth: 25 m, Guan Yang-lian Coll., 18 October 1960; MBM 270560, 1♂, Beibu Gulf, Sta. 6257, depth: 64 m, Fan Zhen-gang Coll., 19 April 1960; MBM 27569, 2ovigerous♀1♂, Beibu Gulf, Sta. 6276, depth: 26 m, Shen Shou-peng Coll., 7 July 1960; MBM 270670, 1ovigerous♀, Beibu Gulf, Sta. 6232, depth: 81 m, Shen Shou-peng Coll., 23 October 1960; MBM 270567, 1ovigerous♀1♂, Beibu Gulf, Sta. 6250, depth: 33.5 m, Sun Fu-zeng Coll., 13 April 1960; MBM 270565, 1ovigerous♀, Beibu Gulf, Sta. 6265, depth: 63 m, 24 October 1960; MBM 270566, 1♂, Beibu Gulf, Sta. 6236, depth: 41.5 m, Sun Fu-zeng Coll., 15 July 1960; MBM 081142, 1♂, East China Sea, Sta. 4081, depth: 81 m, Xu Jie-shan Coll., 2 February 1959; MBM 270529, 1ovigerous♀5♂, Beibu Gulf, Sta. 7603, depth: 43 m, 11 October 1959; MBM 270526, 1ovigerous♀5♂, Beibu Gulf, Sta. 7503, depth: 52 m, 22 August 1962; MBM 270562, 1ovigerous♀, Beibu Gulf, Sta. 6283, depth: 25 m, Fan Zhen-gang Coll., 10 July 1960; MBM 270564, 1ovigerous♀, Beibu Gulf, Sta. 6239, depth: 63.3 m, Fan Zhen-gang Coll., 4 December 1959; MBM 270471, 3ovigerous♀5♂, Beibu Gulf, Sta. 7503, depth: 47 m, Sun Fu-zeng Coll., 15 January 1962; MBM 270140, 1ovigerous♀16♂, Beibu Gulf, Sta. 7405, depth: 60 m, Sun Fu-zeng Coll., 14 January 1962; MBM 270138, 1♂, Beibu Gulf, Sta. 6179, depth: 14 m, Sun Fu-zeng Coll., 17 April 1960; MBM 270133, 1ovigerous♀, Beibu Gulf, Sta. 7404, depth: 58 m, Sun Fu-zeng Coll., 14 January 1962; MBM 270131, 1♂, Beibu Gulf, Sta. 7204, depth: 49 m, Sun Fu-zeng Coll., 22 January 1962; MBM 270130, 1ovigerous♀10♂, Beibu Gulf, Sta. 7304, depth: 54 m, Sun Fu-zeng Coll., 21 January 1962; MBM 270135, 2ovigerous♀1♂, Beibu Gulf, Sta. 7502, depth: 32 m, Sun Fu-zeng Coll., 21 August 1962; MBM 270136, 2ovigerous♀2 ♂, 1 immature specimens, Beibu Gulf, Sta. 7405, depth: 60 m, Sun Fu-zeng Coll., 14 January 1962; MBM 270137, 5♂, Beibu Gulf, Sta. 7404, depth: 58 m, Sun Fu-zeng Coll., 14 January 1962; MBM 270139, 2♂, Beibu Gulf, Sta. 7405, depth: 63 m, 12 October 1962; MBM 270590, 1ovigerous♀4♂, Beibu Gulf, Sta. 6249, depth: 31 m, Zhang Bao-lin Coll., 7 November 1960; MBM 107830, 2ovigerous♀3immature♂, the East China Sea, Sta. V-6, depth: 112 m, Zhang Bao-lin & Xu Feng-shan Coll., 5 July 1976; MBM 107833, 2♂, East China Sea, Sta. V-2, depth: 99 m, Tang Zhi-chan Coll., 5 July 1976; MBM 107832, 1♂1immature specimens, the East China Sea, Sta. V-7, depth: 100 m, Tang Zhi-chan & Xu Jie-shan Coll., 10 October 1975; MBM 107829, 1ovigerous♀, the East China Sea, Sta. III-7, depth: 100 m, Tang Zhi-chan Coll., 27 June 1976; MBM 107827, 1♂, the East China Sea, Sta. VI-4, depth: 139 m, Tang Zhi-chan Coll., 29 August 1976; MBM 270596, 4♂, Beibu Gulf, Sta. 6218, depth: 31 m, Zhang Bao-lin Coll., 4 November 1960; MBM 107828, 1ovigerous♀, the East China Sea, Sta. V-5, depth: 150 m, Tang Zhi-chan & Zhang Bao-lin Coll., 28 August 1976; MBM 270593, 13♂, Beibu Gulf, Sta. 6266, depth: 50.5 m, Shen Shou-peng & Fan Zhen-gang Coll., 13 July 1960; MBM 107836, 5ovigerous♀7♂, the East China Sea, Sta. V-3, depth: 105 m, Tang Zhi-chan Coll., 27 August 1976; MBM 270195, 1♂, Beibu Gulf, Sta. 7604, depth: 58 m, 11 October 1962; MBM 270592, 8ovigerous♀12♂, Beibu Gulf, Sta. 6277, depth: 35 m, Fan Zhen-gang Coll., 9 July 1960; MBM 270196, 1♂, Beibu Gulf, Sta. 7603, depth: 42 m, Sun Fu-zeng Coll., 12 January 1962; MBM 042233, 5 specimens, the East China Sea, Sta. 4082, depth: 86 m, Liu Rui-Yu Coll., October 1959; MBM 158504, 1ovigerous♀, Yezhu Island, depth: 10 m, Zhang Bao-lin Coll., 18 March 1992; MBM 158690, 1♂, Nansha Islands, Sta. 55, depth: 46 m, 8 June 1990; MBM 158776, 1♂, Nansha Islands, Sta. 15, 28 May 1985; MBM 214610, 5♂, Beibu Gulf, Sta. 6236, depth: 47.5 m, Sun Fu-zeng Coll., 18 April 1960.

**Description** Carapace smooth, glabrous. Rostrum about 2.5 times as long as wide at base; acute tip reaching about 2/3 length of visible part of first antennular segment. Orbital hoods slightly shorter than rostrum. Pterygostomial corner produced into acute angle. Cardiac notch well developed.

Telson about 1.4 times as long as wide at base; dorsal surface with two pairs of moderately small spines, inserted approximately at 8/11 and 9/11 length of telson; posterior margin strongly produced, fringed with long setae, posterolateral angle each with two pairs of spines, lateral about 1/2 length of mesial, posterolateral margins not produced.

Antennular peduncle slender, with distally acute stylocerite, latter reaching about 1/4 length of second antennular segment; second segment about 1.5 times as long as wide at base; visible part of first segment longer than second and about twice as long as third. Basicerite with acute distodorsal tooth, distolateral tooth slightly shorter than stylocerite; carpocerite far overreaching end of third antennular segment; scaphocerite blade moderately narrow, reaching 3/4 length of third antennular segment, lateral spine developed, reaching end of antennular peduncle, distinctly shorter than carpocerite.

Third maxilliped far beyond antennular peduncle and carpocerite when extended; antepenultimate segment longest; penultimate segment about 1.5 times as long as wide at base; tip of ultimate segment bearing crown of small four spine-like setae, with transverse rows of long setae along ventral margin.

Major cheliped with dactylus as long as fixed finger. Palm about 2.4 times as long as dactylus, distodorsal of palm with one strong acute projecting, curved to dactylus. Carpus cup-shaped, with setae distodorsally. Merus about twice as long as wide at base, distodorsal margin with one subacute triangular tooth. Minor chela with dactylus and fixed finger slightly compressed, tapering, distal single tooth. Palm slightly longer than fingers. Carpus cup-shaped, proportionally longer that of major cheliped. Merus about four times as long as wide at base, distodorsal margin with one subacute triangular tooth.

Second pereiopod with fingers near as long as palm. Carpus five-segmented, segment ratio subequal to 6: 1: 1: 1: 2, with first segment longer than sum of last four segment; dorsal margin of palm convex.

Third pereiopod moderately stout, with dactylus about 2/7 length of propodus, strongly curved to propodus, biunguiculate, extensor tooth about 1/3 length of flexor tooth, notch between extensor and flexor tooth V-shaped, flexor tooth about 4–5 times as wide at base as extensor tooth. Propodus with nine small spines along ventral margin plus one distal pair of spines near dactylus. Carpus about 1/3 length of merus, with dorsal margin extended into an obtuse tooth, ventral margin with one distal spine. Merus slightly longer than propodus, about 3.5 times as long as wide at base, ventral margin with 4–7 spines. Fourth pereiopod generally similar to third; merus shorter than propodus, ventral margin of propodus with four spines. Fifth pereiopod similar to third and fourth pereiopods, ventral margin of propodus with five transversal rows of spine-like setae and 3 spines.

**Remarks** *S. neomeris* is originally described by De Man (1897) based on two specimens. Banner & Banner (1975) have well discussed the history of this species and synonymized *Synalpheus gravieri* with it. *Synalpheus miscellaneous* De Man, 1909 has be synonymized with *S. neomeris* by Banner & Banner (1985). The present specimens differ from *S. streptodactylus*, *S. iphinoe*, and *S. modestus* by the short rather than long (general in this genus) crown of spines on the tip of third maxilliped. The present abundant specimens are precisely agreed with the description given by Banner & Banner (1975).

**Distribution** China Seas; Suez Canal; Red Sea; eastern Africa; Persian Gulf; Thailand; Japan; Philippines; Indonesia; Australia.

## 2.15.20 *Synalpheus neptunus* (Dana, 1852) (Fig. 2.177)

*Alpheus neptunus* Dana, 1852b: 553, Pl. 35, fig. 5.

*Synalpheus neptunus*–. Coutière, 1909: 87, fig. 53; Banner AH & Banner DM, 1972: 24; Banner DM & Banner AH, 1978: 244; Wang & Sha, 2015: 408.

*Synalpheus neptunus germanus* Banner DM & Banner AH, 1975: 317, fig. 11.

**Material Examined** This species was reported by Banner & Banner (1978) and Liu (2008) from South China Sea, but not found in present specimens.

**Description** Rostrum with tip rounded, slightly overreaching mid-length of first antennular segment. Orbital teeth shorter than rostrum, and narrower than rostrum at base.

Antennular peduncle with stylocerite variable from reaching to 3/4 length of first antennular segment to overreaching distal margin of that segment; three antennular segments nearly subequal in length, with second segment longer than wide. Antenna with basicerite without dorsolateral teeth (or only slightly projecting), with ventrolateral teeth teaching to 3/4 length of second antennular segment; antennal scale with blade reduced or reaching to distal margin of second antennular segment; carpocerite overreaching distal margin of antennular peduncle.

Major cheliped with chela about 2.5 times as long as wide. Finger 1/3 length of palm. Palm with dorsal margin terminating in rounded tubercles. Carpus cup-shaped. Merus about 3.5 times as long as wide, with dorsal margin rounded distally; ventromesial margin unarmed with spines. Ischium unarmed. Minor cheliped with chela 2.4

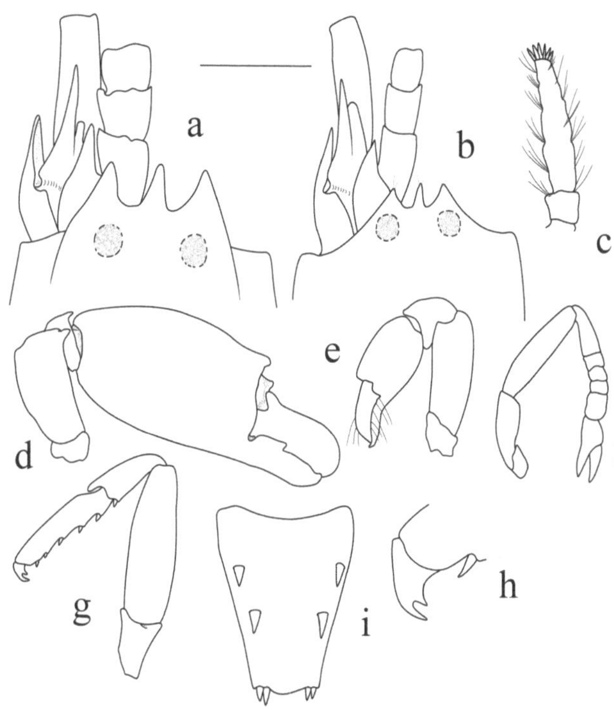


**Fig. 2.177** *Synalpheus neptunus* (Dana, 1852). (Adapted from Banner & Banner, 1975). **a** & **b**, anterior part of carapace, dorsal view; **c**, ultimate of third maxilliped; **d**, major cheliped; **e**, minor cheliped; **f**, second pereiopod; **g**, third pereiopod; **h**, same, distal part of propodus and dactylus; **i**, telson.

times as long as wide. Fingers nearly as long as palm; dactylus spatulate, with acute tip; cutting edges of fingers bearing row of setae. Carpus cup-shaped. Merus similar to that of major cheliped.

Second pereiopod with carpus 5-segments, ratios sub-equal to 10: 2: 2: 2: 5.

Third pereiopod with dactylus biunguiculate, superior unguis nearly as long as inferior unguis, but slightly narrower at base than superior unguis. Propodus shorter than merus, with ventral margin bearing 6–9 spines. Carpus about 1/2 length of merus, with ventral armed with 1–2 spines distally. Merus 3 times as long as wide, unarmed with spines on ventral margin.

Telson about 2.3 times as long as posterior width; dorsal margin bearing two pairs spines, with anterior pair situated on 1/3 length of telson; posterolateral angle not projecting as teeth.

**Distribution** The South China Sea (Hong Kong, Taiwan); Red Sea; Indian Ocean; Indonesia; Japan; Sulu Sea; Philippines; Australian.

### 2.15.21 *Synalpheus nilandensis* Coutière, 1905 (Figs. 2.178 and 2.179)

*Synalpheus nilandensis* Coutière, 1905: 871, Pl. 70, fig. 4–4d; De Man, 1911: 246, fig. 38; Banner DM & Banner AH, 1975: 327, fig. 14; Wang & Sha, 2015: 408, figs. 35–36.

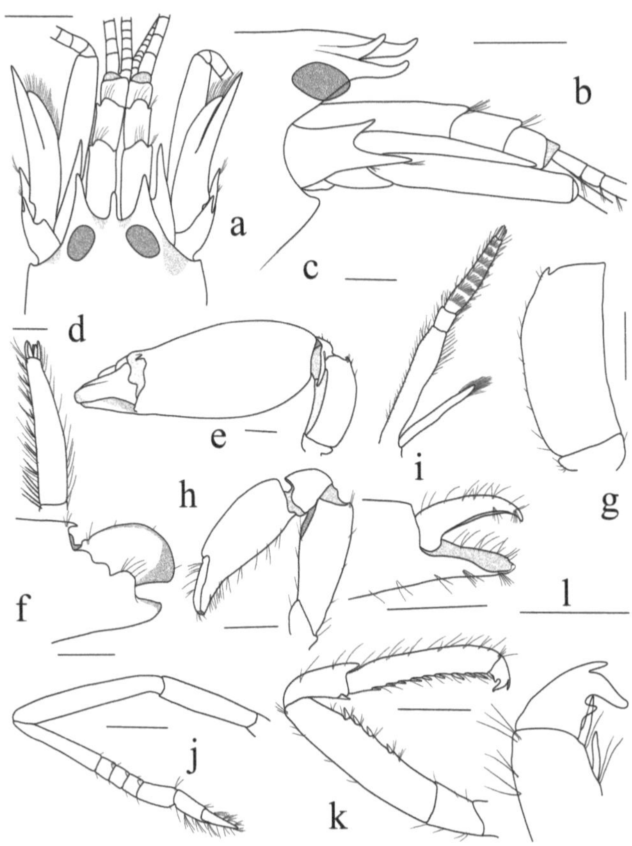


**Fig. 2.178** *Synalpheus nilandensis* Coutière, 1905. MBM 107821, male, CL 5.9 mm, **a**, anterior part of carapace, dorsal view; **b**, same, lateral view; **c**, third maxilliped; **d**, same, ultimate segment; **e**, left major cheliped; **f**, same, fingers; **g**, merus of major cheliped; **h**, right minor cheliped; **i**, same, fingers; **j**, second pereiopod; **k**, third pereiopod; **l**, same, distal part of propodus and dactylus (scale bar: **a**–**c**, **e**–**k** = 1 mm; **d**, **l** = 0.5 mm)

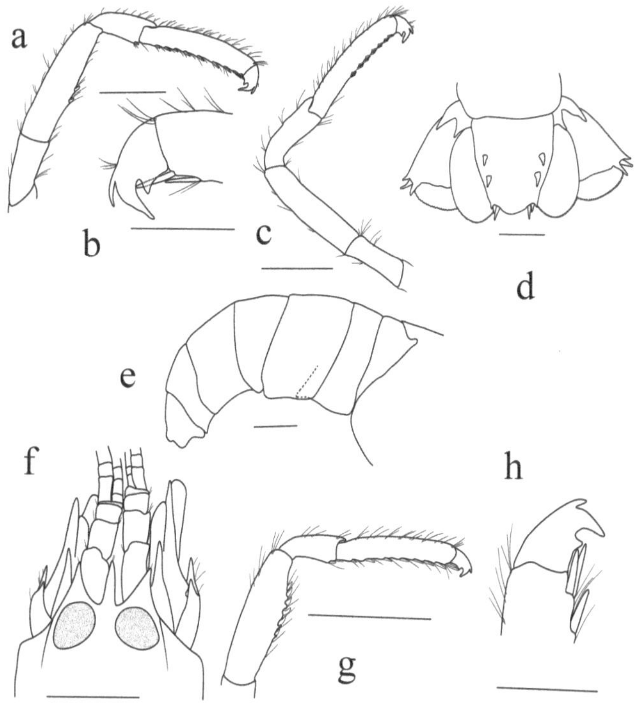


**Fig. 2.179** *Synalpheus nilandensis* Coutière, 1905. MBM 107821, male, CL 5.9 mm, **a**, fourth pereiopod; **b**, same, distal part of propodus and dactylus; **c**, fifth pereiopod; **d**, uropods and telson **e**, abdominal segments, lateral view. MBM 158690, male, CL 6.1 mm, **f**, anterior part of carapace, dorsal view; **g**, third pereiopod; **h**, same, distal part of propodus and dactylus (scale bar: **a**, **c**–**g** = 1 mm; **b**, **h** = 0.5 mm)

*Synalpheus nilandensis* var. *oxyceros* Coutière, 1905: 871, Pl. 70, fig. 5–5a.
*Synalpheus nilandensis* var. *bandaensis* De Man, 1909: 121; 1911: 249, fig. 38a, b.

**Material Examined** MBM 042230, 1♂, CL 4.9 mm, TL 11.2 mm, the East China Sea, Sta. 4081, depth: 74 m, 5 April 1959, 无采集人 Coll. Guang-yu Lin, Yu-lin Liao; MBM 081125, 1♀, 2♂, CL 5.5–6.5 mm, TL 19.3–22.3 mm, the East China Sea, Sta. 4081, depth: 81 m, Xu Jie-shan Coll., 2 February 1959; MBM 081145, 1♂, CL 7.2 mm, TL 19.1 mm, the East China Sea, Sta. 4081, depth: 79 m, Cheng Li-ren Coll., 3 July 1959; MBM 107826, 1♀, the East China Sea, Sta. 4083, 8 December 1959; MBM 042234, 1♀, the East China Sea, Sta. 4081, depth: 79 m, 3 July 1959; MBM 042231, 1♂, the East China Sea, Sta. 4081, depth: 79 m, Zhang Bao-lin Coll., 25 October 1959; MBM 107821, 3♀4♂, the East China Sea, Sta. V-3, depth: 105 m, Tang Zhi-chan Coll., 27 August 1976; MBM 158690, 1♀, 2♂, Nansha Islands, South China Sea, Sta. 55, 16°27′57″N; 111°44′17″E, depth: 46 m, 8 June 1990; MBM 158622, 1♀, Nansha Islands, Sta. E13, 3°53.08.40N, 112°16.22.80E, depth: 50 m, 28 April 1986; 1♀, Nansha Islands, South China Sea, Sta. 55, 16°27′57″N; 111°44′17″E, depth: 46 m, 8 June 1990.

**Description** Carapace smooth, glabrous. Acute tip of rostrum not reaching middle of first antennular segment, strongly upturned. Orbital much longer than rostrum, slightly upturned at tip. Pterygostomial corner produced into acute angle. Cardiac notch well developed.

Telson about 1.2 times as long as wide at base; dorsal surface with two pairs of spines, inserted approximately at 1/3 and 7/11 length of telson;

posterior margin slightly produced, fringed with long setae, posterolateral angle each with two pairs of spines, lateral about 1/2 length of mesial, posterolateral margins each produced into acute small tooth, distinctly shorter than lateral spines.

Antennular peduncle slender, with distally acute stylocerite, overreaching end of first antennular segment; second segment about 1.3 times as long as wide at base; visible part of first segment much longer than second and about 3 times as long as third. Basicerite with acute distodorsal tooth, distolateral tooth shorter than stylocerite, distolateral margin bearing setae; carpocerite overreaching end of third antennular segment; scaphocerite blade moderately narrow, not reaching end of third antennular segment, lateral spine developed, overreaching end of antennular peduncle, sub-equal to carpocerite.

Third maxilliped far beyond antennular peduncle and carpocerite when extended; antepenultimate segment longest; penultimate segment about 1.5 times as long as wide at base; tip of ultimate segment bearing crown of five long spine-like setae, with transverse rows of setae along ventral margin.

Major cheliped with dactylus distinctly longer than fixed finger. Palm about 2.4 times as long as dactylus, distodorsal of palm with one acute projecting. Carpus cup-shaped, with setae distodorsally. Merus about 2.7 times as long as wide at base, distodorsal margin with one small acute triangular tooth. Minor chela with dactylus and fixed finger tapering, distal single tooth. Palm about 1.8 times as long as fingers. Carpus cup-shaped, proportionally longer that of major cheliped. Merus about 3.5 times as long as wide at base, distodorsal margin rounded.

Second pereiopod with fingers slightly longer than palm. Carpus five-segmented, segment ratio subequal to 5.4: 1: 1: 1: 2.2, with first segment slightly longer than sum of last four segments.

Third pereiopod moderately stout, with dactylus curved to propodus, triunguiculate, middle tooth about 3 times as long as extensor tooth, extensor tooth about 1.5 times as long as flexor tooth, notch between extensor and middle tooth V-shaped, middle tooth distinctly much wider at base than extensor tooth, notch between middle tooth and flexor tooth U-shaped, middle tooth wider at base than flexor tooth. Propodus with nine spines along ventral margin plus one distal pair of spines. Carpus about 1/3 length of merus, with dorsal margin extended into an obtuse tooth, ventral margin with one distal spine. Merus longer than propodus, about 4.5 times as long as wide at base, ventral margin with 4 moderately large spines. Fourth pereiopod generally similar to third; merus near as long as propodus, ventral margin with two spines; ventral margin of propodus bearing ten spines. Fifth pereiopod similar to third and fourth pereiopods, ventral margin of carpus without one distal spine; ventral margin of propodus with seven transversal rows of spine-like setae and two spines.

**Remarks** *S. nilandensis* is originally described by Coutière (1905). This species is complex, mainly because of the variation of the anterior region of carapace and the dactylus of the third pereiopod (Banner & Banner 1975). The variations discussed in Banner & Banner (1975) are also shown in the present specimens collected from the East China Sea and the South China Sea. The present specimens collected from the East China Sea are similar to *S. nilandensis forma alpha* in the following characters: the rostrum much shorter than the orbital hoods, the tips of them were upturned; the dactylus of third pereiopod was moderately stout and the middle tooth of it was at right angles to the axis of the dactylus. The present specimens collected from the South China Sea are similar to the original description (Coutière 1905) in the following characters: the distolateral tooth of the basicerite shorter than the stylocerite; the dactylus of the third pereiopod was about 1.5 times as long as wide at base, the extensor tooth slender and reduced, the middle tooth heavy and turning right angle to the axis of the dactylus, the flexor tooth shortest.

**Distribution** China Seas; Red Sea; Maldives and Laccadives; Andaman Sea; Indonesia; Borneo; Hong Kong; Tuamotus Archipelago.

### 2.15.22 *Synalpheus odontophorus* De Man, 1909 (Fig. 2.180)

*Synalpheus odontophorus* De Man, 1909: 113; Banner DM & Banner AH, 1978: 244, fig. 4j–m; Wang & Sha, 2015: 411.

**Material Examined** This species was reported by Banner & Banner (1978) and Liu (2008) from the South China Sea, but not found in present specimens.

**Description** Rostrum triangular, reaching to 2/3 length of second antennular segment; rostral carina present, extending to posterior margin of orbital hoods. Orbital teeth teaching to 1/3 length of rostrum.

Antennular peduncle with short stylocerite reaching to tip of rostrum; second segment 1–1.3 times as long as wide; visible part of first segment nearly as long as third segment, shorter than second segment. Antenna with basicerite bearing dorsolateral teeth about 1/2 length of ventrolateral teeth; antennal scale with blade reaching to mid-length of third antennular segment, shorter than lateral tooth; carpocerite stout, overreaching distal margin of antennular peduncle.

Major cheliped with chela cylindrical, about 3 times as long as wide. Fixed finger with mesial face bearing two triangular teeth. Palm with dorsal margin of mesial face bearing acute tooth distally. Merus 2.2–2.5 times as long as wide, with dorsal margin bearing acute triangular tooth distally. Minor cheliped with chela cylindrical, 3.3 times as long as wide. Fingers about 1/2 length of palm. Merus similar to that of major cheliped, but more slender, about 4.0–4.6 times as long as wide.

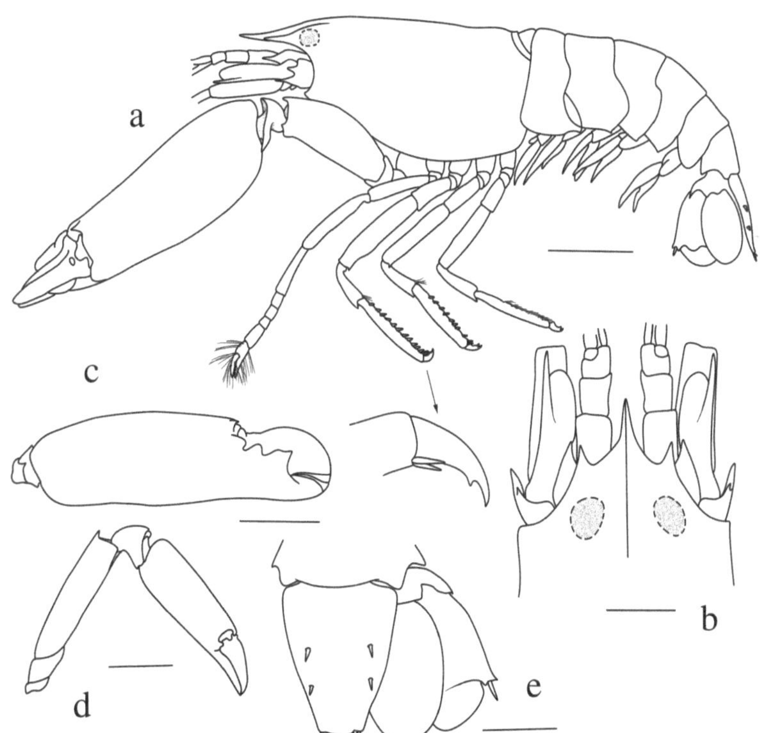


**Fig. 2.180** *Synalpheus odontophorus* De Man, 1909. (Adapted from Miya, 1972) **a**, entire animal, lateral view; **b**, anterior part of carapace, dorsal view; **c**, major chela; **d**, minor cheliped; **e**, right uropod and telson (scale bar: **a** = 2 mm; **b**–**e** = 1 mm)

Second pereiopod with carpus 5-segmented, ratios sub-equal to 10: 2–3: 2–3: 2–3: 4.

Third pereiopod with slender dactylus biunguiculate, inferior unguis much smaller. Carpus with ventral margin bearing 1–2 spines. Merus about 5.4 times as long as wide, with ventral margin terminating in acute tooth.

Telson about 4.0–4.4 times as long as posterior width; dorsal margin bearing two pairs of spines, with posterior pair situated on 1/3 length of telson.

**Distribution** The South China Sea (Hong Kong); Japan; Indonesia.

### 2.15.23 *Synalpheus paraneomeris* Coutière, 1905 (Figs. 2.181 and 2.182)

*Synalpheus paraneomeris* Coutière, 1905: 872, Pl. 71, fig. 7; Banner, 1953: 40, figs. 13, 14; 1956: 331, fig. 6; Miya, 1972: 54, Pl. 9; Banner DM & Banner AH, 1975: 383, fig. 29; 1978: 245; Wang & Sha, 2015: 411, figs 37–38.

*Synalpheus paraneomeris halmaherensis* De Man, 1909a: 122.

*Synalpheus paraneomeris praedabundus* De Man, 1909a: 123.

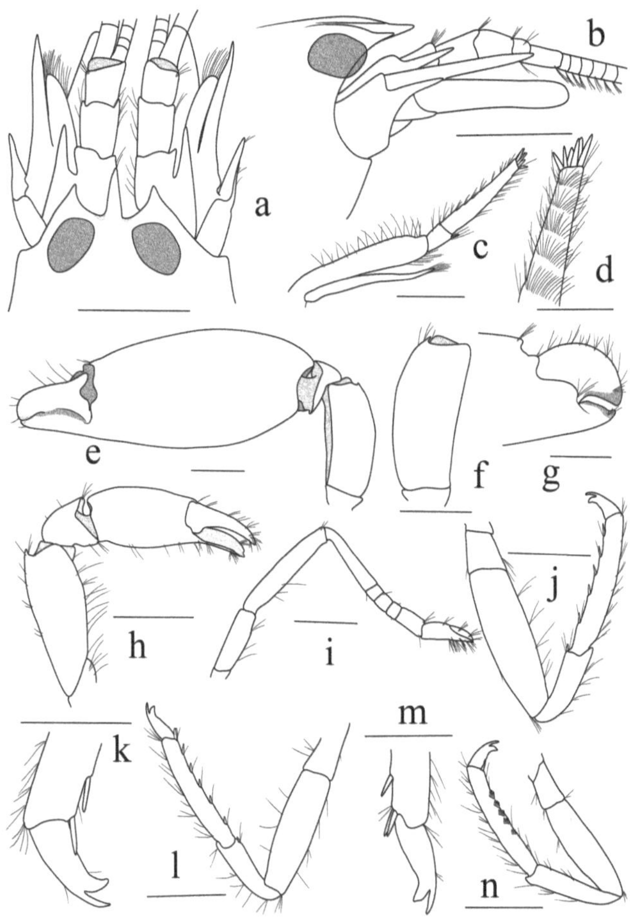


**Fig. 2.181** *Synalpheus paraneomeris* Coutière, 1905. MBM 107746, female, CL 4.1 mm, **a**, anterior part of carapace, dorsal view; **b**, same, lateral view; **c**, third maxilliped; **d**, same, tip; **e**, left major cheliped; **f**, same, merus; **g**, fingers of major chela; **h**, right minor cheliped; **i**, second pereiopod; **j**, third pereiopod; **k**, same, distal part of propodus and dactylus; **l**, fourth pereiopod; **m**, same, distal part of propodus and dactylus; **n**, fifth pereiopod (scale bar: **a**–**c**, **e**–**j**, **l**, **n** = 1 mm; **d**, **k**, **m** = 0.5 mm)

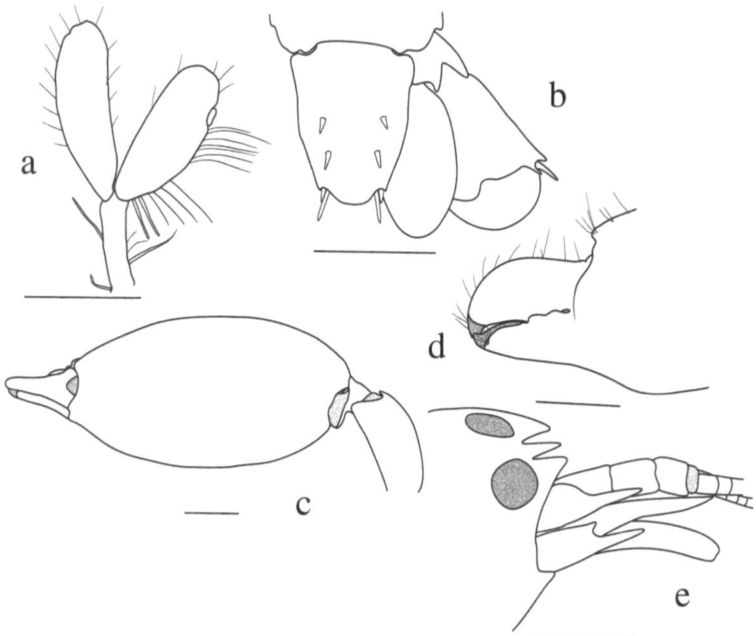


**Fig. 2.182** *Synalpheus paraneomeris* Coutière, 1905. MBM 107746, female, CL 4.1 mm, **a**, second pleopod; **b**, right uropod and telson. MBM 107746, female, CL 3.6 mm, **c**, left major cheliped; **d**, same, fingers. MBM 107746, female, CL 4.5 mm, **e**, anterior part of carapace (scale bar = 1 mm)

*Synalpheus paraneomeris prolatus* De Man, 1911: 241.
*Synalpheus paraneomeris prasalini* Coutière, 1921: 415, Pl. 61, fig. 6.
*Synalpheus paraneomeris seychellensis* Coutière, 1921: 415, Pl. 61, fig. 7.
*Synalpheus townsendi* Coutière, 1909: 32.

**Material Examined** CL 3.2–5.7 mm, TL 6.9–14.3 mm. MBM 158980, 22ovigerous♀37♂, Chenhang Island N-side, Xisha Islands, Meng Zhao-hong & Chen Hui-lian Coll., 18 May 1981; MBM 270026, 1♂, Xisha Islands, 15 April 1975; MBM 157973, 1♂, Jinqing Island, Xisha Islands, Fan Zhen-gang & Xu Jie-shan Coll., 25 April 1958; MBM 107762, 2ovigerous♀, Shi Island, Xisha Islands, Fang Xiu-bin Coll., 10 May 1980; MBM 158852, 1ovigerous♀, Nansha Islands, 14 May 1990; MBM 107748, 2♂, depth: 1–3 m, 26 March 1992; MBM 107750, 1ovigerous♀, Dong Island, Xisha Islands, 28–31 May 1980; MBM 107746, 10ovigerous♀10♂, Yongxing Island, Xisha Islands, Fang Xiu-bin Coll., 11–13 June 1980; MBM 157909, 1♂; MBM 158233, 1ovigerous♀, Dazhou Island, 12 June 1957; MBM 107753, 1♂, Dong Island, Xisha Islands, Fang Xiu-bin Coll., 28–31 May 1980; MBM 107759, 1ovigerous♀, Jinyin Island, Xisha Islands, Fang Xiu-bin Coll., 15 May 1980; MBM 158296, 1♂, Dong Island, Xisha Islands, Ren Xian-qiu Coll., 9 June 1975; MBM 107757, 1ovigerous♀, Yongxing Island, Xisha Islands, Fang Xiu-bin Coll., 11–13 June 1980. MBM 233239, 2♂, Yongxing Island, Xisha Islands, Sta. 3, 26 May 1975; MBM 157802, 1ovigerous♀, Dong Island, Xisha Islands, 28–31 May 1980; MBM 233267, 3ovigerous♀3♂, Dong Island, Xisha Islands, Sta. 5-9, 9 June 1975; MBM 158204, 1♂, Sanya, Li Xin-zheng Coll., 1 March 1997; MBM 157802, 1ovigerous♀; MBM 157995, 1♂, Xisha Islands, Zheng Shu-lian Coll., June 1958; MBM 157827, 6ovigerous♀3♂; MBM 158499, 2♂; MBM 158308, 1ovigerous♀, Shi Island, Xisha Islands, 8 May 1975; MBM 107749, 1ovigerous♀, Chenhang Island, Xisha Islands, Fang Xiu-bin Coll., 18 May 1980; 2ovigerous♀,

Zhubi Reef, Nansha Islands, depth: 2 m, Wang Yong-qiang Coll., 11 May 2004; 1ovigerous♀, Zhubi Reef, Nansha Islands, depth: 1.3 m, Wang Yong-qiang Coll., 4 May 2004; 1ovigerous♀1♂, Zhubi Reef, Nansha Islands, depth: 3 m, Wang Yong-qiang Coll., 6 May 2004; 2ovigerous♀3♂, Zhubi Reef, Nansha Islands, depth: 3 m, Wang Yong-qiang Coll., 10 May 2004; 2ovigerous♀2♂, Zhubi Reef, Nansha Islands, depth: 1.5 m, Wang Yong-qiang Coll., 1 May 2004; MBM 107747, 1♂, Yongxing Island, Xisha Islands, Fang Xiu-bin Coll., 11 June 1980.

**Description** Carapace smooth, glabrous. Rostrum about 2.3 times as long as wide at base; acute tip not reaching end of first antennular segment; orbital hoods shorter than rostrum, wider at base than long. Pterygostomial corner produced into acute angle. Cardiac notch well developed.

Telson about 1.3 times as long as wide at base; dorsal surface with two pairs of spines, inserted approximately at 1/2 and 2/3 length of telson; posterior margin strongly produced, fringed with long setae, posterolateral angle each with two pairs of spines, lateral about 1/3 length of mesial, posterolateral margins each produced into small acute tooth, distinctly shorter than contiguous short spines.

Antennular peduncle moderately slender, with distally acute stylocerite, latter reaching middle of second antennular segment; second segment about 1.5 times as long as wide at base; visible part of first segment longer than second and less twice as long as third. Basicerite without distodorsal tooth, distolateral tooth near as long as stylocerite, overreaching end of first antennular segment; carpocerite far overreaching end of third antennular segment; scaphocerite blade moderately narrow, hardly reaching end of third antennular segment, lateral spine developed, overreaching end of antennular peduncle, much shorter than carpocerite.

Third maxilliped far beyond antennular peduncle and carpocerite when extended; antepenultimate segment longest; penultimate segment about 1.5 times as long as wide at base; tip of ultimate segment bearing crown of five long spine-like setae, with transverse rows of setae along ventral margin.

Major cheliped with dactylus as long as fixed finger, distally slightly curved. Palm about 2.5 times as long as dactylus, distodorsal of palm slightly projecting. Carpus cup-shaped, with setae distodorsally. Merus about 3.5 times as long as wide at base, distodorsal margin with one acute triangular tooth. Minor chela with dactylus and fixed finger tapering, distal single tooth. Palm about 1.3 times as long as fingers. Carpus cup-shaped, proportionally longer that of major cheliped. Merus about 3.6 times as long as wide at base, distodorsal margin with one acute triangular tooth.

Second pereiopod with fingers near as long as palm. Carpus five-segmented, segment ratio subequal to 7: 1.2: 1.5: 1.2: 2.6, with first segment longer than sum of last four segments.

Third pereiopod moderately stout, with dactylus about 1/4 length of propodus, slightly curved to propodus, biunguiculate, extensor tooth slightly longer than flexor tooth, notch between extensor and flexor tooth V-shaped, flexor tooth much wider at base than extensor tooth. Propodus with 4–5 spines along ventral margin plus one distal pair of spines near dactylus. Carpus about 1/2 length of merus, with dorsal margin extended into an obtuse tooth, ventral margin with one distal spine. Merus slightly longer than propodus, about 3.5 times as long as wide at base. Fourth pereiopod generally similar to third; merus near as long as propodus; ventral margin of propodus with 4–5 spines plus one distal pair of spines. Fifth pereiopod similar to third and fourth pereiopods, ventral margin of carpus without distal one spine; ventral margin of propodus with five transversal rows of spine-like setae and one spines.

**Remarks** S. *paraneomeris* is originally described by Coutière (1905). The present specimens show similar variations as showed in Banner & Banner (1975). Miya (1972) has described this species from Japan. the present specimens are slightly difference from the re-description made by Miya (1972) in the following characters: the rostrum more slender and reaching the middle of the second segment of the antennular peduncle rather than the end of the first segment; distal superior margin of the palm in the major chela with one tooth; the ventral margin of the propodus in third pereiopods with 7–8 spines rather than 4–5 spines; two pairs of spines on the posterior lateral angle of the telson larger.

**Distribution** the South China Sea; Maldive Archipelago; Indonesia; Japan (variety?); Philippines; Mariana; Caroline; Marshall; Gilbert; Fiji; Samoa; Line and Hawaiian Islands; Australian.

### 2.15.24 *Synalpheus pescadorensis* Coutière, 1905 (Fig. 2.183)

*Synalpheus pescadorensis* Coutière, 1905: 877, Pl. 73, fig. 15; De Man, 1911: 298, fig. 63; 1926: 341; Tiwari, 1963: 279, fig. 5; Banner DM & Banner AH, 1975: 301, fig. 6; 1978: 245; Wang & Sha, 2015: 414.

**Material Examined** This species was reported by Banner & Banner (1978) and Liu (2008) from South China Sea, but not found in present specimens.

**Description** Rostrum slender, about 3.5 times as long as wide at base, reaching to 1/4 length of first antennular segment. Orbital teeth triangular, nearly as long as rostrum, much wider at base than rostrum.

Antennular peduncle with distally acute stylocerite, not reaching to distal margin of first seg-

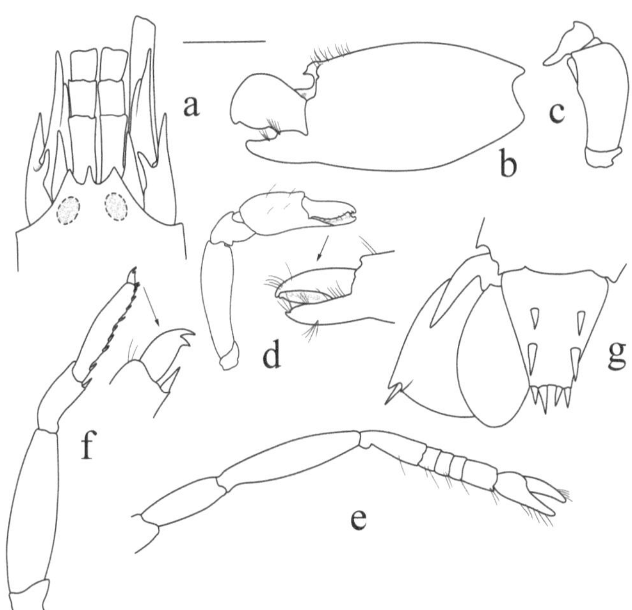


**Fig. 2.183** *Synalpheus pescadorensis* Coutière, 1905. (Adapted from Banner & Banner, 1975) **a**, anterior part of carapace, dorsal; **b**, major chela; **c**, carpus to ischium of major cheliped; **d**, minor cheliped; **e**, second pereiopod; **f**, third pereiopod; **g**, left uropod and telson (scale bar = 1 mm)

ment; visible part of first segment twice as long as second segment; second segment about 1.3 times as long as wide, slightly longer than third segment. Antenna with basicerite bearing dorsolateral teeth reaching to mid-length of first antennular segment, and ventrolateral teeth overreaching distal margin of that segment; antennal scale with narrow blade reaching to distal margin of second antennular segment, much shorter than lateral tooth which overreaching distal margin of antennular peduncle; carpocerite slender, far beyond antennular peduncle.

Major cheliped with chela cylindrical, about 2.3 times as long as wide. Fixed finger slightly longer than dactylus. Tubercle on distal margin of dorsal surface of palm slightly inflated, with acute teeth distally. Merus about twice as long as wide, with dorsal margin rounded distally. Minor cheliped with chela 3 times as long as wide. Dactylus spatulate, terminating in two acute teeth; fixed finger bearing one tooth distally. Merus similar to that of major cheliped, but more slender, about 3.5 times as long as wide.

Second pereiopod with carpus 5-segmented, ratios sub-equal to 10: 2: 2: 2: 5.

Third pereiopod with dactylus biunguiculate, superior unguis slightly longer than inferior unguis, but as wide at base as inferior unguis. Propodus about 3/5 length of merus, with ventral margin bearing 8–9 spines. Carpus about 1/3 length of merus, ventral margin bearing one spine distally. Merus about 3.5 times as long as wide, unarmed with spines on ventral margin.

Telson about 2.3 times as long as posterior width; dorsal surface bearing two pairs of moderately large spines, with posterior pair situated on 1/3 length of telson; posterior margin narrow, nearly straight.

**Distribution** The South China Sea (Taiwan); Maldive Archipelago; Vietnam; Indonesia; Sulu Archipelago; Pescadores; Solomon Islands; Australian; Philippines.

### 2.15.25 *Synalpheus pococki* Coutière, 1989 (Fig. 2.184)

*Synalpheus neomeris* var. *pococki* Coutière, 1898c: 167, fig. 2; Wang & Sha, 2015: 414.
*Synalpheus pococki*–. De Man, 1911: 234, fig. 32.

**Material Examined** This species was reported by Liu (2008) from South China Seas, but not found in present specimens.

**Description** Rostrum slender, about 3.5 times as long as wide at base, overreaching distal margin of first antennular segment. Orbital teeth slightly shorter than rostrum, but much wider at base than rostrum.

Antennular peduncle with distally acute stylocerite reaching to mid-length of second segment; visible part of first segment much longer than second segment; second segment longer than third segment. Antenna with basicerite bearing dorsolateral teeth shorter than ventrolateral teeth; antennal scale with blade shorter than lateral tooth reaching to distal margin of antennular peduncle; carpocerite slightly shorter than antennal scale.

Major cheliped with chela cylindrical, about 2.6 times as long as wide. Fingers about 1/3 length of palm. Palm with dorsal margin terminating in acute tooth. Merus about 3 times as long as wide, with dorsal margin bearing triangular acute tooth distally. Minor cheliped with chela about 4 times as long as wide. Finger about 1/3 length of palm. Carpus cup-shaped, with dorsal margin bearing acute lobes distally. Merus similar to that of major cheliped, but more slender.

Second pereiopod with carpus 5-segmente, with ratios sub-equal to 10: 1: 1: 1: 3.

Third pereiopod with dactylus biunguiculate, about 1/4 length of propodus, superior unguis slightly longer than inferior unguis, but narrower at base than inferior unguis. Propodus bearing one spine proximally, one spine on 3/4 length and distally one. Carpus with dorsal margin terminat-

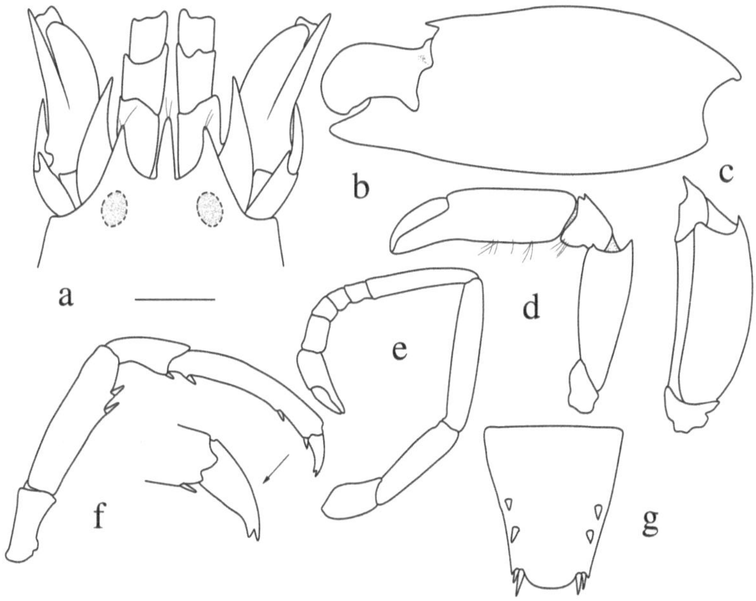


**Fig. 2.184** *Synalpheus pococki* Coutière, 1898. (Adapted from Banner & Banner, 1975). **a**, anterior part of carapace; **b**, major chela; **c**, carpus to ischium of major cheliped; **d**, minor cheliped; **e**, second pereiopod; **f**, third pereiopod; **g**, telson (scale bar = 1 mm)

ing in sub-acute tooth; ventral margin bearing one spine distally. Merus about 4 times as long as wide, with ventral margin armed with 2 spines.

Telson about twice as long as posterior width; posterolateral angle projecting as acute tooth; dorsal margin bearing two pairs of spines, with anterior pair situated on 1/2 length of telson; posterior margin convex, bearing two pairs of posterolateral spines.

**Distribution** The South China Sea (Taiwan); Aru Islands; Australian.

### 2.15.26 *Synalpheus stimpsonii* (De Man, 1888) (Figs. 2.185 and 2.186)

*Alpheus stimpsonii* De Man, 1888b: 513, Pl. 22, fig. 3.

*Synalpheus stimpsonii*–. Banner AH & Banner DM, 1966b: 46, fig. 12; 1968b: 274; Wang & Sha, 2015: 414, figs. 39–40.

*Synalpheus stimpsoni*–. Banner DM & Banner AH, 1975: 292, fig. 2m, 4.

*Alpheus amhoinae* Zehntner, 1894: 202, Pl. 8, fig. 23.

*Synalpheus amhoinae* De Man, 1911: 203, fig. 20; 1922: 26, fig. 13; Banner, 1958: 159, fig. 2.

*Synalpheus consobrinus* De Man. 1911: 204, fig. 21.

*Synalpheus brucei* Potts, 1915: 76, Pl. I, fig. 2; Clark, 1921: 625.

*Synalpheus striatus* Kubo, 1938: 89, figs. 1, 2; Miya, 1972: 47, Pl. 6.

**Material Examined** MBM 157748, 1♀1♂, CL 7.3–10.5 mm, TL 17.2–22.3 mm; MBM 158719, 1♀, CL 7.9 mm, TL 12.5 mm; MBM 157903, 1♂, CL 8.2 mm, TL 18.5 mm, Ren Xian-qiu Coll., 19 April 1975; MBM 158503, 1♀1♂, CL 11.5–12.1 mm, TL 25.9–28.8 mm; MBM 158592, 1 ♂, CL 8.2 mm, TL 18.7 mm, 19 April 1958.

**Diagnosis** Carapace smooth, glabrous. Rostrum slightly depressed, slightly concave in lateral

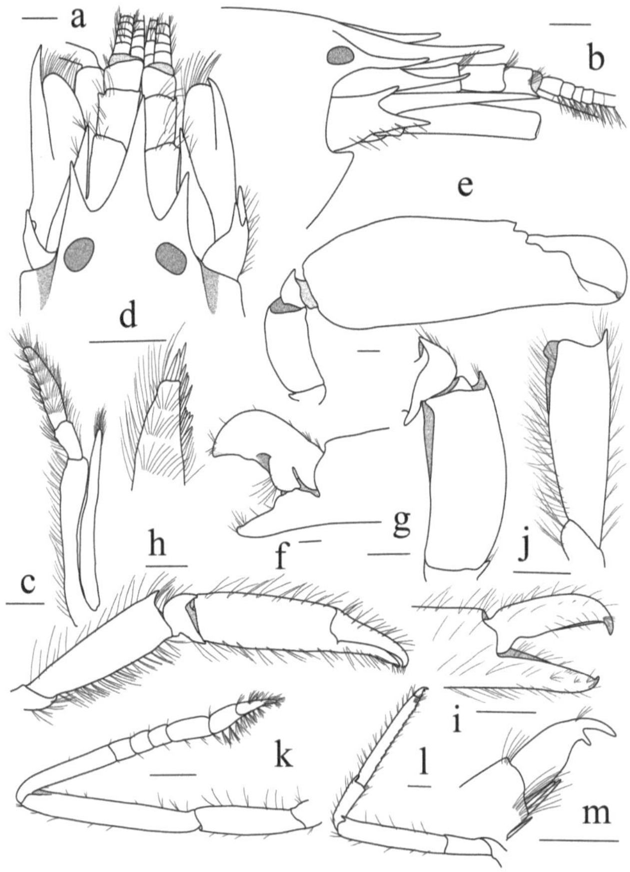


**Fig. 2.185** *Synalpheus stimpsonii* (De Man, 1888). MBM 157903, CL 8.2 mm, male, **a**, anterior part of carapace, dorsal view; **b**, same, lateral view; **c**, third maxilliped; **d**, same, tip; **e**, left major cheliped; **f**, same, fingers; **g**, merus of major cheliped; **h**, right minor cheliped; **i**, same, fingers; **j**, merus of minor cheliped; **k**, second pereiopod; **l**, third pereiopod; **m**, same, distal part of propodus and dactylus (scale bar: **a**–**c**, **e**–**l** = 1 mm; **d**, **m** = 0.5 mm)

view; acute tip reaching end of second antennular segment, with rounded carina, extending to level of eyes. Orbital hoods near half-length of rostrum. Pterygostomial corner produced into blunt acute angle. Cardiac notch well developed.

Telson about 1.3 times as long as wide at base; dorsal surface with two pairs of moderately large spines, inserted approximately at 1/2 and 2/3 length of telson; posterior margin slightly produced, fringed with long setae, posterolateral angle each with two pairs of spines, lateral about 2/3 length of mesial, posterolateral margins not produced.

Antennular peduncle moderately stout, with distally acute stylocerite, latter reaching about 1/3 length of second antennular segment; second segment about 1.3 times as long as wide; visible part of first segment longer than second and about twice as long as third. Basicerite with acute distodorsal tooth, distolateral tooth reaching about 1/2 length of visible part of first segment of antennular, distinctly shorter than stylocerite; carpocerite slightly overreaching end of third antennular segment; scaphocerite blade moderately broad, reaching middle of third antennular segment, lateral spine developed, distally curved inward, reaching end of antennular peduncle, shorter than carpocerite.

Third maxilliped far beyond antennular peduncle and carpocerite when extended; antepenultimate segment longest; penultimate seg-

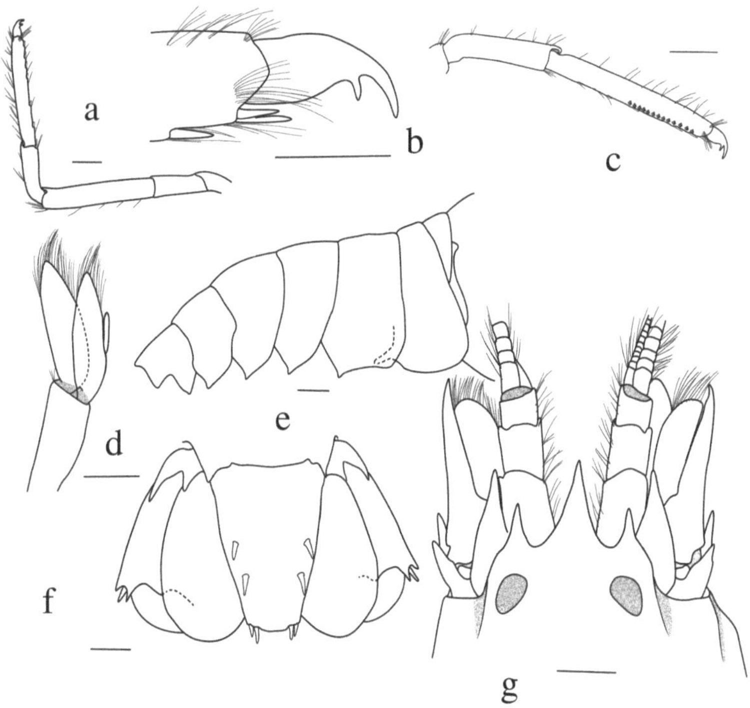


**Fig. 2.186** *Synalpheus stimpsonii* (De Man, 1888). MBM 157903, CL 8.2 mm, male, **a**, fourth pereiopod; **b**, same, distal part of propodus and dactylus; **c**, fifth pereiopod; **d**, second pleopod; **e**, abdominal segments, lateral view; **f**, uropods and telson. Female, CL 7.9 mm, **g**, anterior part of carapace (scale bar: **b** = 0.5 mm; **a**, **c**–**g** = 1 mm)

ment about 1.5 times as long as wide at base; tip of ultimate segment bearing crown of four spine-like setae, with transverse rows of setae along ventral margin.

Major cheliped with chela slightly compressed. Dactylus as long as than fixed finger, distally slightly curved. Palm about 2.2 times as long as dactylus, distodorsal of palm with one obtuse projecting. Carpus cup-shaped, with setae distodorsally. Merus about 3.5 times as long as wide at base, distodorsal margin with acute triangular tooth. Minor chela with dactylus and fixed finger tapering, distal single tooth, margin with fine setae. Palm about 1.6 times as long as fingers, dorsal and ventral margin with fine setae. Carpus cup-shaped, proportionally longer that of major cheliped. Merus about 5.2 times as long as wide at base, distodorsal margin with acute triangular tooth, ventral and dorsal margin bearing fine long setae

Second pereiopod with fingers near as long as palm. Carpus five-segmented, segment ratio subequal to 5.5: 1: 1: 1: 2.1, with first segment slightly longer than sum of last four segment; dorsal margin of palm convex.

Third pereiopod slender, with dactylus slightly curved to propodus, biunguiculate, extensor tooth about twice as long as flexor tooth, notch between extensor and flexor tooth V-shaped. Propodus with ten small spines along ventral margin plus one distal pair of spines near dactylus. Carpus about 1/3 length of merus, with dorsal margin extended into an obtuse tooth, ventral margin with two distal spines. Merus slightly longer than propodus, about 6.6 times as long as wide at base. Fourth pereiopod generally similar to third; ventral margin of propodus with six small spines. Fifth pereiopod similar to third and fourth pereiopods, propodus with distally 13 transversal rows of spine-like setae and three spines on ventral margin.

**Remarks** *S. stimpsonii* is originally described by De Man (1988). This species is variable in following characters: the length of the rostrum; the length of orbital tooth compared to rostrum; the length of first antennular segment compared to the sum of two following segments; the relative length of the distodorsal and distolateral tooth; the proportions of the pereiopods; the dorsal spines and posterolateral margins of telson (Banner & Banner 1975). Those characters were taken variable in *Synalpheus*. The present seven specimens are also variable in the length of the rostrum. Other than the length or relative length of characters mentioned above, the present specimens also differ from Kubo (1938, Pl. 6, fig. D) and Banner & Banner (1966: fig. 12B) in the smaller and obtuse tooth on the distodorsal margin of the palm of major chela.

**Distribution** China Seas; Eastern Africa and Madagascar to Singapore; Thailand; Japan; Philippines; Indonesia; Australia; Marshall; Gilbert; Loyalty Islands.

### 2.15.27 *Synalpheus streptodactylus* Coutière, 1905 (Figs. 2.187, 2.188, and 2.189)

*Synalpheus neomeris streptodactylus* Coutière, 1905: 870, fig. 1.

*Alpheus neomeris* De Man: 734; Coutière, 1905: 869, fig. 1.

*Synalpheus streptodactylus*–. De Man, 1911: 226, Pl. 7, fig. 29; Banner DM & Banner AH, 1975: 362, fig. 23; 1985: 55; Wang & Sha, 2015: 417, figs. 41–43.

*Synalpheus metaneomeris streptodactylus* Coutière, 1921: 414, Pl. 60, fig. 4.

*Synalpheus streptodactylus streptodactylus* Banner AH & Banner DM, 1966a: 157; 1966b: 50, fig. 14; Miya, 1972: 69, Pl. 14.

*Synalpheus streptodactylus hadrungus* Banner AH & Banner DM, 1966a: 158.

*Synalpheus jedanensis*–. Barnard, 1950: 738, fig. 139c–i [Not *S. jedanensis* De Man, 1909].

**Material Examined** CL 3.2–5.7 mm, TL 6.5–12.4 mm. MBM 270607, 1♂, Beibu Gulf, Sta. 6216, depth: 67 m, Sun Fu-zeng Coll., 10 April 1962; MBM 270607, 1♂, Beibu Gulf, Sta. 6216, depth: 67 m, Sun Fu-zeng Coll., 10 April 1962; MBM 270216, 1ovigerous♀, Beibu Gulf, Sta. 6216, depth: 57 m, Sun Fu-zeng Coll., 13 March 1960; MBM 270223, 1♂, Beibu Gulf, Sta. 6252, depth: 56.2 m, Shen Shou-peng Coll., 7 July 1960; MBM 270222, 1non-ovigerous♀, Beibu Gulf, Sta. 6251, depth: 47 m, Fan Zhen-gang Coll., 7 February 1960; MBM 270221, 1non-ovigerous, Beibu Gulf, Sta. 6262, depth: 49.6 m, Fan Zhen-gang Coll., 7 July 1960; MBM 270219, 2♂, Beibu Gulf, Sta. 6273, depth: 46 m, Shen Shou-peng Coll., 25 October 1960; MBM 270220, 1♂, Beibu Gulf, Sta. 6222, depth: 24 m, Fan Zhen-gang Coll., 13 February 1960; MBM 270217, 1♂, Beibu Gulf, Sta. 6249, depth: 30.5 m, Sun Fu-zeng Coll., 13 July 1960; MBM 270227, 1♂, Beibu Gulf, Sta. 7405, depth: 65 m, 12 October 1962; MBM 107815, 1♂, the East China Sea, Sta. III-4, depth: 82 m, Shen Shou-peng Coll., 27 June 1976; MBM 107813, 2♀, Beibu Gulf, Sta. 7405, depth: 42 m, Shen Shou-peng Coll., 12 January 1962; MBM 270197, 1ovigerous♀, the East China Sea, Sta. III-4, depth: 82 m, Shen Shou-peng Coll., 27 June 1976; MBM 081123, 1ovigerous♀, the East China Sea, Sta. 4083, depth: 89 m, Liu Rui-Yu Coll., 24 October 1959; MBM 270415, 1♀, Beibu Gulf, Sta. 7404, depth: 49.6 m, 13 October 1962; MBM 270226, 1♂, Beibu Gulf, Sta. 6262, depth: 51.5 m, Fan Zhen-gang Coll., 8 February 1960; MBM 270201, 1♂, Beibu Gulf, Sta. 6167, depth: 48 m, Sun Fu-zeng Coll., 8 February 1960; MBM 270218, 1♂, Beibu Gulf, Sta. 6243, depth: 93 m, Ma Xiu-tong Coll., 11 December 1959; MBM 270472, 1ovigerous♀, 2♂, Beibu Gulf, Sta. 6182, depth: 18 m, Sun Fu-zeng Coll., 11 February 1960; MBM 270453, 9ovigerous♀, Beibu Gulf, Sta. 6276, depth: 29 m, Zhang Bao-lin Coll., 6 November 1960; MBM 270363, 1♂, Beibu Gulf, Sta. 7302, depth: 30 m, Sun Fu-zeng Coll., 10 April 1962; MBM 270452, 1♂, Beibu Gulf, Sta. 6265, depth: 60 m, Ma Xiu-tong Coll., 1 December 1959;

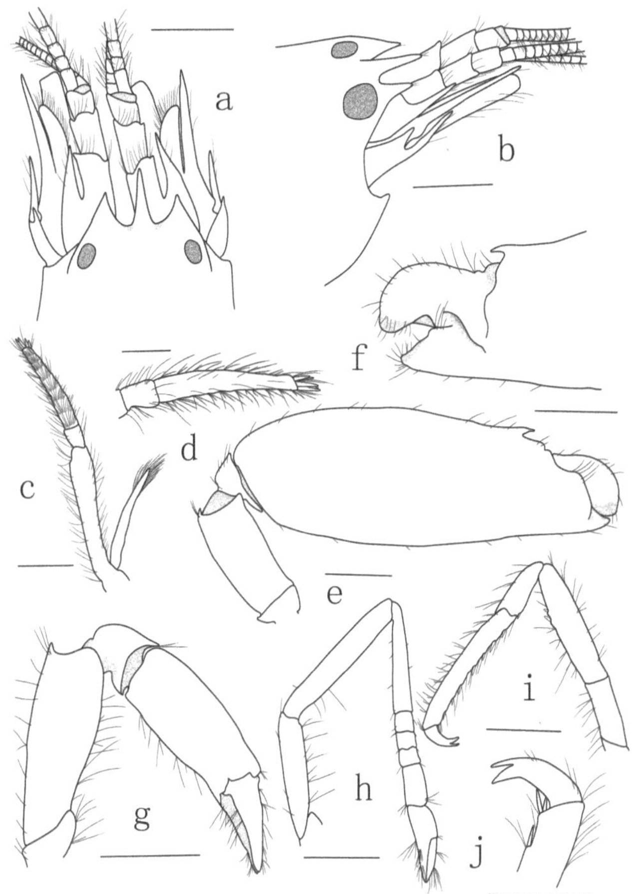


**Fig. 2.187** *Synalpheus streptodactylus* Coutière, 1905. MBM 270223, female, CL 5.1 mm, **a**, anterior part of carapace, dorsal view; **b**, same, lateral view; **c**, third maxilliped; **d**, same, tip; **e**, left major cheliped; **f**, same, distal part of palm and fingers; **g**, right minor cheliped; **h**, second pereiopod; **i**, third pereiopod; **j**, distal part of propodus and dactylus (scale bar: **a**–**i** = 1 mm; **j** = 0.5 mm)

MBM 270637, 1ovigerous♀1♂, depth: 53 m, 22 August 1962; 1♂, the East China Sea, Sta. 4083, depth: 88 m, Cheng Li-ren Coll., 4 July 1959; MBM 081146, 2♂, the East China Sea, Sta. 4084, depth: 100 m, Xu Jie-shan Coll., 1 February 1959; MBM 270424, 1♂, Beibu Gulf, Sta. 6240, depth: 58 m, Fan Zhen-gang Coll., 9 December 1960; MBM 270525, 1ovigerous♀, Beibu Gulf, Sta. 7503, depth: 52 m, Sun Fu-zeng Coll., 22 August 1962; MBM 270134, 1♂, Beibu Gulf, Sta. 7702, depth: 55 m, Sun Fu-zeng Coll., 12 January 1962; MBM 107809, 1ovigerous♀1♂, Nanji island, depth: 15 m, Ren Xian-qiu Coll., 24 October 1979; MBM 107805, 2ovigerous♀2♂, the East China Sea, Sta. V-7, depth: 100 m, Tang Zhi-chan & Zhang Bao-lin Coll., 10 October 1975; MBM 107817, 1♂, the East China Sea, Sta. V-2, depth: 85 m, Tang Zhi-chan & Xu Feng-shan Coll., 4 July 1976; MBM 107818, 2♂, the East China Sea, Sta. V-7, depth: 131 m, Tang Zhi-chan & Liu Rui-Yu Coll., 5 July 1976; MBM 107804, 1♂, the East China Sea, Sta. V-6, depth: 115 m, Liu Rui-Yu Coll., 4 July 1976; MBM 107816, 1ovigerous♀2♂, the East China Sea, Sta. V-10, depth: 110 m, Tang Zhi-chan & Xu Jie-shan Coll., 10 October 1975; MBM 107808, 1ovigerous♀1♂, the East China Sea, Sta. V-4, depth: 99 m, Tang Zhi-chan & Wu Yao-quan Coll., 5 July 1976; MBM 270594, 1ovigerous♀1♂, the

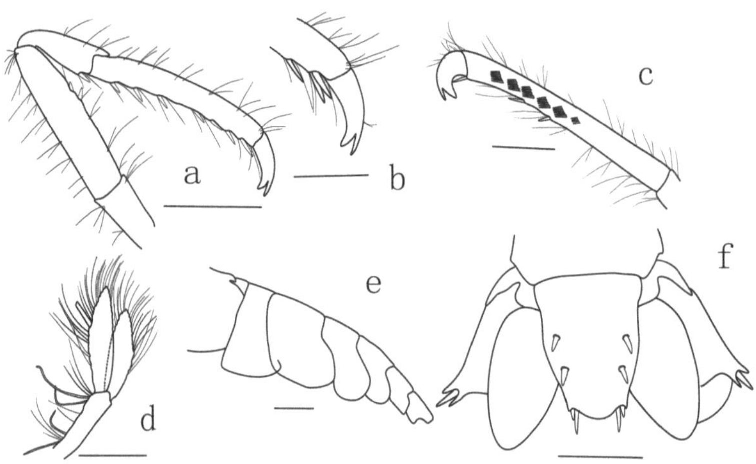


**Fig. 2.188** *Synalpheus streptodactylus* Coutière, 1905. MBM 270223, female, CL 5.1 mm, **a**, fourth pereiopod; **b**, same, distal part of propodus and dactylus; **c**, propodus and dactylus of fifth pereiopod; **d**, second pleopod; **e**, abdominal segments, lateral view; **f**, uropods and telson (scale bar: **a**, **d**–**f** = 1 mm; **b**, **c** = 0.5 mm)

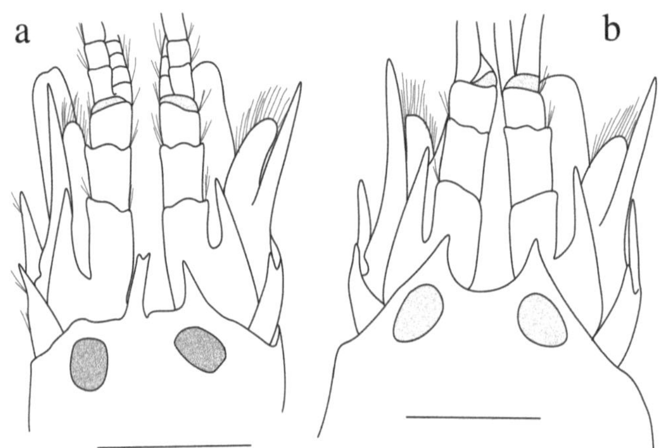


**Fig. 2.189** *Synalpheus streptodactylus* Coutière, 1905. MBM 158375, male, CL 3.1 mm, **a**, anterior part of carapace, dorsal view. MBM 107814, male, CL 4.5 mm, **b**, anterior part of carapace, dorsal view (scale bar = 1 mm)

East China Sea, Sta. 6277, depth: 31 m, Fan Zhen-gang Coll., 17 May 1960; MBM 107810, 2ovigerous♀1♂, the East China Sea, Sta. V-2, depth: 99 m, Tang Zhi-chan Coll., 5 July 1976; MBM 107814, 16ovigerous♀12♂, the East China Sea, Sta. V-3, depth: 105 m, Tang Zhi-chan Coll., 27 August 1976; MBM 270591, 1♂, Beibu Gulf, Sta. 6277, depth: 35 m, Fan Zhen-gang Coll., 9 March 1960; MBM 107807, 2♂, the East China Sea, Sta. V-5, depth: 90 m, Tang Zhi-chan Coll., 11 October 1975; MBM 107806, 2ovigerous♀7♂, the East China Sea, Sta. V-5, depth: 100 m, Tang Zhi-chan & Xu Feng-shan Coll., 10 October 1975; MBM 107820, 1ovigerous♀1♂, the East China Sea, Sta. V-6, depth: 112 m, Tang Zhi-chan & Xu Feng-shan Coll., 5 July 1976; MBM 107812, 1ovigerous♀, the East China Sea, Sta. V-7, depth: 126 m, Tang Zhi-chan & Xu Feng-shan Coll., 21 September 1976; MBM 107811, 2ovigerous♀, the East China Sea, Sta. V-3, depth: 105 m, Zhang Bao-lin Coll., 11 June 1978; 1ovigerous♀, MBM 270588, Beibu Gulf, Sta. 6265, depth: 62 m, Sun Fu-zeng Coll., 25 May 1960; MBM 270609, 1♂, Beibu Gulf, Sta. 7901, depth: 52 m, Sun Fu-zeng Coll., 17

August 1962; MBM 107740, 1ovigerous♀2♂, Xincun, depth: 1–3 m, Liu Rui-Yu Coll., 26 March 1992; 1♂, Yezhu Island, 11 April 1992; MBM 042231, 1♂, the East China Sea, Sta. 4081, depth: 79 m, Zhang Bao-lin Coll., 25 October 1959; MBM 158542, 1♂, Yezhu Island, depth: 97 m, from living millipede, 25 March 1992; MBM 159262, 1♂, Sanya, Dadonghai, E-side, depth: 4–6 m, 1 December 1997; MBM 158375, 1♂, the East China Sea, Sta. V-8, depth: 79 m, Tang Zhi-chan Coll., 10 October 1975; MBM 158076, 1♂, Sanya, Dadonghai, E-side, depth: 6–9 m, Li Xin-zheng Coll., 27 November 1997; MBM 158209, 1♂, Sanya, Dadonghai, Li Xin-zheng Coll., 1 March 1997; MBM 158652, 1♂, Yalong Bay, Liu Rui-Yu Coll.; MBM 270720, 1ovigerous♀, Xinying, 14 January 1956; MBM 158334, 1♂; MBM 159319, 3ovigerous♀2♂, Xincun, depth: 1–3 m, 26 March 1992; MBM 159319, 1ovigerous♀2♂, Xincun, depth: 1–3 m, 21 March 1992; MBM 159279, 1ovigerous♀1♂, Yezhu Island, 27 March 1992; MBM 159278, 1♂; MBM 158223, 1ovigerous♀4♂, Weizhou Island, Liu Rui-Yu Coll., 26 December 1954; MBM 107777, 1♂, Shalao, depth: 12 m, 4 April 1992; MBM 107766, 1ovigerous♀, Yezhu Island, 17, November 1990; MBM 159299, 1♂.

**Description** Carapace smooth, glabrous. Rostrum about 2.7 times as long as wide at base; acute tip not reaching end of first antennular segment. Orbital hoods about 5/8 length of rostrum. Pterygostomial corner produced into acute angle. Cardiac notch well developed.

Telson about 1.4 times as long as wide at base; dorsal surface with two pairs of moderately large spines, inserted approximately at 2/5 and 3/5 length of telson; posterior margin strongly produced, fringed with long setae, posterolateral angle each with two pairs of spines, lateral about 1/3 length of mesial, posterolateral margins each produced into acute small tooth, but distinctly shorter than lateral spines.

Antennular peduncle moderately slender, with distally acute stylocerite, latter reaching about 1/4 length of second antennular segment; second segment about 1.5 times as long as wide at base; visible part of first segment longer than second and about twice as long as third. Basicerite with acute distodorsal tooth, distolateral tooth almost as long as stylocerite; carpocerite overreaching end of third antennular segment; scaphocerite blade moderately broad, reaching 3/4 length of third antennular segment, lateral spine developed, overreaching end of antennular peduncle, slight longer than carpocerite.

Third maxilliped far beyond antennular peduncle and carpocerite when extended; antepenultimate segment longest; penultimate segment about 1.3 times as long as wide at base; tip of ultimate segment bearing crown of five spine-like setae, with transverse rows of setae along ventral margin.

Major cheliped with dactylus slightly longer than fixed finger, distally slightly curved. Palm about 2.8 times as long as dactylus, distodorsal of palm with one acute projecting, slightly curved to dactylus. Carpus cup-shaped, with setae distodorsally. Merus about 2.7 times as long as wide at base, distodorsal margin with one acute triangular tooth. Minor chela with dactylus and fixed finger tapering, distal single tooth. Palm about 1.5 times as long as fingers. Carpus cup-shaped, proportionally longer that of major cheliped. Merus about 3.7 times as long as wide at base, distodorsal margin with one acute triangular tooth.

Second pereiopod with ischium shorter than merus; carpus five-segmented, segment ratio subequal to 5: 1: 1: 1: 1.8, with first segment slightly longer than sum of last four segment; dorsal margin of palm convex; fingers about 1.5 times as long as palm.

Third pereiopod moderately slender, dactylus about 1/4 length of propodus, slightly curved to propodus, biunguiculate, extensor tooth slight shorter than flexor tooth, notch between extensor and flexor tooth V-shaped, flexor tooth about twice as wide at base as extensor tooth. Propodus with seven spines along ventral margin plus one distal pair of spines near dactylus. Carpus about 1/2 length of merus, with dorsal margin extended into an obtuse tooth, ventral margin with one distal spine. Merus slightly shorter than propodus, about 4.5 times as long as wide at base, ventral margin with 2–5 spines. Fourth pereiopod generally similar to third; merus shorter than propodus, ventral margin of propodus with seven spines. Fifth pereiopod similar to third and fourth pereiopods, ventral margin of propodus with six transversal rows of spine-like setae and two spines.

**Remarks** *S. streptodactylus* is original described by Coutière (1905). The variations of this species and the differences between this species and its most morphologically similar species (*S. neomeris*) are well discussed by Banner & Banner (1975). *Synalpheus streptodactyloides* De Man, 1909 was described as one new form by De Man based on single specimen, which was synonymized with *S. streptodactylus* by Banner & Banner (1985). The present abundant specimens are generally agreed with the re-description made by Banner & Banner (1975).

**Distribution** China Seas; Red Sea; eastern and South Africa to Thailand; Indonesia; Japan; Australia to Hawaii; Philippines.

### 2.15.28 *Synalpheus theano* De Man, 1910 (Fig. 2.190)

*Synalpheus theano* De Man, 1910: 296; 1911: 293, fig. 61; Banner AH & Banner DM, 1972: 20, fig. 3f; Banner DM & Banner AH, 1975: 314, fig. 10; Wang & Sha, 2015: 421

*Not Synalpheus theano*–. Banner AH & Banner DM, 1966b: 69, fig. 24 (= *S. neptunus* (Dana, 1852)).

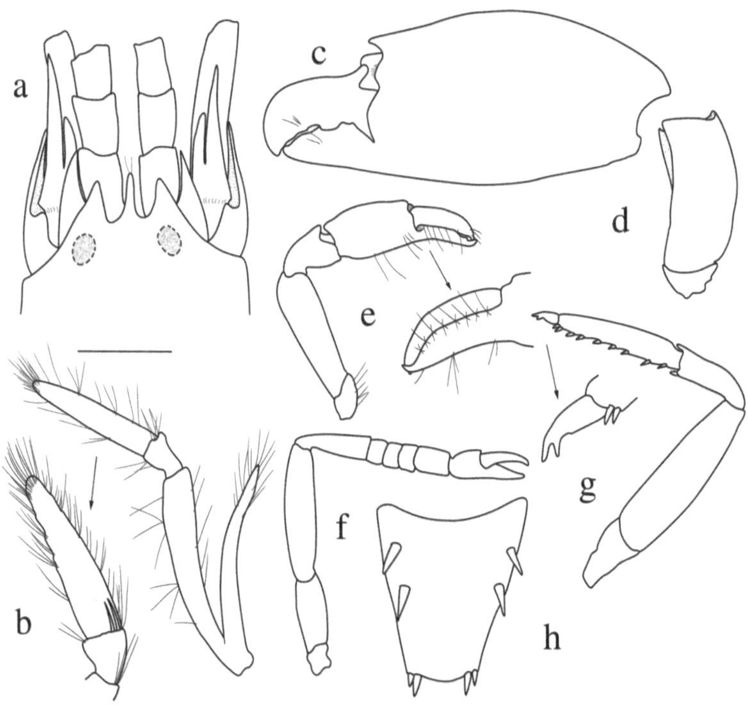


**Fig. 2.190** *Synalpheus theano* De Man, 1910 (Banner & Banner, 1975) **a**, anterior part of carapace, dorsal view; **b**, third maxilliped; **c**, major chela; **d**, merus and ischium of major cheliped; **e**, minor cheliped; **f**, second pereiopod; **g**, third pereiopod; **h**, telson (scale bar = 1 mm)

**Material Examined** This species was reported by Liu (2008) from South China Seas, but not found in present specimens.

**Description** Rostrum overreaching mid-length of first antennular segment. Orbital teeth nearly as long as rostrum, but twice as wide at base as rostrum.

Antennular peduncle with distally acute stylocerite, reaching to distal margin of first segment; visible part of first segment slightly longer than second segment; second segment about 1.3 times as long as wide, longer than third segment. Antenna with basicerite without dorsolateral teeth, with ventrolateral teeth hardly reaching to distal margin of second antennular segment; antennal scale with narrow blade not reaching to distal margin of antennular peduncle, distinctly shorter than lateral teeth; carpocerite slender, far overreaching antennular peduncle.

Major cheliped with chela cylindrical, about 2.4 times as long as wide. Palm with dorsal margin bearing sub-acute large tooth distally. Merus with all margins terminating in acute tooth. Minor chela with chela about 2.4 times as long as wide. Fingers nearly as long as palm, with dactylus spatulate, lateral margin bearing 7–8 long setae; mesial margins of both dactylus and fixed finger bearing row of setae. Carpus cup-shaped. Merus about 3.2 times as long as wide, with dorsal margin rounded distally.

Second pereiopod with carpus 5-segmented, ratios sub-equal to 5: 1: 1: 1: 2.

Third pereiopod with dactylus biunguiculate, superior unguis nearly as long as inferior unguis, but narrower at base than inferior unguis. Propodus shorter than merus, with ventral margin bearing 7 spines and distally one pair. Carpus about 2/5 length of merus, with dorsal margin terminating in sub-acute tooth; ventral margin bearing one spine distally. Merus about 3.7 times as long as wide, ventral margin unarmed with spines.

Telson about 2.2 times as long as posterior width; dorsal surface bearing two pairs of spines, with posterior pair situated on 1/2 length of telson; posterolateral angle not projecting as tooth; posterior margin slightly convex, bearing two pairs of posterolateral spines.

**Distribution** The South China Sea; Indonesia; Singapore; Australian.

### 2.15.29 *Synalpheus tonkinensis* Wang & Sha, 2015 (Figs. 2.191 and 2.192)

*Synalpheus tonkinensis* Wang & Sha, 2015: 421, figs. 44, 45.

**Material Examined** MBM 270602, ♀, CL 5.3 mm, TL 16.5 mm, Beibu Gulf, Sta. 7305, depth: 62 m, Sun Fu-zeng Coll., 21 January 1962; MBM 270602, 1♂, CL 4.8 mm, TL 12.0 mm, Beibu Gulf, Sta. 7305, depth: 62 m, Sun Fu-zeng Coll., 21 January 1962.

**Description** Carapace smooth, glabrous. Rostrum flattened dorsally, with lateral margins slightly convex; rounded tip overreaching 1/2 length of visible part of first antennular segment; tip of rostrum bearing two short setae. Orbital hoods about twice as wide at base as rostrum, length almost equal to rostrum, rounded tip bearing two short setae; notch between orbital hoods and rostrum deep, between V- and U- shaped. Pterygostomial corner produced into blunt acute angle.

Abdominal segments smooth, glabrous, sex bimorphology, female with posterior corner of first pleura broadly rounded, ventral and posterior margins of second pleura longer and wider; male with posterior corner of first pleura acutely produced ventrally. Telson about 1.4 as long as wide at base; lateral margins concave posteriorly; dorsal surface with two pairs of spines, inserted approximately at 1/4 and 1/2 length of telson; posterior margin strongly producing, fringed with long setae, posterolateral angle each with two pairs of spines, lateral about 1/2 length of

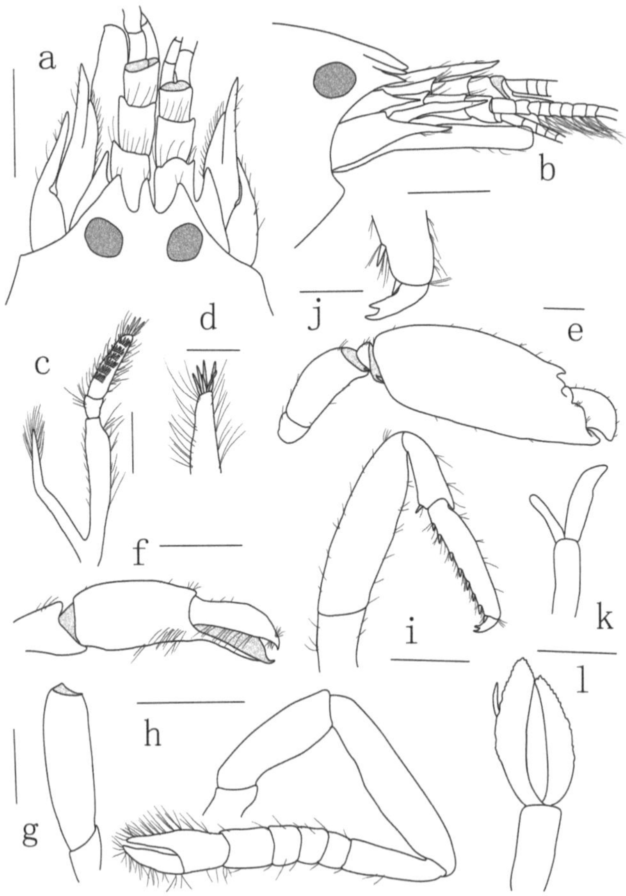


**Fig. 2.191** *Synalpheus tonkinensis* Wang & Sha, 2015. MBM 270602, female, CL 5.3 mm, **a**, anterior part of carapace, dorsal view; **b**, same, lateral view; **c**, third maxilliped; **d**, same, tip; **e**, major cheliped; **f**, minor chela; **g**, merus of minor cheliped; **h**, second pereiopod; **i**, third pereiopod; **j**, same, distal part of propodus and dactylus; **k**, first pleopod; **l**, third pleopod (scale bar: **a**–**i**, **k**, **l** = 1 mm; **j** = 0.25 mm)

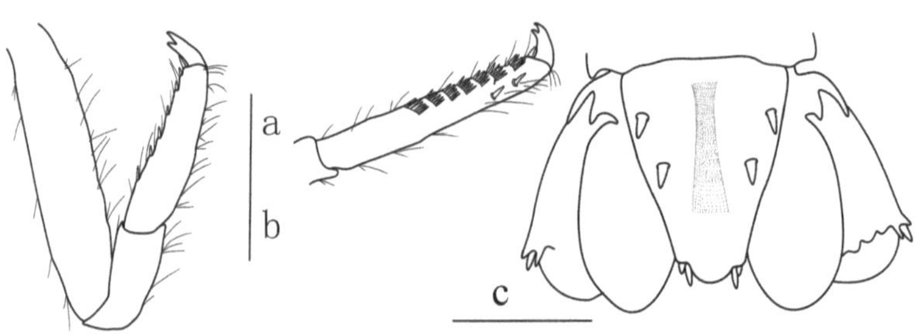


**Fig. 2.192** *Synalpheus tonkinensis* Wang & Sha, 2015. MBM 270602, female, CL 5.3 mm, **a**, fourth pereiopod; **b**, propodus and dactylus of fifth pereiopod; **c**, uropods and telson (scale bar = 1 mm)

mesial, posterolateral margins forming right angles.

Antennular peduncle with distally acute stylocerite, latter hardly reaching distal margin of first antennular segment; second segment distinctly longer than wide; visible part of first segment longer than second and about twice as long as third segment. Antenna with basicerite without distodorsal tooth, distolateral tooth distinctly longer than stylocerite, overreaching middle of second antennular segment; carpocerite beyond end of antennular peduncle by the length of third segment; scaphocerite blade narrow, hardly reaching end of second antennular segment, lateral spine robust, with lateral margin slightly concave, subequal to antennular peduncle.

Mouthparts not dissected. Third maxilliped far beyond antennular peduncle and carpocerite when extended; antepenultimate segment longest, about 5.5 times as long as wide; penultimate segment about 1.8 times as long as wide at base; tip of ultimate segment bearing crown of five-six spin-like long setae, with several brush consisting of six-eight transverse rows of setae along ventral margin.

Major cheliped with chela cylindrical, not a bit compressed or twisted. Dactylus far longer than fixed finger. Palm about 3.2 times as long as dactylus, distodorsal tooth on palm subacute, slightly curved downward to dactylus. Carpus cup-shaped, with long setae distodorsally. Merus three times as long as wide at base, distodorsal margin rounded without any projecting, inferointernal margin with rounded tooth. Minor chela with both dactylus and fixed finger dorsal surface convex, broadened from base to 8/9 length of fingers, terminating in slender, acute, curved teeth, crossing distally, lateral margins of fingers bearing 6–8 long stiff setae that cross in regular fashion. Palm 1.3 times as long as fingers, lateral margin near dactylus bearing patch of stiff setae. Carpus cup-shaped, proportionally longer than that of major cheliped. Merus about two times as long as wide, longer than that of major cheliped, distodorsal margin same with that of major cheliped, without any projecting.

Second pereiopod with fingers 1.3 times as long as palm. Carpus five-segmented, segment ratio subequal to 4: 1: 1: 1: 2, with first segment shorter than sum of last four segments.

Third pereiopod stout, with dactylus around 2/11 propodus length, biunguiculate, moderately slender, extensor tooth slightly longer, thicker at base than flexor tooth, notch between extensor tooth and flexor tooth V-shaped. Propodus with usual seven spines along ventral margin plus one distal pair of spines near dactylus. Carpus about 3/8 merus length, with dorsal margin extended into an obtuse tooth, ventral margin with one short distal spine. Merus longer than propodus, about four times as long as wide, without spines on ventral margin. Fourth pereiopod generally similar to third; but ventral margin of carpus without distal spine. Fifth pereiopod with merus about 3.7 times as long as wide, without spines on ventral margin; carpus about 9/11 merus length, ventral margin without distal spine; propodus as long as merus, with seven brush consisting of five-six transverse rows of short setae along on posterior face, and two mobile spines on ventral margin plus one distal pair of spines near dactylus.

**Remarks** *Synalpheus tonkinensis* is mainly identified by the configuration of the frontal region of the carapace; the shape of the major and minor cheliped and the telson. The differences between this species and its similar species are discussed in Wang & Sha (2015).

**Distribution** Beibu Gulf, the South China Sea.

### 2.15.30 *Synalpheus trispinosus* De Man, 1910 (Figs. 2.193 and 2.194)

*Synalpheus trispinosus* De Man, 1910: 300; 1911: 288, Pl. 12, fig. 58; Banner AH & Banner DM, 1981: 221, figs. 2a–c; Banner DM & Banner AH, 1981: 82; Wang & Sha, 2015: 424, figs. 46, 47.

**Material Examined** MBM 042237, 6♀6♂, CL 5.3–8.6 mm, TL 13.1–21.5 mm, the East China Sea, Sta. 4082, depth: 84 m, 1 July 1959; MBM 042234, 3♀, 3♂, CL 5.1–9.5 mm, TL 13.5–24.1 mm, the East China Sea, Sta. 4082, depth: 86 m, Liu Rui-Yu Coll., 25 October 1959; MBM 107822, 1♂, CL 6.2 mm, TL 14.1 mm, the East China Sea, Sta. V-2, depth: 99 m, Tang Zhi-chan Coll., 5 July 1976; MBM 081144, 1♂, CL 5.6 mm, TL 13.9 mm, the East China Sea, Sta.

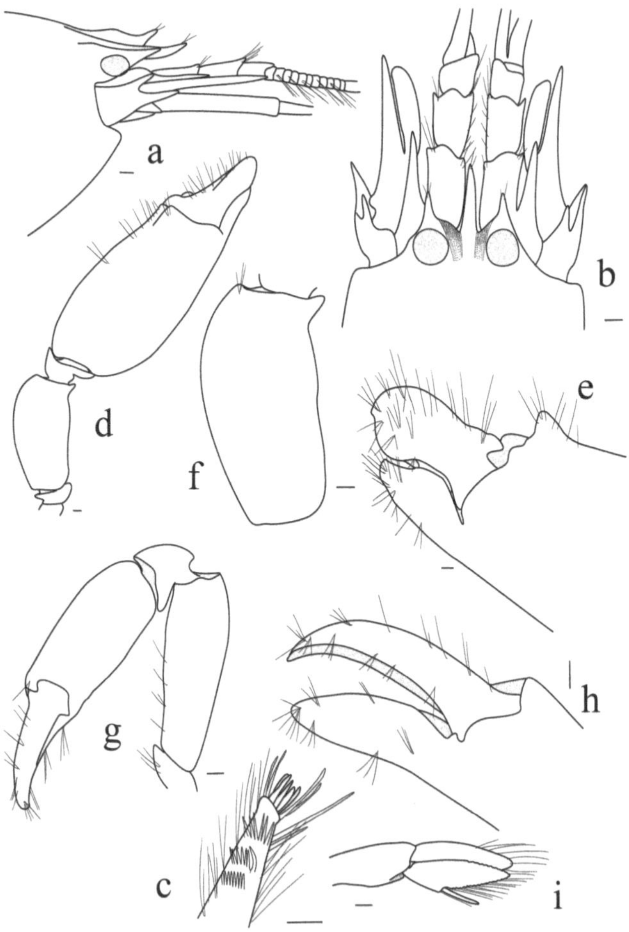


**Fig. 2.193** *Synalpheus trispinosus* De Man, 1910. MBM 042234, female, CL 9.5 mm, **a**, anterior part of carapace, lateral view; **b**, same, dorsal view; **c**, tip of third maxilliped; **d**, major cheliped; **e**, same, distal part of palm and fingers; **f**, merus of major cheliped; **g**, minor cheliped; **h**, same, fingers; **i**, second pleopod (scale bar = 0.25 mm)

4083, depth: 90 m, Cui Yu-hang Coll., 8 December 1959; MBM 081143, 1♀, CL 8.1 mm, TL 19.8 mm, the East China Sea, Sta. 4061, depth: 81 m, 5 December 1959; MBM 270162, 1♀2♂, CL 5.9–9.1 mm, TL 14.3–22.5 mm, the East China Sea, Sta. 4081, depth: 81 m, Xu Jie-shan Coll., 2 February 1959; MBM 042236, 4♀1♂, CL 6.6–8.8 mm, TL 14.9–22.5 mm, the East China Sea, Sta. 4081, depth: 79 m, Cheng Li-ren Coll., 3 July 1959; MBM 042232, 1♀1♂, CL 9.2 mm, TL 20.5 mm, CL 7.1 mm, TL 15.2 mm, the East China Sea, Sta. 4081, depth: 79 m, Zhang Bao-lin & Liu Rui-Yu Coll., 25 October 1959; MBM 107825, 2♂, CL 5.4–7.7 mm, TL 12.1–17.9 mm, the East China Sea, Sta. V-7, depth: 100 m, Tang Zhi-chan & Xu Jie-shan Coll., 10 October 1975; MBM 042238, 1♀, CL 8.4 mm, TL 20.2 mm, the East China Sea, Sta. 4081, depth: 79 m, Cui Yu-hang Coll., 9 December 1959; MBM 107814, 5♀2♂, CL 5.3–7.8 mm, TL 12.5–17.4 mm, the East China Sea, Sta. V-3, depth: 105 m, Tang Zhi-chan Coll., 27 August 1959; MBM 158587, 1♀, CL 8.1 mm, TL 18.3 mm, the East China Sea, Sta. 0108, depth: 94 m, Cui Yu-hang Coll., 9 December 1959.

**Description** Carapace smooth, glabrous. Rostrum slender, about 3.7 times as long as wide at base, with lateral margins slightly convex;

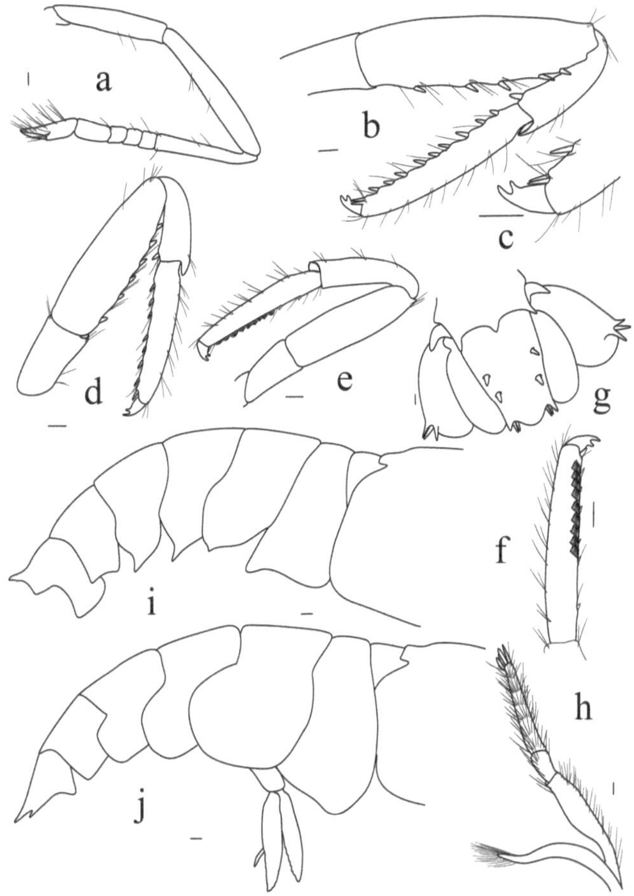


**Fig. 2.194** *Synalpheus trispinosus* De Man, 1910. MBM 042234, male, CL 9.5 mm, **a**, second pereiopod; **b**, third pereiopod; **c**, same, distal part of propodus and dactylus; **d**, fourth pereiopod; **e**, fifth pereiopod; **f**, same, propodus and dactylus; **g**, uropods and telson; **h**, third maxilliped; **i**, abdominal segments, lateral. Female, CL 9.5 mm, **j**, abdominal segments, lateral view (scale bar = 0.25 mm)

rounded tip bearing two short setae, varying from hardly reaching to slightly overreaching end of first antennular segment, tip strongly curved upturned. Orbital hoods distinctly shorter than rostrum, reaching middle of first antennular segment, acute tip strongly curved upturned. Pterygostomial corner produced into blunt acute angle. Cardiac notch well developed.

Abdominal segments smooth, glabrous, posterior margin of sixth pleura with each angle one acute tooth, and an third tooth at middle, slightly shorter. Telson about 1.35 times as long as wide at base; lateral margin slightly concave posteriorly; dorsal surface with two pairs of moderately small spines, inserted approximately at 2/5 and 3/5 length of telson; posterior margin strongly produced, fringed with long setae, posterolateral angle each with two pairs of shorter spines, lateral slightly longer than mesial, posterolateral margins each produced into acute large tooth, slightly longer than lateral spines.

Antennular peduncle slender, with distally acute stylocerite, latter reaching about 1/5 length of second antennular segment, longer than or as long as rostrum; second segment about twice times as long as wide; visible part of first segment longer than second and more than twice as long as third. Antenna with basicerite with acute distodorsal tooth, less 1/2 length of distolateral tooth, latter distinctly shorter than stylocerite; carpocerite slightly beyond end of antennular peduncle; scaphocerite blade moderately narrow, hardly reaching end of third antennular segment, lateral spine developed, with lateral margin slightly concave posteriorly, slightly overreaching end of antennular peduncle, shorter than carpocerite.

Third maxilliped far beyond antennular peduncle and carpocerite when extended; antepenultimate segment longest; penultimate segment twice as long as wide at base; tip of ultimate segment bearing crown of five long spine-like setae, with several transverse rows of setae along ventral margin.

Major cheliped with dactylus distinctly longer than fixed finger, bearing fine setae. Palm about twice as long as dactylus, distodorsal of palm with one large triangular tubercle directed obliquely upturned, bearing fine setae. Carpus cup-shaped, with setae distodorsally. Merus about 3.25 times as long as wide at base, distodorsal margin without triangular tooth, but infero-external and infero-external margin each bearing acute triangular tooth. Minor chela with dactylus and fixed finger tapering, distal single tooth. Palm about 1.1 times as long as fingers. Carpus cup-shaped, proportionally longer that of major cheliped. Merus about 4.5 times as long as wide, longer than that of major cheliped, distodorsal margin without any projecting.

Second pereiopod with fingers sub-equal to palm. Carpus five-segmented, segment ratio subequal to 6: 1: 1: 1: 2, with first segment longer than sum of last four segments.

Third pereiopod moderately stout, with dactylus around 1/8 propodus length, biunguiculate, extensor tooth longer than flexor tooth, and as wide at base as flexor tooth, notch between extensor and flexor tooth U-shaped. Propodus with nine spines along ventral margin plus one distal pair of spines near dactylus, dorsal margin with fine setae. Carpus about 1/3 merus length, with dorsal margin extended into an obtuse tooth, ventral margin with one distal spine. Merus subequal to propodus, about 3.7 times as long as wide at base, with five-nine spines on ventral margin. Fourth pereiopod generally similar to third; ventral margin of merus with six spines; ventral margin of carpus with one distal spine. Fifth pereiopod similar to third and fourth pereiopods, but ventral margin of carpus without distal spine, with merus about 3.3 times as long as wide, ventral margin without spines; carpus about 3/4 merus length, ventral margin without distal spine; propodus as long as merus, with six-nine transversal rows of setae and 3 spines along ventral margin.

**Remarks** *S. trispinosus* is original described by De Man (1910) based on one ovigerous female and one immature specimen collected from southern part of Molo-strait, in a depth of 54–90 m. The original description is simple, and without figures to illustrate. De Man (1911) has re-described this species based on the type specimens, and added some figures to illustrate. *S. trispinosus* is morphologically most similar to *Synalpheus triacanthus*, because of the posterior margin of the sixth abdominal segment bearing teeth. But, this species can be readily distinguished from *S. triacanthus* by the merus of third pereiopod unarmed with spines. The present materials are agreed well with the original description, except for the rostrum variable from hardly reaching to slightly overreaching the end of first antennular segment.

**Distribution** The Ease China Sea; Gulf of Aden and eastern Africa to Philippines and Indonesia.

### 2.15.31 *Synalpheus tumidomanus* (Paul'son, 1875) (Figs. 2.195, 2.196, 2.197, and 2.198)

*Alpheus tumidomanus* Paul'son, 1875: 101, Pl. 13, fig. 2–2i.
*Alpheus tumidomanus gracilimanus* Paul'son, 1875: 102, Pl. 13, fig. 3–3c.
*Synalpheus hululensis* Coutière, 1908: 202.
*Synalpheus maccullochi* Coutière, 1908: 203.
*Synalpheus tumidomanus exilimanus* Coutière, 1909: 10.
*Synalpheus theophane* De Man, 1910: 292.
*Synalpheus anisocheir* Stebbing, 1915: 86, Pl. 87.

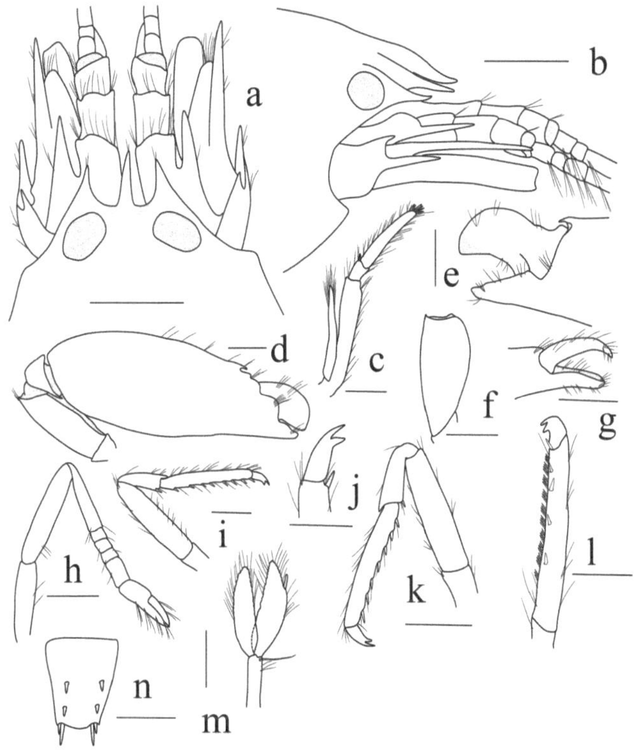


**Fig. 2.195** *Synalpheus tumidomanus* (Paul'son, 1875). MBM 159000, female, **a**, anterior part of carapace, dorsal view; **b**, same, lateral view; **c**, third maxilliped; **d**, major cheliped; **e**, same, distal part of palm and fingers; **f**, merus of minor cheliped; **g**, fingers of minor chela; **h**, second pereiopod; **i**, third pereiopod; **j**, same, distal part of propodus and dactylus; **k**, fourth pereiopod; **l**, propodus and dactylus of fifth pereiopod; **m**, second pleopod; **n**, telson (scale bar: **a–i**, **k**, **m–n** = 1 mm; **j**, **l** = 0.5 mm)

*Synalpheus japonicus* Yokoya, 1936: 133, fig. 3.
*Synalpheus tumidomanus* –. Banner AH & Banner DM, 1968b: 275; Banner DM & Banner AH, 1975: 377, fig. 28; Wang & Sha, 2015: 428, figs. 48–51.

**Material Examined** CL 4.2–6.1 mm, TL 7.5–13.9 mm. MBM 159000, 3ovigerous♀, Sanya, Dadonghai, depth: 2–3 m, Li Xin-zheng Coll., 5 March 1997; MBM 158898, 2♂, Sanya, Dadonghai, depth: 2–3 m, Li Xin-zheng Coll., 5 March 1997; MBM 159158, 1ovigerous♀1♂, Sanya, Dadonghai, depth: 6–9 m, 2 December 1997; MBM 159310, 1♀, Xincun, depth: 1–3 m, 26 March 1992; MBM 158202, 1♂, Sanya, Dadonghai, Li Xin-zheng Coll., 1 March 1997; MBM 159160, 1♂, Sanya, Dadonghai, Li Xin-zheng Coll., 17 November 1997; MBM 270055, 2ovigerous♀2♂, Xizhou island, 30 December 1955; MBM 157827, 2ovigerous♀1♂, Shalao, depth: 6 m, 4 April 1992; MBM 158171, 1♂, Sanya, Dadonghai, Li Xin-zheng Coll., 1 March 1997; 1♂, Xincun, Ding Lan-ping & Wang Yong-qiang Coll., 22 December 2007; MBM 158503, 1♂, Yezhu Island, depth: 10 m, 18 March 1992; MBM 158887, 1ovigerous♀, Sanya, Dadonghai, Li Xin-zheng Coll., 1 March 1997; MBM 158724, 2ovigerous♀1♂, Shalao, depth: 2 m, 4 April 1992; MBM 107824, 10ovigerous♀4♂, Shalao, 5 April 1992; MBM 107778, 1ovigerous♀2♂, Shalao, depth: 6 m, Li Xin-zheng Coll.; MBM 107745, 1♂, Shalao, 7 November 1990; MBM 159158, 1ovigerous♀1♂, Sanya, depth: 6–9 m, 2 December 1997; MBM 156160, 1ovigerous♀, Sanya, Dadonghai, Li Xin-zheng Coll., 17 November 1997; MBM 107784, 3ovigerous♀, Shalao, depth; 12 m, 4 April 1992; MBM 159299,

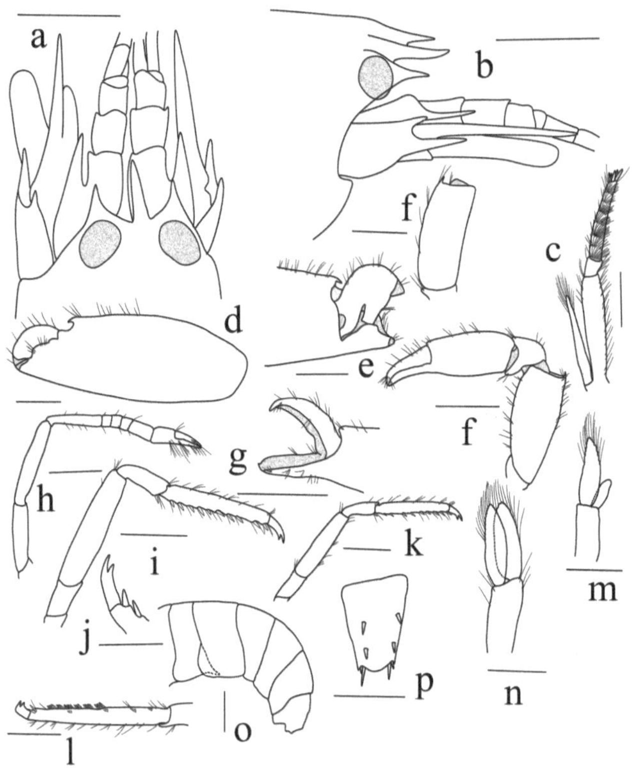


**Fig. 2.196** *Synalpheus tumidomanus* (Paul'son, 1875). MBM 159158, male, **a**, anterior part of carapace, dorsal view; **b**, same, lateral view; **c**, third maxilliped; **d**, major chela; **e**, same, fingers; **f**, minor cheliped; **g**, same, fingers; **h**, second pereiopod; **i**, third pereiopod; **j**, same, distal part of propodus and dactylus; **k**, fourth pereiopod; **l**, propodus and dactylus of fifth pereiopod; **m**, first pleopod; **n**, second pleopod; **o**, abdominal segments, lateral view; **p**, telson (scale bar: **a–f**, **h–i**, **k**, **o–p** = 1 mm; **g**, **j**, **l–n** = 0.5 mm)

2ovigerous♀1♂; MBM 107824, 2ovigerous♀1♂, Shalao, 5 April 1992; 1ovigerous♀2♂, Zhubi Reef, Nansha Islands, depth: 3 m, Wang Yong-qiang Coll., 8 May 2004; 1ovigerous♀1♂, Zhubi Reef, Nansha Islands, depth: 2 m, 8 May 2004; MBM 270163, 1♂, Beibu Gulf, Sta. 6210, depth: 37 m, Fan Zhen-gang Coll., 8 January 1959; MBM 159244 1♂; MBM 158071, 1♀, Dengqing island, Fan Zhen-gang & Xu Jie-shan Coll., 3 May 1958; MBM 159262, 3♂, Sanya, Dadonghai, depth: 4–6 m, 1 December 1997.

**Description** Carapace smooth, glabrous. Rostrum about twice as long as wide at base; acute tip not reaching or slightly overreaching end of first antennular segment. Orbital hoods shorter than rostrum, longer than wider at base. Pterygostomial corner produced into acute angle. Cardiac notch well developed.

Telson about 1.3 times as long as wide at base; dorsal surface with two pairs of spines, inserted approximately at 2/5 and 3/4 length of telson; posterior margin strongly produced, fringed with long setae, posterolateral angle each with two pairs of spines, lateral about 1/3 length of mesial, posterolateral margins each produced into acute tooth, but distinctly shorter than contiguous short spines.

Antennular peduncle moderately slender, with distally acute stylocerite, latter reaching middle of second antennular segment; second segment about 1.5 times as long as wide at base; visible part of first segment longer than second and less

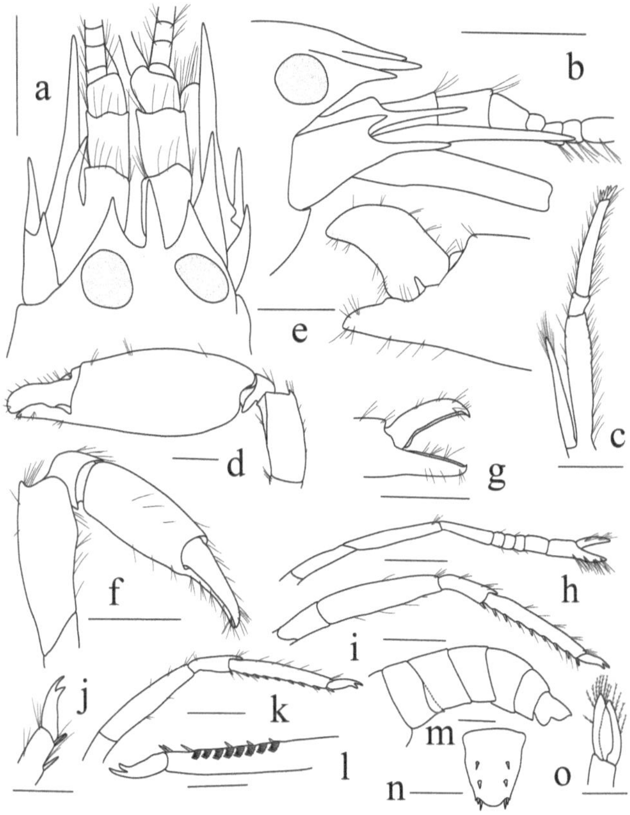


**Fig. 2.197** *Synalpheus tumidomanus* (Paul'son, 1875). MBM 159244, male, **a**, anterior part of carapace, dorsal view; **b**, same, lateral view; **c**, third maxilliped; **d**, major cheliped; **e**, same, fingers; **f**, minor cheliped; **g**, same, fingers; **h**, second pereiopod; **i**, third pereiopod; **j**, same, distal part of propodus and dactylus; **k**, fourth pereiopod; **l**, propodus and dactylus of fifth pereiopod; **m**, abdominal segments; **n**, telson; **o**, second pleopod (scale bar: **a**–**i**, **k**, **m**–**n** = 1 mm; **j**, **l**, **o** = 0.5 mm)

twice as long as third. Basicerite with or without distodorsal tooth, distolateral tooth shorter than stylocerite; carpocerite overreaching end of third antennular segment; scaphocerite blade moderately narrow, reaching end of third antennular segment, lateral spine developed, overreaching end of antennular peduncle, sub-equal to or longer than carpocerite.

Third maxilliped far beyond antennular peduncle and carpocerite when extended; antepenultimate segment longest; penultimate segment about 1.5 times as long as wide at base; tip of ultimate segment bearing crown of five long spine-like setae, with transverse rows of setae along ventral margin.

Major cheliped with dactylus slightly longer than fixed finger, distally slightly curved. Distodorsal of palm with at most an obtuse projecting or strongly acute tooth , slightly curved to dactylus. Carpus cup-shaped, with setae distodorsally. Merus about 3 times as long as wide at base, distodorsal margin with one acute triangular tooth. Minor chela with dactylus and fixed finger tapering, distal single tooth. Palm about 1.3 times as long as fingers. Carpus cup-shaped, proportionally longer that of major cheliped. Merus about 3.6 times as long as wide at base, distodorsal margin with one acute triangular tooth.

Second pereiopod with fingers near as long as palm. Carpus five-segmented, first segment subequal to sum of last four segments.

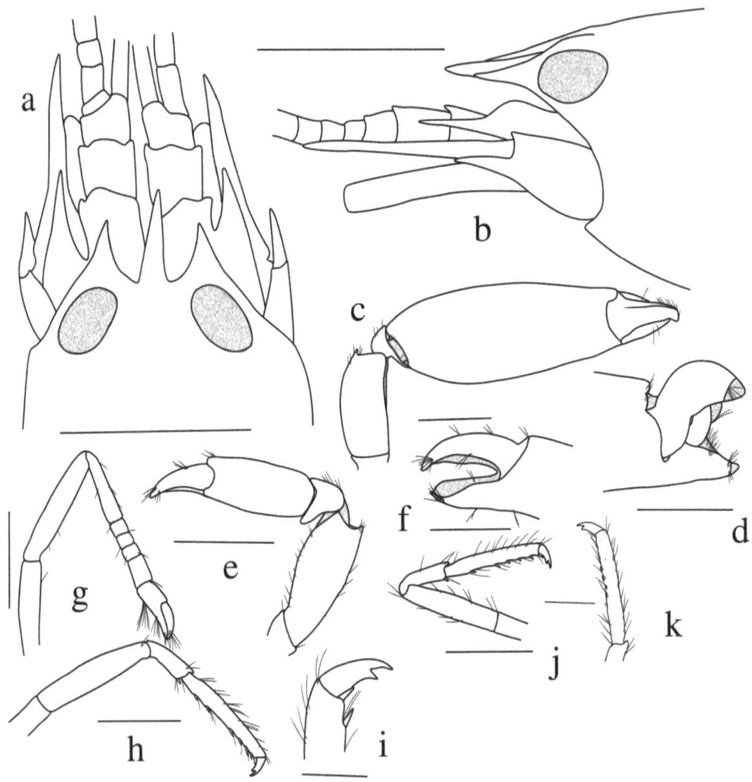


**Fig. 2.198** *Synalpheus tumidomanus* (Paul'son, 1875). MBM 159262, male, **a**, anterior part of carapace, dorsal view; **b**, same, lateral view; **c**, major cheliped; **d**, same, distal part of palm and fingers; **e**, minor cheliped; **f**, same, fingers; **g**, second pereiopod; **h**, third pereiopod; **i**, same, distal part of propodus and dactylus; **j**, fourth pereiopod; **k**, propodus and dactylus of fifth pereiopod, (scale bar: **a–e**, **g–h**, **j** = 1 mm; **f**, **i**, **k** = 0.5 mm)

Third pereiopod moderately stout, with dactylus about 1/4 length of propodus, slightly curved to propodus, biunguiculate, extensor tooth slightly longer than flexor tooth, notch between extensor and flexor tooth V-shaped, flexor tooth wider at base than or sub-wider at base to extensor tooth. Propodus with 6–8 spines along ventral margin plus one distal pair of spines near dactylus. Carpus about 1/3 length of merus, with dorsal margin extended into an obtuse tooth, ventral margin with one distal spine. Merus slightly longer than propodus, about 4 times as long as wide at base. Fourth pereiopod generally similar to third; ventral margin of propodus with 5–7 spines plus one distal pair of spines. Fifth pereiopod similar to third and fourth pereiopods, ventral margin of carpus without distal one spine; ventral margin of propodus with 5–8 transversal rows of spine-like setae and 3–5 spines.

**Remarks** *S. tumidomanus* is rather complex (Banner & Banner 1968). The present specimens are mainly variable in the following characters: the length of the rostrum; the presence or absence of distodorsal tooth of the basicerite; the projecting on distodorsal margin of the major chela; the relative wide at base of the extensor and flexor tooth of the dactylus of last three pereiopods.

**Distribution** The South China Sea; Mediterranean coast of Israel and Red Sea to South Africa, eastward to Japan; Philippines; Indonesia; Australia and across the Pacific to the Phoenix Islands.

## 2.16 Genus *Thuylamea* Nguyên, 2001

**Diagnose** Body elongate, subcylindrical. Rostrum greatly reduced. Orbital hoods absent. Pterygostomian angle rounded. Cardiac notch well accentuated. Dorsal surface of carapace with a faint disk-shaped depression extending from posterior end of rostrum to gastric region; postero-dorsal margin of carapace armed with 2-submedian slightly outward and backward directed crests which are very high, short and with a bluntly rounded top. Antennule with stylocerite elongate. Antenna with basicerite armed with an inner distolateral strong spine, scaphocerite well developed. First pereiopods robust, similar, usually subequal; movable finger laminar, somewhat twisted, basal portion triangular in cross-section; fixed finger slender, tapering into curved acute claw; cheliped carried with dactylus dorso-laterally; coxa with epipod hook-like and 2 setobranchs. Second pereiopod filiform; fingers slightly longer than palm; carpus with 5 articles; coxa with epipod and 2 setobranchs. Ambulatory pereiopods moderately robust; dactyli simple; third and fourth pereiopod respectively

with epipod and 2 setobranchs. Fifth pereiopod with 2 setobranchs and without epipod; outer disto-lateral half of lower margin of propodus with dense transverse rows of setae (Nguyên, 2001).

### 2.16.1 *Thuylamea camelus* Nguyên, 2001 (Figs. 2.199 and 2.200)

*Thuylamea camelus* Nguyên, 2001: 218–228; figs. 1–22; Wang & Sha, 2016b: 1186, figs. 1, 2.

**Material Examined** MBM 158136, 1♀, CL 11.5 mm, TL 35.1 mm, Woneihai, Guangdong, the South China Sea, Liu Rui-Yu Coll., 15 December 1954.

**Description** Carapace smooth, slightly setose; not concealing the eyes anteriorly. Orbital hoods absent. Rostrum small, acute, broadly triangular, not reaching end of eyes; without carina. Pterygostomian angle not produced, obtuse. Cardiac notch well developed. Postero-dorsal margin armed with 2 sub-median slightly outward inclined and backward directed, larger crests, upper and posterior margin rounded and with some setae; a smaller conical crest inserted middle of that two larger crests. Dorsal surface with slight depression. Ventral margin of carapace with some plumose setae.

Pleura of first abdominal somite cover small part of poster-lateral corners of carapace; anterior margin of first abdominal somite tergum only slightly elevated medially in female. Sixth abdominal somite without articulate pleura, anterior dorsal margin depressed medially; sixth somite distinctly longer than fifth, postero-lateral angle acute. Telson distinctly longer than uropodal endopod, about 2.3 times as long as wide at base, lateral margin slightly sinuous; dorsal surface with two pair of spines situated at some distance from lateral margin, anterior and posterior pair situate about 0.36 and 0.62 length of telson, respectively; posterior margin of telson broadly rounded, bearing two pairs of small, slender spines, mesial about twice as long as lateral.

Antennular peduncle moderately slender, setose; first segment longest; second segment about twice as long as wide at base; third segment longer than wide at base; stylocerite not reaching end of first segment of antennular peduncle, acute distally; lateral flagellum biramous. Basicerite bearing strong ventro-lateral tooth and dorsal-lateral tooth at distal margin; scaphocerite oval, anterior margin of blade rounded; distolateral tooth strong, acute, not reaching anterior margin of blade, reaching about end of third segment of antennular peduncle; carpocerite slender, far overreaching end of third segment of antennular peduncle.

Mandible with palp two-segmented; incisor process bearing distally 6 rounded, strong, irregular teeth; molar process well developed. Maxillule with two endites, distal endite with

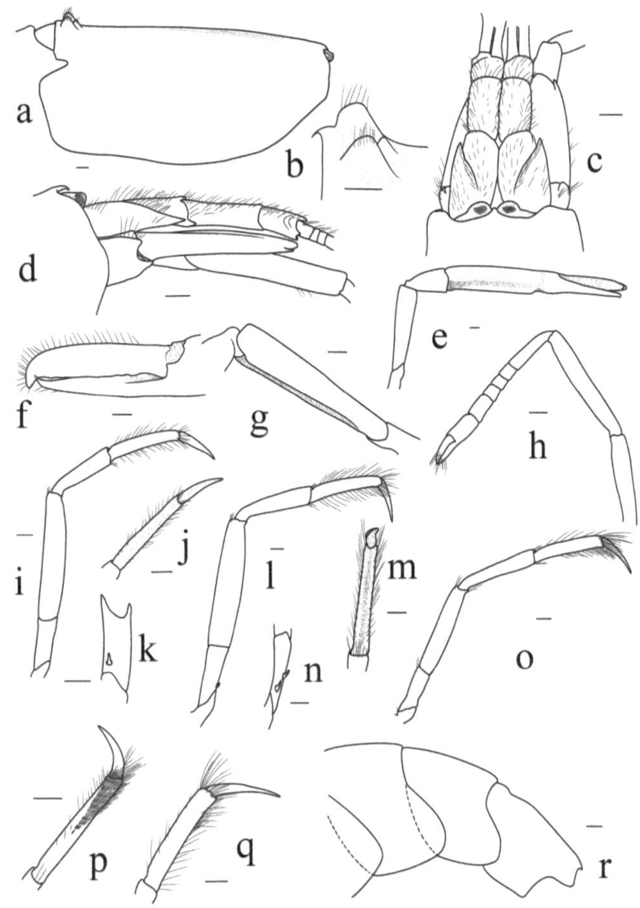


**Fig. 2.199** *Thuylamea camelus* Nguyên, 2001. MBM 158136, female, CL 11.5 mm, **a**, carapace, lateral view; **b**, posterior part of carapace, lateral view; **c**, anterior part of carapace, dorsal view; **d**, same, lateral view; **e**, first cheliped; **f**, fingers of first chela; **g**, merus of first cheliped; **h**, left second pereiopod; **i**, left third pereiopod; **j**, dactylus and propodus of third pereiopod; **k**, ischium of third pereiopod; **l**, left fourth pereiopod; **m**, dactylus and propodus of fourth pereiopod; **n**, ischium of fourth pereiopod; **o**, left fifth pereiopod; **p**, dactylus and propodus of fifth pereiopod; **q**, dactylus and propodus of fifth pereiopod; **r**, posterior three somites of abdominal plate (scale bar = 0.5 mm)

numerous large spine-like setae on inner margin, basal endite with several spine-like setae on distal margins; palp bilobed, lower lobe with one robust seta, upper lobe rounded, with two moderately shorter setae. Maxilla with slender tapering palp, distally with setae; distal endite deeply bilobed, densely setose medially; proximal endite with single lobe, lanceolate, distally with setae; scaphognathite well developed. Basal and coxal endite of first maxilliped with medial margin slightly concave, with densely setae; exopod with well-developed flagellum, distally with long setae. Second maxilliped with dactylar segment narrow, medial margin with moderately short spiniform setae; propodal segment slightly elongate rounded distally, distomedial border with long setae; carpus short; ischium short, basis long, stout; coxa with large epipod; exopod well developed, with long flagellum, distally with long setae. Third maxilliped slender; exopod slightly overreaching end basal segment; antepenultimate segment longest, superior margin with moderately long setae; penultimate segment about 1/2 length of antepenultimate segment; ventral margin of ultimate segment setose, distally without spine-like setae; basipo-

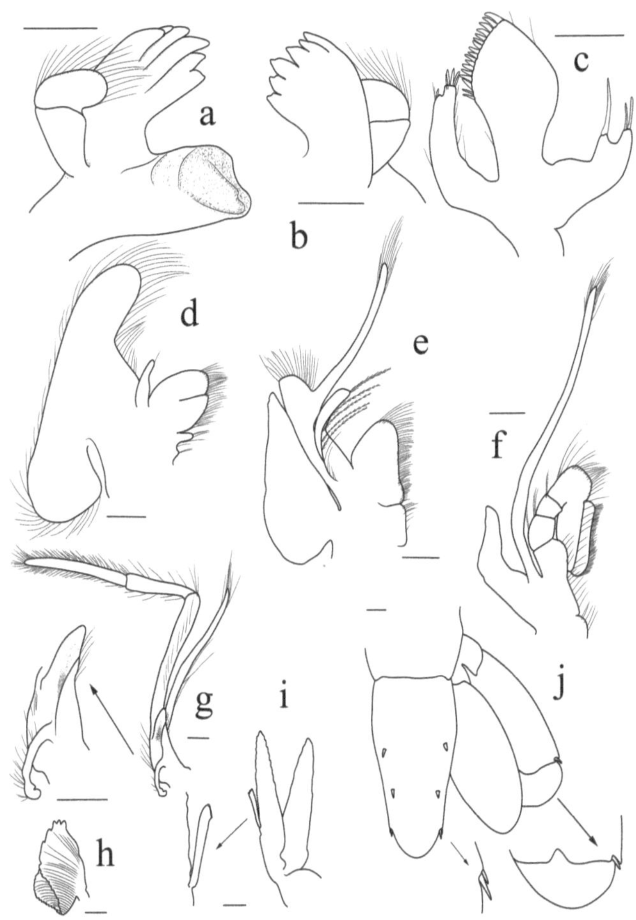


**Fig. 2.200** *Thuylamea camelus* Nguyên, 2001. MBM 158136, female, CL 11.5 mm, **a**, left mandible; **b**, incisor process of mandible; **c**, left maxillule; **d**, left maxilla; **e**, left first maxilliped; **f**, left second maxilliped; **g**, left third maxilliped; **h**, arthrobranch of left third maxilliped; **i**, left second pleopod; **j**, right uropod and telson (scale bar = 0.5 mm)

dite with epipod hook-shaped; arthrobranch normally developed.

First cheliped slender, elongated; cheliped with dactylus dorso-laterally. Chela with palm subcylindrical, distally about 1/4 length with a depressed saddle-shaped notch in lower face; fingers about 3/5 length of palm, dactylus more robust than fixed finger, and slightly longer than fixed finger, cutting edge finely serrated with one or two large teeth in proximal half, distally hooked, dorsal surface with one row of moderately long setae; fixed finger more slender, distally strongly hooked, cutting edge finely dentate. Carpus cup-shaped, dorsal margin with one small hump at base. Merus slender, about 5.3 times as long as wide medially, and widened distally; ischium unarmed.

Second pereiopod moderately stout; chela simple, fingers slightly shorter than palm; carpus 5-segmented, length ratio subequal to 2: 0.5: 0.5: 0.5: 1.1.

Third pereiopod stout, with dactylus simple, slender, curved, more than 1/2 length of propodus, ventral concave. Propodus longer than carpus, ventral margin concave, with row of setae on outer and inner margin, respectively, dorsal margin with one row of setae, distal margin with long setae, unarmed. Carpus about 1/2 length of merus, distal ventral margin with tuff setae, unarmed with spine. Merus about 3 times as long as ischium, unarmed. Ischium armed with one spine on ventral margin. Fourth pereiopod similar to third, ischium ventral margin with two moveable spines; dactylus more slender and curved. Fifth pereiopod more slender and shorter than third and fourth pereiopod, dactylus more slender; propodus about 6.7 times as long as wide; outer distal part of lower margin with several dense transverse rows of normal setae; inner side of lower margin one row of long setae; carpus about as long as merus; merus about 5.2 times as long as wide, unarmed; ischium about 1/2 length of merus, unarmed.

Endopod of second pleopod with appendix interna only in female.

**Remarks** The present specimen slightly differs from the original description by the following characters: the rostrum not overreaching the anterior margin of the cornea; the carapace not covering the eyes; the end margin of first segment of the antennular peduncle smooth instead of sinuous or finely dentate; the incisor process of the mandible bearing 6 acute teeth on the distal margin rather than 5; the anterior dorsal surface only slightly elevated medially, not forming a distinctly dorso-median hump; the ventral margin of the ischium of the fourth pereiopod armed with two spines rather than one, and by the setae on the outer distal part of the fifth pereiopod more longer in the present specimen. Those differences are taken as sexual dimorphism or intraspecific variation, and discussed respectively. The sexual dimorphism is common in the Alpheidae, e.g. *Alpheus tenuicarpus* de Man, 1908 (the arrangement of setae on fingers of the minor chela), *Alpheus inopinatus* Holthuis & Gottlieb, 1958 (the notch on the palm of the minor chela and the arrangement of setae on the fingers of the minor chela), *Athanas dimorphus* Ortmann, 1894 (e.g., the merus of the major cheliped), *Alpheopsis yaldwyni* Banner & Banner, 1973 (the teeth on the cutting edge of the fingers in the major chela). The sexual dimorphism also exists in the *T. camelus*: the end margin of first segment of the antennular peduncle sinuous in male and smooth in female; the setae on the propodus of the posterior three pereiopods longer in female than in male; a distinct elevated hump only existing on the anterior dorsal margin of the first abdominal somite in male, while in the female the anterior dorsal surface of the first abdominal somite only slightly elevated. The shorter rostrum and the visible eyes in the present material maybe indicated that it is not a mature one. Additionally, the number of spine on the ventral margin of the ischium of the fourth pereiopod and the number on teeth on the incisor process of the mandible are taken as intraspecific variation of *T. camelus*.

**Distribution** Guangdong; Vietnam; Singapore.

## References


Alcock A, Anderson ARS (1899) Natural history notes from H. M. Royal Indian marine survey ship Investigator, commander T. H. Heming, R. N., commanding. Series III, No. 2. An account of the deep-sea Crustacea dredged during the surveying season of 1897–98. Ann Magaz Nat Hist 3(7):278–292

Anker A (2003a) Alpheid shrimps from the mangroves and mudflats of Singapore. Part I. Genera *Salmoneus*, *Athanas* and *Potamalpheops*, with the description of two new species (Crustacea: Decapoda: Caridea). Raffles Bull Zool 51:283–314

Anker A (2003b) New records of *Salmoneus* Holthuis, 1955 (Crustacea: Decapoda: Alpheidae) from northern Australia, with description of one new species and remarks on *S. serratidigitus* (Coutière, 1896). Beagle Records Mus Art Galleries N Territory 19:101–117

Anker A (2007) New species and records of alpheid shrimps, genera *Salmoneus* Holthuis and *Parabetaeus* Coutière, from the tropical western Atlantic (Decapoda, Caridea). *Zootaxa* 1653:21–39

Anker A (2010) The shrimp genus *Salmoneus* Holthuis, 1955 (Crustacea, Decapoda, Alpheidae) in the tropical western Atlantic, with description of five new species. In: De Grave S, Fransen CHJM (eds) Contributions to shrimp taxonomy. Zootaxa 2372:177–205

Anker A, Iliffe TM (2000) Description of *Bermudacaris harti*, a new genus, and species (Crustacea: Decapoda: Alpheidae) from anchialine caves of Bermuda. Proc Biol Soc Washington 113(3):761–775

Anker A, Jeng MS (2007) Establishment of a new genus for *Arete borradailei* Coutière, 1903 and *Athanas verrucosus* Banner and Banner, 1960, with redefinitions of *Arete* Stimpson, 1860 and *Athanas* Leach, 1814 (Crustacea: Decapoda: Alpheidae). Zool Stud 46(4):454–472

Anker A, Komai T (2004) Descriptions of two new species of alpheid shrimps from Japan and Australia, with notes on taxonomy of *Automate* De Man, *Coronalpheus* Wicksten and *Bermudacaris* Anker and Iliffe (Crustacea: Decapoda: Caridea). J Nat Hist 38:1895–1914

Anker A, Marin IN (2006) New records and species of Alpheidae (Crustacea: Decapoda) from Vietnam. Part I. Genus *Salmoneus* Holthuis, 1955. Raffles Bull Zool 54(2):295–319

Anker A, Tóth E (2008) A preliminary revision of the *Synalpheus* paraneptunus Coutière, 1909 species complex (Crustacea: Decapoda: Alpheidae). Zootaxa 1915:1–28

Anker A, Jeng MS, Chan TY (2001) Two unusual species of Alpheidae (Decapoda: Caridea) associated with upogebiid mudshrimps in the mudflats of Taiwan and Vietnam. J Crustac Biol 21(4):1049–1061

Anker A, d'Udekem d'Acoz C, Poddoubtchenko D (2005a) Description of a new species of *Alpheopsis* from the Azores, with remarks on *A. africana* Holthuis, 1952 and other species of the *A. trispinosa* (Stimpson, 1860) group (Crustacea, Decapoda, Caridea, Alpheidae). Bull Inst R Sci Nat Belg Biologie 75:97–100

Anker A, Murina GV, Lira C et al (2005b) Macrofauna associated with echiuran burrows: a review with new observations of the innkeeper worm, Ochetostoma erythrogrammon Leuckart and Ruppel, in Venezuela. Zool Stud 44:157–190

Anker A, Komai T, Marin IN (2015a) A new echiuran-associated snapping shrimp (Crustacea: Decapoda: Alpheidae) from the Indo-West Pacific. Zootaxa 3914(4):441–455

Anker A, Pratama IS, Firdaus M, Rahayu DL (2015b) On some interesting marine decapod crustaceans (Alpheidae, Laomediidae, Strahlaxiidae) from Lombok, Indonesia. Zootaxa 3911(3):301–342

Armstrong JC (1941) The Caridea and Stomatopoda of the second Templeton Crocker– American Museum Expedition to the Pacific Ocean. Am Mus Novit 1137:1–14

Audouin V (1826) Explication sommaire des planches de crustaces de l'Egypte et de la Syrie, publiées par Jules-Cesar Savigny, membre de l'Inst.; offrant un exposé des caractères naturels des genres avec la disctinction des espèces. Descrip Egypte Hist Nat 1(4):77–98

Balss H (1914) Beiträge zur Naturgeschichte Ostasiens. Herausgegeben von Dr. F. Dolflein. Ostasiatische Decapoden II. Die Natantia und Reptantia. Abhandlungen der Mathematisch-Physikalischen Klasse der Königlich Bayerischen Akademie der Wissenschaften 10(Supplement 2): 1–101

Balss H (1921) Results of Dr. E. Mjöbergs Swedish scientific expeditions to Australia 1910-13 XXIX. Stomatopoda, Macrura, Paguridea und Galatheidea. Kungliga Svenska Vetenskapsakademiens Handlingar 61(10):1–24

Banner AH (1953) The Crangonidae, or snappping shrimp of Hawaii. Pac Sci 7(1):3–144

Banner AH (1956) Contributions to the knowledge of the alpheid shrimp of the Pacific Ocean. Part I. Collections from the Mariana Archipelago. Pac Sci 10:318–373

Banner AH (1957) Contributions to the knowledge of the alpheid shrimp of the Pacific Ocean Part II. Collection from Arno Atoll, Marshall islands. Pac Sci 11:190–206

Banner AH (1958) Contributions to the knowledge of the alpheid shrimp of the Pacific Ocean. Part III. On a small collection from Onotoa, Gilbert Islands. Pac Sci 12:157–169

Banner AH (1959) Contributions to the knowledge of the alpheid shrimp of the Pacific Ocean. Part IV. Various small collections from the Central Pacific area, including supplementary notes on alpheids from Hawaii. Pac Sci 13:130–155

Banner AH, Banner DM (1964) Contributions to the knowledge of the alpheid shrimp of the Pacific Ocean, IX. Collections from the Phoenix and Line Islands. Pac Sci 18(1):83–99

Banner AH, Banner DM (1966a) Contributions to the knowledge of the alpheid shrimp of the Pacific Ocean. Part X. Collections from Fiji, Tonga, and Samoa. Pac Sci 20:145–188

Banner AH, Banner DM (1966b) The alpheid shrimp of Thailand. Siam Soc Monogr Ser 3:1–168

Banner AH, Banner DM (1967) Contributions to the knowledge of the alpheid shrimps of the Pacific Ocean, XI. Collections from the Cook and Society islands. Occas Pap Bernice P. Bishop Mus 23(12):253–286

Banner AH, Banner DM (1968a) Three new species of the genus *Alpheus* (Decapoda, Alpheidae) from the International Indian Ocean Expedition. Crustaceana 15:141–148

Banner AH, Banner DM (1968b) Contributions to the knowledge of the alpheid shrimp of the Pacific Ocean. Part XII. Collections from the Marshall and Caroline Islands. Micronesica 4(2):261–294

Banner AH, Banner DM (1972) Contributions to the knowledge of the alpheid shrimp of the Pacific Ocean.15.The relationship of *Synalpheus neptunus* (Dana, 1852) to *Synalpheus theano* De Man, 1911, and the establishment of a neotype for *Synalpheus neptunus* (Decapoda, Alpheidae). Crustaceana 23(1):20–27

Banner DM, Banner AH (1973) The alpheid shrimp of Australia. Part I: the lower genera. Rec Aust Mus 28(15):291–382

Banner AH, Banner DM (1974) Contributions to the knowledge of the alpheid shrimp of the Pacific Ocean. Part XVII. Additional notes on the Hawaiian alpheids: new species, subspecies, and some nomenclatorial changes. Pac Sci 28(4):423–437

Banner AH, Banner DM (1975) Contributions to the knowledge of the alpheid shrimp of the Pacific Ocean. Part XVIII: a new species of the genus *Alpheus* from the mouth of the Sepik River, New Guinea. Rec Aust Mus 29:261–266

Banner DM, Banner AH (1975) The alpheid shrimp of Australia. Part 2: the genus *Synalpheus*. Rec Aust Mus 29(12):267–389

Banner DM, Banner AH (1978) Annotated checklist of alpheid and ogyridid shrimp from the Philippine archipelago and the South China Sea. Micronesica 14(2):215–257

Banner AH, Banner DM (1981a) Decapod Crustacea: Alpheidae. In: Résultats des campagnes MUSORSTOM, I: Philippines (18–28 Mars 1976). Collec Mém ORSTOM 91(l):217–235

Banner DM, Banner AH (1981b) Annotated checklist of the alpheid shrimp of the Red Sea and Gulf of Aden. Zool Verh Leiden 190:1–99

Banner DM, Banner AH (1982) The alpheid shrimp of Australia Part III: the remaining alpheids, principally the genus *Alpheus*, and the family Ogyrididae. Rec Aust Mus 34(1):1–357

Banner AH, Banner DM (1983) An annotated checklist of the alpheid shrimp from the western Indian ocean. Travaux et Documents de l'ORSTOM, No. 158. ORSTOM, Bondy, pp 1–164

Banner AH, Banner DM (1984) Old and unreported collections of alpheid shrimp from the Zoologisches Museum, Berlin, principally from Melanesia. Pac Sci 38(1):34–50

Banner DM, Banner AH (1985) The alpheid shrimp of Indonesia, based upon J. G. De Man's "The Decapoda of the Siboga Expedition, Part II, Family Alpheidae." (1911). Mar Res Indones 25:1–79

Barnard KH (1947) Descriptions of new species of South African decapod Crustacea, with notes on synonymy and new records. Ann Magaz Nat Hist, series 11 13 [for 1946]:361–392

Barnard KH (1950) Descriptive catalogue of South African decapod Crustacea (crabs and shrimps). Ann S Afr Mus 38:1–837

Barnard KH (1955) Additions to the fauna list of South African Crustacea and Pycnogonida. Ann S Afr Mus 43:8–107

Barnard KH (1962) New records of marine Crustacea from the East African region. Crustaceana 3(3):239–245

Bate CS (1888) Report on the Crustacea Macrura collected by H.M.S. Challenger during the years 1873–76. In: Murray J (ed) Zoology. Report on the Scientific Results of the Voyage of H.M.S. Challenger During the Years 1873–76 Under the Command of Captain George S. Nares, R.N., F.R.S. and the Late Captain Frank Tourle Thomson, R.N. Wyville Thomson, C. and J. Murray (series eds.), vol 24. Neill and Company, Edinburgh, pp i–xc, 1–942, Plates 1–157

Boone L (1931) A collection of anomuran and macruran Crustacea from the Bay of Panama and the fresh waters of the Canal Zone. Bull Am Mus Nat Hist 63(2):137–189

Boone L (1935) Crustacea and Echinodermata. Scientific results of the world cruise of the yacht Alva, 1931, William K. Vanderbilt, commanding. Bull Vanderbilt Mar Mus 6:1–263

Borradaile LA (1898) On some crustaceans from the South Pacific. Part III. Macrura. Proc Zool Soc Lond 1898:1000–1015

Borradaile LA (1900) On the Stomatopoda and Macrura brought by Dr. Willey from the South Seas. In: Willey A (ed) Zoological results based on material from in New Britain, New Guinea, Loyalty Islands and elsewhere collected during the years 1895, 1896 and 1897, vol 4. University Press, Cambridge, pp 395–428

Bruce AJ (1969) *Aretopsis amabilis* De Man, an alpheid shrimp commensal of pagurid crabs in the Seychelle Islands. J Mar Biol Assoc India 11(1–2):175–181

Bruce AJ (1976) A report on a small collection of shrimps from the Kenya National Marine Parks at Malindi, with notes on selected species. Zool Verh Leiden 145:3–72

Bruce AJ (1987) A new species of alpheid shrimp, *Alpheus bannerorum* from northern Australia. Beagle Rec N Territory Mus Arts Sci 4:61–72

Bruce AJ (1989) A report on some coral reef shrimps from the Philippine Islands. Asian Mar Biol 6:173–192

Bruce AJ (1990) Additions to the marine shrimp fauna of Hong Kong. In: Morton B (ed) Proceedings of the second international marine biological workshop: the marine flora and fauna of Hong Kong and Southern China, Hong Kong, 1986. Hong Kong University Press, Hong Kong, pp 611–648

Bruce AJ, Coombes KE (1997) An annotated check-list of the caridean shrimps (Crustacea: Decapoda) of Darwin Harbour, with descriptions of three new species of *Periclimenes* [Palaemonidae: Pontoniinae]. In: Proceedings of the sixth international marine biological workshop. The marine flora and fauna of Darwin Harbour, Northern Territory, Australia. Museums and Art Galleries of the Northern Territory and the Australian Marine Sciences Association, Darwin, pp 301–337

Cha HK, Lee JU, Park CS, Baik CI, Hong SY, Park JH, Choi JH (2001) Shrimps of the Korean waters. National Fisheries Research Development Institute, Busan, pp 1–188

Chace FA Jr (1955) Notes on shrimps from the Marshall Islands. Proc U S Natl Mus 105(3349):1–22

Chace FA Jr (1962) The non-brachyuran decapod crustaceans of Clipperton Island. Proc U S Natl Mus 113(3466):605–635

Chace FA Jr (1972) The shrimps of the Smithsonian–Bredin Caribbean expeditions with a summary of West Indian shallow–water species (Crustacea: Decapoda: Natantia). Smithson Contrib Zool (98):1–179

Chace FA Jr (1988) The caridean shrimps (Crustacea: Decapoda) of the "Albatross" Philippine Expedition, 1907–1910, Part 5: Family Alpheidae. Smithson Contrib Zool 466:1–99

Clark AH (1921) A monograph of the existing crinoids. U S Natl Mus Bull 82(2):1–795

Coutière H (1897a) Note sur quelques Alphées nouveaux. Bull Mus Hist Nat Paris 3(7):303–306

Coutière H (1897b) Note sur quelques genres nouveaux ou peu connus d'Alphéidés, formant la sousfamille des Alphéopsidés. Bull Mus Hist Nat Paris 2: 380–386. (imprint: 1896)

Coutière H (1897c) Note sure *Betæus Jousseaumei*, nouvelle espèce d'Alphée de la Mer Rouge. Bull Mus Hist Nat 2:236–237. (imprint 1896)

Coutière H (1897d) Note sur quelques espèces du genre *Alpheus* du Musée de Leyde. Notes Leyden Mus 19:195–207

Coutière H (1898a) Note sur quelques formes nouvelles d'alpheides voisines de *A. Bouvieri* A. M. -Edwards (Crust.). Bull Soc Entomol France 1898(5):131–134

Coutière H (1898b) Note sur quelques alpheides nouveaux de la collection du British Museum (Crust). Bull Soc Entomol France 1898(6):149–152, 2 figs

Coutière H (1898c) Note sur quelques cas de regeneration hypotypique chez *Alpheus* (Crust.). Bull Soc Entomol France 1898(12):248–250

Coutière H (1898d) Notes sur la faune des recifs madreporiques de Djibouti. Bull Mus Natl Hist Nat Paris 4(4):195–198

Coutière H (1899) Les "Alpheidae" morphologie externe et interne, formes larvaires, bionomie. Theses presenteesala Faculte des Sciences de Paris, Ser. A, No. 321 No. d'ordre 980. 559 pp

Coutière H (1900) Note sur une collection d'Alpheidae provenant detroit de Torres. Bull Mus Natl Hist Nat Paris 6(8):411–414

Coutière H (1902) Sur quelques espèces nouvelles du genre Automate De Man. Bull Mus Natl Hist Nat Paris, pp 337–342

Coutière H (1903) Note sur quelques Alpheidæ des Maldives et Laquedives. Bull Soc philomathique Paris (9) 5(2):72–90

Coutière H (1905) Les Alpheidae. In: Stanley Gardiner J (ed) The fauna and geography of the Maldive and Laccadive archipelagoes 2(4):852–921, Plates. 70–87

Coutière H (1908) Sur quelques nouvelles espèces d'Alpheidae. Bull Soc philomathique Paris (9) 10:191–216

Coutière H (1909) The American species of snapping shrimps of the genus *Synalpheus*. Proc U S Natl Mus Smithson Inst Washington 36(1659):1–93

Coutière H (1921) No. X.–Les espèces d'Alpheidae rapportées de l'Océan Indien par M.J. Stanley Gardiner. Trans Linnean Soc Lond, Series 2, Zool 17(4):413–428

Crosnier A, Forest J (1966) Crustaces decapodes: Alpheidae. In Campagne de la Calypso dans le Golfe de Guinee et aux lies principe, Sao Tome et Annobon (1956), et campagne aux lies du Cap Vert (1959). XXVII (Fasc. 7). Ann Inst Océanogr 44:199–314

Cui DL, Sha ZL (2014) A new record of a species of *Alpheus* (Decapoda: Caridea: Alpheidae) from China. Chin J Oceanol Limnol 32(5):1123–1127

Dana JD (1852a) Conspectus crustaceorum, &c. Conspectus of the Crustacea of the exploring expedition under Capt. C. Wilkes, U.S.N. Macroura. Proc Acad Nat Sci Phila 6:10–28

Dana JD (1852b) Crustacea. Part I. United States Exploring Expedition. During the years 1838, 1839, 1840, 1841, 1842. Under the command of Charles Wilkes, U.S.N. Vol. 13. C. Sherman, Philadelphia. 685 pp

Dana JD (1855) Atlas. Crustacea. United States Exploring Expedition. During the years 1838, 1839, 1840, 1841, 1842. Under the command of Charles Wilkes, U.S.N. Vol. 13–14. C. Sherman, Philadelphia. 27 pp, Plates 1–96

De Grave S, Fransen CHJM (2011) Carideorum catalogus: the recent species of the dendrobranchiate, stenopodidean, procarididean and caridean shrimps (Crustacea: Decapoda). Zool Mededelingen Leiden 89(5):195–589

De Haan W (1833–1850) Crustacea. In: von Siebold, P. F., Fauna Japonica sive Descriptio Animalium, quae in Itinere per Japoniam, Jussu et Auspiciis Superiorum, qui Summum in India Batava Imperium Tenent, Suspecto, Annis 1823-1830 Collegit, Notis, Observationibus et Adumbrationibus Illustravit: i–xxxi, ix–xvi, 1–243.

De Man JG (1888a) Report on the podophthalmous Crustacea of the Mergui Archipelago, collected for the Trustees of the Indian Museum, Calcutta, by Dr. John Anderson F.R.S., Superintendent of the Museum. J Linnean Soc (Zool) 22:1–305, Plates 1–19

De Man JG (1888b) Bericht über die von Herrn Dr. J. Brock im indischen Archipel gesammelten Decapoden und Stomatopoden. Arch Naturgeschichte 53(1):289–600, Plates 11–22a

De Man JG (1890) Carcinological studies in the Leyden Museum. No. 4. Notes Leyden Mus 12:49–126

De Man JG (1897) Bericht über die von Herrn Schiffscapitan Storm zu Atjeh, an den westlichen Küsten von Malakka, Borneo und Celebes sowie in der Java-See gesammelten Decapoden und Stomatopoden. Fünfter Thiel. Zool Jahrb Abt Syst Geogr Biol Thiere 9:725–790, Plates 12–14

De Man JG (1898) Note sur quelques especes du genre *Alpheus* Fabr. appartenant a la section dont *l'Alpheus Edwardsi* Aud. est le representant. Mém Soc Zool France 11:309–325, pl. 4

De Man JG (1902) Die von Herrn Professor Kükenthal im Indischen Archipel gesammelten Dekapoden und Stomatopoden. In: W. Kükenthal Ergebnisse einer zoologischen Forschungsreise in den Molukken und Borneo, im Auftrage der Senckenbergischen naturforschenden Gesellschaft 25: 465–929

De Man JG (1907) On a collection of Crustacea, Decapoda and Stomatopoda, chiefly from the Inland Sea of Japan; with descriptions of new species. Trans Zool Soc Lond 9(11):387–454

De Man JG (1908) Diagnoses of new species of macrurous decapod Crustacea from the "Siboga-expedition". Notes Leyden Mus 30(14):98–112

De Man JG (1909a) Diagnoses of new species of macrurous decapod Crustacea from the "Siboga-Expedition". Tijdschrift der Nederlandse Dierkundige Vereeniging 9(2):99–125
De Man JG (1909b) Note sur quelques espèces du genre *Alpheus* Fabr., appartenant au groupe brevirostris de M. Mém Soc Zool France 22:146–164, Plates 7–8
De Man JG (1909c) Description of a new species of the genus *Alpheus* Fabr. from the Bay of Batavia. Proc Zool Soc Lond 1909:663–666, Plate 70
De Man JG (1910) Diagnoses of new species of macrurous decapod Crustacea from the «Siboga-Expedition», V. Tijdschrift der Nederlandsche Dierkundige Vereeniging 11:287–319
De Man JG (1911) The Decapoda of the Siboga Expedition, Part II Family Alpheidae. Siboga Exped Monogr 39a1:18–464, Plates XXIII
De Man JG (1920) Diagnoses of some new species of Penaeidae and Alpheidae with remarks on two known species of the gneus *Penaeopsis* A. M.-Edw. from the Indian Archipelago. Zool Mededeelingen Leiden 5(3):103–109
De Man JG (1922) The Decapoda of the Siboga Expedition. Part V. On a collection of macrurous decapod Crustacea of the Siboga Expedition, chiefly Penaeidae and Alpheidae. Siboga Exped 39a 4(5):1–51
De Man JG (1924) On a collection of macrurous decapod Crustacea, chiefly Penaeidae and Alpheidae from the Indian Archipelago. Arch Nat 90(1):1–60
De Man JG (1926) Beschreibung zweier Arten von Decapoda Macrura von der Insel Buka (Salomoninseln). Mitteilungen aus dem Zoologischen Mus Berlin 12(2):341–345
De Man JG (1929) On a collection of decapod and stomatopod Crustacea from Pulau Berhala, an islet situated in the Straits of Malacca. Bijdragrn tot de Dierkunde 26:1–26
Didderen K, Fransen CHJM, De Voogd NJ (2006) Observations on sponge-dwelling colonies of *Synalpheus* (Decapoda, Alpheidae) of Sulawesi, Indonesia. Crustaceana 79(8):961–975
Duffy JE, Macdonald KS (1999) Colony structure of the social snapping shrimp *Synalpheus filidigitatus* in Belize. J Crustac Biol 19(2):283–292
Edmondson CH (1925) Marine zoology of tropical central Pacific. Crustacea. Bulletin of the Bernice P. Bishop Mus 27:3–62
Fabricius JC (1775) Systema entomologiae sistens insectorum classes ordines, genera, species adiectis synonymis, locis, descriptionibus, observationibus. Flensburgi et Lipsiae (Flensburg and Leipzig):1–832
Felder DL, Chaney AH (1979) Decapod fauna of Seven and One-Half Fathom Reef, Texas: species composition, abundance and diversity. Contrib Mar Sci 22:1–29
Fishelson L (1966) Observations on the littoral fauna of Israel. V. On the habitat and behaviour of *Alpheus frontalis* H. Milne-Edwards (Decapoda, Alpheidae). Crustaceana 11(1):98–104
Forest J, Guinot D (1958) Sur une collection de crustaces decapodes des cotes d'lsrael. Bull Sea Fish Res Station Israel 15:4–16
Gravely FH (1930) The Alpheidae of Krusadai Island. Bull Madras Govern Mus 1(2):77–79
Green A. (1972) A report of the fifth expedition to the Easter and Pelsart Croup of Houtman's Abrolhos, Aug. 22–28, 1970 made by the students of Aquinas College, Manning, W.A. 80 pp
Guérin-Méneville FE (1838) Crustacés, Arachnides et Insectes. In: Duperrey LI (ed) Voyage autour de monde, execute par ordre du Roi, sur la corvette de sa Majeste, la Coquille, pendant les annees 1822, 1823, 1824 et 1825. vol. 2 (pt 2, div. 1), Paris, 319 pp
Gurney R (1927) Report on the larvae ofthe Crustacea Decapoda. In Zoological results of Cambridge Expedition to the Suez Canal, 1924. Trans Zool Soc Lond 22(2):231–286
Hay WP (1917) Preliminary descriptions of five new species of crustaceans from the coast of North Carolina. Proc Biol Soc Washington 30:71–74
Hayashi KI (1995) Prawns, shrimps and lobsters from Japan (85). Family Alpheidae – Genera *Automate* and *Salmoneus*. Aquabiology (Tokyo) 17(6):520–524
Hayashi KI (1998a) Prawns, shrimps and lobsters from Japan (100). Family Alpheidae-Genus *Alpheus* (9). Aquabiology 116(20):209–216
Hayashi KI (1998b) A new genus and a new species of alpheid shrimp (Decapoda, Caridea) from Japan. Zoosystema 20:229–238
Hayashi K.I. (2002). A new species of the genus *Athanas* (Decapoda, Caridea, Alpheidae) living in the burrows of mantis shrimp. Crustaceana, 75(3/4): 395–404.
Hayashi KI, Nagata M (2002) Identity of *Alpheus digitalis* De Haan, 1844 and description of a new closely related species from the northwestern Pacific (Decapoda: Caridea: Alpheidae). Crustac Res 31:73–90
Heller C (1861) Beitrage zur crustaceen-fauna des Rothen Meeres. Sitzungsberichte der Mathematisch-Naturwissenschaftlichen Klasse der Kaiserlichen Akademie der Wissenschaften, Wien 44(2):241–295
Heller C (1865) Crustaceen. Reise der österreichischen Fregatte Novara um die Erde in den Jahren 1857, 1858, 1859 unter den Befehlen des Commodore B. von Wüllerstorf-Urbair. Zoologischer Theil. Zweiter Band. III. Abteilung, vol. 2, 280 pp
Hilgendorf F (1879) Die von Herrn W. Peters in Mocambique gesammelten Crustaceen. Monatsberichte der Königlichen Preussischen Akademie des Wissenschaften zu Berlin 1879:782–851
Holthuis LB (1958) Contributions to the knowledge of the Red Sea. nos. 8-9. Crustacea Decapoda from the Northern Red Sea (Gulf of Aqaba and Sinai peninsula). I. Macrura. Bull Sea Fish Res Station Haifa 17:1–40
Holthuis LB (1961) On the dates of publication of the crustacean plates in Duperry's "Voyage Autour du Monde ... sur ... la Coquille". Crustaceana 3(2):168–169
Holthuis LB, Gottlieb E (1958) An annotated list of the decapod Crustacea of the Mediterranean coast of

Israel, with an appendix listing the Decapoda of the eastern Mediterranean. Bull Res Council Israel Sect B Zool 7B(1–2):1–126

Holthuis LB, Sakai T (1970) Ph. F. Von Siebold and Fauna Japonica. A history of early Japanese Zoology, i–xviii+ part I, 1–132. part II, 207–323

Igarashi T (1969) A list of marine decapod crustaceans from Hokkaido, deposited at the Fisheries Museum, Faculty of Fisheries, Hokkaido University, I. Macrura Contr Fish Mus Fac Fish Hokkaido Univ 11:1–15

Jacquotte R (1964) Notes de faunistique et de biologie marines de Madagascar II – Decapodes nageurs associes aux echinodermes dans la region de Tuléar (sud-ouest de Madagascar). Recueil des Travaux de la Station marine d'Endoume 32(48):179–182

Jeng MS (1997) Studies on the land and aquatic decapod crustacean fauna of the Kenting National Park (II). Conservation Research Report, 96. Kenting National Park, Taiwan

Jeng MS, Chang KH (1985) Snapping shrimps (Crustacea: Decapoda: Alpheidae) of Taiwan. Bull Inst Zool Acad Sin 24:241–256

Johnson DS (1962) A synopsis of the Decapoda Caridea and Stenopodidea of Singapore, with notes on their distribution and a key to the genera of Caridea occurring in Malayan waters. Bull Mus Hist Nat Singapore 30:44–79

Kazmi MA (1974) A new species of the genus *Alpheus* from the Bay of Bengal (Decapoda, Carisea). Crustaceana 27(2):170–174

Kemp S (1915) Crustacea Decapoda, Fauna of the Chilka Lake. Mem Indian Mus 5(3):199–326

Kensley B (1969) Decapod Crustacea from the south-west Indian Ocean. Ann S Afr Mus 52:149–181

Kim HS (1977) Macrura. Illustrated Flora and Fauna of Korea 19:1–414

Kim W (1998) *Chelomalpheus koreanus*, a new genus and species of snapping shrimp from Korea (Crustacea: Decapoda: Alpheidae). Proc Biol Soc Washington 111(1):140–145

Kingsley JS (1878) A synopsis of the North American species of the genus *Alpheus*. Bulletin of the United States Geological and Geographical Survey of the Terri-tories, F. V. Hayden, United States Geologist-in-charge 4(1):189–199

Kubo I (1936a) A description of a new alpheoid shrimp from Japan. J Imp Fish Inst 31(2):43–46

Kubo I (1936b) Two new littoral macrurous crustaceans from Japan. J Imp Fish Inst 31(2):47–54

Kubo I (1938) A new snapping shrimp belonging to the genus *Synalpheus*. Annotationes Zoologicae Japonenses 17(1):89–92

Kubo I (1940) Notes on the Japanese shrimps of the genus *Athanas* with a description of one new species. Annotationes Zoologicæ Japonenses 19(2):99–106

Kubo I (1942) On a new snapping shrimp, *Athanas kominatoensis*. Zool Mag Tokyo 54(2):82–85

Kubo I (1951) Some macrurous decapod Crustacea found in Japanese waters, with descriptions of four new species. J Tokyo Univ Fish 38(2):259–289

Lanchester WF (1901) On the Crustacea collected during the "Skeat" Expedition to the Malay Peninsula, together with a note on the genus Actaeopsis. Part I.-Brachyura, Stomatopoda, and Macrura. Proc General Meetings Sci Business Zool Soc Lond 1901(2):534–574

Ledoyer M (1970) Étude systématique et remarques écologiques sur les Caridae recueillis principalement dans les biotopes de substrat meuble régions de Tuléar et de Nosy-Be. Ann Univ Madagascar (Sci Nat Math) 7:121–157

Lewinsohn C, Holthuis LB (1964) New records of decapod Crustacea from the Mediterranean coast of Israel and the eastern Mediterranean. Zool Mededelingen 40(8):45–63

Liu R.Y. (1995). Economic shrimps and prawns of Northern China. Marine Biological Institute of Academy Science, Beijing, i–iii+73pp. (in Chinese)

Liu RY (2008) Checklist of Marine Biota of China Sea. Science Press, Beijing, pp 712–716. (in Chinese)

Liu RY, Lan JY (1980) On a collection of the genus Alpheus (Crustacea Decapoda) from the Xisha Islands, Guangdong Province, China. Studia Marina Sinica 17:77–115. (in Chinese)

Marin I (2010) On the presence of the symbiotic alpheid shrimp *Aretopsis amabilis* in the coastal waters of Nhatrang Bay, Vietnam. Mar Biodivers Records 3:e74

Marin I (2014) An unusual habitat for hermit crab-associated alpheid shrimp *Aretopsis amabilis* De Man, 1910 (Crustacea: Decapoda: Alpheidae): a possible evidence of socially monogamous species? Arthropoda Selecta 23(3):265–268

Miers EJ (1875) On some new or undescribed species of Crustacea from Samoa Islands. Ann Mag Nat Hist IV 16(95):341–344

Miers EJ (1879) On a collection of Crustacea made by Capt. H.C. St. John, R.N., in the Korean and Japanese Seas. With an appendix by Capt. H.C. St. John. Proc Zool Soc Lond 1879:18–61

Miers EJ (1881) On a collection of Crustacea made by Baron Hermann-Maltzan at Gorce island, Senegambia. Ann Mag Nat Hist (5)8:204–220, 259–281, 364–377, Plates 13–16

Miers EJ (1884) Crustacea. In: Report of the zoological collections made in the Indo-Pacific Ocean during the voyage of H.M.S. "Alert", 1881–2, pp 178–322, 513–575, Plates 18–35, 46–52

Milne-Edwards H (1837) Histoire naturelle des crustaces, comprenant I'anatomie, la physiologie et la classification de ces animaux 2:1–532

Miya Y (1972) The Alpheidae (Crustacea, Decapoda) of Japan and its adjacent waters. Part I. Publ Amakusa Mar Biol Lab 3(1):23–101

Miya Y (1974) The Alpheidae (Crustacea, Decapoda) of Japan and its adjacent waters. Part II. Publ Amakusa Mar Biol Lab 3(2):103–195

Miya Y (1984) Alpheid Shrimps from the Truks, Ponape and Majuro Atoll (Crustacea, Decapoda). Proc Jpn Soc Syst Zool 27:67–100

Miya Y (1997) *Stenalpheops anacanthus*, new genus, new species (Crustacea, Decapoda, Alpheidae) from the Seto Inland Sea and the Sea of Ariake, South Japan. Bulletin of the Faculty of Liberal Arts. Nagasaki University. Nat Sci 38(1):145–161
Miya Y, Miyake S (1968) Redefinition of the genus *Batella* (Crustacea, Decapoda, Alpheidae), with description of a new species from Kyushu, Japan. Ohmu 1(5):113–120
Miyake S (1998) Japanese Crustacean Decapods and Stomatopods in Color, vol. I. Macrura, Anomura and Stomatopoda. Second Printing. Hoikusha, Osaka, vii+261 pp. (in Japanese)
Miyake S, Miya Y (1966) On a new species and a new record of alpheid shrimps from Japan. J Fac Agric Kyushu Univ 14(1):133–141
Miyake S, Miya Y (1967) A rare alpheid shrimp *Aretopsis amabilis* De Man from the Ryuku Islands (Decapoda, Crustacea). J Fac Agric Kyushu Univ 14(2):267–273
Motoh H (1972) A faunal list of the macruran Decapoda from Nanao Bay, Ishikawa Prefecture, middle Japan. Bull Ishikawa Prefectural Mar Cult Station 2:29–83
Nakazawa K, Kubo I (1947) Illustrated encyclopedia of the fauna of Japan, vol 2. Tokyo. 5213 pp
Nguyên VX (2001) A new alpheid shrimp (Crustacea: Decapoda: Alpheidae) from South Vietnam. Zool Mededelingen Leiden 75(13):217–228
Nishimura S, Suzuki K (1971) Common seashore animals of Japan in color. Hoikusha, Osaka. 169 pp. (in Japanese)
Nobili G (1901) Decapodi e Stomatopodi Eritrei del Museo Zoologico dell'Università di Napoli. Annuario del Museo Zoologico della R.Università di Napoli 1(3):1–20
Nobili G (1906a) Fauna carcinologique de la Mer Rouge. Decapodes et Stomatopodes. Ann Sci Nat 9 série 4:1–347
Nobili G (1906b) Crustaces decapodes et stomatopodes. Mission J. Bonnier et Ch. Perez (Golfe Persique, 1901). Bull Jpn Soc Sci Fish 40:13–159
Nobili G (1907) Ricerche sui Crostacei della Polinesia. Decapodi, Stomatopodi, Anisopodi e Isopodi. Mem Reale Acad Sci Torino Serie 2(57):351–430
Oliveira MV, Costa-Souza AC, Guimaraes FJ, Almeida AO, Baeza JA (2015) Observations on the life history of a rare shrimp, *Salmoneus carvachoi* (Crustacea: Caridea: Alpheidae), a possible simultaneous hermaphrodite. Mar Biodivers Records 8:e141
Olivier AG (1811) Palemon. Palaemon. In: Diderot & d'Alembert (premiers) (eds) Endyclopedie methodique, ou par ordre de matieres ... Histoire naturelle. Insectes 8:652–667. Chez Panckonshe, Paris
Ortmann AE (1890) Die Unterordnung Natantia Boas: Die Decapoden-Krebse des Strassburger Museums, mit besonderer Berücksichtung der von Herrn Dr. Döderlein bei Japan und bei den Liu-Kiu-Inseln gesammelten und z. Z. im Strassburger Museum aufbewahrten Formen. I. Zoologische Jahrbücher. Abteilung für Systematik, Geographie und Biologie der Thiere 5:437–542, Plates XXXVI–XXXVII
Ortmann AE (1893) Decapoden und Schizopoden. Ergebnisse der Plankton-Expedition der Humbolt-Stiftung im Atlantischen Ozean, 1889(2): 1–120
Ortmann A.E. (1894). Crustaceen. In: Semon (ed) Zoologische Forschungsreisen in Australien und dem malayischen Archipel. Denkschriften der Medicinisch-Naturwissenschaftlichen Geselschaft zu Jena 8:3–80
Paul'son O. (1875). Studies on Crustacea of the Red Sea with notes regarding other seas. Part I. Podophthalmata and Edriophthalmata (Cumacea). (In Russian, English translation by the Israel Program for Scientific Translations, Jerusalem, 1961). S.V. Kul'zhenko, Kiev. xiv, 144 pp, Plates 1–22
Pearson J. (1905). Report on the Macrura collected by Professor Herdman at Ceylon In 1902. Report to the Government of Ceylon on the Pearl Oyster Fisheries of the Gulf of Manaar, Supplementary Report 4:65–92
Pearson J (1911) Ceylon Crustacea. Part 1. Notes on the Alpheidae. Spolia Zeylan 7:169–186
Potts F.A. (1915). The fauna associated with crinoids of a tropical coral reef: with especial reference to its color variation. Papers from the Department of Marine Biology, Carnegie Institution of Washington 8(212):73–96
Rafinesque CS (1815) Analyse de la Nature, ou Tableau de l'Univers et des Corps Organisés. L'Imprimerie de Jean Barravecchia, Palermo. 224 pp
Ramadan MM (1936) Report on a collection of Stomatopoda and Decapoda from Ghardaqa, Red Sea. Bull Fac Sci Egypt Univ 6:1–43
Randall JW (1839) Catalogue of the Crustacea brought by Thomas Nuttall and J. K. Townsend, from the west coast of North America and the Sandwich Islands, with descriptions of such species as are apparently new, among which are included several species of different localities, previously existing in the collection of the Academy. Proc Acad Nat Sci Phila 8(1):106–147
Rathbun MJ (1901) The Brachyura and Macrura of Porto Rico. Bull U S Fish Commission 20(2):1–127, pls. 1–2. (for 1900)
Richters F (1880) Decapoda. In: Mobius K (ed) Beitrage zur Meeresfauna der Insel Mauritius und der Seychellen bearbeitet von K Mobius, F. Richters und E. von Martens nach Sammlungen, angelegt auf einer Reise nach Mauritius, vi+352 pp
Sakamoto T, Hayashi KI (1977) Prawns and shrimps collected from the Kii Strait by small type trawlers. Bull Jpn Soc Sci Fish 43:1259–1268
Sha ZL, Liu JY (2007) Study on Alpheidae (Crustacea, Decapoda) of China seas, genus *Athanas* Lecah. Acta Zootaxon Sin 32(4):749–755
Sha ZL, Liu JY (2008) Studies on genus *Arete* Stimpson of Alpheidae (Crustacea, Decapoda) from China seas. Acta Zootaxon Sin 33(1):7–9
Stebbing TR (1915) South African Crustacea (Part VIII. of S.A. Crustacea, for the Marine Investigations in South Africa). Ann S Afr Mus 15:57–104. Plates 13–25
Stimpson W (1860) Prodromus descriptionis animalium evertebratorum, quae in Expeditione ad Oceanum

Pacificum Septentrionalem, a Republica Federata missa, Cadwaladaro Ringgold et Johanne Rodgers Ducibus, observavit et descripsit. Pars VIII. Crustacea Macrura. Proc Acad Nat Sci Phila 12:22–47
Suzuki H (1971) On some commensal shrimps found in the western region of Sagami Bay. Res Crustacea 4(5):1–31
Tattersall WM (1921) Report on the Stomatopoda and Macrurous Decapoda collected by Mr. Cyril Crossland in the Sudanese Red Sea. J Linnean Soc Lond Zool 34:345–398
Tirmizi NM, Kazmi MA (1969) The occurrence of *Alpheus inopinatus* Holthuis & Gottlieb, 1958, in the Arabian Sea. (Decapoda, Alpheidae). Crustaceana 16(1):99–100. fig. 1
Tiwari KK (1963) Alpheid Shrimps (Crustacea: Decapoda: Alpheidae) of Vietnam. Ann Fac Sci Saigon 2:269–362
Tiwari KK (1964) Diagnosis of two new species of alpheid shrimps from Vietnam. Crustaceana 7(4):313–315
Utinomi H (1956) Coloured illustrations of seashore animals of Japan. Hoikusha, Osaka. 168 pp
Wang YR, Sha ZL (2015) A review of the genus *Synalpheus* (Crustacea: Decapoda: Caridea: Alpheidae) from China seas. Zool Syst 40(4):357–435
Wang YR, Sha ZL (2016a) A review of the genus *Alpheopsis* (Decapoda, Caridea, Alpheidae) from the seas of China. Crustaceana 89(9):1099–1113
Wang YR, Sha ZL (2016b) Redescription of the type species of the genus *Thuylamea* Nguyên, 2001 (Decapoda, Caridea, Alpheidae), based on a female specimen. Crustaceana 89(10):1185–1191
Wang YR, Sha ZL (2016c) Taxonomic Notes on the genus *Salmoneus* Holthuis, 1955 (Decapoda, Caridea, Alpheidae) from the South China Sea. Crustaceana 89(14):1585–1609
Wang YR, Sha ZL (2017a) Description of two new species of the genus *Automate* De Man, 1888 (Crustacea: Decapoda: Caridea) from the South China Sea. Zootaxa 4238(1):30–42
Wang YR, Sha ZL (2017b) The caudal appendix as an important character to identify various species in the genus *Stenalpheops* (Decapoda, Alpheidae). Crustaceana 90(13):1615–1640
White A (1847) List of the specimens of Crustacea in the collection of the British Museum:i–viii, 1–143. British Museum, London
Yaldwyn JC (1971) Preliminary descriptions of a new genus and twelve new species of natant decapod Crustacea from New Zealand. Rec Dominion Mus 7(10):85–94
Yamaguchi T, Baba K (1993) Crustacean specimens collected in Japan by Ph. F. von Siebold and H. Bürger and held by the National Natuurhistorisch Museum in Leiden and other museums. Ph. F. von Siebold and Natural History of Japan, Crustacea. The Carcinological Society of Japan, Tokyo, pp 145–570
Yang HJ, Kim CH (1998) Zoeal stages of *Alpheus brevicristatus* De Haan, 1849 (Decapoda, Caridea, Alpheidae) with a key to the first zoeal larvae of three Korean *Alpheus* species. Kor J Biol Sci 2:187–193
Yokoya Y (1927) Notes on two alpheoid shrimps from Japan. J Coll Agric Tokyo Imp Univ 9:171–176, Plate 7
Yokoya Y (1930) Macrura of Mutsu Bay. Report on the Biological Survey of Mutsu Bay. 16. Sci Rep Tohoku Univ 5:525–548
Yokoya Y (1933) On the distribution of decapod crustaceans inhabiting the continental shelf around Japan, chiefly based upon the materials collected by S. S. Sôyô–Maru, during the year 1923–1930. J Coll Agric Tokyo Imp Univ 12(1):1–226
Yokoya Y (1936) Some rare and new species of decapod crustaceans found in the vicinity of the Misaki marine biological station. Jpn J Zool 7:129–146
Yu SC (1930) Deux nouvelles crevettes de Chine. Bull Soc Zool Fr 55:454–462
Zehntner L (1894) Voyage de MM. M. Bedot et C. Pictet dans l'Archipel Malais. Crustacés de l'Archipel Malais. Revue Suisse de Zoologie et Annales du Musée d'Histoire Naturelle de Genève 2:135–214, Plates VII–IX

Kolibri 360 GmbH
Ferdinand-Jühlke-Straße 7
99095 Erfurt, Deutschland
produktsicherheit@kolibri360.de